Lecture N
Computer Science

Lecture Notes in Computer Science

Vol. 270: E. Börger (Ed.), Computation Theory and Logic. IX, 442 pages. 1987.

Vol. 271: D. Snyers, A. Thayse, From Logic Design to Logic Programming. IV, 125 pages. 1987.

Vol. 272: P. Treleaven, M. Vanneschi (Eds.), Future Parallel Computers. Proceedings, 1986. V, 492 pages. 1987.

Vol. 273: J.S. Royer, A Connotational Theory of Program Structure. V, 186 pages. 1987.

Vol. 274: G. Kahn (Ed.), Functional Programming Languages and Computer Architecture. Proceedings. VI, 470 pages. 1987.

Vol. 275: A.N. Habermann, U. Montanari (Eds.), System Development and Ada. Proceedings, 1986. V, 305 pages. 1987.

Vol. 276: J. Bézivin, J.-M. Hullot, P. Cointe, H. Lieberman (Eds.), ECOOP '87. European Conference on Object-Oriented Programming. Proceedings. VI, 273 pages. 1987.

Vol. 277: B. Benninghofen, S. Kemmerich, M.M. Richter, Systems of Reductions. X, 265 pages. 1987.

Vol. 278: L. Budach, R.G. Bukharajev, O.B. Lupanov (Eds.), Fundamentals of Computation Theory. Proceedings, 1987. XIV, 505 pages. 1987.

Vol. 279: J.H. Fasel, R.M. Keller (Eds.), Graph Reduction. Proceedings, 1986. XVI, 450 pages. 1987.

Vol. 280: M. Venturini Zilli (Ed.), Mathematical Models for the Semantics of Parallelism. Proceedings, 1986. V, 231 pages. 1987.

Vol. 281: A. Kelemenová, J. Kelemen (Eds.), Trends, Techniques, and Problems in Theoretical Computer Science. Proceedings, 1986. VI, 213 pages. 1987.

Vol. 282: P. Gorny, M.J. Tauber (Eds.), Visualization in Programming. Proceedings, 1986. VII, 210 pages. 1987.

Vol. 283: D.H. Pitt, A. Poigné, D.E. Rydeheard (Eds.), Category Theory and Computer Science. Proceedings, 1987. V, 300 pages. 1987.

Vol. 284: A. Kündig, R.E. Bührer, J. Dähler (Eds.), Embedded Systems. Proceedings, 1986. V, 207 pages. 1987.

Vol. 285: C. Delgado Kloos, Semantics of Digital Circuits. IX, 124 pages. 1987.

Vol. 286: B. Bouchon, R.R. Yager (Eds.), Uncertainty in Knowledge-Based Systems. Proceedings, 1986. VII, 405 pages. 1987.

Vol. 287: K.V. Nori (Ed.), Foundations of Software Technology and Theoretical Computer Science. Proceedings, 1987. IX, 540 pages. 1987.

Vol. 288: A. Blikle, MetaSoft Primer. XIII, 140 pages. 1987.

Vol. 289: H.K. Nichols, D. Simpson (Eds.), ESEC '87. 1st European Software Engineering Conference. Proceedings, 1987. XII, 404 pages. 1987.

Vol. 290: T.X. Bui, Co-oP A Group Decision Support System for Cooperative Multiple Criteria Group Decision Making. XIII, 250 pages. 1987.

Vol. 291: H. Ehrig, M. Nagl, G. Rozenberg, A. Rosenfeld (Eds.), Graph-Grammars and Their Application to Computer Science. VIII, 609 pages. 1987.

Vol. 292: The Munich Project CIP. Volume II: The Program Transformation System CIP-S. By the CIP System Group. VIII, 522 pages. 1987.

Vol. 293: C. Pomerance (Ed.), Advances in Cryptology – CRYPTO '87. Proceedings. X, 463 pages. 1988.

Vol. 294: R. Cori, M. Wirsing (Eds.), STACS 88. Proceedings, 1988. IX, 404 pages. 1988.

Vol. 295: R. Dierstein, D. Müller-Wichards, H.-M. Wacker (Eds.), Parallel Computing in Science and Engineering. Proceedings, 1987. V, 185 pages. 1988.

Vol. 296: R. Janßen (Ed.), Trends in Computer Algebra. Proceedings, 1987. V, 197 pages. 1988.

Vol. 297: E.N. Houstis, T.S. Papatheodorou, C.D. Polychronopoulos (Eds.), Supercomputing. Proceedings, 1987. X, 1093 pages. 1988.

Vol. 298: M. Main, A. Melton, M. Mislove, D. Schmidt (Eds.), Mathematical Foundations of Programming Language Semantics. Proceedings, 1987. VIII, 637 pages. 1988.

Vol. 299: M. Dauchet, M. Nivat (Eds.), CAAP '88. Proceedings. 1988. VI, 304 pages. 1988.

Vol. 300: H. Ganzinger (Ed.), ESOP '88. Proceedings, 1988. VI, 381 pages. 1988.

Vol. 301: J. Kittler (Ed.), Pattern Recognition. Proceedings, 1988. VII, 668 pages. 1988.

Vol. 302: D.M. Yellin, Attribute Grammar Inversion and Source-to-source Translation. VIII, 176 pages. 1988.

Vol. 303: J.W. Schmidt, S. Ceri, M. Missikoff (Eds.), Advances in Database Technology – EDBT '88. X, 620 pages. 1988.

Vol. 304: W.L. Price, D. Chaum (Eds.), Advances in Cryptology – EUROCRYPT '87. Proceedings, 1987. VII, 314 pages. 1988.

Vol. 305: J. Biskup, J. Demetrovics, J. Paredaens, B. Thalheim (Eds.) MFDBS 87. Proceedings, 1987. V, 247 pages. 1988.

Vol. 306: M. Boscarol, L. Carlucci Aiello, G. Levi (Eds.), Foundations of Logic and Functional Programming. Proceedings, 1986. V, 218 pages. 1988.

Vol. 307: Th. Beth, M. Clausen (Eds.), Applicable Algebra, Error-Correcting Codes, Combinatorics and Computer Algebra. Proceedings, 1986. VI, 215 pages. 1988.

Vol. 308: S. Kaplan, J.-P. Jouannaud (Eds.), Conditional Term Rewriting Systems. Proceedings, 1987. VI, 278 pages. 1988.

Vol. 309: J. Nehmer (Ed.), Experiences with Distributed Systems. Proceedings, 1987. VI, 292 pages. 1988.

Vol. 310: E. Lusk, R. Overbeek (Eds.), 9th International Conference on Automated Deduction. Proceedings, 1988. X, 775 pages. 1988.

Vol. 311: G. Cohen, P. Godlewski (Eds.), Coding Theory and Applications 1986. Proceedings, 1986. XIV, 196 pages. 1988.

Vol. 312: J. van Leeuwen (Ed.), Distributed Algorithms 1987. Proceedings, 1987. VII, 430 pages. 1988.

Vol. 313: B. Bouchon, L. Saitta, R.R. Yager (Eds.), Uncertainty and Intelligent Systems. IPMU '88. Proceedings, 1988. VIII, 408 pages. 1988.

Vol. 314: H. Göttler, H.J. Schneider (Eds.), Graph-Theoretic Concepts in Computer Science. Proceedings, 1987. VI, 254 pages. 1988.

Vol. 315: K. Furukawa, H. Tanaka, T. Fujisaki (Eds.), Logic Programming '87. Proceedings, 1987. VI, 327 pages. 1988.

Vol. 316: C. Choffrut (Ed.), Automata Networks. Proceedings, 1986. VII, 125 pages. 1988.

Vol. 317: T. Lepistö, A. Salomaa (Eds.), Automata, Languages and Programming. Proceedings, 1988. XI, 741 pages. 1988.

Vol. 318: R. Karlsson, A. Lingas (Eds.), SWAT 88. Proceedings, 1988. VI, 262 pages. 1988.

Vol. 319: J.H. Reif (Ed.), VLSI Algorithms and Architectures – AWOC 88. Proceedings, 1988. X, 476 pages. 1988.

Vol. 320: A. Blaser (Ed.), Natural Language at the Computer. Proceedings, 1988. III, 176 pages. 1988.

Vol. 321: J. Zwiers, Compositionality, Concurrency and Partial Correctness. VI, 272 pages. 1989.

Vol. 322: S. Gjessing, K. Nygaard (Eds.), ECOOP '88. European Conference on Object-Oriented Programming. Proceedings, 1988. VI, 410 pages. 1988.

Vol. 323: P. Deransart, M. Jourdan, B. Lorho, Attribute Grammars. IX, 232 pages. 1988.

Lecture Notes in Computer Science

Edited by G. Goos and J. Hartmanis

378

J.H. Davenport (Ed.)

EUROCAL '87

European Conference on Computer Algebra
Leipzig, GDR, June 2–5, 1987
Proceedings

Springer-Verlag
Berlin Heidelberg New York London Paris Tokyo Hong Kong

Editor

James H. Davenport
University of Bath, School of Mathematical Sciences
Claverton Down, Bath BA2 7AY, UK

CR Subject Classification (1987): I. 1, J. 2, G. 4

ISBN 3-540-51517-8 Springer-Verlag Berlin Heidelberg New York
ISBN 0-387-51517-8 Springer-Verlag New York Berlin Heidelberg

Printing and binding: Druckhaus Beltz, Hemsbach/Bergstr.
2145/3140-543210 – Printed on acid-free paper

Preface

This volume contains the papers presented at EUROCAL '87, the sixth in a series of International Conferences on Computer Algebra, held in Europe with proceedings published by Springer–Verlag in this series. Previous conferences were: EUROSAM '79 (Marseille; vol. **72**), EUROCAM '82 (Marseille; vol. **144**), EUROCAL '83 (Kingston-on-Thames; vol. **162**), EUROSAM '84 (Cambridge; vol. **174**) and EUROCAL '85 (Linz; vol. **203** and **204**).

EUROCAL '87 took place from the second to the fifth of June 1987, at the Karl-Marx University of Leipzig, GDR. There were invited lectures in the mornings, many of which are reproduced in the early part of this volume, and contributed papers and abstracts in the afternoons. There were also demonstrations of various software systems, a poster session and numerous meetings and informal encounters that cannot be repesented in a formal record, but which contributed greatly to the success of the conference. In addition, two papers presented at the Conference actually appeared elsewhere: *Operational semantics of behavioural covers based on narrowing* by H. Reichel, which appeared in Springer Lecture Notes in Computer Science 332, pp. 235–248, and *What is the rank of the Demjanenko matrix?* by H. G. Folz and H. G. Zimmer, which appeared in J. Symbolic Computation **4** (1987) pp. 53–67.

The General Chairman of the Conference was Prof. Dr. W. Lassner, of the Mathematics Department of the Karl-Marx University. To him and his colleagues is due the success of the local arrangements. Prof. Dr. N.J. Lehmann's international experience was invaluable in organising the many facets of a large intercontinental conference. Generous sponsorship was provided by the Robotron organisation, and by the Rank Xerox group. Like any scientific conference, the quality is assured by the many referees who read the papers, and whose comments often greatly improve the presentation of the conference. To all of these people the Programme Committee extends its sincere thanks. I personally am grateful to many colleagues, notably Prof. J.P. Fitch, for their assistance in the preparation of these proceedings.

Bath, April 1989 J.H. Davenport

Contents

Invited Papers

Computer algebra in physical research of Joint Institute for Nuclear Research 1
R. N. Fedorova, V. P. Gerdt, N. N. Govorun, V. P. Shirikov

Complexity of quantifier elimination in the theory of ordinary differential equations 11
D. Yu. Grigor'ev

Groups and polynomials 26
G. C. Smith

Symbolic computation in relativity theory 34
M. A. H. MacCallum

A zero structure theorem for polynomial-equations-solving and its applications 44
W. Wu (Abstract)

Some algorithms of rational function algebra 45
S. A. Abramov

Applications and Systems

The computer algebra system SIMATH 48
R. Böffgen, M. A. Reichert

Converting SAC-2 code to LISP 50
L. Langemyr

Computer algebra system for continued fractions manipulation 52
V. Tomov, M. Nisheva, T. Tonev

Computing a lattice basis from a system of generating vectors 54
J. Buchmann, M. Pohst

Expression optimization using high-level knowledge 64
M. P. W. Mutrie, B. W. Char, R. H. Bartels

CATFACT: Computer algebraic tools for applications of catastrophe theory 71
R. G. Cowell, F. J. Wright

Computer algebra application for investigating integrability of nonlinear evolution systems . 81
V. P. Gerdt, A. B. Shabat, S. I. Svinolupov, A. Yu. Zharkov

Computer classification of integrable seventh order MKdV-like equations 93
V. P. Gerdt, A. Yu. Zharkov

Symbolic computation and the finite element method 95
J. P. Fitch, R. G. Hall

Application of Lie group and computer algebra to nonlinear mechanics 97
D. M. Klimov, V. M. Rudenko, V. F. Zhuravlev

Hierarchical symbolic computations in the analysis of large-scale dynamical systems 107
J. Paczyński

SCHOONSCHIP for computing of gravitino interaction cross sections in n=2 supergravity . 116
N. I. Gurin

Creation of efficient symbolic-numeric interface 118
N. N. Vasiliev

Automatic generation of FORTRAN-coded Jacobians and Hessians 120
P. van den Heuvel, J. A. van Hulzen, V. V. Goldman

Laplace transformations in REDUCE 3 132
C. Kazasov

REDUCE 3.2 on iAPX 86/286-based personal computers 134
T. Yamamoto, Y. Aoki

Some extensions and applications of REDUCE System . . . 136
M. Spiridonova
Infinite structures in SCRATCHPAD II . . . 138
W. H. Burge, S. M. Watt
Application of a structured LISP system to computer algebra . . . 149
J. Smit, S. H. Gerez, R. Mulder
Number-theoretic transforms of prescribed length . . . 161
R. Creutzburg, M. Tasche
A Hybrid algebraic-numeric system ANS and its preliminary implementation . . . 163
M. Suzuki, T. Sasaki, M. Sato, Y. Fukui
The calculation of QCD triangular Feynman graphs in the external gluonic field using REDUCE-2 system . . . 172
L. S. Dulyan
Computer algebra application for determining local symmetries of differential equations . . . 174
R. N. Fedorova, V. V. Kornyak
Trace calculations for gauge theories on a personal computer . . . 176
J. Ranft, H. Perlt
Evaluation of plasma fluid equations collision integrals using REDUCE. . . . 178
R. Liska, D. Drska (Abstract)
Computerised system of analytic transformations for analysing of differential equations . . . 179
V. L. Katkov, M. D. Popov
Integral equation with hidden Eigenparameter solver: REDUCE and FORTRAN in tandem . . . 186
E. Shablygin
Combinatorial aspects of simplification of algebraic expressions . . . 192
A. Ya. Rodionov, A. Yu. Taranov
Dynamic program improvement . . . 202
P. D. Pearce, J. P. Fitch
Computer algebra and numerical convergence . . . 204
K.-U. Jahn
Computer algebra and computation of Puiseux expansions of algebraic functions. . . . 206
V. P. Gerdt, N. A. Kostov, Z. T. Kostova
Boundary value problems for the Laplacian in Euclidean space solved by symbolic computation . . . 208
F. Brackx, H. Serras
The methods for symbolic evaluation of determinants and their realization in the planner-analytic system . . . 216
V. A. Eltekov, V. B. Shikalov
Transformation of computation formulae in systems of recurrence relations . . . 223
E. V. Zima
DIMREG - The package for calculations in the dimensional regularization with 4-dimensional γ^5-matrix in quantum field theory . . . 225
V. A. Ilyin, A. P. Kryukov
CTS - Algebraic debugging system for REDUCE programs . . . 233
A. P. Kryukov, A. Ya. Rodionov
Applications of computer algebra in solid modelling . . . 244
A. Bowyer, J. H. Davenport, P. S. Milne, J. A. Padget, A. F. Wallis
Implementation of a geometry theorem proving package in SCRATCHPAD II . . . 246
K. Kusche, B. Kutzler, H. Mayr
Collision of convex objects . . . 258
B. Roider, S. Stifter

Polynomial Algorithms

Solving algebraic equations via Buchberger's algorithm 260
S. R. Czapor
Primary ideal decomposition 270
H. Kredel
Solving systems of algebraic equations by using Gröbner bases 282
M. Kalkbrener
Properties of Gröbner bases under specializations 293
P. Gianni
The computation of polynomial greatest common divisors over an algebraic number field . . 298
L. Langemyr, S. McCallum
An extension of Buchberger's algorithm to compute all reduced Gröbner bases of a polynomial ideal 300
K.-P. Schemmel
Singularities of moduli spaces 311
B. Martin, G. Pfister
Radical simplification using algebraic extension fields 313
T. J. Smedley
Hermite normal forms for integer matrices 315
R. J. Bradford
Mr Smith goes to Las Vegas: Randomized parallel computation of the Smith Normal Form of polynomial matrices 317
E. Kaltofen, M. S. Krishnamoorthy, B. D. Saunders
Fonctions symétriques et changements de bases 323
A. Valibouze
Complexity of standard bases in projective dimension zero 333
M. Giusti
Gröbner bases for polynomial ideals over commutative regular rings 336
V. Weispfenning
Some algebraic algorithms based on head term elimination over polynomial rings 348
T. Sasaki
Algorithmic determination of the Jacobson radical of monomial algebras 355
T. Gateva-Ivanova
A recursive algorithm for computation of the Hilbert polynomial 365
M. V. Kondratéva, E. V. Pankratév
An affine point of view on minima finding in integer lattices of lower dimensions 376
B. Vallée
A combinatorial and logical approach to linear-time computability 379
P. Scheffler, D. Seese
Complexity of computation of embedded resolution of algebraic curves 381
J. P. G. Henry, M. Merle
Polynomial factorization: An exploration of Lenstra's algorithm 391
J .A. Abbott, J. H. Davenport

Advanced Algorithms

A matrix-approach for proving inequalities 403
A. Ferscha
Using automatic program synthesizer as a problem solver: Some interesting experiments . . 412
P. Návrat, Ľ. Molnár, V. Vojtek
Strong splitting rules in automated theorem proving 424
M. Baaz, A. Leitsch

Towards a refined classification of geometric search and computation problems 426
T. Fischer
Matrix-Padé fractions 438
G. Labahn, S. Cabay
Computation of generalized Padé approximants 450
G. Németh, M. Zimányi
A critical pair criterion for completion modulo a congruence 452
L. Bachmair, N. Dershowitz
Shortest paths of a disc inside a polygonal region 454
G. Werner
Rabins width of a complete proof and the width of a semialgebraic set 456
T. Recio, L. M. Pardo
Practical aspects of symbolic integration over Q(x) 463
D. M. Gillies, B. W. Char
Integration: Solving the Risch differential equation 465
J. A. Abbott
Computation and simplification in Lie fields 468
J. Apel, W. Lassner
A package for the analytic investigation and exact solution of differential equations 479
T. Wolf
An algorithm for the integration of elementary functions 491
M. Bronstein

Index of authors 498

COMPUTER ALGEBRA
IN PHYSICAL RESEARCH OF JINR

R.N. Fedorova, V.P. Gerdt, N.N.Govorun, V.P. Shirikov
Laboratory of Computing Techniques and Automation
Joint Institute for Nuclear Research
Head Post Office, P.O. Box 79, Moscow, USSR

I. Originally computers were designed for numeric computation. Computer structure and instruction system were directed just at this application field, though instructions from the very beginning include those ones for logical operations, that allowed one to exceed the bounds of numeric information processing.

Against a breakdown of predominant numeric methods the separate papers [1-3] on computer application to algebraic formula manipulations began to appear about 30 years ago. However, the two reasons delayed the development of computer algebra programs. First, unsufficient training of experts and unpreparedness of their scientific fields for computer algebra usage. Secondly, difficulties of computer adaptation to formula manipulations, since a user was obliged by himself to train a computer to perfom (i.e. to design a compiler,or an interpreter) algebraic transformations and only after that to use it for solving his problem. Therefore it was necessary to combine in one person two different professions system and applied programmer.

It is also necessary to note one more not the least of the factors connected with unsufficient core memory and slow performance of the first generation computers for algebraic manipulations.

Years went by, computer possibilities are rapidly extended from generation to generation, with increasing the efficiency of its usage in different applied fields.

All that furthers the appearance of needs in algebraic manipulations by computer which could already be effectively enought implemented in serial computers. It was getting more and more Soviet and foreign papers on computer algebra (see,for example,reviews [4-7]).Algorithmic and program methods were developed. Languages and compilers for symbolic information processing were created. Among them LISP was the most widely spreaded.

Many early and some present-day CAS had been written in a code or assembler language, sometimes including elements of such numerical languages as FORTRAN,ALGOL,PL/1 and others.It raised an immobility of such a system. An availability of high-level symbolic languages like LISP allowed one create mobile systems.

Just LISP underlies of the most developed and universal CAS, for example, widely-distributed systems REDUCE and MACSYMA. Such a system requires large computer resourses itself, i.e. typically 1 megabyte computer memory to say nothing of a problem to be solved. Howewer, the big universal systems could relatively easy be adopted at another computer. Moreover those CAS permit an extension by the use of addition to a program written in LISP or in "sourse" language for a given system.

Inspite of the appearance of those and other powerful universal CAS [7] (SMP, SCRATCHPAD-II, MAPLE, etc.) the process of creation, development and usage of special purpose CAS is in progress. In its own field such a system could in a sufficiently full measure satisfy the user's requirements. At the same time the special purpose systems gave, as a rule, the most effective algorithms and optimum internal representation for data, corresponding to the mathematical expressions and operations from the field of specialization for a given CAS. Therefore the specialized systems are usually much more effective in computer resourses then the powerful universal CAS.

As the stricking example of the specialized CAS it should be noted the system SCHOONSCHIP [8] which was developed more than 20 years ago (the first SCHOONSCHIP version has been created by Dutch physicist M.Veltman in 1965) and intended for computations in quantum field theory. Inspite of its "middle-aged" SCHOONSCHIP is till now beyond comparison in high energy physics. Just by means of its usage the record (with respect to amount of computation) results in quantum field theory were obtained. Below some of such computation will be briefly described.

In JINR up to now the most part of the problems connected with computer algebra application is solved with the help of SCHOONSCHIP. Among others such general purpose CAS as REDUCE and FORMAC are more widely used.

II. Comparing computer algebra development in the USSR and abroad one can note the same nature of its basic stages:

1. Appearence of separate papers which were pioneers in the field.

2. Creation of tens of specialized and general purpose CAS written in assembler languages or algorithmic languages FORTRAN, ALGOL, PL/1,

LISP and others.

3. Intensive development of the algorithmic base of the present-day CAS and in first place of the general purpose systems. Appearance of new, considerably improved versions of such systems (REDUCE, ANALITIK, SCRATCHPAD).Creation of developed computer algeebra software for mini- and microcomputers.

A shift should be noted in time of coming either stage in the USSR and abroad. If the first papers on computer algebra appeared in our country and abroad approximately at the same time (first stage), then mass creation of CAS in the USSR (second stage) has begun in fact more later. It became possible only with appearance of the BESM-6 and then-serial ES computers. Now, several tens of CAS were created in the USSR (see review [6] and also proceedings of the conferences on computer algebra held in the Soviet Union [9-15]).

Among the Soviet special purpose CAS, ones intended to mechanical problems, are predominated [6]. While from general purpose systems: AVTO-ANALITIC [16],SIRIUS-SPUTNIK [17],AUM [18],ANALITIK [19] (see also [6]) the later is most widely used. In contrast to other Soviet CAS the language ANALITIK has been implemented by hardware, initially for MIR computers, then for the special processor SM-2410, which is a part of the two-processors complex SM-1410, and for the special processors ES-2680 distined for ES computers. Those special processors interprete by hardware the language ANALITIK-79 (SM-2410) and ANALITIK-82 (ES-2680).

It is remarcable that such an approach to support by hardware of a high-level language started to be developed in the Institute of Cybernetics of the Ukrainian Akademy of Sciences as early as the sixties.For some time past one can see a sharp rise of interest to the symbolic processors as abroad where the greatest attention was paid to the different LISP and PROLOG dialects, as also in the Soviet Union. Among home works it should be noted the investigations on creation of the symbolic processor which are carried on at the Institute of Applied Mathematics of the USSR Academy of Sciences [20].

REDUCE is the most widely-distributed foreign CAS in the USSR. It is used in different institutions for solving the scientific and applied problems. With the help of JINR REDUCE was adapted to ES computers in more than 50 Soviet institutions. Soviet papers based on REDUCE usage are sufficiently enough represented in references [10-15].

III. Let us go on to abrief description of basic works on computer algebra carried out at the Joint Institute for Nuclear Research.

The first investigations on computer realizations of symbolic mathematical operations were carried out at the JINR as long ago as early 1960s.

Kim Ze Phen in 1963 has created the program for definite integration of some class of rational functions.

At the same time H. Kaizer has developed computer algorithms and program [21] for the algebra of Dirac γ-matrices.

In 1964 V.I. Sharonov has done extention of ALGOL-60 to make some formula manipulations and, in particular, to calculate the trace of the γ-matrices product [22].

The next step in computer algebra investigations at JINR began in the middle of 1970s after the first CAS SCHOONSCHIP had been obtained. It was in 1975 and created the favourable conditions for making investigations in computer algebra more active. In 1976 there were the first successful attempts [23,24] to use SCHOONSCHIP in quantum field theory.

Now JINR has 12 different CAS [25] for the ES-1060, ES-1061, CDC-6500 and BESM-6 computers (see the table):

ES-1060, ES-1061	CDC-6500	BESM-6
REDUCE 2, 3.2	REDUCE 2	AVTO-ANALITIK
SCHOONSCHIP	SCHOOSCHIP	UPP
FORMAC 73	CLAM	SAVAG
CAMAL	SYMBAL	GRATOS
ASHMEDAI		
AMP		

All systeems for the BESM-6 computer are Soviet ones and are described in refs [6,10-16].

An implementation of CAS has encountered great difficulties because of unsufficient computer memory, differences between operating systems, adaptation of computer dependent parts of CAS and so on.

IV. Anouther group of works is connected with the development of CAS to extend the field of its application in JINR. Some of such works are following:

1) Improvement for interface between SCHOONSCHIP aand FORTRAN for symbolic-numeric computations [26].

2) Development of the algorithm for virtual memory control in case of compiled LISP function in order to improve usage of REDUCE for the

CDC-6500 computer [27].

3) Creation of general mathematical packages to extend possibilities of CAS for the following problems:

- solving by power series method of an ordinary differential equation of the form (REDUCE) [28]

$$y'' + p(x)y' + q(x)y = 0 ,$$

where $p(x)$ and $q(x)$ are rational function in x ,

- construction of the determining equations for finding Lie-Backlund symmetries of differential equations (FORMAC, REDUCE) [29,30],

- determination of the Lie algebra of point and contact symmetries of differential equations (REDUCE) [31],

- classification of integrable scalar nonlinear evolution equations (FORMAC) [32],

$$u_t = F(u, u_1, \ldots u_n) \text{ , } u \equiv u(x,t) \text{ , } u_i \equiv d^i u/dx^i \text{ .}$$

- investigating the integrability of nonlinear evolution system of the form (FORMAC) [33]

$$\bar{u}_t = \Lambda \bar{u}_n + F(x, \bar{u}, \bar{u}_1, \ldots \bar{u}_{n-1}), \quad \bar{u} = (u^1, \ldots u^m) \text{ , } \bar{u}_i \equiv d^i \bar{u}/dx^i \text{ ,}$$

where $\Lambda = diag(\lambda_1, \ldots \lambda_m)$, $\lambda_i \in \mathbb{C}$, $\lambda_i \neq 0$, $\lambda_i \neq \lambda_j$ $(i \neq j)$

- computation of symbolic determinants (SCHOONSCHIP) [34].

4) Creation of special packages for high energy physics

- calculation of the one- and two-loop Feynman integrals by the methods of dimensional regularization (SCHOONSCHIP) [34],

- construction of renormalized coefficient functions of Feynman diagrams in scalar theories (SCHOONSCHIP) [35],

- realization of Feynman diagram technique for virton-quark model (Standard LISP) [36],

-simplification of polynomials in Pauli σ-matrices (SHOONSCHIP) [34]. All the packages listed above form a core of general and special users libraries for REDUCE and SCHOONSCHIP [34] and also for FORMAC.

V. All the work on development of computer algebra systems and methods are closely connected with the scientific program of JINR. CAS are used in such fields of physics and mathematics as

- theoretical hight energy physics,
- physics of atomic nucleus,
- statistical mechanics,
- quantum mechanics,
- electrodynamics of charges particles in accelerators,

- nonlinear problems of theoretical and mathematical physics,
- experimental high energy physics,

and others. Let us consider very briefly some of computer algebra applications in JINR.

The most traditional field of computer algebra application at the JINR is multiloop computation in quantum field theory based on CAS SCHOONSCHIP. As was shown by home and foreign practice it is the most suitable for such problems. Multiloop calculations in gauge and supersymmetric theories [23,37,38] were of great importance in analisys of their renormalization properties. As to computation of three-loop divergences in quantum chromodynamics [37] carried out in 1980 till now it is record. As a result of these computations,a number of universal programs [35,39,40] have been developed giving a possibility to automatize cumbersome algebraic manipulation at separate steps of Feynman diagram technique.

In other group of works (see ref. [41] and its bibliography) in the process of ten years usage of SCHOONSCHIP an effective method for solving a number of problems in theoretical high energy physics was developed. In these works a line has been successively realized, directed to the total algorithmization by means of SCHOONSCHIP of a computational procedure for elementary particles cross-sections,taking into consideration a contribution of a big number of high-order diagrams. By corresponding authors efforts the general algorithm had been created for all computation of the chain "matrix-element" → "total cross-section" that is trace calculations, removal of ultraviolet and infra-red divergences, multiple integrations and other complicated transformations. The algorithm was used for solving the physical problems,connected with an analisis of experimental data from the combined large-scale experiments of JINR and CERN (European Organization for Nuclear Research).

Development of applied software for CAS and in the first place for general-purpose systems REDUCE and FORMAC was greatly stimulated by its applications to nonlinear problems of theoretical and mathematical physics intensively studied at the Joint Institute. Among them are :

- investigation of the nonlinear resonanses influence on charged particle motion in cyclic accelerators, using the asymptotic method by Krylov and Bogoliubov [42,43];
- construction of the general solution of the Chew-Low nonlinear dispersion equations for low-energy πN-scattering [44];
- investigation of nonlinear evolution equations,which at present give

rise to a great interest owing to their soliton solutions [32,33,45];
- group analysis of differential equations; finding the system of determining equations for the Lie aalgebra of point and contact symmetries [29,30].

To solve these problems a number of effective algorithms and universal program have been developed [28-34].The programs [29,31,32] were included in a widely-spreaded CPC program library.

Computer algebra is used in the JINR not only for theoretical and mathematical problems but also in actual experimental investigations in high energy physics. For example, in papers [46] on the base of the new method of information registration from multiwire proportional chambers [47] a number SCHOONSCHIP and REDUCE programs are created. These programs are used for the development of the principal schemes of data compression devices, majority coincidence schemes and for devices realized the switch function. The latter is represented by the special polynomial,which is the element of Galua GF(2^m) field. Using the methods of algebraic coding theory a number of symbolic algorithms for coding and decoding of compressed experimental data are developed. These algorithms and programs were used for designing special processors to registrate nuclear interactions.

VI. To conclude we note a close collaboration of JINR with many scientific and educational institutions in the USSR and socialist countries in the field of computer algebra. JINR passed a number of CAS and packages developed at the Joint Institute to scientific institutions of the Soviet Union, Bulgaria, Hungary, Czechoslovakia, the German Democratic Republic and Viet-Nam. Among the Soviet receivers are such large-scale centres as Moscow and Leningrad State Universities, Institute of Nuclear Physics (Novosibirsk), Institute of Hight Energy Physics (Moscow),Institute for Nuclear Research (Moscow), Steklov Institute of Mathematics (Leningrad Brach), Institute of Geophysics (Kiev), Institute of Physics (Minsk) and many others. Numerous common investigations in the field of computer algebra are well enough presented in Proceedings of International Conferences held at Dubna in 1979 [10], 1982 [12] and 1985 [25].

REFERENCES

1. Kahrimanian H.G. (1953). Analytical Differentiation by a Digital Computer.MA Thesis,Temple University,Philadelphia.
Nolan J. (1953).Analytical Differentiation on a Digital Computer.MA Thesis, MIT, Cambridge, Massachusetts.
2. Kantorovich L.V. (1957). On one mathematical symbolism suited for computations by computer. *DAN SSSR*,V.113,p.738 (in Russian).
3. Shurygin V.A.,Yanenko N.N. (1961).On Computer Realization of Algebraic Differential Algorithms .In: problems of Cybernetics,No.6 Fizmatgiz,Moscow,p.33 (in Russian).
4. Barton D.,Fitch J.P. (1972). A Review of Algebraic Manipulative Programs in Physics. *Rep. Prog. Phys.*,v. 35,p. 235.
5. Gerdt V.P.,Tarasov O.V.,Shirkov D.V. (1980). Analytic Calculations on Digital Computers for Applications in Physics and Mathematics. *Sov.Phys. Usp.* 23 (1),p.59.
6. Grosheva M.V. et al. (1983). Computer Algebra Systems (Analytic Application Packages).Informator No.1, Keldysh Inst. of Appl. Math. Moscow (in Russian).
7. Calmet J.,van Hulzen J.A.(1983).Computer Algebra Systems & Computer Algebra Applications. In: Computer Algebra. Symbolic and algebraic Computation (eds. Buchberger B., Collins G.E., Loos R.),2-nd ed., Springer-Verlag ,Vienna, p.221.
8. Strubbe H. (1974). Manual for SCHOONSHIP a CDC 600/7000 Program for Symbolic Evaluation of Algebraic Expressions. *Comp. Phys. Comm.*,v. 8,p. 1.
9. Computational Mathematics and Techniques.No. III (1972),Krarkov (in Russian).
10.Proceedings of the International Conference on Computer Algebra and its Application in Theoretical physics (1980). JINR, D11-80-13, Dubna.
11.All-Union Conference on Compilation Methods. Theses of Reports (1981). Novosibirsk (in Russian).
12.Proceedings of the (2nd) International Conference on Computer Algebra and its Application in Theoretical Physics (1983). JINR, D11-83-511, Dubna.
13.Theory and Practice of Automatized Computer Algebra Systems. Theses of Reports (1984), Vilnius (in Russian).
14.Computer Algebra Systems and Mechanics. Theses of Reports. (1984), Gorky (in Russian).
15.Proceedings of the (3-nd) International Conference on Computer Algebra and its Application in Theoretical Physics (1985). JINR, D11-85-791, Dubna.
16.Arais E.A.,Yakovlev N.E. (1985). Automation of Analytic Computations in Scientific Research. Nauka, Novosibirsk (in Russian).
17.Akselrod I.R.,Belous L.F. (1981). SIRIUS-SPUTNIK - New Version of Computer Algebra System. In Ref.[11],p. 160 (in Russian).
18.Kalinina N.A.,Pottosin I.V.,Semenov A.L. (1983). Universal Computer Algebra System AUM. In Ref. [12],p.7 (in Russian).
19.Klimenko V.P.,Pogrebinsky S.B.,Fishman Yu.S. (1983). Software Development of MIR Computers for Solving of Mathematical and Applied Problems by Analytic Methods. In: Ref. [12],p. 132 (in Russian).
20.Eisymont L.K.,Platonova L.N. (1983). Choice and Estimation of Basic Language for Symbolic Processor. In: Ref. [12],p. 19 (in Russian).
21.Kaiser H.J. (1963).Trace Calculation on Electronic Computer. *Nucl. Phys.*,v. 43,p. 620.
22.Sharonov V.I. (1964). An Algorithmic Language for Manipulation of Words Based on ALGOL-60. JINR, No. 1668, Dubna (in Russian).
23.Tarasov O.V.,Vladimirov A.A. (1976).Two-loop Renormalization of the Yang-Mills Theory in an Arbitrary Gauge. JINR, E2-10079, Dubna.

24.Bardin D.Yu.,Fedorenko O.M.,Shumejko N.M. (1976). Exact Calculation of the Lowest Order Electromagnetic Correction to the Elastic Scattering of Particles with Spin 0 and 1/2. JINR, P2-10114, Dubna (in Russian).
25.Kim Khon Sen,Kruglova L.Yu,Rostovsev V.A.,Fedorova R.N. (1985). Computer Algebra Systems in JINR, Experience of their Installation, Development and Usage. In: ref. [15], p. 13 (in Russian).
26.Bobyleva L.V.,Fedorova R.N.,Shirikov V.P. (1978). Computer algebra system SCHOONSCHIP for CDC-6500 Computer and Experience of its Exploitation in JINR. In Proceedings of the International Meting on Programming and Mathematical Methods for Solving the Physical Problems. JINR, D10,11-11264, Dubna (in Russian).
27.Rostovsev V.A. (1983). Utilization of Secondary Memory in Computer Algebra Systems. In: Ref. [12], p. 107 (in Russian).
28.Gerdt V.P.,Zharkov A.Yu. (1983). REDUCE - Package for Solving Ordinary Differential Equation. In: ref [12], p. 171 (in Russian).
29.Fedorova R.N.,Kornyak V.V. (1986). Determination of Lie-Backlund Symmetries of Differential Equations Using FORMAC.*Comp. Phys. Comm.* v.39, p. 93.
30.Fedorova R.N.,Kornyak V.V. (1987). A REDUCE Program for Calculation of Determining Equations of Lie-Backlund Symmetries of Differential Equations. JINR, R11-87-19, Dubna (in Russian).
31.Eliseev V.P.,Fedorova R.N.,Kornyak V.V. (1985).A REDUCE Program for Determining Point and Contact Lie Symmetries of Differential Equation. *Comp. Phys. Comm.* v.36,p. 383.
32.Gerdt V.P.,Shvachka A.b.,Zharkov A.Yu. (1985). FORMINT - a Program for Clasification of Integrable Nonlinear Evolution Equations.*Comp. Phys. Comm.* v.34,p. 303.
Gerdt V.P.,Shvachka A.B.,Zharkov A.Yu. (1985). Computer Algebra Application for Classification of Integrable Nonlinear Evolution Equations. *J. Symb. Comp.*,v. 1,p. 101.
33.Gerdt V.P.,Shabat A.B.,Svinolupov S.I.,Zharkov A.Yu.(1987).Computer Algebra Application for Investigating Integrability of Nonlinear Evolution Systems. JINR, E5-87-40, Dubna. See also this volume.
34.Bogolubskaya A.A.,Gerdt V.P.,Tarasov O.V. (1985). About Library Complectation of SCHOONSCHIP and REDUCE Systems. In: ref.[15], p.82 (in Russian).
35.Tarasov O.V. (1978). The Construction of Renormalized Coeffitient Functions of Feynman Diagrams by Computer, JINR,E2-11573, Dubna.
36.Raportirenko A.M. (1985). VIRTON - a Problem Oriented LISP-Package. In: Ref. [15],p. 72 (in Russian).
37.Tarasov O.V.,Vladimirov A.A.,Zharkov A.Yu. (1980).The Gell-Mann-Low Function of QCD in the Three-Loop Approximation.*Phys.Lett.* v.93B,p. 429.
38.Vladimirov A.A.,Tarasov O.V. (1980). Three-Loop Calculations in Non-Abelian Gauge Theories. JINR, E2-80-483, Dubna, 1980.
39.Tarasov O.V. (1980).A Program for Computation of One-, Two- and Three-Loop Feynman Diagrams in Gauge Theories. In: Ref.[10],p. 150 (in Russian).
40.Tarasov O.V. (1983). An Effective Program for Computation of Three-Loop Complanar and Non-Complanar Feynman Diagrams. In: Ref. [12],p. 214 (in Russian).
41.Akhundov A.A.,Baranov S.P.,Bardin D.Yu.,Rimann T. (1985). Computer Algebra Systems Application to Exact Calculations in the Theory of Electroweak Interactions.In: Ref. [15],p. 382 (in Russian).
42.Amirkhanov I.V.,Zhidkov E.P.,Zhidkova I.E. (1983). On Investigation by the Method of Averaging of the Resonance $2\nu_z-\nu_x=1$ and its Influence on Motion of Particles in Cyclic Accelerators. In: Ref. [12],p. 223 (in Russian).

43. Amirkhanov I.V., Zhidkov E.P., Zhidkova I.E. (1985). Research of Non-Linear Resonance Effect on Stability of Charge Particle Motion Using Computer Algebra System REDUCE. In: Ref. [15], p. 361 (in Russian).

44. Gerdt V.P. (1980). Local Construction of the General Solution of the Chew-Low Equations by Computer. In: Ref. [10], p. 159 (in Russian).
Gerdt V.P., Zharkov A.Yu. (1983). An Iteration Sheme for Constracting the General Solution of the Chew-Low Equations using REDUCE-2. In: Ref. [12], p. 232 (in Russian).

45. Gerdt V.P., Shvachka A.B., Zharkov A.Yu. (1984). Classification of Integrable Hight-Ordeer KdV-Like Equations. JINR, P5-84-489, Dubna (in Russian).
Gerdt V.P., Zharkov A.Yu. (1986). Computer Classification of Integrable Seventh Order MKdV Like Equations. JINR, P5-86-371, Dubna (in Russian).

46. Gaidamaka R.I., Nikityuk N.M., Shirikov V.P. (1983). Computer Algebra and Complex of Programs for Constructing of Data Compression and Transformation Devices in Nuclear-Physical Experiments. In: Ref. [12], p. 246 (in Russian).
Alexandrov I.N., Gaidamaka R.I., Nikityuk N.M. (1985). Computer Algebra Application to Computation of Logical Schemes and Special Processors In: Ref. [15], p. 295 (in Russian).

47. Nikityuk N.M., Radzhabov P.S., Shafranov M.D. (1978). A New Method of Information Registration from Multiwire Proportional Chambers. *Nucl. Instr. and Meth.*, v. 155, p. 485.

COMPLEXITY OF QUANTIFIER ELIMINATION IN THE THEORY OF ORDINARY DIFFERENTIAL EQUATIONS

D.Yu.Grigor'ev

Leningrad Department of Mathematical V.A.Steklov Institute of Academy of Sciences of the USSR, Fontanka 27, Leningrad, 191011, USSR

Introduction

Let a formula of the first-order theory of ordinary differential equations be given

$$Q_1 u_1 \dots Q_n u_n (\Omega) \qquad (1)$$

where $Q_1, \dots, Q_n$ are quantifiers (either universal or existential), Ω is a quantifier-free formula containing as atomic subformulas of the kind $(f_i = 0)$, $1 \leq i \leq N$. Here $f_i \in \mathbb{Z}\{u_1, \dots, u_n, v_1, \dots, v_m\}$ are differential polynomials (relatively to differentiating over variable X), indeterminates $u_1, \dots, u_n$ are connected, $v_1, \dots, v_m$ are free (remind, see e.g. [7, 9], that the differential ring $\mathbb{Z}\{u_1, \dots, u_n, v_1, \dots, v_m\}$ is generated as a polynomial ring over $\mathbb{Z}[X]$ by the derivatives $u_s, u_s^{(1)}, u_s^{(2)}, \dots; v_t, v_t^{(1)}, v_t^{(2)}, \dots$ for $1 \leq s \leq n$, $1 \leq t \leq m$). Denote by $ord_{u_s}(f_i)$ the maximal order of derivatives $u_s, u_s^{(1)}, u_s^{(2)}, \dots$ of the indeterminate u_s, occuring in the differential polynomial f_i. Suppose that $ord_{u_s}(f_i) \leq r$, $ord_{v_t}(f_i) \leq r$ for all $1 \leq s \leq n$, $1 \leq t \leq m$, then one can consider f_i as a (usual) polynomial from a ring $\mathbb{Z}[X, u_1, u_1^{(1)}, \dots, u_1^{(r)}, \dots, u_n, u_n^{(1)}, \dots, u_n^{(r)}, v_1, v_1^{(1)}, \dots, v_1^{(r)}, \dots, v_m, v_m^{(1)}, \dots, v_m^{(r)}]$. Assume also that the degree $deg(f_i)$ of the polynomial f_i as an element of the latter ring (relatively to all the indeterminates) is less than d, and finally, that the absolute value of any (integer) coefficient of the polynomial f_i does not exceed 2^M.

In [7] a quantifier elimination method in the first-order theory of ordinary differential equations is described, which allows for a given formula of the kind (1) to produce equivalent to it quantifier-free formula. Here and further we consider the equivalence of the formulas over the differential closure of the quotient field $Z\langle v_1,\dots,v_m\rangle$ of the ring $Z\{v_1,\dots,v_m\}$ (see [7 , 9]). However, the working time of the method from [9] is nonelementary (in Kalmar sense), in particular, it cannot be bounded from above by any finite tower of exponential functions (one can consider the working time on RAM or on any other polynomially equivalent computational model, e.g. Turing machine). The main result of the present paper is the following theorem, in which a quantifier elimination algorithm is designed with an elementary complexity bound (see also [3]).

THEOREM. There is an algorithm which for a given formula of the kind (1) of the first-order theory of ordinary differential equations produces an equivalent to it quantifier-free formula of this theory of the form

$$\bigvee_{1\le i\le \mathcal{N}} \Big(\mathop{\&}_{1\le j\le \mathcal{K}} (g_{i,j}=0)\ \&\ (g_{i,0}\neq 0)\Big) \tag{2}$$

where $g_{i,j}\in Z\{v_1,\dots,v_m\}$ are differential polynomials, within time polynomial in $M(Nd)^{m^n c^{r2^n}}$ for a suitable constant $c>1$. Moreover, for the parameters of the polynomials $g_{i,j}$ hold the following bounds: $ord_{v_t}(g_{i,j})\le r2^n$; $\mathcal{N},\mathcal{K},deg(g_{i,j})\le (Nd)^{O(m^n c^{r2^n})}=\mathcal{M}$ and the absolute value of every (integer) coefficient of a polynomial $g_{i,j}$ is less than $2^{M\mathcal{M}}$.

The method from [9] contains two subroutines, transforming a system of differential equations in a certain disjunction of systems. The first subroutine is applied in the case, when informally speaking, for some distinguished indeterminate its derivative of the maximal order occurs at least in two polynomials. As a result of executing the first subroutine each obtained system has at most one polynomial containing this derivative. The second subroutine consists in splitting a system and decreasing the order of the distinguished indeterminate. Just executing the first subroutine leads in [9] to nonelementary complexity bound. In the present paper transforming (instead of the first subroutine) to a disjunction of systems such that each of then contains at most one polynomial, in which occurs the derivative of the maximal order of the distinguished indeterminate is going in a quite another way, based on the constructing the greatest common divisor of a family of one-variable polynomials with parametric coefficients (lemma 1 in section 1), apparently, interesting itself. The proof of lemma 1 is similar to the construction from [2], but on the other hand the direct application of the result from [2] (see also

[4]) yields a worse complexity bound than in lemma 1. In section 2 a modification of the subroutine from [9] of splitting a system and decreasing the order of the distinguished indeterminate is exposed, then a quantifier elimination algorithm and its complexity analysis are exhibited.

In [9], moreover a quantifier elimination method for the first-order theory of partially differential equations is described. It would be interesting to clarify, whether there exists such a method with elementary complexity? This problem is connected (see [9]) with estimating in an effective version of Hilbert's theorem on Idealbasis.

1. Constructing the greatest common divisor of a family of one-variable polynomials with parametric coefficients

We present the main result of this section (lemma 1) in a more general form than it is necessary for the main theorem, namely for the polynomials with the coefficients from a field F finitely generated over a prime subfield (cf. [1, 2, 4, 5]).

Thus $F = H(T_1,\dots,T_e)[\eta]$ where either $H=\mathbb{Q}$ or $H=\mathbb{F}_p$, i.e. H is a prime subfield, $T_1,\dots,T_e$ are algebraically independent over the field H, an element η is algebraic separable over the field $H(T_1,\dots,T_e)$, let $\varphi(Z)\in H(T_1,\dots,T_e)[Z]$ be its minimal polynomial. Each polynomial $f\in F[X_1,\dots,X_n]$ can be uniquely (up to a factor from H^*) represented in a form $f = \sum_{0\le i<\deg_Z(\varphi);\, i_1,\dots,i_n} (a_{i,i_1,\dots,i_n}/b)\eta^i X_1^{i_1}\cdots X_n^{i_n}$, where $a_{i,i_1,\dots,i_n}, b\in H[T_1,\dots,T_e]$ and $\deg(b)$ is the least possible. Define the degree $\deg_{T_1,\dots,T_e}(f) = \max_{i,i_1,\dots,i_n}\{\deg_{T_1,\dots,T_e}(a_{i,i_1,\dots,i_n}), \deg_{T_1,\dots,T_e}(b)\}$.

The size $l(\alpha)$ for $\alpha\in H$ is defined as its bit-size in the case $H=\mathbb{Q}$ and as $\log_2(p)$ when $H=\mathbb{F}_p$. Denote by $l(f)$ the maximum of the sizes of all the coefficients (from the field H) of the polynomials $a_{i,i_1,\dots,i_n}, b$ at the monomials of variables $T_1,\dots,T_e$.

For the functions $g_1>0, g_2>0, \dots, g_s>0$ we write $g_1 \le \mathcal{P}(g_2,\dots,g_s)$ if for a suitable polynomial P an inequality $g_1\le P(g_2,\dots,g_s)$ is valid.

Consider some polynomials $h_0,h_1,\dots,h_K\in F[X_1,\dots,X_n,Y]$ and assume that the following bounds are true:

$$\deg_{T_1,\dots,T_e,Z}(\varphi)<d_1;\ \deg_{X_1,\dots,X_n,Y}(h_i)<d_0;\ \deg_{T_1,\dots,T_e}(h_i)<d_2;\ l(\varphi)\le M_1;\ l(h_i)\le M_2 \quad (3)$$

for every $0 \leq i \leq K$. Introduce a notation $h_i = \sum_j h_{i,j} Y^j$ where the polynomials $h_{i,j} \in F[X_1, \ldots, X_n]$. Denote by $\bar{F}$ an algebraic closure of the field F.

LEMMA 1. There is an algorithm which for given polynomials h_0, $h_1, \ldots, h_K$ yields such two families of polynomials $g_{q,t} \in F[X_1, \ldots, X_n]$, $\varphi_q \in F_1[X_1, \ldots, X_n, Y]$ for $1 \leq q \leq N_1$, $0 \leq t \leq N_2$ that

a) quasiprojective varieties $\mathcal{V}_q = \{x \in \bar{F}^n : g_{q,1}(x) = \ldots = g_{q,N_2}(x) = 0;\ g_{q,0}(x) \neq 0\}$ for $1 \leq q \leq N_1$ form a decomposition of an open (in Zariski topology) set $\bar{F}^n \setminus \{x \in \bar{F}^n : h_{i,j}(x) = 0$ for all $1 \leq i \leq K$ and $j\}$;

b) for each $1 \leq q \leq N_1$ the following two varieties coincide:

$$\{(x,y) \in \bar{F}^n \times \bar{F} = \bar{F}^{n+1} : h_1(x,y) = \ldots = h_K(x,y) = 0;\ h_0(x,y) \neq 0\} \cap (\mathcal{V}_q \times \bar{F}) =$$
$$= \{(x,y) \in \bar{F}^{n+1} : \varphi_q(x,y) = 0\} \cap (\mathcal{V}_q \times \bar{F})$$

, and besides the leading coefficient $lc_Y(\varphi_q)$ is distinguished from zero everywhere on $\mathcal{V}_q$.

The running time of the algorithm can be estimated by a certain polynomial in $K, M_1, M_2, (d_1 d_2)^e, d_0^{n+e}$. Finally, the following bounds on the parameters of the polynomials are fulfilled:

$$\deg_{X_1,\ldots,X_n,Y}(\varphi_q),\ \deg_{X_1,\ldots,X_n}(g_{q,t}) \leq \mathcal{P}(d_0);$$

$$\deg_{T_1,\ldots,T_e}(\varphi_q),\ \deg_{T_1,\ldots,T_e}(g_{q,t}) \leq d_2 \mathcal{P}(d_1, d_0) \tag{4}$$

$$l(\varphi_q),\ l(g_{q,t}) \leq (M_1 + M_2 + (e+n)\log d_2)\mathcal{P}(d_1, d_0);\ N_1, N_2 \leq K\mathcal{P}(d_0^n).$$

REMARK. 1) The property b) shows that one can consider φ_q as a kind of the greatest common divisor of the polynomials $h_1, \ldots, h_K$ (under the condition $h_0 \neq 0$) considered in a variable Y on the quasiprojective variety $\mathcal{V}_q$;

2) The properties a), b) are still correct if to replace $\bar{F}$ by an arbitrary algebraically closed field containing F.

Proof of Lemma 1. For any $1 \leq i \leq K$, $0 \leq j < d_0$ consider a quasiprojective variety $\mathcal{U}_{i,j} = \{x \in \bar{F}^n : h_{1,d_0-1}(x) = \ldots = h_{1,0}(x) = h_{2,d_0-1}(x) = \ldots = h_{2,0}(x) = \ldots = h_{i,d_0-1}(x) = \ldots = h_{i,j+1}(x) = 0;\ h_{i,j}(x) \neq 0\}$.
Obviously $\bigcup_{i,j} \mathcal{U}_{i,j} = \bar{F}^n \setminus \{x \in \bar{F}^n : h_{i,j}(x) = 0$ for all $1 \leq i \leq K$ and $j\}$. Introduce a notation $\tilde{h}_{i,j} = \sum_{0 \leq \beta \leq j} h_{i,\beta} Y^\beta$. The system under consideration

$$h_1 = \ldots = h_K = 0; \quad h_0 \neq 0 \tag{5}$$

is equivalent to a disjunction (over all $1 \leq i \leq K$, $0 \leq j < d_0$) of the following systems $\tilde{h}_{i,j} = h_{i+1} = \ldots = h_K = h_{1,d_0-1} = \ldots = h_{1,0} = h_{2,d_0-1} = \ldots = h_{2,0} = \ldots = h_{i,d_0-1} = \ldots = h_{i,j+1} = 0;\ h_0 h_{i,j} \neq 0$. Fix for the time being $1 \leq i \leq K$, $0 \leq j \leq d_0$ and consider a system

$$\tilde{h}_{i,j} = h_{i+1} = \ldots = h_K = 0; \quad h_0 \neq 0. \tag{6}$$

Introduce new variables Y_1, Y_0 . For every point $x = (x_1, \ldots, x_n)$ an element y satisfies the system $(6)_x$ (which is obtained from the system (6) by substituting the coordinates $x_1, \ldots, x_n$ instead of $X_1, \ldots, X_n$ respectively) iff there exists y_1 such that $\tilde{h}_{i,j}(x,y) = h_{i+1}(x,y) = \ldots = h_K(x,y) = y_1 h_0(x,y) - 1 = 0$. Consider the following homogeneous relatively to the variables Y_1, Y, Y_0 polynomials: $\bar{h}_\ell(X_1, \ldots, X_n, Y_1, Y, Y_0) = Y_0^{\deg_Y(h_\ell)} h_\ell(X_1, \ldots, X_n, Y/Y_0)$ for $i < \ell \leq K$; $\bar{h}_i(X_1, \ldots, X_n, Y_1, Y, Y_0) = Y_0^j \tilde{h}_{i,j}(X_1, \ldots, X_n, Y/Y_0)$; $\bar{h}_0(X_1, \ldots, X_n, Y_1, Y, Y_0) = Y_0^{\deg_Y(h_0)+1}(Y_1/Y_0\, h_0(X_1, \ldots, X_n, Y/Y_0) - 1)$.
Notice, besides, that for arbitrary point $x \in \mathcal{U}_{i,j}$ the systems $(5)_x$ and $(6)_x$ are equivalent.

Consider some field F_1, a point $x = (x_1, \ldots, x_n) \in \bar{F}_1^n$ and a homogeneous system of equations (in variables Y_1, Y, Y_0):

$$\bar{h}_i(x, Y_1, Y, Y_0) = \bar{h}_{i+1}(x, Y_1, Y, Y_0) = \ldots = \bar{h}_K(x, Y_1, Y, Y_0) = \bar{h}_0(x, Y_1, Y, Y_0) = 0. \tag{7$_x$}$$

Suppose that $h_{i,j}(x) \neq 0$. Then the system $(6)_x$ has a finite number of solutions. If $y \in \bar{F}_1$ is a solution of $(6)_x$ then a point $(1/h_0(x,y) : y : 1) \in \mathbb{P}^2(\bar{F}_1)$ of the projective space is a solution of the system $(7)_x$. Conversely, if $(y_1 : y : y_0) \in \mathbb{P}^2(\bar{F}_1)$ is a solution of $(7)_x$ and $y_0 \neq 0$ then y/y_0 is a solution of $(6)_x$ and apart from that $y_1/y_0 = 1/h_0(x, y/y_0)$; if $y_0 = 0$ then $y = 0$ since $lc_Y(\tilde{h}_{ij}) = h_{ij}$. Thus, the system $(7)_x$ has a finite number of solutions in $\mathbb{P}^2(\bar{F}_1)$, and moreover all these solutions, may be except (1:0:0), correspond bijectively to the solutions of the system $(6)_x$ (provided that $h_{i,j}(x) \neq 0$).

In the sequel we need a certain construction from [8]. Let $g_0, \ldots, g_{t-1} \in F_1[Y_0, \ldots, Y_m]$ be homogeneous polynomials of degrees $\gamma_0 \geq \ldots \geq \gamma_{t-1}$ respectively. Introduce the variables $U_0, \ldots, U_m$ algebraically independent over the field $F_1(Y_0, \ldots, Y_m)$ and a polynomial $g_t = Y_0 U_0 + \ldots + Y_m U_m \in F_1(U_0, \ldots, U_m)[Y_0, \ldots, Y_m]$, its degree $\gamma_t = 1$. Set $D = (\sum_{1 \le \ell \le \min\{t-1, m\}} (\gamma_\ell - 1)) + \gamma_0$. Consider a linear over $F_1(U_0, \ldots, U_m)$ mapping $\mathfrak{A}: \mathcal{H}_0 \oplus \cdots \oplus \mathcal{H}_t \to \mathcal{H}$ where $\mathcal{H}_\ell$ (respectively $\mathcal{H}$) is a space of homogeneous in the variables $Y_0, \ldots, Y_m$ polynomials over the field $F_1(U_0, \ldots, U_m)$ of the degree $D - \gamma_\ell$ (respectively D) for $0 \le \ell \le t$, namely $\mathfrak{A}(f_0, \ldots, f_t) = f_0 g_0 + \ldots + f_t g_t$. Fix some numeration of monomials of degrees $D - \gamma_0, \ldots, D - \gamma_t, D$, respectively and write down the operator $\mathfrak{A}$ in the coordinates corresponding to this numeration, we obtain the matrix A of size $\binom{m+D}{m} \times \sum_{0 \le \ell \le t} \binom{m+D-\gamma_\ell}{m}$. The matrix A can be represented in a form $A = (A^{(num)}, A^{(for)})$ where the submatrix $A^{(num)}$ (called a numerical part of A) contains $\sum_{0 \le \ell \le t-1} \binom{m+D-\gamma_\ell}{m}$ columns and its entries belong to F_1; the submatrix $A^{(for)}$ (called a formal part of A) contains $\binom{m+D-1}{m}$ columns, and its entries are linear forms in the variables $U_0, \ldots, U_m$ over F_1.

PROPOSITION 1. ([8]). a) A system $g_0 = \ldots = g_{t-1} = 0$ has a finite number of solutions in $\mathbb{P}^m(\bar{F}_1)$ iff the rank $rg(A) = \binom{m+D}{m}$;

denote $r = \binom{m+D}{m}$ and assume in the items b), c), d) that $rg(A) = r$;

b) all $r \times r$ minors of the matrix A generate a principal ideal, whose generator $R \in F_1[U_0, \ldots, U_m]$ is also their greatest common divisor;

c) homogeneous relatively to the variables $U_0, \ldots, U_m$ form R equals to the product $R = \prod_{1 \le \varkappa \le D_1} L_\varkappa$ where $L_\varkappa = \sum_{0 \le \alpha \le m} \xi_\alpha^{(\varkappa)} U_\alpha$ is a linear form, moreover $(\xi_0^{(\varkappa)} : \ldots : \xi_m^{(\varkappa)}) \in \mathbb{P}^m(\bar{F}_1)$ is a solution of the system $g_0 = \ldots = g_{t-1} = 0$ and the number of occurences in the product R of the forms proportional to $L_\varkappa$ coincides with the multiplicity of the solution $(\xi_0^{(\varkappa)} : \ldots : \xi_m^{(\varkappa)})$ of the system $g_0 = \ldots = g_{t-1} = 0$ $(1 \le \varkappa \le D_1)$;

d) let Δ be a nonsingular $r \times r$ submatrix of the matrix A, containing $rg(A^{(num)})$ columns in the numerical part $A^{(num)}$ (one can easily see that such a submatrix exists), then $\det(\Delta)$ coincides with R up to a factor from F_1^*, besides the degree $\deg(R) = D_1 = r - rg(A^{(num)})$.

Let us apply the described construction to a system $(7)_{(X_1, \ldots, X_n)}$ taking $F_1 = F(X_1, \ldots, X_n)$, $m = 2$, we get a matrix A with the entries from the ring $F[X_1, \ldots, X_n, U_1, U, U_0]$. According to proposition 1a) the rank $rg(A_x) = r = \binom{D+2}{2}$ for any point x, provided that

$h_{i,j}(x) \neq 0$ (recall that r is the number of the rows of A).

Define a variant of Gaussian algorithm (VGA) as a succession of pairs of indices $(\alpha_0, \beta_0), (\alpha_1, \beta_1), \ldots, (\alpha_\rho, \beta_\rho)$, herein $\alpha_\lambda \neq \alpha_\gamma, \beta_\lambda \neq \beta_\gamma$ for $\lambda \neq \gamma$. VGA determines a succession of matrices $A^{(0)} = A, A^{(1)}, \ldots, A^{(\rho+1)}$. Introduce the notation $A^{(l)} = (a^{(l)}_{\alpha,\beta})$, then $a^{(l+1)}_{\alpha,\beta} = a^{(l)}_{\alpha,\beta} - a^{(l)}_{\alpha_l,\beta} a^{(l)}_{\alpha,\beta_l} / a^{(l)}_{\alpha_l,\beta_l}$ for $\alpha \neq \alpha_0, \ldots, \alpha_l$ distinguished from $\alpha_0, \ldots, \alpha_l$ and $a^{(l+1)}_{\alpha_s,\beta} = a^{(l)}_{\alpha_s,\beta}$ for $0 \leqslant s \leqslant l$, apart from that, the leading entry $a^{(l)}_{\alpha_l,\beta_l} \neq 0$ for all $0 \leqslant l \leqslant \rho$. Then $a^{(l)}_{\alpha,\beta_s} = 0$ for every $0 \leqslant s \leqslant l-1$ unless $\alpha = \alpha_t$ for a certain $0 \leqslant t \leqslant s$. Provided that $\alpha \neq \alpha_0, \ldots, \alpha_{l-1}$ and $\beta \neq \beta_0, \ldots, \beta_{l-1}$ denote by $\Delta^{(l)}_{\alpha,\beta}$ the determinant of $(l+1) \times (l+1)$ submatrix of the matrix A formed by the rows $\alpha_0, \ldots, \alpha_{l-1}, \alpha$ and the columns $\beta_0, \ldots, \beta_{l-1}, \beta$. It is well known that $a^{(l)}_{\alpha,\beta} = \Delta^{(l)}_{\alpha,\beta} / \Delta^{(l-1)}_{\alpha_{l-1},\beta_{l-1}}$ (see e.g. [6]).

We produce a sequence of VGA $\Gamma_1, \Gamma_2, \ldots$ and a corresponding sequence of linearly independent over F polynomials $P_1, P_2, \ldots \in F[X_1, \ldots, X_n][U_1, U, U_0]$. Moreover, VGA Γ_s is applicable correctly to a matrix $A_{\tilde{x}}$ for any point $\tilde{x}$ from (possibly empty) quasi-projective variety $\mathcal{W}_s = \{x \in \bar{F}^n : 0 = P_t(x, U_1, U, U_0) \in \bar{F}[U_1, U, U_0]$ for all $1 \leqslant t < s$ and $0 \neq P_s(x, U_1, U, U_0)\}$. Besides $\bigcup_s \mathcal{W}_s \supset \mathcal{U}_{i,j}$ (see the beginning of the proof of lemma 1).

Considering a current VGA Γ_t we utilize the introduced above notations (in particular, applying Γ_t to A yields a succession $A^{(0)} = A, A^{(1)}, \ldots$). Assume that $\Gamma_1, \ldots, \Gamma_s; P_1, \ldots, P_s$ are already produced $(s \geqslant 0)$. Then as Γ_{s+1} we take VGA satisfying the condition that for every $l \geqslant 0$ the index β_l of the leading entry (α_l, β_l) of the matrix $A^{(l)}$ is the least possible such that $\beta_l > \beta_{l-1}$ and the polynomials $P_1, \ldots, P_s, \prod_{0 \leqslant t \leqslant l} \Delta^{(t)}_{\alpha_t,\beta_t}$ are linearly independent over F. Assume that it is impossible to continue the succession $(\alpha_0, \beta_0), \ldots, (\alpha_{\rho_{s+1}}, \beta_{\rho_{s+1}})$ with fulfilment of the formulated condition. Then we take this succession as VGA Γ_{s+1} and set the polynomial

$$P_{s+1} = \prod_{0 \leqslant t \leqslant \rho_{s+1}} \Delta^{(t)}_{\alpha_t,\beta_t} \tag{8}$$

If each entry of A is linearly dependent over F with $P_1, \ldots, P_s$ then we terminate the process of producing $\Gamma_1, \ldots, \Gamma_s$ without producing Γ_{s+1}.

Observe that if $\rho_{s+1} < r-1$ then $\mathcal{W}_{s+1} \cap \mathcal{U}_{i,j} = \emptyset$. Indeed, let a point $x \in \mathcal{W}_{s+1} \cap \mathcal{U}_{i,j}$. By induction on $0 \leqslant l \leqslant \rho_{s+1}$ we deduce the equality $(a^{(l+1)}_{\alpha,\beta})_x = (\Delta^{(l+1)}_{\alpha,\beta})_x / (\Delta^{(l)}_{\alpha_l,\beta_l})_x = 0$

for any $\beta_l < \beta < \beta_{l+1}$ and $\alpha \neq \alpha_0, \dots, \alpha_l$ since $(\Delta^{(l)}_{\alpha_l,\beta_l})_x \neq 0$ for $x \in \mathcal{W}_{s+1}$ (see (8)), and on the other hand $(\Delta^{(l+1)}_{\alpha,\beta})_x = 0$, otherwise we could take $\beta_{l+1} \leq \beta$, that contradicts to the choice of β_{l+1}. This implies that $(a^{(l+1)}_{\alpha,\beta})_x = 0$ when $\beta < \beta_{l+1}$ and $\alpha \neq \alpha_0, \dots, \alpha_l$, taking into account that the executed elementary transformations with the rows in VGA keep this property. Analogously $(a^{(\rho_{s+1}+1)}_{\alpha,\beta})_x = 0$ for $\beta > \beta_{\rho_{s+1}}$ and $\alpha \neq \alpha_0, \dots, \alpha_{\rho_{s+1}}$. Therefore, $rg(A_x) = \rho_{s+1} + 1 < r$ that contradicts to the relation $x \in \mathcal{U}_{i,j}$ and to proposition 1a) (see the system $(7)_x$).

Now we show the inclusion $\mathcal{U}_{i,j} \subset \bigcup_s \mathcal{W}_s$. Let $x \in \mathcal{U}_{i,j}$, then $rg(A_x) = r$, hence $(a_{\alpha,\beta})_x \neq 0$ for suitable α, β. Suppose that $x \notin \bigcup_s \mathcal{W}_s$, this means that $0 = P_1(x, U_1, U, U_0) = P_2(x, U_1, U, U_0) \cdots$, therefore the entry $a_{\alpha,\beta}$ cannot be linearly dependent with the polynomials $P_1, P_2, \dots$ that contradicts to the condition of terminating the process of producing $\Gamma_1, \Gamma_2, \dots$

Let us prove that for any point $x \in \mathcal{W}_{s+1} \cap \mathcal{U}_{i,j}$ the polynomial R_x, corresponding to the matrix A_x according to proposition 1b), coincides with the minor $\delta_1 \Delta_{s+1}(x) = \delta_1 \Delta^{(r-1)}_{\alpha_{r-1},\beta_{r-1}}(x) \neq 0$ (here and further $\delta_1 \in F^*$, $\bar{\delta}_1, \bar{\delta}_2, \dots \in \bar{F}^*$), arising from VGA Γ_{s+1}. Indeed, consider such a unique λ that the cell $(\alpha_{\lambda-1}, \beta_{\lambda-1})$ is located in the numerical part $A^{(num)}$ and the cell $(\alpha_\lambda, \beta_\lambda)$ is located in the formal part $A^{(for)}$, then $rg(A^{(num)}) = \lambda$ since $(a^{(\lambda)}_{\alpha,\beta})_x = 0$ for any $\beta < \beta_\lambda$ and $\alpha \neq \alpha_0, \dots, \alpha_{\lambda-1}$ by virtue of the proved above. Hence $\delta_1 \Delta_{s+1}(x) = R_x$ in force of proposition 1d), besides $r - \lambda = \deg_{U_1,U,U_0}(\Delta_{s+1})$, denote $D_2 = r - \lambda$.

Represent $\Delta_{s+1} = \sum_{0 \leq \omega \leq D_2} E^{(\omega)}_{s+1} U_0^{D_2 - \omega}$ where $E^{(\omega)}_{s+1}(X_1, \dots, X_n) \in F[X_1, \dots, X_n, U_1, U]$. Introduce quasiprojective varieties $\mathcal{W}^{(\omega)}_{s+1} = \{x \in \mathcal{W}_{s+1} : 0 = E^{(0)}_{s+1}(x) = \dots = E^{(\omega-1)}_{s+1}(x) \in \bar{F}[U_1, U];\ 0 \neq E^{(\omega)}_{s+1}(x)\}$. Then $\mathcal{W}^{(\omega_1)}_{s+1} \cap \mathcal{W}^{(\omega_2)}_{s+1} = \emptyset$ for $\omega_1 \neq \omega_2$ and $\mathcal{W}_{s+1} = \bigcup_{0 \leq \omega \leq D_2} \mathcal{W}^{(\omega)}_{s+1}$. For any point $x \in \mathcal{W}_{s+1} \cap \mathcal{U}_{i,j}$, proposition 1c) and the proved above entail the equality $\Delta_{s+1}(x) = \prod_{\varkappa} L_\varkappa^{c_\varkappa}$ where linear forms $L_\varkappa = \zeta_1^{(\varkappa)} U_1 + \zeta^{(\varkappa)} U + \zeta_0^{(\varkappa)} U_0$ correspond bijectively to the solutions $(\zeta_1^{(\varkappa)} : \zeta^{(\varkappa)} : \zeta_0^{(\varkappa)}) \in \mathbb{P}^2(\bar{F})$ of the system $(7)_x$, therefore for each point $x \in \mathcal{W}^{(\omega)}_{s+1} \cap \mathcal{U}_{i,j}$ the form $E^{(\omega)}_{s+1}(x) \mid \Delta_{s+1}(x)$ coincides with the product $\bar{\delta}_1 \prod_\mu L_\mu^{c_\mu}$ of all linear forms L_μ, for which $\zeta_0^{(\mu)} = 0$. Then $\zeta^{(\mu)} = 0$ for every such index μ according to the proved before proposition 1, hence $E^{(\omega)}_{s+1}(x) = \bar{\delta}_2 U_1^\omega$.

Write $E^{(\omega_1)}_{s+1} = \sum_{0 \leq \gamma \leq \omega_1} E^{(\omega_1)}_{s+1,\gamma} U_1^{\omega_1 - \gamma} U^\gamma$ where $E^{(\omega_1)}_{s+1,\gamma} \in$

$F[X_1,\dots,X_n]$, then $E^{(\omega)}_{s+1}(x)=E^{(\omega)}_{s+1,0}(x)\,U_1^{\omega}$ for $x\in\mathcal{W}^{(\omega)}_{s+1}\cap\mathcal{U}_{i,j}$. Then $\Delta_{s+1}(x)/E^{(\omega)}_{s+1}(x)$ coincides with the product $\bar{\delta}_3\prod_\nu L_\nu^{c_\nu}$ of all linear forms L_ν, for which $\zeta_0^{(\nu)}\neq 0$, when $x\in\mathcal{W}^{(\omega)}_{s+1}\cap\mathcal{U}_{i,j}$, in particular a relation $E^{(\omega)}_{s+1}(x)\mid E^{(\omega_1)}_{s+1}(x)$ is valid in the ring $\bar{F}[U_1,U]$ for arbitrary $\omega_1\geqslant\omega$. Therefore $E^{(\omega_1)}_{s+1,\gamma}(x)=0$ if $\omega_1-\gamma<\omega$, thus $\Delta_{s+1}(x)/E^{(\omega)}_{s+1}(x)=$

$$\frac{1}{E^{(\omega)}_{s+1,0}(x)}\sum_{\omega\leqslant\omega_1\leqslant D_2}U_0^{D_2-\omega_1}\sum_{0\leqslant\gamma\leqslant\omega_1-\omega}E^{(\omega_1)}_{s+1,\gamma}(x)\,U_1^{\omega_1-\gamma-\omega}U^{\gamma}\in\bar{F}[U_1,U,U_0].$$

So, a polynomial $(\Delta_{s+1}(x)/E^{(\omega)}_{s+1}(x))\,(0,-1,Y)\in\bar{F}[Y]$ coincides with the product $\bar{\delta}_4\prod_\nu(Y-y_\nu)^{c_\nu}$, where $y_\nu=\zeta^{(\nu)}/\zeta_0^{(\nu)}$ ranges over all the solutions of the system $(6)_x$ (see the claim proved before proposition 1), for each point $x\in\mathcal{W}^{(\omega)}_{s+1}\cap\mathcal{U}_{i,j}$.

For the fixed indices i,j,m,s denote the quasiprojective variety $\mathcal{P}_q^{(1)}=\mathcal{W}^{(\omega)}_{s+1}\cap\mathcal{U}_{i,j}$ (this yields the polynomials $g^{(1)}_{q,t_1},g^{(2)}_{q,t_2}\in F[X_1,\dots,X_n]$ such that $\mathcal{P}_q^{(1)}=\{x\in\bar{F}^n:\ \&_{t_1}(g^{(1)}_{q,t_1}(x)=0)\ \&\ \bigvee_{t_2}(g^{(2)}_{q,t_2}(x)\neq 0)\}$; fixing a certain t_3 we obtain pairwise disjunctive quasiprojective varieties $\mathcal{P}_q=\{x\in\bar{F}^n:\ \&_{t_1}(g^{(1)}_{q,t_1}(x)=0)\ \&\ \&_{t_2<t_3}(g^{(2)}_{q,t_2}(x)=0)\ \&\ (g^{(2)}_{q,t_3}(x)\neq 0)\}$ and the required in lemma 1a) polynomials $g_{q,t}$). Thereupon set a polynomial $\varphi_q=\sum_{\omega\leqslant\omega_1\leqslant D_2}Y^{D_2-\omega_1}E^{(\omega_1)}_{s+1,\omega_1-\omega}(-1)^{\omega_1-\omega}\in F[X_1,\dots,X_n,Y]$. Then for $x\in\mathcal{P}_q$ the equality $\varphi_q(x)=(E^{(\omega)}_{s+1,0}(x)\,\Delta_{s+1}(x)/E^{(\omega)}_{s+1}(x))\,(0,-1,Y)$ is true, this implies the required in lemma 1b) coincidence of the varieties

$\{(x,y)\in\bar{F}^{n+1}: h_1(x,y)=\ldots=h_K(x,y)=0,\ h_0(x,y)\neq 0\}\cap(\mathcal{P}_q\times\bar{F})=$

$\{(x,y):\tilde{h}_{i,j}(x,y)=h_{i+1}(x,y)=\ldots=h_K(x,y)=0;\ h_0(x,y)\neq 0\}\cap(\mathcal{P}_q\times\bar{F})=$

$\{(x,y):\varphi_q(x,y)=0\}\cap(\mathcal{P}_q\times\bar{F})$. Moreover, the leading coefficient $lc_Y(\varphi_q)=E^{(\omega)}_{s+1,0}$ does not vanish anywhere on $\mathcal{P}_q$. Evidently $\bigcup_q\mathcal{P}_q=\{x\in\bar{F}^n: h_{i,j}(x)\neq 0$ for some $1\leqslant i\leqslant K$ and $j\}$.

It remains to check the bounds (4) and the running time of the algorithm. Taking into account that Δ_{s+1} is a minor of the matrix A, the polynomial P_{s+1} (see (8)) is a product of not more than $r\leqslant\mathcal{P}(D)\leqslant\mathcal{P}(d_0)$ minors of A and involving bounds (3) one can deduce the following bounds: $\deg_{X_1,\dots,X_n,U_1,U,U_0}(\Delta_{s+1})$,

$\deg_{X_1,\dots,X_n,U_1,U,U_0}(P_{s+1}),\ \deg_{X_1,\dots,X_n}(g_{q,t}),\ \deg_{X_1,\dots,X_n,Y}(\varphi_q)\leqslant\mathcal{P}(d_0);$

$\deg_{T_1,\dots,T_e}(\Delta_{s+1}),\ \deg_{T_1,\dots,T_e}(P_{s+1}),\ \deg_{T_1,\dots,T_e}(g_{q,t}),\ \deg_{T_1,\dots,T_e}(\varphi_q)\leqslant d_2\mathcal{P}(d_1,d_0);$

$l(\Delta_{s+1}),\ l(P_{s+1}),\ l(g_{q,t}),\ l(\varphi_q)\leqslant(M_1+M_2+(e+n)\log d_2)\,\mathcal{P}(d_1,d_0).$

Since $P_1,P_2,\dots$ are linearly independent over F, one concludes that the number of them does not exceed $\mathcal{P}(d_0^n)$, hence $1 \leq q \leq N_1 \leq K\mathcal{P}(d_0^n)$, $0 \leq t \leq N_2 \leq K\mathcal{P}(d_0^n)$; the bounds (4) are ascertained. From (4) one can infer $\mathcal{P}(K, M_1, M_2, (d_1d_2)^e d_0^{n+e})$ bound on the running time of the algorithm, because this is a bound on bit-sizes of all the intermediate polynomials in the calculations, and also a bound on the number of executed with them arithmetic operations, that completes the proof of lemma 1.

2. Splitting subroutine and quantifier elimination algorithm

Before describing the splitting subroutine, we ascertain the following lemma 2 allowing under relevant conditions to decrease the order of a system of differential equations. Let $g_0, g_1, \dots, g_\gamma, f_0, f_1, \dots, f_K \in \mathbb{Q}\{u, u_1, \dots, u_n\}$ be differential polynomials. Assume that the bounds $ord_u(g_\beta) \leq r-t$; $ord_u(f_i) \leq r$; $ord_{u_j}(g_\beta), ord_{u_j}(f_i) \leq R$; $deg(g_\beta), deg(f_i) < d$; $l(g_\beta), l(f_i) \leq M$

are valid for any $0 \leq \beta \leq \gamma$, $0 \leq i \leq K$, $1 \leq j \leq n$ where deg (here and further) denotes the degree relatively to all the indeterminates $X, u, u^{(1)}, \dots, u^{(r)}, u_1, \dots, u_1^{(R)}, \dots, u_n, \dots, u_n^{(R)}$.

LEMMA 2. For given $g_0, \dots, g_\gamma, f_0, \dots, f_K$ one can produce such differential polynomials $\hat{f}_0, \hat{f}_1, \dots, \hat{f}_K \in \mathbb{Q}\{u, u_1, \dots, u_n\}$ that a system

$$g_0 = g_1 = \dots = g_\gamma = f_1 = \dots = f_K = 0; \qquad f_0 \frac{\partial g_0}{\partial u^{(r-t)}} \neq 0 \tag{9}$$

is equivalent in the ring $\mathbb{Q}\{u, u_1, \dots, u_n\}$ to a system $g_0 = g_1 = \dots = g_\gamma = \hat{f}_1 = \dots = \hat{f}_K = 0$; $\hat{f}_0 \frac{\partial g_0}{\partial u^{(r-t)}} \neq 0$. Besides, the following bounds: $ord_u(\hat{f}_i) \leq r-t$; $ord_{u_j}(\hat{f}_i) \leq R+t$; $deg(\hat{f}_i) \leq \mathcal{P}(d,t)$; $l(\hat{f}_i) \leq (M+nR+r)\mathcal{P}(d,t)$ are true for any $0 \leq i \leq K$, $1 \leq j \leq n$. Finally, the time of producing $\hat{f}_0, \dots, \hat{f}_K$ can be estimated by $\mathcal{P}(K, M, (dt)^{n(R+t)+r})$.

Proof. Observe that for every $s \geq 1$ a derivative $g_0^{(s)} = u^{(r-t+s)}\left(\frac{\partial g_0}{\partial u^{(r-t)}}\right) - Q_s$, where a differential polynomial $Q_s \in \mathbb{Q}[X, u, \dots, u^{(r-t+s-1)}, u_1, \dots, u_1^{(R+s)}, \dots, u_n, \dots, u_n^{(R+s)}]$. Obviously, $deg(g_0^{(s)}) < d$; $l(g_0^{(s)}) \leq M + O(s \log d)$.

Define a weight of a monomial $X^\beta \prod_{l \leq r} (u^{(l)})^{d_l} \prod_{s,j} (u_j^{(s)})^{\beta_{j,s}}$ as $wgt = \sum_{r-t+1 \leq l \leq r} d_l(l-r+t)$, the weight of a differential polynomial is defined as the maximum of the weights of its monomials. Evidently, $wgt(g_0^{(s)}) \leq s$.

Assume by recursion that for a certain $0 \leq s < t$ differential polynomials $f_{i,s} \in \mathbb{Q}[X, u, u^{(1)}, \dots, u^{(r-s)}, u_1, \dots, u_1^{(R+t)}, \dots, u_n, \dots, u_n^{(R+t)}]$ are already produced such that the system (9) is equivalent to a system $g_0 = g_1 = \dots = g_\gamma = f_{1,s} = \dots = f_{K,s} = 0;\ f_{0,s} \frac{\partial g_0}{\partial u^{(r-t)}} \neq 0$. For the base of the recursion $(s=0)$ we set $f_{i,0} = f_i$. Let $wgt(f_{i,s}) \leq W_s;\ deg(f_{i,s}) < D_s;\ l(f_{i,s}) \leq M_s$. Fix some $0 \leq i \leq K$. Obviously $deg_{u^{(r-s)}}(f_{i,s}) \leq W_s/(t-s)$. Substitute in $f_{i,s}$ instead of $u^{(r-s)}$ a quotient of differential polynomials $Q_{t-s} / \left(\frac{\partial g_0}{\partial u^{(r-t)}}\right)$ and in the obtained expression eliminate a denominator multiplying the expression on $\left(\frac{\partial g_0}{\partial u^{(r-t)}}\right)^{deg_{u^{(r-s)}}(f_{i,s})}$, as a result we get a differential polynomial $f_{i,s+1} \in \mathbb{Q}[X, u, \dots, u^{(r-s-1)}, u_1, \dots, u_n^{(R+t)}]$. Clearly a system $g_0 = \dots = g_\gamma = f_{1,s+1} = \dots = f_{K,s+1} = 0;\ f_{0,s+1}\left(\frac{\partial g_0}{\partial u^{(r-t)}}\right) \neq 0$ is equivalent to $g_0 = \dots = g_\gamma = f_{1,s} = \dots = f_{K,s} = 0;\ f_{0,s}\left(\frac{\partial g_0}{\partial u^{(r-t)}}\right) \neq 0$ and hence is equivalent to (9). Finally, set $\hat{f}_i = f_{i,t}$.

Taking into account inequality $wgt(Q_s) \leq t-s = wgt(u^{(r-s)})$ one can deduce that after the described substitution the weight does not increase, in other words $W_{s+1} \leq W_s$. Moreover $deg(f_{i,s+1}) \leq D_s + d \cdot deg_{u^{(r-s)}}(f_{i,s}) \leq D_s + d \frac{W_s}{t-s}$. Besides $l(f_{i,s+1}) \leq M_s + (M + O(s \log d))(W_s/(t-s)) + (n(R+t)+r) \log(D_{s+1})$. Since $wgt(f_i) \leq W_0 < dt$, we conclude that $deg(\hat{f}_i) = O(d^2 t \log t);\ l(\hat{f}_i) = O(Mdt \log t + dt^2 \log d \log t + (n(R+t)+r)t \cdot \log(dt))$. Finally, the algorithm produces $f_{i,s+1}$ starting with $f_{i,s}$ in time $\mathcal{P}(M_{s+1}, D_{s+1}^{n(R+t)+r})$. Lemma 2 is proved.

Now we proceed to describing a splitting subroutine of a system of the kind

$$g = h_1 = \dots = h_\ell = 0; \quad h_0 \neq 0 \tag{10}$$

where $g, h_i \in \mathbb{Q}[X, u, \dots, u^{(r)}, u_1, \dots, u_n^{(R)}]$, apart from that $0 \leq ord_u(g) = p < r$. Write $g = \sum_{0 \leq \alpha \leq \gamma} g_\alpha (u^{(p)})^\alpha$, herein $ord_u(g_\alpha) \leq p-1$.

The system (10) is equivalent to the dijunction of the following formulas (11), (12) (we call this equivalence a splitting of the system (10) and g a splitted polynomial):

$$\bigvee_{0 \leq \beta \leq \gamma-1} \left(\left(g = \frac{\partial g}{\partial u^{(p)}} = \dots = \frac{\partial^\beta g}{\partial (u^{(p)})^\beta} = h_1 = \dots = h_\ell = 0 \right) \& \left(h_0 \frac{\partial^{\beta+1} g}{\partial (u^{(p)})^{\beta+1}} \neq 0 \right) \right) \tag{11}$$

$$(g_0 = \dots = g_\gamma = h_1 = \dots = h_\ell = 0) \& (h_0 \neq 0). \qquad (12)$$

Let differential polynomials $f_0, \dots, f_K \in \mathbb{Q}[X, u, \dots, u^{(r)}, u_1, \dots, u_n^{(R)}]$ satisfy the following bounds: $\deg(f_i) < d$, $\ell(f_i) \leq M$ for every $0 \leq i \leq K$. Denote $f_i = \sum_s f_{i,s} (u^{(r)})^s$ where $ord_u(f_{i,s}) \leq r-1$. Consider a formula

$$\Omega_1 = ((f_1 = \dots = f_K = 0) \& (f_0 \neq 0)) \qquad (13)$$

Our nearest goal is to design an algorithm producing a quantifier-free formula equivalent to a formula $\exists u(\Omega_1)$. Apply to (13) lemma 1, taking the derivative $u^{(r)}$ as the variable Y and X, $u, \dots, u^{(r-1)}, u_1, \dots, u_n^{(R)}$ as $X_1, \dots, X_n$ respectively. It yields differential polynomials $g_{q,t} \in \mathbb{Q}[X, u, \dots, u^{(r-1)}, u_1, \dots, u_n^{(R)}]$, $\psi_q \in \mathbb{Q}[X, u, \dots, u^{(r-1)}, u^{(r)}, u_1, \dots, u_n^{(R)}]$ such that formula (13) is equivalent to the disjunction of the following formulas (14), (15):

$$\bigvee_q (\mathop{\&}_{t \geq 1} (g_{q,t} = 0) \& (\psi_q = 0) \& (g_{q,0} \neq 0)) \qquad (14)$$

$$\mathop{\&}_{i \geq 1; s} (f_{i,s} = 0) \& (f_0 \neq 0) \qquad (15)$$

To any system $\mathop{\&}_{t \geq 1} (g_{q,t} = 0) \& (\psi_q = 0) \& (g_{q,0} \neq 0)$ (from (14)) the algorithm applies the splitting subroutine, considering this system as (10) and an arbitrary $g_{q,t}$ as a splitted polynomial, provided that $ord_u(g_{q,t}) \geq 0$. Thereupon the algorithm applies repeatedly the splitting subroutine to all the obtained systems of the kind (12) taking them as the system (10) (without taking ψ_q as a splitted polynomial). In a similar way the algorithm applies the splitting subroutine to the system (15) taking it as (10) and an arbitrary polynomial among $\{f_{i,s}\}_{i \geq 1; s}$ as a splitted one, provided that the indeterminate u occurs in it, and continues applying repeatedly the splitting subroutine to all the obtained systems of the kind (12) taking them as (10).

Observe that the algorithm is unable at some step to apply the splitting subroutine to a certain obtained system of the kind (12) iff either the system is of the form $\Omega_2 = \mathop{\&}_{t \geq 1; E} (g_{q,t,E} = 0) \& (\psi_q = 0) \& (g_{q,0} \neq 0)$ (i.e. it arises from (14)) where $g_{q,t} = \sum_E g_{q,t,E} u^{\varepsilon_0} (u^{(1)})^{\varepsilon_1} \dots (u^{(r-1)})^{\varepsilon_{r-1}}$, herein $g_{q,t,E} \in \mathbb{Q}[X, u_1, \dots, u_n^{(R)}]$ and $E = (\varepsilon_0, \varepsilon_1, \dots, \varepsilon_{r-1})$ is a multiindex, or the system arises from (15) and is of the form $\Omega_3 = \mathop{\&}_{i \geq 1; s; E} (f_{i,s,E} = 0) \& (f_0 \neq 0)$.

We claim that a formula $\exists u\,(\Omega_2)$ is equivalent to the following disjunction

$$\bigvee_{E_0} \Big(\mathop{\&}_{t \geq 1; E} (g_{q,t,E} = 0) \,\&\, (g_{q,0,E_0} \neq 0) \Big) \tag{16}$$

Consider some $u_1, \ldots, u_n$ and denote by $K = \mathbb{Q}\langle u_1, \ldots, u_n \rangle$ the differential field generated by them. Let (16) be true. Take $u, u^{(1)}, \ldots, u^{(r-1)}$ to be algebraically independent over the field K, then $g_{q,0}(X, u, u^{(1)}, \ldots, u^{(r-1)}, u_1, \ldots, u_n^{(R)}) \neq 0$ hence $0 \neq lc_{u^{(r)}}(\varphi_q) \in K[u, u^{(1)}, \ldots, u^{(r-1)}]$ by virtue of lemma 1b) and involving remark 2) just after lemma 1. Consider an irreducible over the field $K(u, u^{(1)}, \ldots, u^{(r-1)})$ divisor $\tilde{\varphi}_q \in K[u, \ldots, u^{(r-1)}, u^{(r)}]$ of the polynomial φ_q. Then take u satisfying the single relation $\tilde{\varphi}_q(u, u^{(1)}, \ldots, u^{(r-1)}, u^{(r)}) = 0$ (so, u is an element of the differential factor-ring $K\{u\}/(\tilde{\varphi}_q)$ without divisors of zero, see [7]). The equality $0 = \varphi_q(u, u^{(1)}, \ldots, u^{(r-1)}, u^{(r)}) \in K\{u\}/(\tilde{\varphi}_q)$ proves the claim.

Analogously and even easier, a formula $\exists u\,(\Omega_3)$ is equivalent to the following disjunction

$$\bigvee_{E_0} \Big(\mathop{\&}_{i \geq 1; s; E} (f_{i,s,E} = 0) \,\&\, (f_{0,E_0} \neq 0) \Big) \tag{17}$$

For every obtained (after executing splitting subroutine at some step) formula of the kind (11) the algorithm applies lemma 2 to each its disjunctive term (for a given β we take $g_0 = \dfrac{\partial^\beta g}{\partial (u^{(p)})^\beta}$, see (9)). It yields the differential polynomials $\hat{h}_{0,\beta}, \hat{h}_{1,\beta}, \ldots, \hat{h}_{\ell,\beta} \in \mathbb{Q}[X, u, u^{(1)}, \ldots, u^{(p)}, u_1, \ldots, u_n^{(R+r-p)}]$ such that (11) is equivalent to the following disjunction

$$\bigvee_{0 \leq \beta \leq \gamma - 1} \Big(\Big(g = \frac{\partial g}{\partial u^{(p)}} = \ldots = \frac{\partial^\beta g}{\partial (u^{(p)})^\beta} = \hat{h}_{1,\beta} = \ldots = \hat{h}_{\ell,\beta} = 0 \Big) \,\&\, \Big(\hat{h}_{0,\beta} \frac{\partial^{\beta+1} g}{\partial (u^{(p)})^{\beta+1}} \neq 0 \Big) \Big). \tag{18}$$

To any disjunctive term from the formular (18) the described process is again applied taking this term as a formula of the kind (13) etc. It completes the description of the algorithm producing a quantifier-free formula equivalent to a formula $\exists u\,(\Omega_1)$ (see (13)). Notice that the terminal systems (in which the indeterminate u does not occur) are of the form (16) or (17).

Now we estimate the number of systems obtained by the described above algorithm from (13) and their parameters. Observe that for any intermediate system, occuring in it differential polynomials belong to a ring $\mathbb{Q}[X, u, \dots, u^{(\rho)}, u_1^{(1)}, \dots, u_1^{(R+r-\rho)}, \dots, u_n, \dots, u_n^{(R+r-\rho)}]$ for some ρ, hence the polynomials occuring in systems of forms (16), (17) belong to $\mathbb{Q}[X, u_1, \dots, u_1^{(R+r)}, \dots, u_n, \dots, u_n^{(R+r)}]$ From system (13) lemma 1 yields $\kappa d^{c_1(nR+r)}$ disjunctive terms in formulas (14), (15) (here and below $c_1, c_2, \dots$ are suitable constants). The degree of each polynomial occuring in (14), (15) is less than d^{c_2} by lemma 1. The subroutine splitting disjunctive terms in (14), (15) produces $\kappa d^{c_3(nR+r)}$ systems of kinds (11), (12). For any disjunctive term from (11) lemma 2 gives a system (18) with the polynomials of degrees less than $(d(r-\rho))^{c_4}$.

Basing on these speculations one can prove by induction on $0 \leqslant \rho \leqslant r$ that after repeatedly applying the described process yields in the whole not more than $\kappa(dr)^{c_5^{\rho}(nR+r)}$ intermediate systems of the form $\tilde{g}_{1,\rho} = \dots = \tilde{g}_{\tilde{s},\rho} = 0$, $\tilde{g}_{0,\rho} \neq 0$ where $\tilde{g}_{\alpha,\rho} \in \mathbb{Q}[X, u, u^{(1)}, \dots, u^{(r-\rho)}, u_1, \dots, u_1^{(R+\rho)}, \dots, u_n, \dots, u_n^{(R+\rho)}]$, besides $\tilde{s} \leqslant \kappa(dr)^{c_5^{\rho}(nR+r)}$, $\deg(\tilde{g}_{\alpha,\rho}) \leqslant (dr)^{c_6^{\rho}}$, $l(\tilde{g}_{\alpha,\rho}) \leqslant (M+nR)(dr)^{c_6^{\rho}}$. Thus, the formula $\exists u(\Omega_1)$ (see (13)) is equivalent to a disjunction of $\kappa d^{c_7^{r} nR}$ systems of the kind $g_1 = \dots = g_s = 0$, $g_0 \neq 0$ (see (16), (17)), where $g_\alpha \in \mathbb{Q}[X, u_1, \dots, u_1^{(R+r)}, \dots, u_n, \dots, u_n^{(R+r)}]$, moreover $s \leqslant \kappa d^{c_7^{r} nR}$, $\deg(g_\alpha) \leqslant d^{c_7^{r}}$, $l(g_\alpha) \leqslant (M+nR)d^{c_8^{r}}$. The time required to produce this quantifier-free disjunction is less than $\mathcal{P}(\kappa M d^{c_8^{r} nR})$.

Describe now a procedure reducing a quantifier-free formula of the kind Ω, see (1) in the introduction, to disjunctive normal form. We utilize the notations from (1) and consider $\{f_1, \dots, f_N\}$ occuring in Ω as (usual) polynomials in $(n+m)(r+1)+1$ variables. For any subset $I \subset \{1, \dots, N\}$ we name $\mathcal{B} = \&_{i \in I}(f_i = 0) \& \&_{i \in \{1,\dots,N\} \setminus I}(f_i \neq 0)$ an elementary $\{f_1, \dots, f_N\}$-formula. Corollary 1 to theorem 2 [6] implies a bound $(\sum_{i \in I} \deg f_i)^{(n+m)(r+1)+1} < (dN)^{(n+m)(r+1)+1}$ on the number of elementary $\{f_1, \dots, f_N\}$-formulas, determining nonempty quasiprojective varieties (in a space $\overline{F}_1^{(n+m)(r+1)+1}$ for arbitrary algebraically closed field $\overline{F}_1$), such elementary formulas we call nontrivial. One can find all of them successively. Assuming that for some $0 \leqslant t < N$ all nontrivial elementary $\{f_1, \dots, f_t\}$-formulas $\mathcal{B}_1 = \&_{i \in I_1}(f_i = 0) \& \&_{i \in \{1,\dots,t\} \setminus I_1}(f_i \neq 0)$ are found, one tests, which among two elementary $\{f_1, \dots, f_t, f_{t+1}\}$-formulas $\mathcal{B}_1 \& (f_{t+1} = 0)$ and $\mathcal{B}_1 \& (f_{t+1} \neq 0)$ are nontrivial involving the algorithm from [1] (see also [2, 4, 5]). Thus, one can list all nontrivial elementary $\{f_1, \dots, f_N\}$-formulas in time $\mathcal{P}(M, N^{(n+m)r}, d^{(n+m)^2 r^2})$ ([1]).

Next, for each nontrivial elementary $\{f_1, \dots, f_N\}$-formula $\mathcal{B}$ the procedure detects, whether it is consistent with Ω, replacing in Ω every atomic subformula ($f_i=0$) by its truth value from $\mathcal{B}$. The obtained formula is true iff $\mathcal{B}$ is consistent with Ω. Then Ω is equivalent to the disjunction of all the consistent with Ω nontrivial elementary $\{f_1,\dots,f_N\}$-formulas.

The quantifier elimination algorithm repeatedly applies to (1) alternatively described two procedures of eliminating one quantifier (see (13) and after it) and reducing a quantifier-free formula to disjunctive normal form and yields formula (2). The bounds on parameters of formula (2) and on the working time of the algorithm (see the theorem) one can prove by induction on n.

References

1. A.L.Chistov, D.Yu.Grigor'ev. Solving systems of algebraic equations in subexponential time I, II. - Preprints LOMI, E-9-83, E-10-83, Leningrad, 1983.
2. A.L.Chistov, D.Yu.Grigor'ev. Complexity of quantifier elimination in the theory of algebraically closed fields. - Lect.Notes Comput. Sci., 1984, v.176, p.17-31.
3. D.Yu.Grigor'ev. Complexity of quantifier elimination in the theory of ordinary differential equations. - Proc.VIII All-Union conf. Math.Logic, Moscow, 1986, p.46 (in Russian).
4. D.Yu.Grigor'ev. Computational Complexity in Polynomial Algebra. - Proc.International Congress of Mathematicians, 1987, Berkeley.
5. D.Yu.Grigor'ev, A.L.Chistov. Fast decomposition of polynomials into irreducible ones and the solution of systems of algebraic equations. - Soviet Math.Dokl., 1984, v.29, p.380-383.
6. J.Heintz. Definability and fast quantifier elimination in algebraically closed fields. - Theor.Comput.Sci., 1983, v.24, p.239-278.
7. E.R.Kolchin. Differential algebra and algebraic groups. - Academic Press, 1973.
8. D.Lazard. Resolution des systemes d'equations algebriques. - Theor. Comput.Sci., 1981, v.15, p.77-110.
9. A.Seidenberg. An elimination theory for differential algebra. - Univ.of Calif.Press, 1956, v.3, N 2, p.31-66.

Groups and Polynomials

Geoff C Smith
School of Mathematical Sciences
University of Bath
Claverton Down
Bath
Avon BA2 7AY ENGLAND

Introduction

In this tutorial lecture we will address ourselves to the subject of Galois Theory. This is a substantial area of modern algebra, and in the course of a single lecture, it is impossible to give the subject any more than a distant overview. The lecture will conclude with some remarks about possible applications of Galois Theory to Computer Algebra. As many of you will know, Evariste Galois (1811-1832) led a short and rather unhappy life. The tense political situation in France affected his career, such as it was, adversely. He even spent several months in gaol in consequence of his political activities. He died as a result of wounds received in a duel. It is clear that the pretext for the fatal conflict was "a woman"; it does not seem likely that the conflict was actually a consequence of the intense republican versus royalist rivalry in which he was deeply involved. This is not a talk about the life and times of Galois, though that would be a fascinating topic. He is, of course, in some ways a romantic figure; when one combines the dramatic events of his life with the difficulty he had in achieving recognition before his death, it is not surprising that the man is still a figure of considerable interest. However, his true significance is his fole in initiating modern algebra.

His early death is to be regretted — to put it mildly. One is reminded of the Chinese benediction

"May you live in uninteresting times" [ff]

The modern casting of the theory is very different from the times when Galois was writing. The very notion of a Group is due to Galois, and the language of modern mathematics had not yet been formulated. The works of Galois were not published until well after his death — in 1846 by Liouville [G]. The theory was subsequently advanced by the work of C Jordan [J]. The more concrete ways of thinking mathematically prevalent in those days may have some considerable appeal to this particular audience, so I commend to you especially the excellent text by Edwards [E], where he develops the theory in a constructive fashion. There are presumably texts in other languages which do a similar job. The Russian text by Postkinov [Po] presumably fills this niche in its untranslated form. Perhaps there is a more recent work which does the job as well. It is of course perfectly obvious that this talk is no substitute for a proper study of the theory, and some standard texts are included in the references [A], [E], [Po], [S].

Theory

The following theorem is known to well educated schoolchildren. We are so familiar with it that perhaps, in mature reflection, we do not give it the respect that it deserves. We shall prove it very carefully, in a manner which will generalize.

Theorem: If $F(X)$ is a polynomial in the variable X and has coefficients which are real numbers, then non-real roots of F occur in (complex conjugate) pairs.

It follows easily from results in elementary analysis, and the Fundamental Theorem of Algebra, that the number of non-real roots of $F(X)$ (counting multiplicities) is even. One simply distinguishes two cases depending on the parity of the degree of F. This however is a bad proof (not wrong, but bad). The more insightful proof runs as follows:

Let $g: \mathbf{C} \to \mathbf{C}$ denote complex conjugation, so $g(x+iy) = x - iy$.

One checks that g defines a field automorphism of **C**, that is to say: For all complex numbers w and z, we have $g(w+z) = g(w) + g(z)$ and $g(wz) = g(w)g(z)$, and moreover $g(1) = 1$, so g is certainly a field monomorphism (1-1 and structure preserving). One notices also that g composed with itself is the identity map, so g is a bijection, and so g is indeed an automorphism of **C**, the complex numbers. What complex numbers are fixed by g? Clearly **R**, the set of real numbers. The fact that this fixed set is itself a field is no accident — we will return to this issue later.

Suppose that α is any complex root of F, so $F(\alpha) = 0$. This is an equation in **C**, so we may apply the automorphism g to both sides. Write

$$F(X) = f_0 + f_1 X^1 + \dots\dots + f_n X^n \quad \text{(and each } f_i \text{ is real)}$$

so

$$F(\alpha) = f_0 + f_1 \alpha^1 + \dots\dots + f_n \alpha^n = 0$$

and therefore

$$g(F(\alpha)) = g(f_0) + g(f_1) g(\alpha)^1 + \dots\dots\ g(f_n) g(\alpha)^n = g(0)$$

since g is a field automorphism of the complex numbers. However, each f_i is actually real, so is fixed by g. The previous equation actually says:

$$f_0 + f_1 g(\alpha)^1 + \dots\dots + f_n g(\alpha)^n = F(g(\alpha)) = 0.$$

Thus if α is a root of F, then so is $g(\alpha)$. It might be tempting to conclude that the proof were finished, since when α is not real then α and $g(\alpha)$ are distinct and are interchanged by g. In fact there is a little more to say; we must allow for the possibility of multiple roots. We deal with this as follows; $\alpha + g(\alpha)$ and $\alpha g(\alpha)$ are both real (being fixed by g) so the polynomial

$M(X) = \alpha g(\alpha) - (\alpha + g(\alpha))X + X^2$ has real coefficients.

We divide $M(X)$ into $F(X)$ as real polynomials to obtain a quotient $H(X)$ with remainder $R(X)$ of degree at most one. By the remainder theorem α is a root of the real polynomial $R(X)$. If α is not real this forces $R(X)$ to be the zero polynomial, and so $F(X) = M(X).H(X)$. The roots of F are precisely those of M and H — counting multiplicities. Now we decide that all along we were actually inducting on the degree of F, and the proof is really complete.

The rather laboured exposition of the proof of the theorem above was intentional. The length of the proof is rewarded by its generality. Suppose K is any field, and $F(X)$ is any polynomial with coefficients in K, then there is L, an overfield of K containing a root of $F(X)$. This is by no means obvious in general, indeed if we were ignorant of the properties of complex numbers, it would not even be obvious in the case that the field K were actually **R**. Nonetheless, such a field may be built readily from the data K and $F(X)$. By an inductive procedure we may even build a field S containing a full complement of roots of $F(X)$. Regarded as a polynomial in $S[X]$, the ring of polynomials with coefficients in S, the polynomial $F(X)$ will factorize into linear polynomials.

The field S that one builds is actually a smallest overfield of K in which the polynomial $F(X)$ splits into linear factors. No proper subfield M of S containing K will have the property that $F(X)$ splits into linear factors in the ring $M[X]$. For this reason the field S is called a SPLITTING FIELD for the polynomial $F(X)$ over the field K. It is not obvious that this construction is in any sense unique, but in fact it is (in a strong technical sense). Indeed, the construction of S is "sufficiently unique" that we often employ the definite article and refer to S as THE splitting field of $F(X)$ over K. This abuse is often harmless, is sometimes useful, and is never to be forgiven.

The Galois Group G of the extension of fields $K \leq S$ is defined to be the set of automorphisms of the field S which happen to leave fixed the elements of the field K. We endow G with a group operation defined by composition of maps.

Let us return temporarily to the case where K is **R**, the reals, and S is **C** the complex numbers (which are in fact a splitting field for the polynomial $X^2 + 1$ over **R**). One readily calculates that there are exactly 2 automorphisms of **C** which leave **R** fixed elementwise, the identity map and g, complex conjugation. The Galois Group of this field extension is thus of isomorphism type $\mathbf{C}_2$, the cyclic group of order 2. Notice also that **C** may be regarded as a vector space over the field of scalars **R** in a natural way, and that as such the dimension of **C** is also 2. This is not an accident. Notice also that the set of elements of **C** fixed by all elements of G is precisely **R**. This too is no accident.

There are certain obstructions to the theory when K, the original field, is infinite but of non-zero characteristic. For such fields it is possible to have a polynomial in $K[X]$ which is irreducible (ie does not factorize into polynomials of smaller degree) and yet in any splitting field of the polynomial there will be repeated roots. For fields of characteristic zero, and for finite fields, this phenomenon never happens. This trivial fact simplifies the theory greatly and so from now on we shall specialize to the case where K is either of characteristic zero or finite. In the jargon of the subject, all field extensions will be separable extensions. We remind the reader that (up to isomorphism) there is a unique field of order q, where q is any prime power, and moreover any finite field must have prime power order. The field of order q is denoted $GF(q)$. Finite fields are sometimes known as Galois fields.

The reader may care to think of K as being the rational numbers from now on. Since in computing the word real actually means rational perhaps it is sensible to concentrate on this case!

Given a field K (the rationals?) and a polynomial $F(X)$ in $K[X]$ then we can construct a splitting field for $F(X)$ called S. If we are thinking of K being the rationals, we may as well choose this field S as being the appropriate subfield of the complex numbers — the smallest field containing both **Q** and the roots of $F(X)$ in **C**, or, if you prefer, the field generated by **Q** and these roots. The fact that we may as well do this is actually neither here nor there. The existence of the ambient field **C** is an irrelevance. From the Computer Algebraists' point of view the existence of **C** is so much moonshine — not being able to represent the reals on a machine means trying to represent **C** is hopeless. Representing certain subfields of **C** is another matter entirely. This is reasonably straightforward. The psychological straightjacket of working inside **C** is not present when working with finite fields. If we all conspired to carry around with us a copy of the algebraic closure of $GF(7)$ we could do all our Galois Theory of finite fields of characteristic 7 inside this field. We don't, and it doesn't matter.

We need some notation. Suppose $K \leq L$ is a field extension and M is a field such that $K \leq M \leq L$. We say that M is an intermediate field. Given any field extension we may regard the larger field as a vector space with scalars in the smaller field. The dimension of this vector space is usually written $| L : K |$, and is called the degree of the extension. Notice that the larger field comes first. We say that the extension is finite or infinite as this number is finite or infinite. Infinite field extensions exist — $| \mathbf{R} : \mathbf{Q} |$ is infinite as a cardinality argument shows, but for us this problem will not arise. Splitting fields are finite extensions of the original field. Suppose that M is an intermediate field between K and L, as above. We have the following result:

$$|L : M| |M : K| = |L : K|$$

This formula is suggestive; the connection between this and the theory of groups is not merely notational. If, for a moment, K, M and L were groups, and $|L : M|$ were interpreted as the number of left or right cosets of M in L, the equation would hold good.

Suppose S is the splitting field for some polynomial in $K[X]$. This determines G, the group of all automorphisms of S fixing K elementwise.

Definition: G is the Galois group of the extension $K \leq L$.

G is determined up to group isomorphism by the polynomial $F(X)$, so one may speak of the Galois Group of the polynomial (though the field K must be understood).

Suppose H is a subgroup of G. We denote by $fd(H)$ the set of elements of S which are left fixed by each element of H. One checks easily that $fd(H)$ is actually a field, and of course $K \leq fd(H) \leq S$.

Suppose, on the other hand, that M is an intermediate field of between K and S. We can consider $gp(M)$, the set of elements of G which fix elementwise the whole of M. This is by definition the Galois Group of the extension $M \le S$. Note that S is still the splitting field of the polynomial $F(X)$ thought of as an element of $M[X]$.

The group of automorphisms of an algebraic structure (or any structure for that matter) is clearly an object of considerable theoretical interest, but from the Computer Algebra point of view, this group will be of limited relevance unless there is some particularly tractable concrete realisation of it — just as the group of automorphisms of a finite dimensional vector space can be viewed as a collection of non-singular matrices.

Happily there is a natural way to describe Galois groups. The Galois group G of $F(X) \in K[X]$ will permute the roots of $F(X)$. The reason for this is exactly the same as in our laboured example from secondary school mathematics. If there are n distinct roots of $F(X)$ then we have a group homomorphism from G to Σ_n, the symmetric group on n letters. In fact this homomorphism has trivial kernel; the abstract definition of G belies its natural concrete representation as a permutation group. It is in this earthy fashion that Galois would have thought of his groups.

We are now in a position to state the Fundamental Theorem of Galois Theory:

Let K be a field, and $F(X)$ a polynomial with coefficients in K. Let S be a splitting field for $F(X)$ over K, and G be the Galois Group of the extension $K \le S$.

(a) $|S : K|$ is finite and $|S : K| = |G|$.

(b) If M is an intermediate field between K and S then $fd(gp(M)) = M$.

(c) If H is a subgroup of G then $gp(fd(H)) = H$.

(d) If M is an intermediate field between K and S then $gp(M)$ is the Galois group of the extension $M \le S$.

(e) If H is a subgroup of G then $fd(H) \le S$ will be an extension with Galois Group H.

(f) If N is a normal subgroup of G then the field $fd(N)$ will be a splitting field of some polynomial with coefficients in K, and the Galois Group of the extension $K \le fd(N)$ will be (naturally) isomorphic to the quotient group G/N.

(g) If P is an intermediate field between K and S which happens to be the splitting field of some polynomial with coefficients in K, then $gp(P)$ will be a normal subgroup of G.

(h) If H is a subgroup of G then $|H| = |S : fd(H)|$ and $|G : H| = |fd(H) : K|$.

(i) If M is an intermediate field between K and S then $|S : M| = |gp(M)|$ and $|M : K| = |G : gp(M)|$.

This is rather a lot of information to absorb at a glance. What it means is this; there are inclusion reversing mutually inverse bijections between the set of subgroups of G and the collection of intermediate fields between K and L, these maps being gp and fd. Information about degrees of field extensions manifests itself inside the group G as information about subgroup orders and indices and vice versa. Information about whether field extensions of K are splitting fields of polynomials with coefficients in K manifests itself in G as information about whether the corresponding subgroups are normal and vice versa.

Working with intermediate fields is problematic, but working with subgroups of a finite group is relatively straightforward. The theorem renders the study of field extensions a much easier process. Before discussing the possible uses of this theory in Computer Algebra, we cannot pass this opportunity to mention the most celebrated application of the theory. A polynomial of degree 5 with rational coefficients may not be "soluble by radicals", that is to say it may not be possible to express the roots of the polynomial in terms of the coefficients using the operations of addition, subtraction, multiplication, division and the extraction of nth roots. That is a result of Abel/Ruffini, not as is sometimes erroneously stated, Galois. The result of Galois is much more powerful; it asserts that a given polynomial will be "soluble by radicals" if and only if the Galois group of the polynomial possesses a certain structural condition — and for obvious reasons a group with this structure is called a soluble group. The theorem of Abel/Ruffini follows easily since there exist polynomials of degree 5 with Galois Group the symmetric group on 5 letters. This group is not soluble.

It is widely suspected that every possible isomorphism type of finite group will arise as the Galois Group of some polynomial with rational coefficients, but this seems to be a deep problem however, and no proof can be expected shortly. Most of the sporadic finite simple groups have been shown to be such Galois groups, including even the Fischer-Griess Monster by Thompson [T1], [T2]. Much important work in this area has been pioneered by Heinrich Matzat [M1], [M2], [M3].

There are various ways in which the study of Galois groups could benefit Computer Algebra. We draw the readers attention to only a few. As often happens, there are ways of working modulo primes which algorithmic mathematicians may exploit. Suppose p is a prime number and $F(X)$ is a polynomial with integral coefficients; let $GF(p)$ denote the finite field with p elements. The map from $Z[X]$ to $GF(p)[X]$ obtained by reducing coefficients modulo p is a ring homomorphism. The image of $F(X)$ we call $F_p(X)$, and this polynomial has Galois Group G_p.

(i) Suppose $F(X)$ is an polynomial with integer coefficients and without repeated roots (not a problem — take the g.c.d. of $F(X)$ with its derivative to check that $F(X)$ is squarefree). $F(X)$ will be irreducible (in $\mathbf{Q}[X]$) if and only if G is acts transitively on the roots of $F(X)$. Unless p is one of finitely many "bad primes" — a set which one may calculate, the Galois Group G_p will embed as a subgroup of the permutation group G. The presence of an n-cycle in G_p will then force $F(X)$ to be irreducible. The reader unfamiliar with the theory of finite permutation groups is urged to consult [Pa] or [Wi], or some comparable text.

(ii) A more subtle version of (i). Examine the cycle-lengths for the permutation groups G_p for various good primes p. If $F(X)$ is irreducible we may be able to deduce this fact by forcing the existence of an n-cycle in G.

(iii) Suppose you wish to know how many factors $F(X) \in Z[X]$ has in $\mathbf{Q}[X]$. This turns out to be the average number of linear factors that $F(X)$ has in $GF(p)[X]$. Thus one can reduce mod various primes and work in the tractable rings $GF(p)[X]$. There is even no need to make the procedure probabilistic — one may calculate the "error term", and so know how many primes to use. This result may not appear to be anything to do with Galois Theory. The proof, by Weinberger, relies upon Galois Theory [We].

(iii) Computer Algebra `a la Davenport. The editor of these proceedings and his students have been working on the development of Computer Algebra systems which manipulate algebraic numbers symbolically rather than as ordinary complex numbers. This allows complete precision, and for applications where the limits of computer arithmetic accuracy are too low, provides a way of obviating the difficulty. The nature of this process force Galois Theoretic questions and methods upon developers of the system. In particular, the use of norms is important [ABD].

(iv) Solvability by radicals is in polynomial time [LM]. This interesting result of Landau and Miller is obtained by investigating the permutation representation of the Galois Group of the polynomial on its roots. Questions of blocks and primitivity arise. Serious readers of this paper should know some permutation group theory. In [DST] a practical improvement to the algorithm is suggested for use in the event that it actually be implemented.

Much theoretical and practical work has been done by McKay and incorporated into the computer algebra system MAPLE. For recent and interesting ideas concerning galois theory algorithms, we refer the reader to the forthcoming paper of John Dixon [D].

A note on Language:

Speakers of American English do not use the word "soluble" in a mathematical context, but only in connection with sugar etc. Some British English speakers follow the American practice, but others, myself included, use "solvable" in the context of decision theory, but "soluble" when refering to the structure of groups.

[A] Artin E., Galois Theory, Notre Dame (1942) (reprinted many times)

[ABD] Abbott J.A., Bradford R.J. and Davenport, D.H., Proceedings of 1986 Symposium a Symbolic and Algebraic Computation, Symsac '86, (1986) [ISBN 0-8979/199-7]

[DST] Davenport J.H., Smith Geoff C. and Trager B., Remarks on Solvability by Radicals is in Polynomial Time (to appear)

[D] Dixon John D., Computing Subfields in Algebraic Number Fields (to appear)

[E] Edwards Harold M., Galois Theory, Graduate Texts in Mathematics 101, Springer-Verlag (1984) [ISBN 0-387-90980]

[ff] ffitch J.P., verbal communication

[G] Galois E., Œuvres mathématiques d'Evariste Galois, J. Math. Pures et Appl. (1) 11 (1846) 381-444
(also published by Gauthier-Villars, Paris, (1897))

[J] Jordan C., Traité des substitutions et des 'equations algébriques, Gauthiers-Villars, Paris (1870)

[L] Lagrange J.L., Reflexions sur la resolution algébrique des équations, Nouveaux mémoires de l'Academie royale des Sciences et Belles-Lettres de Berlin (1770-1771)
(also Œuvres, vol 3, 205-421)

[LM] Landau S. and Miller G.L., Solvability by Radicals is in Polynomial Time, J. Comput. System. Sci. 30(1985) no 2 179-208

[M1] Matzat B. Heinrich, Konstruction von Zahl- und Funktionkorpern mit vorgegebener Galoisgruppe, J. Reine Agnew Math. 349(1984) 179-220

[M2] Matzat B. Heinrich, Realisierung endlicher Gruppen als Galoisgruppen, Manuscripta Math. 51(1985) no 1-3 253-265

[M3] Matzat B. Heinrich, Zum Einbettungsproblem der algebraischen Zahlentheorie mit nicht abelschem Kern, Invent.Math 80(1987) no 2 365-374

[Po] Postkinov M.M., Fundamentals of Galois Theory, Noordhof (1962) (translated from Russian)

[Pa] Passman D., Permutation Groups, Benjamin (1968)

[S] Stewart Ian, Galois Theory, Chapman & Hall (1973) [ISBN 0-412-108003]

[T1] Thompson John G., PSL_3 and Galois Groups over Q, Proceedings of the Rutgers group theory year 1983-84 309-319, Cambridge University Press

[T2] Thompson John G., Some finite groups which appear as Gal L/K where K cntndin $Q(\mu_n)$, J. Algebra 89 (1984) no 2 437-499

[We] Weinberger P.J., Finding the Number of Factors of a Polynomial J. Algorithms 5 (1984) 180-186

[Wi] Wielandt H., Permutation Groups, Academic Press (1986), (translated from German)

Symbolic computation in relativity theory

M.A.H. MacCallum
School of Mathematical Sciences
Queen Mary College
London E1 4NS, U.K.
E-mail: mm@maths.qmc.ac.uk

Abstract

After a brief introduction to the theory of relativity, the history and current status of computer algebra programs for this field are reviewed (though full system descriptions are not given). Their applications are described in outline, and the "equivalence problem" is discussed as a more detailed example. Recent developments are highlighted. Finally some prospects and problems are mentioned.

1 Introduction: the theory of general relativity

General relativity was one of the commonest early fields of application of computer algebra systems and remains a major application area. The principal reason for this is the heavy computation required to get from a metric to its curvature in general relativity. To explain the significance of such computations, I must attempt a thumbnail sketch of the theory.

The physical motivation can be explained by considering an experimenter in a freely falling sealed laboratory. At some point in the laboratory's plunge towards the Earth, the experimenter holds out his or her arms and releases two objects. As the laboratory falls, so will the objects, and since they are pulled towards the centre of the Earth, they will appear to the experimenter to move slowly towards one another (they would meet at the centre of the Earth, if the journey could continue that far). Thus these objects initially moving on parallel tracks in space-time (motionless in space and moving at the same rate through time) eventually meet.

An analogous phenomenon occurs on the surface of the Earth. Two paths heading due North from different points of the equator (i.e. on initially parallel tracks) will meet at the North Pole. This does not happen on a flat surface, which suggests that the effect should be related to a concept of curvature. A mathematical formulation of this concept which is closely related to the previous ideas is obtained by considering transport of a pointer round a loop, moving it so that at any two neighbouring points it lies in parallel directions (both the previous examples can be described like this, using "curved triangles" as the loops). The initial and final directions of the pointer, before and after its transport round the loop, will differ by an amount related to the magnitude of the curvature.

To describe this more quantitatively, it is helpful to introduce some coordinate system in terms of which the pointer, when at the point with coordinates x^i $(i = 1, 2, ...n)$, can be represented as a vector with components v^i. The parallel vector at the neighbouring point at $x^i + dx^i$ can be written as

$$v^i - \Gamma^i{}_{kj} v^k dx^j$$

where the convention is used that repetition of an index (e.g. k) implies summation over all its possible values. Γ is called the connection. Then by estimating the integral of this round the loop using Picard's method, one can show that for a small loop the difference between the initial and final positions of the pointer is given (to a first approximation) by $R^i{}_{jkl} A^{kl}/2$ where $A^{kl} = \int x^k dx^l$ represents the area of the loop as projected into the (x^k, x^l) plane and

$$R^i{}_{jkl} = \partial\Gamma^i{}_{jl}/\partial x^k - \partial\Gamma^i{}_{jk}/\partial x^l + \Gamma^i{}_{sk}\Gamma^s{}_{jl} - \Gamma^i{}_{sl}\Gamma^s{}_{jk}.$$

We usually think of two vectors as parallel if they point in the same direction in Euclidean space. This can be described by saying two vectors are equal if the two pairs of ends of the vectors have the same separation. The appropriate generalization, which leads to Riemannian geometry, is to define lengths and angles from the "line-element"

$$ds^2 = g_{ij} dx^i dx^j$$

where g_{ij} is called the metric. (This formula would be just Pythagoras' theorem in Euclidean space.) The resulting formula for parallelism is

$$\Gamma^i{}_{jk} = g^{im}(g_{mj,k} + g_{mk,j} - g_{jk,m})/2$$

where g^{im} is the matrix inverse of g_{ij} and the commas signify taking partial derivatives (so $f_{,i} = \partial f/\partial x^i$).

Finally to fix the amount of curvature produced by a gravitating object we need to specify equations between the curvature and the matter present, and this is done, in general relativity, by using the Ricci tensor defined by $R_{ij} = R^k{}_{ikj}$ in equations of the form $R_{ij} = \kappa\Theta_{ij}$, where κ is a coupling constant whose numerical value depends on the units used, and Θ_{ij} is given by the energy and momentum of the matter.

2 Development and status of computer algebra systems in relativity

One can easily see the strong incentive to use computer algebra in relativity by comparing the equations above with those of Newtonian gravity theory: there are 10 dependent variables (the components of g_{ij}) instead of 1 (the Newtonian potential ϕ), non-orthogonal curvilinear coordinates are in common use, and the field equations are of 8th degree rather than linear in the dependent variables. The net result is that instead of a single equation with three terms for the dependent variable we have 10 equations, 6 of which have 13280 terms and the rest 9990 terms in the metric components. (These numbers themselves were checked by computer algebra, the first to do this being Lars Hörnfeldt and Richard Pavelle.) Although other ways of formulating the equations (e.g.

the use of differential forms as in EXCALC, mentioned below, or the Clifford algebra methods used in the GEOCALC system presented by Tombal et al. at Leipzig) reduce these numbers, the basic difficulties remain.

It should be noted that the design of computer algebra systems for relativity has some special features which distinguish the resulting systems from general purpose systems. The first of these is the need for efficient storage and retrieval of tensor components, making full use of the symmetries. For example, the Riemann tensor R_{ijkl} has 256 components (in 4 dimensions) but only 20 of them are linearly independent. Thus one needs combinatorial algorithms to handle the symmetries. In the early systems, these were "hard-wired" into the code; modern systems provide more convenient means to define new tensors' symmetries.

A second point is that the dimensions of the spaces used is small (4 for general relativity, and not more than 10 or so for most of its extensions). Algorithms with good asymptotic properties for large n are not necessarily good for small n (after all, n is never in practice infinite) and attention has to be given to the small n performance.

Moreover, systems for relativity make little use, in many cases, of integration, polynomial factorization and so on, so these facilities and the necessary support for them can be dispensed with, producing a more efficient and smaller system. On the other hand differentiation and very good control of substitutions (superior, for example, to that provided by REDUCE and MACSYMA) are needed, the latter being usable as a substitute for factorization and g.c.d.'s in simplification.

With these requirements, the hardware constraints of the 1960s led people to build their own specialized systems for relativity. Many such systems were discussed in the survey by d'Inverno [1] but almost all of these are effectively defunct. Nevertheless, there still is and will continue to be a demand for smaller systems until the present more advanced technologies have become readily available world-wide. On the other hand, even the smaller systems are becoming more sophisticated (e.g. the use of differential forms in the system GRG discussed at Leipzig by Obukhova et al).

Of those 1960s systems only two are still in regular use. One is the CAMAL system developed by John Fitch [2], and the other, which survives in a disguised form, is Ray d'Inverno's LAM. The latter was developed in a series of versions for different machines such as ALAM, CLAM and ILAM for Atlas, CDC and IBM mainframes, was incorporated into SHEEP, originally written in MACRO-10 for Digital Equipment machines, and finally grew into the present-day SHEEP 2, written in Standard Lisp. SHEEP is largely the work of Inge Frick, with contributions from Jan Åman, myself and others: it is rarely used in bare form as most applications need either the package of programs for component-wise calculations called CLASSI, whose principal author is Jan Åman, or the indicial tensor package STENSOR written by Lars Hörnfeldt. (The distinction here is that CLASSI can, for example, calculate R_{12} in terms of, say, $g_{01}, g_{02}, ...etc$, while the output of STENSOR consists of equations exactly like the formulae in the above Introduction, e.g. $\Gamma^i{}_{jk} = g^{im}(g_{mj,k} + g_{mk,j} - g_{jk,m})/2$, with the indices left purely symbolic. However, the capabilities of the two packages overlap to some extent, and in particular STENSOR can produce SHEEP/CLASSI programs defining new tensors.) CLASSI and STENSOR began life in the late 1970s, and are still being actively developed.

The only substantial special purpose system written subsequently is Krasinski's ORTOCARTAN [3]. The program SCHOONSCHIP [4], developed for quantum field theory calculations, has also been used in quantum gravity. Apart from these cases, most applications of computer algebra in relativity in recent years have made use of one of the general purpose algebra systems. This is becoming an increasingly viable option as the power and storage capacity of machines grows, and I now consider the available packages.

From the 1960s onwards a number of packages were written in FORMAC (see [1]), but this system is now obsolete in North America and Western Europe. MACSYMA has, among its distribution software, two packages developed in the late 1970s and used, for instance, by Pavelle. One of these, CTENSR, is for tensor component calculations in coordinate bases and the other, ITENSR, is for indicial tensor calculations.

There are a considerable number of relativity packages written in REDUCE: many authors developed their own packages, some of them quite effective. (A simple one is among the REDUCE test suite.) Most of these have been superseded in flexibility and power (though not necessarily in efficiency) by the EXCALC package (see [5]). Another possibly interesting approach is to build on top of REDUCE a system with facilities like those of SHEEP, and there are two projects to do this, one of which, Dautcourt's GENRE, was presented at Leipzig (the other is RSHEEP, a single image containing both REDUCE and SHEEP, built by Jim Skea at QMC).

MAPLE has a growing number of relativity packages, and these are under active development at the University of Waterloo. For instance there is a package to handle Newman-Penrose formalism, called NP. Moreover, some of the same programs have also been converted to muMATH [6], for which other more specialized programs, e.g. for handling the Ernst equations which arise from metrics with special symmetry, also exist.

These packages have been reviewed in more detail in various past works [7,8,9] and I do not propose to repeat all the details now. Let me make, though, the important point that the main systems - SHEEP (including CLASSI and STENSOR), REDUCE (and EXCALC), and MAPLE - all run perfectly well on desktop machines costing less than US $5000, in some cases notably less. It is this development which can bring such tools to the desk of every relativist.

It is important to recognise that a good system for CA in GR these days should provide more than just curvature calculations. Unfortunately, a proper explanation for non-experts of the nature of the desirable extra facilities would take several pages, so I can do no more here than mention some words which will mean little to non-relativists. An easy way to define new tensors, algorithms for Petrov and Ricci tensor classifications, ways to deal with splitting into sub-manifolds (for 3+1 formalism or Kaluza-Klein), and calculus of variations are among the features one would wish to develop and which do already exist in some of the present systems.

To conclude this section, I would like to highlight some particular recent developments. EXCALC is a program for exterior calculus in close-to-textbook format written in REDUCE by Eberhard Schrüfer. It can handle a wide range of problems in differential geometry and is coming into use for relativity [5]; its textbook-like notation makes it very easy to use. I am sure it will become a popular package.

An important area of application of computer algebra systems in relativity is at the interface with numerical relativity, a special example of the general development of interfaces between algebraic and numerical computing (for which see e.g. [10]). Numerical relativity is a major supercomputer user for exactly the same reasons as computer algebra is used for analytic computation in relativity, and is just begining to be able to do fully four-dimensional (three space and one time) calculations. Computer algebra is already playing an important part in making this possible (see the articles by Tsvi Piran and Takashi Nakamura in [11]), and its role will no doubt increase as work on the symbolic/numerical interface progresses. (It is worth noting that as well as the generation of code, symbolic computation can be used to analyse numerical schemes, e.g. for stability and convergence, as was done in [12], in which the identification of a singularity in a numerical scheme was actually carried out using MACSYMA.)

The third development I wish to highlight is the range of facilities offered by STENSOR. It can handle many types of non-commuting objects, deal with splittings of manifolds into submanifolds (e.g. Kähler geometry, Kaluza-Klein theory), substitute for sums (e.g to make efficient use of the identity $\sin^2 x + \cos^2 x = 1$), use disk files to manage large calculations by 'bucketing' (a simple hashing), and in particular has very clever methods for dealing with sums of terms with apparently different dummy indices, including the exploitation of symmetries.

In its handling of commutation rules, it has a range of different classifications. Objects may be defined to be non-commuting, never-commuting, never-commuting but commuting with specified objects, or as always commuting. These can be used, for example, to define the octonians and perform calculations with them. A special category is available for differential operators of various types. These facilities make the system very flexible for use in areas such as classical and quantum mechanics (where Poisson brackets and operator commutator relations must be handled) and in supersymmetry, supergravity and superstrings, the theories of modern physics in which almost all the categories of commutation rule just described arise.

Like the commutation rules, the splitting rules are very important to modern theories of gravity, and unified theories including gravity. Almost all of these involve spaces of dimension greater than 4 in which subspaces have different properties and eventually different physical roles. Kaluza-Klein methods split the dimensions into the space-time and internal dimensions of the matter field, while Kähler spaces, which are complex and can be split into real and imaginary parts, are used not only directly as models of space-time but also in complex generalizations and abstractions which aim to provide new ways of approaching the unification of the forces of nature (a problem for which supergravity represents the end-point of an evolution of the standard quantum field theory). To achieve the handling of these problems, STENSOR allows the introduction of several different index types for the different subspaces.

The sum substitutions are an interesting area mathematically. For example, STENSOR's trigonometric simplifier can turn the expansion of $(1-\cos^2 x)^5(1-\sin^2 x)^5(\cos^2 x+\sin^2 x)^5$ into the more natural $\sin^{10} x \cos^{10} x$ (91 terms reduced to 1) fairly rapidly by a recursive scan of neighbouring possible values of the powers of $\sin x$ and $\cos x$. An application in the type of geometrical problem of use in gravity has been in the calculation of the curvature of n-dimensional ellipsoids in $(n+1)$-dimensional space. However, the procedure used, though clearly effective, has not been strictly proved to be optimal, and this suggests a problem for future research. The difficulty arises because the criterion for simplicity here is the number of summed terms, and although this clearly gives rise to a well-determined ordering of expressions, this ordering does not, under multiplication, obey the necessary rules for the application of the theorems of the Gröbner basis approach.

The recognition of terms with apparently different "dummy" indices (repeated indices implying summation over all values) is another impressive feature (it involves a rather complicated and subtle ordering and substitution recursion): what this means is that the system can recognise $A^{ijk}B_{kl}$ and $A^{ijm}B_{ml}$ as the same (and of course can also handle much more complicated examples of this type, as well as the problem of ordering of indices to take symmetries into account). An amusing example of the combination of this and the previous feature (sum substitution) is the discovery of the identity

$$R^{ijkl}R_{ij}{}^{mn}R_{klmn} = 4R^{ijkl}R_{ik}{}^{mn}R_{jmln}$$

from the symmetries of the Riemann tensor and the identity $R_{ijkl} + R_{iljk} + R_{iklj} = 0$. The discovered identity is not obvious, and choosing the right manipulations to prove it by hand is tricky.

3 Applications of computer algebra in relativity

There are three main ways of approaching a theory of mathematical physics (assuming we are not ambitious enough to be constructing one of our own) namely to prove general theorems about it, to apply (analytic or numerical) approximation schemes, and to find (necessarily very special) exact solutions. The first of these is meritorious but very difficult, the second is what most applied mathematicians do although it frequently suffers, in the case of non-linear theories like GR, from lack of a rigorous basis for the scheme used, and the third, because it takes exact account of the non-linearities, may be useful in providing simple models, in generating comparisons to test against the approximation schemes (see e.g. [13]), and in other ways, despite the very special character of the solutions.

CA has been used in all these approaches, amd many references can be found in earlier reviews [8,9]. Here for brevity I will simply indicate some of the ways in which the systems can be used.

For exact solutions one can verify results for a given metric; one can use computer algebra as a tool when seeking new solutions; and one can compute additional properties such as Petrov type or symmetry group. The verification methods should be flexible enough to work in coordinates and in constant and variable frame bases. As a tool for finding solutions, systems have been used both as a 'workpad' while carrying out integrations by hand, and in constructive methods where these exist, e.g. for applying the transformations of Ernst potentials which give new solutions with commuting Killing vectors. The range of additional properties one could want is considerable, but there are already algorithms for Petrov and Segre types, for dimension of isometry and isotropy groups, and their Lie algebra structure, and for the unique characterization of a solution (see below). Further aspects of the search for properties are the investigation of exact solutions of variants of Einstein's theory to test their experimental viability, the study of singularities and horizons and the study of physical fields, e.g. quantum field theory, in a given space-time.

On the level of approximations, one can do perturbation calculations, and thus for example study stability and oscillations of stars or the growth of cosmological fluctuations. One can also investigate asymptotic properties such as the cosmological far-future or the approach to singularities. Another type of perturbative calculation one can do is the calculation of scatterings in quantum gravity. Finally, one can use CA with numerical methods as mentioned above in two ways, to generate the necessary equations, or to investigate the properties of numerical schemes.

General theoretical calculations such as derivation of field equations from a Lagrangian, indicial equation calculations, and so on, are carried out too, mainly with systems like ITENSR or STENSOR (whereas, in the case of SHEEP, the applications mentioned in the previous paragraphs would use CLASSI). These can be useful in studies of general theorems.

I would like to emphasize that such applications are equally available for theories other than GR, and many of them have been used in such cases as quantum gravity, supergravity, string theory or alternative classical theories of gravity. For example, I was recently able to help colleagues investigating representations of superalgebras by providing a simple REDUCE program [14].

4 The equivalence problem

This particular application is concerned with the characterization and identification of solutions of the Einstein equations. Although we may never be able to find a general solution, many particular solutions are known. They are found by simplifying the equations in various ways, e.g. by imposing symmetry or some special algebraic structure [15]. This leads to the basic problem of recognising the same metric in two coordinate systems, or of characterizing a metric uniquely in a coordinate-free way. In fact the method starts by doing the second of these, which for a sufficiently differentiable metric is an algorithmic procedure, and then prescribing how to compare two metrics, which is a well-defined problem but has no algorithmic solution.

Finding the basic method is a purely mathematical problem. Having achieved that, one has to find algorithms. Then one must provide a practical implementation, and finally one should study the physical interpretation problem that arises because the local gravitational field properties must be entirely coded in the invariant characterization. Thus the work mixes pure mathematics, software engineering, and physics.

The pure mathematical parts have been reviewed elsewhere [16] [17]. The problem has a long history going back to Christoffel, and the reason the comparison of metrics is not formally decideable is that it involves determining the existence of solutions of a set of simultaneous equations, which may be non-linear or perhaps transcendental, and even for comparatively simple classes of functions no algorithm for such a problem is possible.

It would be inappropriate in a review for the present audience to go into full details of the method, but I will give a statement of it motivated by heuristic arguments. The basic idea, which relativists would quickly spot, is to use invariants, e.g. to equate, in two spaces, the values of the Ricci scalar $R = g^{ij}R_{ij}$. Unfortunately the example of plane waves and flat space (in both of which all scalar polynomial invariants of the curvature vanish) shows that such invariants are not sufficient to distinguish two spaces.

This shows us we need tensor components, but these depend on the choice of vector basis as well as the space-time position. Thus we must work on a frame bundle over space-time, and try to provide a unique characterization of location in this larger space. To do so we consider the Riemann tensor and its (covariant) derivatives, since these give the only possible invariants of a Riemannian geometry (extension to other problems which can be thought of in terms of a connection or a set of connections is quite direct [17]). Now it is easy to see that if the (p+1)-th derivative depends on the derivatives up to the p-th, all subsequent derivatives also depend on the first p (by repeated differentiation of the relation for the (p+1)-th and back-substitution). Thus we will at worst need to differentiate up to $p = d$ where d is the dimension of the frame bundle considered.

However, using coordinates in 4 dimensions gives a frame bundle of dimension 20, and the derivatives of the Riemann tensor up to the 20th have 29 320 310 074 020 terms of which 79 310 are independent. An important reduction is made by using constant frames (i.e. a basis of vectors with constant scalar products, a generalization of the idea of unit vectors) in which case the dimension is 10 and the number of possible independent terms 8690: this was first noted by Cartan, and its application proposed by Brans. A further improvement (also based on ideas of Cartan [17]) was given by Karlhede who suggested that the dimension of the frame bundle to be dealt with could be reduced by fixing the frame at each step of calculation so that the Riemann tensor and its derivatives are at each step in a canonical form. Working this way one can easily show (for the GR case) that at most the seventh derivative is required (with a maximum of 3156 independent terms), and experience so far suggests that no more than the third derivatives (at worst 430 terms) are really needed.

Thus the procedure is to start with an order of differentiation $q = 0$. Then:

1. Calculate the derivatives of the Riemann tensor up to the q-th.
2. Find the canonical form of the Riemann tensor.
3. Fix the frame as far as possible by this canonical form, and note the residual frame freedom (isotropy group).
4. Find the number of independent functions of space-time position in the components of the Riemann tensor and its derivatives in canonical form, and record that.
5. If the isotropy group and number of independent functions are the same as at the previous step, stop,
or otherwise (or if $q = 0$) increment q by 1 and go to step 1.

The space-time is then characterized by the canonical form used, the successive isotropy groups and independent function numbers, and the values of the non-constant components of the Riemann tensor and derivatives which are found. To compare two space-times one can first compare the discrete properties such as the sequence of isotropy groups, and only if those all match does one have to check whether the set of equations obtained by equating corresponding components of the Riemann tensor and its derivatives in canonical form have a solution.

The implementation of these steps has been the major impetus behind the development of CLASSI. Step 1 for $q = 0$ is straightforward, but steps 2 and 3 then need careful investigation of the Petrov and Segre classifications and efficient methods of locating the canonical frames. They have been implemented but the details are not yet published. Steps 4 and 5 (for any q) are again straightforward, but returning to step 1 with $q = 1$ leads to the necessity of characterizing a minimal set of derivatives on which the others depend. This again has been done [18] and makes the whole procedure workable, but further effort is going into consideration of steps 2 and 3 at $q = 1$ and higher q, and into finding the least upper bound on the required q.

5 Problems and prospects

Of course, in discussing the possible future of symbolic computing in gravity theory one must remember that one person's prospect is another person's problem, but I would like to mention a few areas of possible improvement.

The first is that improvements to the user interface will make systems much easier to use in what I call "calculator mode", i.e. where all the user has to do is type a simple command interactively, without having to write programs or submit batch jobs, and with the convenience of an easily intelligible graphical input and output presentation. This applies to computer algebra systems in general, and in that context I would add that we need high-volume low-cost high-functionality systems to match Hearn's price/performance predictions for hardware: none exist so far.

At the other extreme we must be asking what are the useful limits of computational power. In a recent extension to 5 dimensions of his calculations of the number of terms in the Ricci tensor, Lars Hörnfeldt found the diagonal components contain 263 598 terms each. To do this calculation proved impossible even with 48Mb processes on a High Level Hardware Orion, without the use of bucketing. With bucketing, 1000 files containing 40Mb of data were generated. While the result is impressive, one cannot help wondering how useful it will be to go further (except to use as an argument to discourage people working on higher-dimensional field theories).

I do expect a further development of integrated use of symbolic, numerical and graphical systems in relativity, and I expect to see development of systems for analysis as well as generation of numerical methods. I also expect fuller functionality of the GR packages in various systems (for example, the MAPLE programs at present behave as stand-alone programs rather than an integrated system like SHEEP, and the REDUCE programs lack easy ways of defining new tensors - except for indexed forms in EXCALC - and expressing their symmetries). I also anticipate further improvements in ways of dealing with the special features I mentioned at the start of section 2, i.e. of all the old problems.

References

[1] R.A. d'Inverno, "Algebraic computing in general relativity", *Gen. Rel. Grav.*, **6**, 567 (1976).

[2] J.P. Fitch, "An algebraic manipulator", Ph.D. thesis, Cambridge University (1971).

[3] A. Krasinski, "The program ORTOCARTAN for applications in Einstein's relativity theory", *Lect. Notes Comp. Sci.* **204**, 159-160, Springer-Verlag, Berlin and Heidelberg (1985).

[4] H. Strubbe, "Development of the SCHOONSCHIP program", *Comp. Phys. Commun.* **18**, 1-5 (1979).

[5] E. Schrüfer, F. Hehl & J.D. McCrea, "Application of the REDUCE package EXCALC to the Poincare gauge field theory", *Gen. Rel. Grav.*, **19**, 197 (1987).

[6] C.C. Dyer, S. Allen & J.S. Harper, "The MuTENSOR computer system for general relativity", in *Abstracts of the 11th international conference on general relativity and gravitation, Stockholm, Sweden*, ed. B. Laurent and K.Rosquist (1986).

[7] R.A. d'Inverno, "A review of algebraic computing in general relativity", in *General relativity and gravitation: one hundred years after the birth of Albert Einstein, vol. 1*, ed. A. Held, Plenum Press, New York and London, 491-537 (1980).

[8] M.A.H. MacCallum, "Algebraic computing in general relativity", in *Classical general relativity*, ed. W.B. Bonnor, J.N. Islam and M.A.H. MacCallum, Cambridge University Press, London and New York, 145-171 (1984).

[9] M.A.H. MacCallum, "Algebraic computing in relativity", in *Dynamical spacetimes and numerical relativity*, ed. J.M. Centrella, Cambridge University Press, London and New York, 411-445 (1986).

[10] J.A. van Hulzen, "Program generation aspects of the symbolic-numeric interface", in *Proceedings of the international conference on computer algebra and its applications in theoretical physics*, ed. N.N. Govorun, Dubna (1985).

[11] M.A.H. MacCallum (ed.), "*General Relativity and Gravitation (Proceedings of the 11th International conference on general relativity and gravitation)*", Cambridge University Press, London and New York (1987).

[12] J.W. Eastwood & W. Arter, "Interpretation of disruptions in tokamaks", *Phys. Rev. Lett.* **57**, 2528-2531 (1986).

[13] J.M. Centrella, S.L. Shapiro, C.R. Evans, J.F. Hawley & S.A. Teukolsky, "Test-bed calculations in general relativity", in *Dynamical spacetimes and numerical relativity*, ed. J.M. Centrella, Cambridge University Press, London and New York, 326-344 (1986).

[14] J.W.B. Hughes & R.C. King, "A conjectured character formula for atypical irreducible modules of the Lie superalgebra $sl(m/n)$", *J. Phys. A* **20**, L1047-52 (1987).

[15] D. Kramer, H. Stephani, M.A.H. MacCallum and E. Herlt, "*Exact solutions of Einstein's field equations*", Deutscher Verlag der Wissenschaften, Berlin, and Cambridge University Press, London and New York (1980).

[16] M.A.H. MacCallum, "Computer-aided classification of exact solutions in general relativity", in *Gravitational collapse and relativity*, ed. T. Nakamura and H. Sato, World Scientific, Singapore (1986).

[17] N. Kamran, "Contributions to the study of the equivalence problem of Élie Cartan and its applications to partial and ordinary differential equations", *Princeton University preprint* (1988).

[18] M.A.H. MacCallum and J.E. Åman, "Algebraically independent n-th derivatives of the Riemann curvature spinor in a general space-time", *Class. Quant. Grav.* **3**, 1133-1141 (1986).

A Zero Structure Theorem for Polynomial-Equations-Solving and its Applications

Wu Wen-tsun
Institute of Systems Science
Academia Sinica

Abstract

Let PS be a finite set of non-zero polynomials in variables $X_1, \ldots, X_n$ with coefficients in a field K of characteristic 0. Denote by K'–Zero(PS) or simply Zero(PS) the totality of zeros in some extension field K' of K of polynomials in PS. A certain polynomial set CHS of special type can then be formed from PS in an algorithmic manner for which the zeros of CHS will have a close relation with Zero(PS) expressed in some form to be called Zero Structure Formula. This formula is at the basis of a general method of solving systems of polynomial equations. Programs have been accordingly implemented on some microcomputer of 1MB memory. Experiments for comparing the efficiency of this new method and some known methods of equations-solving have been made. Besides, some direct method of finding the resultant of two polynomials with multivariate polynomial coefficients is introduced and comparison with the well-known Collins PRS methods is also made.

SOME ALGORITHMS OF RATIONAL FUNCTION ALGEBRA

S.A.Abramov
Acad. of Sciences of USSR, Computing Center

This lecture is concerned generally with linear difference equations

$$R(x+t)+l_{t-I}R(x+t-I)+\ldots+l_0R(x)=F(x). \qquad (I)$$

with F(x) as rational function. For the coefficient field there exist the algorithms of arithmetic operation and element equivalence recognition . This field should be suitable in terms of the following definition: a field K of a characteristics 0 is suitable, if an algorithm exists that finds the greatest integer root of an equation P(x)=0 for any P(x) K[x]; here an integer n is identified with ne, where e is the identity of field K. The field $\mathbb{Q}$ of rational numbers is an example of a suitable field. A simple extension K(θ) of a suitable field, algebraic or trancendental, is a suitable field. As a consequence any field of algebraic function of the finite number of variables $x_I,\ldots,x_n$ over $\mathbb{Q}$ is suitable.

An algorithm has been proposed [2] for recognizing a rational solution (i.e. a solution in K(x)) of the equation (I), if one exists, to construct all set of the rational solutions. In the theory of differential equations the powers of various factors in denominator f(x) of the right hand side F(x) are very important. For our theory, however, the greatest h for which the polynomials f(x) and f(x+h) possess a common divisor is more important. This value is denoted by Dis F(x). The following propositions are proved:

a) The value Dis F(x) may be evaluated for any $F(x)\in K(x)$, where K is suitable field.

b) Let R(x) satisfy (I) and let $l_0\neq 0$, then Dis R(x)=Dis F(x)-t.

c) Let R(x) satisfy (I) and Dis R(x)<c for some integer c. Then for R(x) one may derive linear difference equation with rational right hand side and coefficients in K which contains only the terms $b_mR(x+mc)$ (here $b_m\in K$).

In this algorithm one does not need to analyze the roots of characteristic equation and to use the method of indefinite coefficient.

In the absence of rational solution it may be desirable to isolate a rational component from the solution so that other component will satisfy the more simple differece equation. An algorithm for isolating the rational component in the first order equation has been described in detail [3]. The second component of the solution satisfies an equation with the same left hand side but with denominator of the lowest possible degree in the right hend side. This algorithm again requires no factorization of functions over the simplest fractions. Analogous algorithm but more complex for the second order equation has been described [4]. Solution of the problem in this case is based on the proven theorem given here in the following simplified version.

Let $l_0, l_I \in \mathbb{Q}, l_0 \neq 0$. Then an algorithm exists which for any polynomial $f(y) \in \mathbb{Q}[y]$ may recognize such rational non-negative integer n that $f(y) \equiv uy^n \pmod{y^2 + l_I y + l_0}$. If the answer is positive this algorithm gives all pairs (u,n).

Summation of rational functions has been examined in detail. Finding $\sum_{x=p}^{q} F(x)$ obviously reduces to solving the equation

$$R(x+I)-R(x)=F(x). \qquad (2)$$

The above-mentioned general algorithm may be improved for equation (2). To isolate the rational component in (2) the algorithm expands $\sum_{x=p}^{q} F(x)$ in form $A(p,q)+ \sum_{x=p}^{q} S(x)$, where $A(p,q) \in K(p,q)$, $S(x) \in K(x)$ and the denominator of S has the lowest possible degree. It is analogous to Ostrogradsky - Hermite s method for rational function integration.

An algorithm has been described [5] for expanding the rational function in terms with separation of varibles. Tne coefficient field may be arbitrary. The simplest problem is the following: if possible, expand the given rational function R(x,y) as $S_I(x)T_I(y) +\ldots+ S_m(x)T_m(y)$, where $S_i(x) \in K(x)$, $T_i(y) \in K(y)$, $i=I, \ldots, m$. The general case: let $R(x_I,\ldots,x_n) \in K(x_I,\ldots,x_n)$, and let the set of variables $x_I,\ldots,x_n$ be partitioned into τ nonoverlapping subsets

$$\{x_I,\ldots,x_i\}, \{x_{i_I+I},\ldots,x_{i_2}\}, \ldots, \{x_{i_{\tau-I}+I},\ldots, x_n\} \; ; \qquad (3)$$

is it possible to expand $R(x_I,\ldots,x_n)$ over products of rational functions with coefficients in K and any product satisfying two conditions:

a) for any factor there is a subset in (3) of which this factor is independent;

b) if a factor depends on at least one variable of some subset in (3), then all other factors in the product are independent of the variables of this subset.

When separation of the variables is possible, it should be done.

This problem is concerned with reducing the multiplicity of a rational function sum. For example, for the sum

$$\sum_{i=0}^{n} \sum_{j=0}^{n} \sum_{k=0}^{n} \sum_{l=0}^{n-j} \sum_{m=i}^{n} R(i,j,k,l,m)$$

the whole set of varables $\{ i,j,k,l,m \}$ is partitioned into subsets $\{ i,m \}$, $\{ j,l \}$, $\{ k \}$. If one could expand tne function R(i,j,k,l,m) over above-mentioned products, the initial fivefold sum would reduce its multiplicities. In this case one may derive the following expressions:

$$\sum_{i=0}^{n} \sum_{m=i}^{n} \sum_{k=0}^{n} F(i,m,k) \sum_{j=0}^{n} \sum_{l=0}^{n-j} G(j,l) +$$

$$+ \sum_{k=0}^{n} V(k) \sum_{j=0}^{n} \sum_{l=0}^{n-j} W(j,l) + \sum_{j=0}^{n} \sum_{m=j}^{n} T(i,m),$$

ect. The problem of separation of variables arises not only in multiple sums or integrals, but also in solving the differential equations, in investigation of the operator kernel degeneracy. The algorithm is described as recursive. Its application requires no complete factorization of polynomials and expansion of rational function over tne simplest fractions.

It may be noted finally that the algorithm of summing the rational functions, that of a search of rational solution of equation (I) and also the analogy of Ostrogradsky - Hermite's method has been derived and published in I97I-I975, i.e. before M.Karr, J.Cohen, J.Katcoff and R.W.Cosper's intrinsic studies.

REFERENCES

I. Abramov S.A. On the summation of rational functions. - USSR Comput. Math. Math. Phys., I97I, II, No. I.
2. Abramov S.A. Solution of linear finite-difference equations with constant coefficients in the field of rational functions. - USSR Comrut. Math. Phys., I974, I4, No. 4.
3. Abramov S.A. The rational component of the solution of a first-order linear recurrence relation with a rational right side.- USSR Comput. Math. Math. Phys., I975, I5, No. 4.
4. Abramov S.A. Second-order finite-difference equations with constant coefficients in the rational function field. - USSR Comput. Math. Math. Phys., I977, I7, No. 3.
5. Abramov S.A. Separation of variables in rational functions. - USSR Comput. Math. Math. Phys., I985, 25, No. 9.

THE COMPUTER ALGEBRA SYSTEM SIMATH

R. Böffgen and M.A. Reichert
Departement of Mathematics
University of Saarbrücken
D–6600 Saarbrücken

SIMATH (that is, SINIX–mathematics) is a computer algebra system which is being developed at the Departement of Mathematics of the University of Saarbrücken on a Siemens Personal Computer MX 2.

What are our intentions in introducing another new system? How does it differ from already existing systems? We think that it would be desirable to develop a general portable computer algebra system having two important features. First it should be based on a well-known programming language (like "C"), so that the user is able to work with it right away in an efficient manner, and second, it should be easily extendable to special applications (for example, in constructive number theory) by the user himself.

The complete SIMATH–system operates under the conventions of "C". There are two distinct ways of working with SIMATH.

First, the calculator SIMCALC gives an easy access to the system. All SIMATH–functions are available in an interactive mode. Thus the calculator can be used for intermediate calculations, in case it is not worth the effort of writing a new program. SIMCALC is equipped with some useful help features; it allows recording, tracing of execution times and state of storage, as well as syntax testing of the input lines. Furthermore, it is possible to display previous input lines on the screen and to re–edit them for the purpose of repeated execution.

Second, calls of SIMATH functions can be utilized in programs written in "C" or in FORTRAN.

System initialization and administration is carried out under the user level, so that system handling causes no problems at all. Similarly to SAC-2, for internal data representation, SIMATH is based on a list system. Emphasis lies on high speed arithmetic algorithms, required for higher applications of SIMATH in constructive number theory, algebraic geometry and group theory. For example, in order to compute the Mersenne–prime number $2^{132049} - 1$ by a "general" exponentiation–algorithm (that is, without shift operations for the powers of 2), SIMATH needs only a running time of 6.6 seconds on the MX 2 machine. This number consists of 39751 decimal digits!

SIMATH is constructed as an open system with accessible data–structures and source–codes. This facilitates

- the adaption of available general algorithms to the special problems of the user,
- the addition of user–designed algorithms to any section of the system.

The hierarchical structure of the program collection can be utilized by the user as the need arises of extending the system for the purpose of other applications. This can be done starting from an appropriate level of the hierarchy.

The computer algebra system SIMATH consists of

- the programming language "C";
- a list processing with automatic garbage collector and dynamic space;
- fundamental software libraries
  - for list administration, sorting and searching
  - for multiple precision arithmetic containing standard algorithms over the integers, the rationals, over residue class rings, prime fields or arbitrary finite fields, for example gcd, chinese remainder theorem, primality testing and integer factorization.
  - for multivariate polynomial arithmetic over various rings and fields with a collection of useful routines (evaluation, substitution, resultant, discriminant, Hensel–Lemma lifting, factorization over the rationals, over prime fields or arbitrary finite fields), with polynomials given either in sparse or in dense representation.
  - for matrix arithmetic over the above–mentioned structures as well as over polynomial rings.
- a special library in constructive number theory with algorithms concerning
  - elliptic curves: minimal model of a curve, arithmetic of rational points, Néron–Tate–height
  - algebraic number fields: decomposition of primes, integral bases
  - congruence function fields: integral bases,continued fraction expansions, unit groups, class groups;
- software libraries to be implemented by the user for special applications;
- extensive tools for tracing and testing;
- the calculator SIMCALC which makes all functions available in an interactive mode.

To date about 500 algorithms of SIMATH are available — including an exhaustive documentation. They can be obtained from the SIMATH Group:

SIMATH Group
Fachbereich 9 Mathematik
Universität des Saarlandes
D – 6600 Saarbrücken

Prof.Dr. H.G. Zimmer
Dipl.-Math. M.A.Reichert
Dipl.-Math. R.Böffgen
Dipl.-Math. B.Weis

Converting SAC–2 Code to Lisp

Lars Langemyr*
NADA, Royal Institute of Technology, S–100 44 Stockholm, Sweden.

Introduction and Motivation

The SAC-2 computer algebra system is designed to perform operations on multivariate polynomials over a variet of computational domains. It was developed by Collins and Loos [2], and evolved from SAC-1. The SAC-2 systen consists of a series of modules all written in ALDES and each module in the series adds a functional capability to th system. In the distributed version of SAC-2 a translator for the ALDES language to ANSI FORTRAN [5] is included

We have used the SAC-2 system and the ALDES FORTRAN translator for algorithm development under the UNIX operating system. It has been noted that a *lisp* environment where the same algebraic algorithms as in SAC- existed would be nice for algorithm development. Using the FORTRAN-based system we have observed the followin problems: • If we run a program that contains an error, we normally get an illegal memory reference or a bu error signal which aborts the program and produces a core dump. The cause of the error can then be determine by using a low level core dump analyzer (adb). • If high level debuggers (dbx) are used we can access the FORTRA program code and the integers representing lists in SAC-2. It is a tedious process to examine pointer structure when represented as integers. The trace feature of the ALDES-translator makes it possible to print easy-readabl list structures, but this process is not interactive. • We have also observed that the time for translating, compilin and linking for one single routine using the FORTRAN translator can be quite long.

We therefore suggest to convert the SAC-2 library to *lisp* code to use it interactively in a *lisp* environment. W choose to use *Common-LISP* [6] since it has the advantage of having become more or less standard. It would b convenient if the translation could be automatically performed by a program. A number of advantages would b obtained: • The ALDES description language could be used to develop new routines, and sites where the FORTRA version of ALDES is used could still use these newly developed routines. • Error signals would be caught by th *lisp* system, and would trigger debugging routines. • We could use the *lisp*-debugger which is often built int *Common-LISP* systems (however not standardized). If lists are represented with *lisp* lists within the translate system we would be able examine these list structures interactively. • The vast knowledge incorporated into th thoroughly documented SAC-2 library would be automatically translated to *lisp*. • The SAC-2 subroutines coul be called from any *lisp* program to perform algebraic manipulation tasks. The major computer algebra system on the market are not designed with this goal in mind. • If one wants to make a small change to one routin one would not have to recompile and link, which can be quite tedious using FORTRAN. Often it is much quicker t load (and perhaps compile) a single *lisp* function instead. Considering these advantages we have implemented translator from the ALDES language to *Common-LISP*. Recently Loos [4] has extended implementation ALDES t the full ASCII character set. Publication ALDES, however, has not been changed. We refer to the new version implementation ALDES as ALDES-2. A version of the *Common-LISP* translator has been adapted to ALDES-2.

Implementation

We will indicate how the translation is performed. The syntax diagram of ALDES is available in the SAC-2 manua The main idea is to translate each ALDES routine to a single *lisp* prog form, where we let statement labels in ALDE correspond to tags in the tagbody form implicit in a prog form. When variables are used they are inserted into th prog variable list, if not already declared in the header of the algorithm. We preserve the semantics of the ALDE return statement by using the return-from form of *Common-LISP*. An ALDES subroutine may return sever values. This can be implemented in *Common-LISP* by using the multiple-value-return feature. Consider a ALDES algorithm with the following header: DEMO(r,A;s,c,L). r and A are inputs and s,c and L are output Below we show how the following code fragment from within the algorithm DEMO is translated.

```
for i:=1,...,r do {                 (FOR |i| 1 |r| (PROGN
     if PDEG(a) # 0 then return;      (IF (NOT (EQL (PDEG |a|) 0))
     a:=PLDCF(a) };                     (RETURN-FROM DEMO (VALUES |v| |c| L)))
                                      (SETQ |a| (PLDCF |a|))))
```

*Supported by: Swedish Board for Technical Development. *Internet:* larsl%nada.kth.se@uunet.uu.net

[1]UNIX is a trademark of AT&T Bell Laboratories.

ote that the code fragment is given in the new implementation language ALDES-2. The *lisp* code to the right ove is returned by the translator. The `for` loop is then translated to the *Common-LISP* primitives `loop` and turn. We have had some problems with the semantics of the SAFE declaration. In ALDES a SAFE declaration les a variable for the garbage collector by storing the variable in a FORTRAN common block instead of on the ıck. This can be used to avoid interpreting integers as pointers to list structures. In *lisp* this cannot happen so have decided to just ignore SAFE declarations. If the fact that SAFE variables are stored in common blocks is ed to gain global access to these variables the ignored SAFE declarations may cause problems. SAFE declarations e also used to insert optimizations in the code manually. This use of the SAFE declaration causes no semantic oblems in *lisp*.

erging the Low-Level Systems

ie SAC-2 library contains subsystems that deals with basic I/O, list processing and long integer arithmetic. One ajor problem with the design of the *lisp* version of the SAC-2 library is to decide what parts of these subsystems e to be replaced with the corresponding *lisp* facilities. The basic I/O system of our *lisp* version of ALDES has been tained, and only the most low level functions have been coded in *lisp*. We have decided that the list processing bsystem of SAC-2 is to be replaced by the corresponding *lisp* list handling functions. This is done by a library *lisp* macros which are inserted into in the *lisp* code at compile time. We thereby obtain one of our major design als in that the list structures can be easily inspected with *lisp* debugging systems. After communication with os [4] we have also decided to allow the use of both SAC-2 long integer arithmetic and of the bignum routines in *ommon-LISP*, depending on the choice of the user. This is because we want to allow access to the source code the long integer arithmetic. This has the disadvantage that some parts of the code of the SAC-2 had to be anged. This is because the test for *list* property of an integer in the SAC-2 library, is identical to the test for the teger being greater than the number β, which is the basis for the long integer arithmetic.

ew Developments

ıce [3] some things have changed. The lisp translator has also been coded in *Common-LISP* only. We have cided to assume that *Common-LISP* is the only *lisp* dialect in which to use the system. Thus the design als that the code generated should be easily ported to other *lisp* system has been relaxed. This is because is *lisp* implementation heavily relies on the efficient implementation of the `multiple-value-return` feature of *ommon-LISP* and also because we believe that *Common-LISP* is becoming more or less standard.

onclusions

e have tested the system in Extended Common-LISP 2.0 from Franz Inc. [1], with a `sun3/75`. We have noted at the efficiency is similar to the FORTRAN version, at least if `lisp` bignums are used. The performance seems to slightly slower when the arithmetic subsystem is used to perform long integer arithmetic.

We believe that this implementation is valuable due to the obtained interactiveness of the system, and because e algorithmic information is automatically available for groups that are *lisp*-oriented. Some other benefits are entioned in the introductory section.

e author is grateful to Rüdiger Loos for inviting him to Tübingen and for many useful and stimulating discussions during the visit.

eferences

Extended Common Lisp User Guide. Franz Lisp Inc., release 2.0 edition, April 1987.

G. E. Collins. ALDES and SAC–2 now available. *SIGSAM bull.*, 12(2):19, 1980.

L. Langemyr. Converting SAC-2 code to lisp. *SIGSAM bull.*, 20(4):11–13, December 1986.

R. G. K. Loos. 1988. Personal communication.

R. G. K. Loos. The algorithm description language ALDES. *SIGSAM bull.*, 14(1):15–39, 1976.

G. Steel. *Common LISP.* Digital Press, 1984.

COMPUTER ALGEBRA SYSTEM FOR CONTINUED FRACTIONS MANIPULATION

V.Tomov, M.Nisheva, T.Tonev
Institute of Mathematics with Computer Centre,
Bulgarian Academy of Sciences
Acad. G.Bonchev str., Bl. 8, Sofia 1113, Bulgaria

Continued fractions are used in solving important problems of mathematical analysis, mathematical physics, number theory,probability theory and many other areas. For the automation of some labour-consuming operations that are typical in the manipulation of continued fractions,a special purpose computer algebra system has been developed. It can find analytical solutions of the following problems: power series expansion of a rational function; finite continued fraction representation of a rational function; rational function representation of a finite continued fraction; construction of continued fractions corresponding to power series; construction of Pade approximants of power series. The system includes also some tools for computing approximate values of the poles of a complex function given by a power series. The formulations and the algorithms for solving the listed problems are described in [2] and [3].

The language of the system is developed with the purpose of providing some assistance for users who are not professional programmers. The user queries and the results are written in a form similar to the conventional one in mathematics. Possibilities for manipulation of the following object types are provided: numbers (whole, rational, complex - rational, decimal fractions); univariate polynomials; univariate rational functions; truncated power series; finite continued fractions whose partial numerators and denominators are univariate polynomials.

The commands of the user language are described in [1]. They enable to make queries for performing the main operations of the system, for printing or displaying the results and for performing some additional actions: evaluation; substitution; manipulation of a file of results; releasing the memory used for expressions that are not necessary; printing or displaying information about the used objects and about the user language commands.

The system consists of two main parts: a managing module and a processing module. The processing module includes programs realizing the algorithms for solving the problems listed above. The managing module reads the next command of the user program (the next user enquiry), recognizes and analyses it and addresses the necessary programs of the processing module or displays information about the detected errors.

The system is implemented in the ALDES language. Some basic symbolic manipulation programs of the SAC-2 system and of several special purpose computer algebra systems developed at the Institute of Mathematics with Computer Centre are used. The system can run on a computer with 512 K bytes memory.

The authors thank the staff of the Laboratory of Informatics with Computer Centre at the University of Sofia for the support during the development of the system.

REFERENCES

1. Nisheva, M. User's language of a computer algebra system for continued fractions manipulation. In "Mathematics and mechanics and their application in science and practice", Sofia, 1985, pp. 120-127 (in Bulgarian).

2. Nisheva, M., T.Tonev. A system for continued fractions manipulation. Complex Analysis and Applications'85, Sofia, Publishing House of the Bulgarian Academy of Sciences, 1986, pp. 482-488.

3. Nisheva, M., T.Tonev. Main possibilities of a system for continued fractions manipulation. Proc. of the Int. Conf. on Computer - Based Scientific Research (Plovdiv, 1984), pp. 398-404 (in Russian).

Computing a lattice basis from a system of generating vectors

Johannes Buchmann and Michael Pohst
Mathematisches Institut
Universität Düsseldorf
4000 Düsseldorf
FRG

Abstract

In this paper we describe how the LLL-algorithm can be used to compute a basis of a lattice L in $\mathcal{R}^n$ from a system of k generating vectors and a lower bound for the lengths of the non zero vectors in L. The algorithm which we present is proved to be polynomial time in $n + k$ and the size of the input data. The algorithm is applied to the problem of finding multiplicative relations between units of algebraic number fields. Numerical results show that our method works very efficiently.

1 Introduction

Let $\mathbf{b}_1, \ldots, \mathbf{b}_k$ be vectors in $\mathcal{R}^n$ different from $\mathbf{0}$. In this paper we discuss the problem of computing a basis of the lattice

$$\mathbf{L} = \sum_{j=1}^{k} \mathcal{Z}\mathbf{b}_j.$$

The algorithm which was proposed for this purpose in Hastad, Just, Lagarias and Schnorr [4] seems to be of theoretical value only since it requires exact computations with real numbers and can therefore not be implemented on a computer. The performance of the modified LLL-algorithm of Pohst [6] will be investigated in a subsequent paper by the authors.

Here we present an algorithm which is based on an idea of Lenstra, Lenstra and Lovasz [5]. Assuming that we know bounds

$$M \le \min\{\|\mathbf{v}\| : \mathbf{0} \ne \mathbf{v} \in \mathbf{L}\} \tag{1}$$
$$B \ge \max\{\|\mathbf{b}_j\| : 1 \le j \le k\} \tag{2}$$

(where $\| \ \|$ denotes the Euclidean norm on $\mathcal{R}^n$) we prove that our algorithm computes an integral transformation which changes the generating system $\mathbf{b}_1, \ldots, \mathbf{b}_k$ into a basis of $\mathbf{L}$ in $O((k+n)^6 \log(B/M))$ binary operations. We discuss applications to unit computations in algebraic number fields and present numerical examples.

2 Preliminary results

Finding a basis of $\mathbf{L}$ requires to determine relations

$$\sum_{j=1}^{k} m_j \mathbf{b}_j = \mathbf{0} \tag{3}$$

with $m_j \in \mathcal{Z}$ for $1 \le j \le k$ and $\sum_{j=1}^{k} |m_j| \ne 0$. Any such integer vector $\mathbf{m} = (m_1, \ldots, m_k)^t \in \mathcal{Z}^k$ is called a **relation** for $\mathbf{b}_1, \ldots, \mathbf{b}_k$.

Since we cannot compute with real input vectors directly, we introduce rational approximations. We choose $q \in \mathcal{Z}^{>0}$ and we put

$$\hat{b}_{ij} = < 2^{q-1} b_{ij} > \qquad \text{for } 1 \le i \le n, 1 \le j \le k,$$
$$\hat{\mathbf{b}}_j = (\hat{b}_{1j}, \ldots, \hat{b}_{nj})^t \qquad \text{for } 1 \le j \le k$$

where b_{ij} is the i-th coordinate of $\mathbf{b}_j$ and $< >$ denotes the nearest integer function. Then we have

$$\|2^{-q+1}\hat{\mathbf{b}}_j - \mathbf{b}_j\| \leq \sqrt{n}2^{-q} \qquad \text{for } 1 \leq j \leq k, \tag{4}$$

and therefore

$$\|\hat{\mathbf{b}}_j\| \leq 2^{q-1}B + \sqrt{n}/2 \qquad \text{for } 1 \leq j \leq k. \tag{5}$$

Proposition 2.1 *Let* $\mathbf{m} = (m_1, \ldots, m_k)^t \in \mathbb{Z}^k$, $M^* = \sum_{j=1}^k |m_j|$ *and assume that*

$$2^{q-1} > \sqrt{n}M^*/M. \tag{6}$$

Then $\mathbf{m}$ *is a relation for* $\mathbf{b}_1, \ldots, \mathbf{b}_k$ *if and only if*

$$\|\sum_{j=1}^{k} m_j \hat{\mathbf{b}}_j\| \leq \sqrt{n}M^*/2. \tag{7}$$

Proof: If $\mathbf{m}$ is a relation then we have

$$\|\sum_{j=1}^{k} m_j \hat{\mathbf{b}}_j\| = 2^{q-1}\|\sum_{j=1}^{k} m_j(2^{-q+1}\hat{\mathbf{b}}_j - \mathbf{b}_j)\| \leq \sqrt{n}M^*/2.$$

Here we have used (4). If (7) holds then it follows again from (4) that

$$\|\sum_{j=1}^{k} m_j \mathbf{b}_j\| \leq \|\sum_{j=1}^{k} m_j(\mathbf{b}_j - 2^{-q+1}\hat{\mathbf{b}}_j)\| + 2^{-q+1}\|\sum_{j=1}^{k} m_j \hat{\mathbf{b}}_j\| \leq \sqrt{n}2^{-q+1}M^* < M.$$

□

Using Proposition 2.1 we can check whether a given $\mathbf{m} \in \mathbb{Z}^k$ is a relation. The precision constant q has to be chosen according to (6). By means of the next statement the dependency of q on $\mathbf{m}$ can be removed.

Proposition 2.2 *Let* r *be the dimension of* $\mathbf{L}$. *Then there are k-r linearly independent relations* $\mathbf{m}_1, \ldots, \mathbf{m}_{k-r}$ *of* $\mathbf{b}_1, \ldots, \mathbf{b}_k$ *with*

$$|m_{i,j}| \leq C = (B\gamma_r^{1/2}/M)^r \qquad \text{for } 1 \leq i \leq k, \quad 1 \leq j \leq k-r$$

where $m_{i,j}$ *is the i-th coordinate of* $\mathbf{m}_j$, *and* γ_r^r *is the r-th Hermite constant.*

Proof: Without loss of generality, we assume that $\mathbf{b}_1, \ldots, \mathbf{b}_r$ are linearly independent. We let $\mathbf{v}_1, \ldots, \mathbf{v}_r$ be a basis of $\mathbf{L}$ and we put $V = (\mathbf{v}_i^t \mathbf{v}_j)_{1 \le i,j \le r})$. Then there is a decomposition $V = W^t W$ with $W \in GL(r, \mathcal{R})$. We denote the columns of W by $\mathbf{w}_1, \ldots, \mathbf{w}_r$ and we let $\mathbf{L}'$ be the lattice generated by those columns. The mapping

$$\begin{array}{ccc} \mathbf{L} & \rightarrow & \mathbf{L}' \\ \mathbf{v} = \sum_{j=1}^r x_j \mathbf{v}_j & \mapsto & \Phi(\mathbf{v}) = \sum_{j=1}^r x_j \mathbf{w}_j \end{array}$$

is an isometry between $\mathbf{L}$ and $\mathbf{L}'$. We put $\mathbf{b}'_j = \Phi(\mathbf{b}_j)$ for $1 \le j \le k$ and

$$m_{i,j-r} = \begin{cases} \det(\mathbf{b}'_1, \ldots, \mathbf{b}'_{i-1}, \mathbf{b}'_j, \mathbf{b}'_{i+1}, \ldots, \mathbf{b}'_r) / \det \mathbf{L}' & \text{for } 1 \le i \le r, r < j \le k, \\ \det(\mathbf{b}'_1, \ldots, \mathbf{b}'_r) / \det \mathbf{L}' & \text{for } r+1 \le i = j \le k, \\ 0 & \text{otherwise.} \end{cases}$$

The $m_{i,j}$ are either 0 or indices of sublattices of $\mathbf{L}$ and therefore, they are rational integers. Since $m_{i,i} \ne 0$ for $r < i \le k$ but $m_{i,j} = 0$ for $r < i, j \le k$, $i \ne j$ the vectors $\mathbf{m}_1, \ldots, \mathbf{m}_{k-r}$ are in fact linearly independent. Moreover, we have by Cramer's rule

$$\Phi\left(\sum_{i=1}^k m_{i,j} \mathbf{b}_i\right) = \mathbf{0} \qquad \text{for } 1 \le j \le k - r.$$

Hence, we have found $k - r$ linearly independent relations. Finally, we have $\det \mathbf{L}' \ge (M/\gamma_r^{1/2})^r$ and using Hadamard's inequality, we get the bounds on $m_{i,j}$. □

3 The algorithm

We define

$$\tilde{\mathbf{b}}_j = (\mathbf{e}_j^t, \hat{\mathbf{b}}_j^t)^t \qquad \text{for } 1 \le j \le k$$

where $\mathbf{e}_j$ denotes the j-th unit vector of $\mathcal{Z}^k$. Then $\tilde{\mathbf{b}}_1, \ldots, \tilde{\mathbf{b}}_k$ are linearly independent integer vectors and therefore they form a basis of the lattice

$$\tilde{\mathbf{L}} = \bigoplus_{j=1}^k \mathcal{Z} \tilde{\mathbf{b}}_j.$$

We apply the LLL-reduction algorithm [5] to this basis, the result being the new basis $\tilde{\mathbf{c}}_1, \ldots, \tilde{\mathbf{c}}_k$. We write

$$\tilde{\mathbf{c}}_j = (\mathbf{m}_j^t, \hat{\mathbf{c}}_j^t)^t \qquad \text{for } 1 \leq j \leq k$$

with $\mathbf{m}_j = (m_{1,j}, \ldots, m_{k,j})^t \in \mathbb{Z}^k$ and $\hat{\mathbf{c}}_j = (\hat{c}_{1,j}, \ldots, \hat{c}_{n,j})^t \in \mathbb{Z}^n$. We also define

$$M_j^* = \sum_{i=1}^{k} |m_{i,j}| \qquad \text{for } 1 \leq j \leq k.$$

Since by [5] (1.12)

$$\|\tilde{\mathbf{c}}_j\| \leq 2^{(k-1)/2} \lambda_j(\tilde{\mathbf{L}}),$$

where $\lambda_j(\tilde{\mathbf{L}})$ denotes the j-th successive minimum of the lattice $\tilde{\mathbf{L}}$, we can expect to find relations among the $\mathbf{b}_j$ provided that q was chosen sufficiently big. More precisely we have

Theorem 3.1 *Let r be the dimension of $\mathbf{L}$ and assume that*

$$2^q > (\sqrt{nk} + 2)\tilde{M} 2^{(k-1)/2} / M \tag{8}$$

with $\tilde{M} = (k\sqrt{n}/2 + \sqrt{k})C$, C as in Proposition 2.2. Then $\mathbf{m}_1, \ldots, \mathbf{m}_{k-r}$ are linearly independent relations for $\mathbf{b}_1, \ldots, \mathbf{b}_k$ and the vectors

$$\mathbf{c}_j = \sum_{i=1}^{k} m_{i,j} \mathbf{b}_i \qquad \text{for } k - r + 1 \leq j \leq k$$

form a basis of $\mathbf{L}$.

Proof: By Proposition 2.2, the first $k - r$ successive minima of $\tilde{\mathbf{L}}$ are bounded from above by $\tilde{M}$. Hence we have by (1.12) in [5]

$$\|\tilde{\mathbf{c}}_j\| \leq 2^{(k-1)/2} \tilde{M} \qquad \text{for } 1 \leq j \leq k - r. \tag{9}$$

On the other hand, we will prove that

$$\|\tilde{\mathbf{c}}\| > 2^{(k-1)/2} \tilde{M} \tag{10}$$

for every $\tilde{\mathbf{c}} = (\mathbf{m}^t, \hat{\mathbf{c}}^t)^t \in \tilde{\mathbf{L}}$ for which $\mathbf{m} = (m_1, \ldots, m_k)^t$ is not a relation. Those two inequalities (9) and (10) clearly show that $\mathbf{m}_1, \ldots, \mathbf{m}_{k-r}$ are relations. In order to prove (10) we note that $\|\tilde{\mathbf{c}}\|^2 = \|\mathbf{m}\|^2 + \|\hat{\mathbf{c}}\|^2$. If $\|\mathbf{m}\| > 2^{(k-1)/2}\tilde{M}$ then we are done. Otherwise we have

$$\begin{aligned} \|\hat{\mathbf{c}}\| &\geq 2^{q-1}\|\sum_{j=1}^{k} m_j \mathbf{b}_j\| - 2^{q-1}\|\sum_{j=1}^{k} m_j(\mathbf{b}_j - 2^{-q+1}\hat{\mathbf{b}}_j)\| \\ &\geq 2^{q-1}M - \sqrt{nk}\tilde{M}2^{(k-1)/2}/2 \\ &> 2^{(k-1)/2}\tilde{M}. \end{aligned}$$

Finally, the independency of the relations $\mathbf{m}_1, \ldots, \mathbf{m}_{k-r}$ follows from the fact that the determinant of the matrix with columns $\mathbf{m}_1, \ldots, \mathbf{m}_k$ is ± 1 and for the same reason, the vectors $\mathbf{c}_{k-r+1}, \ldots, \mathbf{c}_k$ form a basis of $\mathbf{L}$. □

Remarks 3.1 *1. If the dimension r of $\mathbf{L}$ is unknown we have to replace C by $C' = (\gamma_\beta^{1/2} B/M)^\beta$, $\beta = \min(n, k)$.*

2. In practice, the described algorithm yields the necessary $k - r$ relations for q much smaller than required by (8). When using the algorithm we therefore choose q according to our experience from extensive calculations. Then we apply Proposition 2.1 to identify the relations.

Theorem 3.2 *An integer transformation which changes the input vectors $\mathbf{b}_1, \ldots, \mathbf{b}_k$ into a basis of $\mathbf{L}$ can be computed in $O((n+k)^6 \log(B/M))$ bit operations.*

Proof: Since the core of our algorithm is LLL-reduction we can use the running time estimate for this algorithm given in Proposition 1.26 of [5]. The dimension of our lattice is $n + k$ and we only need to estimate the lengths of the input vectors $\tilde{\mathbf{b}}_1, \ldots, \tilde{\mathbf{b}}_k$ which is done in (5) and (8). □

4 Application to the computation of units

Let $\mathcal{F}$ be an algebraic number field of degree $n = s + 2t$ where s, t denote the number of real, complex conjugates of $\mathcal{F}$, respectively. For any $\mathcal{Z}$-order $\mathcal{O}$ of $\mathcal{F}$ the structure of its unit group $\mathcal{O}^\times$ is given by Dirichlet's Theorem:

$$\mathcal{O}^\times = \mathrm{TU}(\mathcal{O}) \times \langle \epsilon_1 \rangle \times \ldots \times \langle \epsilon_r \rangle. \tag{11}$$

Here $r = s + t - 1$, $\mathrm{TU}(\mathcal{O})$ is the torsion subgroup of $\mathcal{O}^\times$ and the generators of the infinite cyclic groups $\langle \epsilon_i \rangle$ are called **fundamental units**. The determination of a full system of fundamental units is one of the most important and most difficult tasks of computational algebraic number theory. The methods of the preceding sections are especially well suited in this context since in that case lower bounds for the shortest non-zero vector of the considered lattice can be calculated.

Usually one proceeds in two steps. In step one, units $\eta_1, \ldots, \eta_k$ (k sufficiently large) are computed which generate a subgroup of finite index of $\mathcal{O}^\times$. In step two, fundamental units are calculated from the η_i. Hence we need to determine units $\phi_1, \ldots, \phi_r$ in $\mathcal{O}^\times$ satisfying

$$G = \langle \mathrm{TU}(\mathcal{O}), \eta_1, \ldots, \eta_k \rangle = \mathrm{TU}(\mathcal{O}) \times \langle \phi_1 \rangle \times \ldots \langle \phi_r \rangle. \tag{12}$$

The final step of computing fundamental units from the independent units $\phi_1, \ldots, \phi_r$ will not be discussed in this paper.

To diminish the number of generators of G (for $k > r$) we apply the methods of the preceding section to the lattice

$$\mathbf{L} = \mathbf{Log}\,(\mathcal{O}^\times) \tag{13}$$

where $\mathbf{Log}$ denotes the mapping of $\mathcal{O}^\times$ into the logarithmic space :

$$\begin{array}{ccl} \mathcal{O}^\times & \to & \mathcal{R}^r \\ \epsilon & \mapsto & (\log|\epsilon^{(1)}|^{c_1}, \ldots, \log|\epsilon^{(r)}|^{c_r})^t. \end{array} \tag{14}$$

Here the $\epsilon^{(i)}$ denote the conjugates of ϵ and

$$c_i = \begin{cases} 1 & \text{for} \quad i \le s \\ 2 & \text{for} \quad i > s. \end{cases}$$

Then a lower bound $M \le \min\{\|\mathbf{v}\| : \mathbf{0} \ne \mathbf{v} \in \mathbf{L}\}$ needs to be computed. It can be obtained as follows: Let $\mathbf{Log}\,\eta_{i_1}, \ldots, \mathbf{Log}\,\eta_{i_r}$ be linearly independent and subject to $\|\mathbf{Log}\,\eta_{i_1}\| \le \ldots \le \|\mathbf{Log}\,\eta_{i_r}\|$ for suitable indices $1 \le i_1, \ldots, i_r \le k$. That linear independence can be checked by the Gram-Schmidt orthogonalization procedure, for example. By Hadamard's inequality,

$$\det \mathbf{L} \le M \|\mathbf{Log}\,\eta_{i_2}\| \cdots \|\mathbf{Log}\,\eta_{i_r}\|. \tag{15}$$

Hence, all we need is a lower bound for det $\mathbf{L}$. Note that the estimate in (15) can still be improved by using the vectors of a LLL-reduced basis of the lattice generated by $\mathbf{Log}\,\eta_{i_1}, \ldots, \mathbf{Log}\,\eta_{i_r}$ instead of those vectors themselves. The best known method for computing lower bounds for det $\mathbf{L}$ works as follows (see [7], V.6): First determine for $1 \le i \le r$ numbers M_i^* as large as possible such that there are at most i units η in $\mathcal{O}$ with $\sum_{j=1}^{n} |\eta^{(i)}|^2 \le M_i^*$ whose $\mathbf{Log}$-vectors are linearly independent and define

$$M_i = \frac{n}{4}(\log(\frac{M_i^*}{n} + ((\frac{M_i^*}{n})^2 - 1)^{1/2}))^2.$$

Then

$$\det \mathbf{L} \ge (\gamma_r^{-r} M_1 \cdots M_r \frac{1}{n} 2^t)^{1/2}. \tag{16}$$

5 Numerical results

Using the method described in [1] we have computed for each of the 889 totally real quartic fields of discriminant below 100000 given in [2] four units ($\neq \pm 1$) which generate a subgroup of finite index in the unit group $\mathcal{O}^\times$ of the maximal order $\mathcal{O}$. Since the free rank of this unit group is 3 there must be a relation

$$\eta_1^{k_1}\eta_2^{k_2}\eta_3^{k_3}\eta_4^{k_4} = \pm 1. \tag{17}$$

For each quadruple of units we have computed a relation (17) by using the methods of the preceding sections.

Example 5.1 *$\mathcal{F}$ is the field generated by a root ρ of the totally real quartic polynomial $f(x) = x^4+2x^3-8x^2+x+3$ which is of discriminant $\mathcal{D} = 69805$. This discriminant is square free. Therefore the discriminant of $\mathcal{F}$ equals $\mathcal{D}$ and $1, \rho, \rho^2, \rho^3$ is an integral basis of $\mathcal{F}$. The units*

$$\begin{aligned}
\eta_1 &= 4 + \rho \\
\eta_2 &= 2 + 4\rho - 3\rho^2 \\
\eta_3 &= 1 - \rho \\
\eta_4 &= 8 + 13\rho - 5\rho^2 - 2\rho^3
\end{aligned}$$

generate a subgroup of finite index in the unit group $\mathcal{O}^\times$. By the method of section 4 we found the lower bound 0.58 for the lenghts of the non zero vectors in the unit lattice $\mathbf{Log}\,\mathcal{O}^\times$. *Using this bound we computed the relation*

$$\eta_1^3\eta_2\eta_3^5\eta_4 = 1$$

which means that

$$\langle -1, \eta_1, \eta_2, \eta_3 \rangle = \langle -1, \eta_1, \eta_2, \eta_3, \eta_4 \rangle.$$

Here we had to choose the constant $q = 2$.

In each case we have computed the minimum q-value which yields a relation between the units. Here is a table that shows the frequency $f(q)$ with which each q occured.

q	1	2	3	4	5
f(q)	618	135	53	67	16

The computation of the relations took 290 seconds on an ATARI 1040 ST personal computer.

References

[1] J. Buchmann, *On the computation of units and class numbers by a generalization of Lagranges's algorithm*, J. Number Theory **26** (1987), 8–30.

[2] J. Buchmann and D. Ford, *On the computation of totally real fields of small discriminant*, Math. Comp., to appear.

[3] U. Fincke and M. Pohst, *Improved methods for calculating vectors of short length, including a complexity analysis*, Math. Comp. **44** (1985), 463–471.

[4] J. Hastad, B. Just, J.C. Lagarias and C.P. Schnorr, *Polynomial time algorithms for finding integer relations among real numbers*, Proceedings STACS 86.

[5] A.K. Lenstra, H.W. Lenstra Jr. and L. Lovasz, ***Factoring polynomials with rational coefficients***, Math. Ann. **261** (1982), 515–534.

[6] M. Pohst, ***A modification of the LLL–algorithm***, J. Symb. Comp. **4** (1987),123–127.

[7] M. Pohst und H. Zassenhaus, ***Algorithmic algebraic number theory***, Cambridge University Press, to appear.

Expression Optimization Using High-Level Knowledge

Mark P.W. Mutrie
Bruce W. Char
Richard H. Bartels*
Department of Computer Science
University of Waterloo
Waterloo, Ontario
Canada N2L 3G1

Abstract

Combining symbolic algebra with numerical computation has become an effective way of solving many scientific and engineering problems. One of the difficulties in practice is producing concise, efficient, stable code from the large expressions generated in symbolic algebra. Most of the existing optimization techniques are applied after an operation or algorithm has been performed. We introduce techniques using high-level knowledge which are to be applied while the output expressions are generated.

1 Introduction

1.1 Rationale for Needing Optimization

Symbolic algebra and numerical computing techniques can be combined to solve problems in engineering an physics [Wir80, Wan86], but expressions generated using symbolic algebra systems for subsequent use in numeri codes tend to be very large [SR84, Bre84]. Translating these expressions directly into a computing language ca lead to disappointing results. The code may be inefficient and numerically unstable. Furthermore, improvin the code by hand is impractical because of its size. Consequently, an interface between symbolic and numerica computing facilities should include procedures for optimizing the number of arithmetic operations in the formula and tools for analyzing the numerical stability of the generated code. The term optimization is used in the sens of *improving* the arithmetic operation count and compactness of the code rather than "the best possible".

Most of the existing optimization techniques are applied to the large output expressions *after* a procedure ha been performed. The process applied to the input to produce the output expression is usually well-understooc although perhaps not by the user. We discuss an approach to expression optimization which uses knowledge abou the operation or algorithm being performed, and the structure of the input expressions to generate concise outpu formulae that can be evaluated efficiently.

In Section 3 we present three examples in which optimization using high-level knowledge has been applie The examples involve differentiation, Taylor series approximation, and integration operations. Within the pro cess of this knowledge-based approach, we use some of the existing structure-determining and structure-reducin optimization techniques on small components of the larger problem. A brief summary of existing optimizatio techniques with references is given in Section 2.

2 Survey of Existing Optimization Techniques

Using Built-in Procedures The built-in procedures in symbolic algebra languages like Reduce, Macsyma an Maple can be used interactively to improve the quality of many codes. Steinberg and Roache [SR84] and Cahi and Reeder [CR85] used built-in Macsyma procedures to interactively analyze and produce improved FORTRA code for specific applications in physics.

*This work was supported by grants A5471 and A5471 of the Natural Sciences and Engineering Research Council of Canada. Secon author's present address: Department of Computer Science, University of Tennessee, Knoxville Tennessee U.S.A. 37996-1301.

xpression Analysis and Compression Hulshof and van Hulzen developed a Reduce package for interac- ᵣely analyzing and restructuring expressions [vHH82] and a package for expression compression [HvH85]. After alyzing the expression(s), compression techniques using local factorization [Coo82] and controlled expansions sed on work done by Brenner can be applied to produce a more compact representation.

ommon-Subexpression Searching van Hulzen developed a common-subexpression searching procedure in educe [vH81, vH83], based on Breuer's algorithm [Bre69], that performs a common-subexpression search on ultivariate polynomials. This approach to common-subexpression searching has also been implemented in Maple.

The **optimize** procedure in the Maple (version 4.0) library, developed by M. Monagan, makes use of "remem- r" tables [CGGW85, page 186], which are based on hashing, to identify subexpressions which are complete btrees of an expression's parse tree. Only subexpressions that Maple simplifies as identical syntactic structures e recognized as common. In $(x+y+z)(x+y+f)$, $x+y$ is not recognized as a common component since it is ly a part of the sum structure, whereas it is recognized in $(x+y)\exp(5(x+y))$.

Macsyma's `OPTIMIZE` command also employs a heuristic based upon hashing. During the descent of an ex- ession tree, a hash code is formed for all subexpressions at a given level. A collision in the hash codes triggers check for a common subexpression. Thus `OPTIMIZE` makes two passes through the expression – one to find and ark common subexpressions, and another to form the list of expressions for the result.

Macsyma's and Maple's optimization procedures, which we will refer to as "syntactic-optimization procedures", th make use of hashing to discover occurrences of common subexpressions in linear time. If expression-analysis d compression techniques are applied to an expression first, the result of a common-subexpression search is ually more compact than if only a common-subexpression search is performed.

valuation of Polynomials Given the polynomial $p(x) = a_o + a_1x + a_2x^2 + \ldots + a_nx^n$, Horner's rule is suggested an economical way of evaluating the polynomial. Horner's rule is not always optimal [Fat69]; methods for econditioning the coefficients can further improve the operation count. For example, preconditioning based on lynomial chaining, an extension of addition chaining, is given in Knuth [Knu81, pages 446-505]. These methods tend to multivariate polynomials. Horner's rule and preconditioning methods very often give good results, but eir numerical stability depends on the polynomial's coefficients [MW67].

ther Techniques for Optimization The procedure `FORT`, due to Harten [Har84], uses the built-in functions `TIMIZE` and `HORNER` to produce improved FORTRAN code.

Brenner [Bre84] describes techniques for simplying large algebraic expressions which include segmenting the lculation in parts, and analyzing the structure so that rational simplification can be applied at appropriate points the computation. By segmenting, he is able to reuse some intermediate results in the symbolic simplification ocess. His approach is implemented in Macsyma in a procedure called `LTAB`.

Paul Wang [Wan86] describes several techniques for improving the code produced in his symbolic system for the tomatic generation of numerical programs for finite element analysis. These techniques, which include automatic termediate-expression labelling, exploiting the symmetry of a problem using generated functions and subroutines, d interleaving formula derivation with code generation, could be useful in other applications. Intermediate- pression labelling is an illustrative example of Wang's methods. This method generates machine-created labels r certain intermediate expressions. These intermediate results are saved on an association list, which is checked fore each computation to avoid reevaluating expressions. For an expression not on the association list, a label generated and the expression-label pair is added to the list. For an expression already on the association list, e label is returned. This technique is much like Maple's option remember [CGGW85, page 169], but a label presenting the result is returned instead of the actual result.

An Alternative Approach Using High-Level Knowledge

'e present an approach which uses knowledge about the operations, and the structure of the input expressions these operations, to produce a compact representation of the output. This technique does not necessarily eserve the original structure of the problem, but does exploit it to a great extent. Since the approach depends knowledge about the operations being performed, we apply it to three examples – differentiation, Taylor series proximation, and integration – to illustrate how it works. These are representative of the (small number of) sic operations that all symbolic algebra systems use to create and transform expressions.

3.1 Differentiation

Solving systems of nonlinear equations is just one problem where differentiation can be an important part of the solution process. Symbolic differentiation of formulae can often produce large expressions that are difficult to understand and expensive to evaluate. However, the components of gradients and Jacobians tend to contain many common subexpressions. This observation has been exploited in developing software for the automatic differentiation of computer code [Spe80, Ked80, Ral81, KKRS86, Gri88, Iri84] and also for generating concise and efficient code using symbolic algebra systems [NC79].

When using a symbolic algebra system to generate the derivatives, we could apply a common subexpression search on the result to generate a more concise representation for inclusion in numeric code. On the other hand, by considering the differentiation operation itself, we can observe that certain differentiation rules, such as the chain rule, generate particular structures in the solution expressions. We can take advantage of the knowledge we have about the operation being performed and about the structure of the input expressions to eliminate the need for a common subexpression search on the generated expressions.

Consider computing the Jacobian of the following nonlinear problem [Ric83, page 241]:

$$\begin{aligned} h1 &:= \sin(xy+1)/(1+x+y)-1 \\ h2 &:= \tan(xy+2)/(1-x+y)-2. \end{aligned}$$

The Jacobian is:

$$\begin{bmatrix} \frac{\cos(xy+1)y}{1+x+y} - \frac{\sin(xy+1)}{(1+x+y)^2} & \frac{\cos(xy+1)x}{1+x+y} - \frac{\sin(xy+1)}{(1+x+y)^2} \\ \frac{\sec^2(xy+2)y}{1-x+y} - \frac{\tan(xy+2)}{(1-x+y)^2} & \frac{\sec^2(xy+2)x}{1-x+y} - \frac{\tan(xy+2)}{(1-x+y)^2} \end{bmatrix}.$$

We could then perform a Breuer-algorithm based common-subexpression search on the elements of this matrix which would find 16 common structures. As an alternative we could exploit our knowledge of the differentiation process, and the structure of the problem. In the given problem we can use our knowledge of the division rule

$$\frac{d}{dx}\left(\frac{u}{v}\right) = \frac{v\frac{du}{dx} - u\frac{dv}{dx}}{v^2} = \frac{1}{v}\frac{du}{dx} - \frac{u}{v^2}\frac{dv}{dx}.$$

The denominator v is a common subexpression in the final result.

In the given problem, the following structures were identified using Maple's op procedure to extract the components of the internal representation of the input expressions.

$$\begin{aligned} t1 &= xy \\ u1 &= \sin(t1+1) \\ v1 &= 1+x+y \\ u2 &= \tan(t1+2) \\ v2 &= 1-x+y \end{aligned}$$

The component $x+y$ could also be identified if a Breuer-algorithm-based common-subexpression search was used.

For lack of space, we will just consider calculating the $(1,1)$ and $(2,1)$ elements of the Jacobian corresponding to the derivatives of $h1$ and $h2$ with respect to x. We first calculate the derivatives of the components that we identified above. For assignment variable *var*, its derivative with respect to x is assigned to the variable *varp*.

$$\begin{aligned} t1p &= y \\ u1p &= \cos(t1+1)t1p \\ v1p &= 1 \\ u2p &= \sec^2(t1+2)t1p \\ v2p &= -1 \end{aligned}$$

The $(1,1)$ and $(2,1)$ elements of the Jacobian are given by

$$\begin{aligned} j11 &= \frac{1}{v1}u1p - \frac{u1}{v1^2} \\ j21 &= \frac{1}{v2}u2p + \frac{u2}{v2^2} \end{aligned}$$

ıere simple intermediate expressions representing simple constants have been eliminated. The components, their rivatives, and the result can be saved as elements of an expression table. The above approach could be used to nerated the entire Jacobian; this approach produces an improved description of the output expressions, finding similar number of common subexpressions as a Breuer-algorithm-based search, without requiring a search upon mpleting the differentiation process.

To further illustrate how this strategy can be used, consider taking the derivative of a product of 3 terms u, v, d w which are functions of the variable x. Using the product rule, the derivative is

$$\frac{d(uvw)}{dx} = uv\frac{dw}{dx} + uw\frac{dv}{dx} + vw\frac{du}{dx}.$$

y identifying the operation we are performing and the existing factors in the product, we see that the result ntains the common subexpressions u, v and w. If the number of factors is greater than two, the product rule for fferentiation will produce common subexpressions (as long as the components are independent of each other). ırthermore, a factor, u, and its derivative, $\frac{du}{dx}$, often contain common components; after $\frac{du}{dx}$ has been computed, e could consider performing a common-subexpression search on the smaller problem comprised of the expression and any factors identified in carrying out the differentiation $\frac{du}{dx}$, rather than doing a systematic search on the tire result afterwards.

If we consider a product of 4 terms, we can determine a more efficient representation than would be expected om just applying the product rule directly. Using the product rule:

$$\frac{d(uvwy)}{dx} = uvw\frac{dy}{dx} + uvy\frac{dw}{dx} + uwy\frac{dv}{dx} + vwy\frac{du}{dx}$$

here the original factors in the product u, v, w, and y are now common subexpressions. uv and wy could identified as common subexpressions *afterwards* using a search based on Breuer's algorithm, but would not found using syntactic optimization. On the other hand, we can take advantage of the structure during the fferentiation process and eliminate the need for a common-subexpression search on the final result. Consider

$$\frac{d(uvwy)}{dx} = \frac{d(uv)(wy)}{dx} = uv\frac{d(wy)}{dx} + wy\frac{d(uv)}{dx} = uv\left(w\frac{dy}{dx} + y\frac{dw}{dx}\right) + wy\left(u\frac{dv}{dx} + v\frac{du}{dx}\right).$$

) and wy were identified as significant structures from the beginning of the process using knowledge about the eration and the expression. Assuming the components u, v, w, and y are evaluated once, the direct use of the oduct rule requires 12 multiplications and 3 additions; using knowledge-based optimization, 8 multiplications and multiplications are needed. This improvement has been achieved without the need for a Breuer-algorithm-based mmon-subexpression search. Greater improvements in efficiency and compactness of the code would be seen in rge problems where one uses knowledge about several differentiation rules, and all the structures identified in e input expressions.

Similar strategies can be established for other differentiation rules involving exponentials, integer powers, and wers which are functions of x. Using the differentiation operation with this code improvement strategy includes eps to identify existing common subexpressions, factors in a product, the number of such factors, and the use of vision and exponentiation in the input expressions, and then, steps to take advantage of these structures *within* e differentiation process. The derivatives of the identified components can be calculated, common subexpressions each component and its derivative can be identified, and the required derivatives can be constructed from signed variable names representing the common subexpressions.

2 Taylor Series Approximation

he standard form of the Taylor series for $f(x)$ about the point $x = a$ is given by

$$f(x) = f(a) + (x-a)f'(a) + \frac{(x-a)^2}{2!}f''(a) + \frac{(x-a)^3}{3!}f'''(a) + \ldots + \frac{(x-a)^n}{n!}f^{(n)}(a) + \ldots$$

here is two sources for improvement. Firstly, we can take advantage of the common subexpression $\theta = (x-a)$. condly, we can exploit the common subexpressions produced by the differentiation operation. For example, in e Taylor series expansion of $\exp(y/(1-x)+a)$ about $x = 0$, each term in the result contains the expression p$(-y+a)$. Both sources of common subexpressions are identified without the need for a common-subexpression arch.

Subsequent operations on the Taylor series approximation will have to deal with the common subexpression repeatedly. In a symbolic context, Maple already uses this fact by only storing the common components once in its internal data representation of the expression, and using pointers to the single representation of the common component. Furthermore, Maple's `taylor` procedure uses a remember table [CGGW85, page 186] to avoid recomputing the common pieces of the Taylor expansion.

The proposal that is being made here goes further; in order to describe large output expressions effectively to the user in the symbolic algebra system, and in order to generate efficient code for a numeric environment, the output should be generated using, for example, the high-level-knowledge approach described above, in a form where the common subexpressions are described only once as a side relation [Hea84].

3.3 Integration

Consider applying this strategy to integrals calculated using the Risch algorithm [Ris69]. Liouville's theorem states that if the integral of an elementary function $f(x) \in K(x, \theta_1, \theta_2, \ldots \theta_n)$ is an elementary function then it can be expressed in the form

$$\int f(x)dx = v_0 + \sum_{i=1}^{k} c_i \log v_i$$

where θ_i, c_i and v_i belong to a tower of algebraic, logarithmic or exponential extension fields, or their algebraic closure, of the base differential field, $K(x)$.

Consider two examples; in the first integral, I_1, the extension field used in the Risch algorithm is $\phi = \ln(x-1)$ and in the second integral, I_2, the extension field is $\omega = \ln(x)$.

$$\begin{aligned}
I_1 &= \int \frac{(x^2-x)\ln^5(x-1) - (x^2-x)\ln^4(x-1) + 2\ln^3(x-1) - 8\ln(x-1) + 8}{(x-1)\ln^4(x-1) - (x-1)\ln^3(x-1)} dx \\
&= \frac{4}{(x-1)\phi^2} + \ln(\phi - 1) + \ln(\phi) + \frac{x^2-1}{2}\phi - \frac{x^2+2x}{4} \qquad \text{where } \phi = \ln(x-1) \\
I_2 &= \int \frac{x\ln^2(x)+1}{x\ln(x)} dx \\
&= x\ln(x) - x + \ln(\ln(x)) \\
&= x\omega - x + \ln(\omega) \qquad \text{where } \omega = \ln(x).
\end{aligned}$$

The extension fields act as common subexpressions in the indefinite integral. The final result can be expressed in terms of variables, θ_i representing the extension fields, and a set of side relations describing the θ_i subexpressions. We observe that the integral can be described more concisely simply by taking advantage of our knowledge about the Risch algorithm without having to do a common-subexpression search on the result that is produced. Further investigation is needed to determine if this optimization approach can be used to identify common subexpressions that are not algebraic extensions and if this approach can be used with other procedures used for integration.

4 Concluding Remarks

We have proposed an approach to expression optimization that differs from most existing techniques, which are applied after the output has been generated. We take advantage of our knowledge of the operation and the structure of the input to this operation to produce concise, efficient code. The approach was illustrated using three operations as examples. It should not be necessary to apply this approach to a large number of operations; most of the preprocessing occurring in symbolic algebra systems for numeric codes appears to use only a few key algebraic operations such as differentiation, integration, and power series. The proposed optimization scheme need only be used in a small number of operations, in which the optimization strategies that can be exploited are easily identifiable, in order to achieve code improvement for a significant range of problems.

Future research includes determining the class of operations with which this approach is useful and comparing the cost and effectiveness of this approach with existing optimization techniques. Since all the optimization techniques improve the representation of the large expressions but do not necessarily generate the true optimum, the investigation should determine how much improvement is achieved by applying each method, for what type of problems each method is best suited, and when the use of several techniques in sequence is effective. Another issue that must be addressed is the generation of expressions that are numerically stable as well as efficient to evaluate.

eferences

re69] Melvin A. Breuer. Generation of Optimal Code for Expressions via Factorization. *Communications of the ACM*, 12(6):333–340, June 1969.

re84] Richard L. Brenner. Simplifying Large Algebraic Expressions by Computer. In V. Ellen Golden, editor, *Proceedings of the 1984 Macsyma User's Conference*, pages 50–109. General Electric, Schenectady, New York, July 1984.

GGW85] B. W. Char, K. O. Geddes, G. H. Gonnet, and S. M. Watt. *Maple User's Guide*. WATCOM Publications Ltd., Waterloo, Ontario, Canada, 4 edition, 1985.

oo82] Cook Jr., G. O. *Development of a Magnetohydrodynamic Code for Axisymmetric, High-Beta Plasmas with Complex Magnetic Fields*. PhD thesis, Lawrence Livermore National Laboratory, University of California, Livermore, 1982. Also available as report from Lawrence Livermore National Laboratory, Livermore.

R85] Kevin Cahill and Randolph Reeder. Using MACSYMA to Write Long FORTRAN Codes for Simplicial-Interpolative Lattice Gauge Theory. Research Report DOE-UNM-85/3, Department of Physics and Astronomy, University of New Mexico, Department of Physics and Astronomy, University of New Mexico, Albuquerque, New Mexico 87131, July 1985.

at69] Richard J. Fateman. Optimal Code for Serial and Parallel Computation. *Communications of the ACM*, 12(12):694–695, December 1969.

ri88] Andreas Griewank. Book review of *Numerical Derivatives and Nonlinear Analysis*. *SIAM Review*, 30(2):327–329, June 1988.

ar84] Leo P. Harten. Using Macsyma to Generate (Somewhat) Optimized FORTRAN Code. In V. Ellen Golden, editor, *Proceedings of the 1984 Macsyma User's Conference*, pages 524–544. General Electric, Schenectady, New York, July 1984.

ea84] Anthony C. Hearn. Structure: The Key to Improved Algebraic Computation. In *The Second Symposium on Symbolic and Algebraic Computation by Computers, RIKEN, Japan, 21-22 August 1984*, pages 18–1 — 18–16. RSYMSAC, 1984.

vH85] B. J. A. Hulshof and J. A. van Hulzen. An Expression Compression Package for REDUCE based on Factorization and Controlled Expansion. Twente University of Technology, Department of Computer Science, Enschede, the Netherlands, 1985.

i84] Masao Iri. Simultaneous computation of functions, partial derivatives and estimates of rounding errors — complexity and practicality —. *Japan Journal of Applied Mathematics*, 1:223–252, 1984.

ed80] G. Kedem. Automatic Differentiation of Computer Programs. *ACM Transactions on Mathematical Software*, 6(2):150–165, June 1980.

KRS86] Harriet Kagiwada, Robert Kalaba, Nima Rosakhoo, and Karl Spingarn. *Numerical Derivatives and Nonlinear Analysis*, volume 31 of *Mathematical Concepts and Methods in Science and Engineering*. Plenum Press, Inc., 1986.

nu81] D. E. Knuth. *The Art of Computer Programming*, volume 2. Addison-Wesley Publishing Company, Reading, Massachusetts, 2nd edition, 1981. Seminumerical Algorithms.

W67] C. Mesztenyi and C. Witzgall. Stable Evaluation of Polynomials. *Journal of Research of the National Bureau of Standards - B. Mathematics and Mathematical Physics*, 71B(1):11–17, January-March 1967.

C79] E. Ng and B. W. Char. Gradient and Jacobian Computation for Numerical Applications. In V. Ellen Golden, editor, *Proceedings of the 1979 Macsyma User's Conference*, pages 604–621. NASA, Washington, D.C., June 1979.

al81] Louis B. Rall. *Automatic Differentiation: Techniques and Applications*. Lecture Notes in Computer Science 120. Spring-Verlag, Berlin, Germany, 1981.

[Ric83] John R. Rice. *Numerical Methods, Software, and Analysis*, chapter 3, pages 33–58. McGraw-Hill Inc., New York, IMSL Reference edition, 1983.

[Ris69] R. H. Risch. The Problem of Integration in Finite Terms. *Transactions of the American Mathematical Society*, 139:167–189, May 1969.

[Spe80] Bert Speelpenning. *Compiling Fast Partial Derivatives of Functions Given by Algorithms.* PhD thesis, Department of Computer Science, University of Illinois at Urbana-Champaign, Urbana, Illinois 61801, January 1980.

[SR84] Stanly Steinberg and Patrick Roache. Using Vaxima to Write Fortran Code. In V. Ellen Golden, editor, *Proceedings of the 1984 Macsyma User's Conference*, pages 1–22. General Electric, Schenectady, New York, July 1984.

[vH81] J. A. van Hulzen. Breuer's grow factor algorithm in computer algebra. In *Proceedings of the 1981 ACM Symposium on Symbolic and Algebraic Computation*, pages 100–104, 1981.

[vH83] J. A. van Hulzen. Code Optimization of Multivariate Polynomial Schemes: A Pragmatic Approach. In J. A. van Hulzen, editor, *Computer Algebra, Eurocal 83 (European Computer Algebra Conference, London, England, March 1983*, pages 286–300, Berlin, 1983. Springer-Verlag.

[vHH82] J. A. van Hulzen and B. J. A. Hulshof. An Expression Analysis Package for REDUCE. *SIGSAM Bulletin*, 16(4):32–44, 1982.

[Wan86] Paul S. Wang. FINGER: A Symbolic System for Automatic Generation of Numerical Programs in Finite Element Analysis. *Journal of Symbolic Computation*, 2(3):305–316, September 1986.

[Wir80] Michael C. Wirth. *On the Automation of Computational Physics.* PhD thesis, University of California, Davis, October 1980. Also available as Lawrence Livermore National Laboratory, Livermore, Report UCRL-52996 (October 1980).

CATFACT: Computer Algebraic Tools for Applications of Catastrophe Theory

R. G. Cowell and F. J. Wright
School of Mathematical Sciences, Queen Mary College
University of London, Mile End Road, LONDON E1 4NS, UK

ABSTRACT

We describe the current state of a package, written in REDUCE, that is being developed to solve the following problems that arise in applications of elementary catastrophe theory. For an input unfolding of some singularity, the *recognition problem* is to find a set of topological invariants that fix the equivalence class of the singularity. If the modality invariant is less than 3 then normal forms for unfoldings are known. The recognition algorithm employs the Buchberger Algorithm for Gröbner bases modified to the *local* requirements of singularity theory. The *mapping problem* is to find the taylor polynomial, up to any desired degree, of the right-equivalence that transforms the given unfolding into its normal form.

1. *INTRODUCTION*

Existing uses of computer algebra in applications of catastrophe theory were surveyed at EUROCAL'85 by Millington and Wright (1985), and a new project was outlined. The main thread of that project is now nearing completion, and it is our purpose in this paper to give some details of its implementation in REDUCE. The mathematical details underlying the algorithms are presented elsewhere (Cowell 1989; Cowell and Wright 1989a, 1989b).

Elementary catastrophe theory (*ECT*) was developed by Thom (1972) from ideas in differential topology and dynamical systems theory, motivated largely by a desire to be able to apply mathematical modelling to the irregular, asymmetrical systems that arise in biology. However, it has been fruitfully applied in areas ranging from quantum mechanics to sociology (Zeeman 1977; Poston and Stewart 1978; Stewart 1981, 1982). We are concerned with applications at the quantitative end of this spectrum.

Mathematically, *ECT* provides a local classification of the critical or stationary point structure of *typical families* of real-valued (easily extended to complex-valued) functions of the form

$$\phi : \mathbf{R}^{n+r} \otimes \mathbf{R}^K \to \mathbf{R},$$

where each member of the family is regarded as a function of an $(n+r)$-dimensional *state variable* $S \in \mathbf{R}^{n+r}$, and the family is parametrized by a K-dimensional *control variable* $c \in \mathbf{R}^K$. Families are classified into *right-equivalence* classes, such that ϕ is right-equivalent to ϕ' (denoted $\phi \sim \phi'$) if there exists a diffeomorphism $S' : \mathbf{R}^{n+r} \to \mathbf{R}^{n+r}$ parametrized by the control variables $c \in \mathbf{R}^K$ such that

$$\phi(S; c) = \phi'(S'(S; c); c).$$

This classification is useful in modelling any system that is governed by minima, maxima or stationary points in general of some kind of potential function, if the system can be considered to depend also on external parameters (representing perhaps knobs on an instrument, members of an ensemble, etc.). We assume that ϕ is given explicitly and algebraically by some modelling process.

The first problem (the recognition problem) in applying *ECT* to analyse any model is to recognise the right-equivalence class to which ϕ belongs. The second problem (the mapping problem) which arises in some applications, particularly those in which one seeks not only topological but also (geo)metrical information, is to determine the right-equivalence that relates ϕ to the standard representative of its equivalence class, the so called *normal form.* This problem is in general insoluble, but fortunately it suffices in applications to find an approximation to the right-equivalence, usually its taylor polynomial to some (fairly small) degree. This is consistent with the local nature of *ECT*, and is justified at length by Wright and Cowell (1987).

A family ϕ is classified in terms of its main *singularity*, which is the function for which the largest number of critical points have coalesced at a point taken as the origin of $\mathbf{R}^{n+r} \otimes \mathbf{R}^K$. The family is then said to *unfold* the singularity. The classification is performed in terms of a set of quantities that are invariant under right-equivalence transformations. Millington and Wright (1985) discussed two possible approaches to constructing right-equivalences; we have proceeded by generalising that of Wright and Dangelmayr (1985). Hence we construct the right-equivalence step by step as a composition of right-equivalences, each of which systematically removes terms from the taylor expansion of ϕ that do not appear in its normal form. Eventually only higher-order terms are left, which can be simply discarded (as proved by Cowell and Wright 1989a, 1989b).

In summary, we aim to automate the construction and application of a parametrized diffeomorphism $S(s, q; c)$ such that

$$\phi(S(s, q; c); c) = \phi_M(s; c) + \sum_{\rho=1}^{r} e_\rho q_\rho^2 + O(c^{M+1}),$$

where r is the rank (which may be zero) of the singularity at the origin of $\mathbf{R}^{n+r} \otimes \mathbf{R}^K$, $e \in \mathbf{R}^r$ are constants, $q \in \mathbf{R}^r$ are *inessential* (or Morse) state variables and $s \in \mathbf{R}^n$ are *essential* state variables. We call the function

$$\phi_M(s; c) = \Psi(s) + \sum_{p=0}^{P} u_p(c)\psi_p(s)$$

the *M-reduced form* of ϕ, where $\Psi(s)$ is the normal form of the singularity (without Morse terms), $\{\psi_p(s) \mid p = 0, ..., P\}$ is an unfolding basis for $\Psi(s)$ and the unfolding functions $u_p : \mathbf{R}^K \to \mathbf{R}$ are polynomials of degree M in the control variables c that all vanish at $c = 0$.

2. *THE* CATFACT *PACKAGE*

CATFACT – Computer Algebraic Tools For Applications of Catastrophe Theory – is a package which is designed to automate the solution of the recognition and mapping problems for unfoldings of singularities with modality ≤ 2, which is all that have been completely classified (by Arnol'd 1974). At present the package can perform the following tasks (after expanding an input unfolding as a multivariate polynomial if necessary). 1) Calculate the corank of the singularity and if necessary implement the splitting lemma to reduce the number of independent variables to that of the corank. 2) Test the singularity for determinacy up to some pre-assigned maximum value. 3) If the singularity is found to be finitely determined evaluate its codimension. 4) If the singularity is simple (modality 0) find an approximate diffeomorphism to reduce the unfolding to normal form. (A simple singularity can be identified from a knowledge of its determinacy, corank and codimension without directly evaluating its modality. We do not have an efficient algorithm for evaluating the modality of a singularity.) We believe that most of the algorithms in the package developed so far are new. The programs are written in REDUCE 3.2 (about 200K of code) so they should be widely portable. Our version is built on Cambridge Lisp, and runs under the Berkeley UNIX 4.2 operating system on a High Level Hardware ORION super-micro-computer.

The package has been developed in a modular fashion in order to facilitate extentions to incorporate higher-modal singularities. An overview of the package is provided in Figure 1 below. At present the first two columns of the 'flow chart' have been developed. The third is currently under construction but is complete for simple singularities of corank two.

3. *THE REDUCTION PROCESS*

Let s and c denote respectively state and control variables of the unfolding ϕ (which is a singularity if there are no control variables), not necessarily in a normal form but expressed as a truncated taylor series. After the input of the unfolding, the first step is to perform a linear coordinate transformation to yield a new unfolding in which all the terms quadratic in s and independent of c are diagonalised; this simultaneously yields the corank of the singularity. The quadratic variables of the transformed singularity (if present) are labelled by the coordinates q. The remaining inessential variables retain the symbols s, but with re-numbered indices. If the determinacy k of the singularity is not known, then a test is performed to see if k is less than some integer k_{det} (input together with the unfolding). The test employs a generalisation of Siersma's trick for corank 1 and 2 singularities, together with a result of Bruce, du Plessis and Wall (1987) concerning affine algebraic groups. (We have been unable to prove that our algorithm will always work, although it has done so in all cases tested so far. Briefly, the difficulty is that Bruce, du Plessis and Wall give conditions for the *existence* of a certain nilpotent vector field, but our *constructive* algorithm might not detect such a vector field for some singularity (if it exists) because the result obtained from the algorithm may be sensitive to the order in which a set of equations are generated and reduced.) If a value $k < k_{det}$ is found then the program continues, otherwise it stops with an error message.

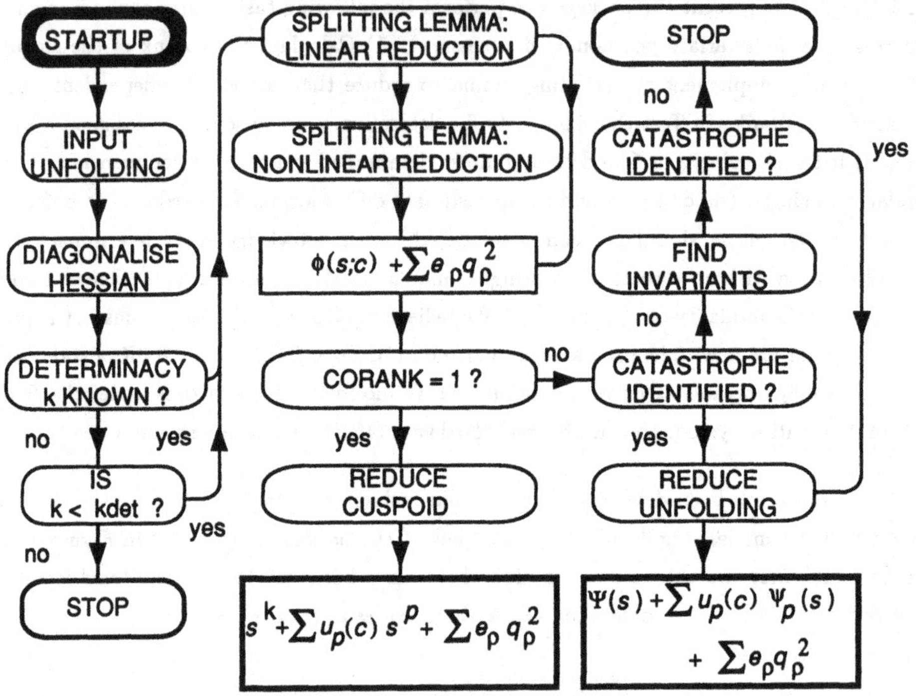

Figure 1. Stages employed in CATFACT for classifying and reducing unfoldings to normal form.

The next phase is to simplify the unfolding by removing any inessential or Morse variables. The program employs a simplification of Poston and Stewart's (1978) algorithm for implementing the splitting lemma. It can be shown that, for a singularity whose quadratic terms are diagonalised, a linear coordinate transformation of the Morse variables (but generally having an inhomogeneous part nonlinear in the essential variables) is sufficient for the splitting stage of the reduction if only the M-reduced form and not the transformation that produces it is required (Cowell and Wright, 1989b). Thus this linear transformation is performed. It is left as an option for the user to employ a subsequent nonlinear transformation if the reduction transformation to some finite degree in all the state variables is sought, because nonlinear transformations generally require a lot of computer time. At this stage the unfolding is split into a 'core unfolding', which is dependent upon a set of essential state variables only, and a sum of quadratic terms involving the inessential state variables.

If the corank is unity then the unfolding is readily identified as belonging to the cuspoid family and is reduced to normal form, to some pre-assigned degree M in the control variables, by using a variation of the algorithm of Wright and Dangelmayr (1985). For example, the unfolding

$$\phi(s;c) = s_1^2 + s_2^4 + c(s_1 + s_1 s_2 + s_1^4),$$

where s_1, s_2 are state variables, and c is a control variable, yields the following cuspoid normal form to degree 7 in c :

$$\Phi(s,q;c) = s_1^4 - c^2/4 + c^5/16 + c^7/32 - (c^2/2 - c^5/4 - 5c^7/128)s_1 \\ - (c^2/4 - 3c^5/8 - c^7/128)s_1^2 + q_1^2 + O(c^8).$$

If the corank is greater than unity, then the package moves into the third column, which at present is incomplete but is planned to perform as follows. If the normal form of the singularity is known a reduction is performed; if not, other invariants (for example modality) are calculated in order to classify the singularity. If it is verified that the singularity belongs to Arnol'd's list, which occurs if the modality ≤ 2, a sequence of transformations of the essential variables is performed which reduce the singularity to normal form.

The current experimental version of CATFACT is not pre-compiled. An interactive session typically generates the following dialogue, in which user input (underlined) is read using REDUCE's symbolic level 'XREAD' function.

```
1:  IN CATFACT $
Input number of state variables :  2;
Input number of control variables :  1;
Input degree in control variables for reduced form :  2;
The default notation for the state variables is stored in an array s, with
elements s(1) = s1, s(2)=s2, ...  etc, except when there is only one state
variable, in which case s(1) = s is employed.
Do you wish to change this notation ?  (Y or N) Y
    Default notation for s(1) = s1
    Input new notation :  X;
    Default notation for s(2) = s2
    Input new notation :  Y;
The default notation for the control variables is stored in an array c, with
elements c(1) = c1, c(2)=c2, ...  etc, except when there is only one control
variable, in which case c(1) = c is employed.
Do you wish to change this notation ?  (Y or N) N
Input upper bound on strong determinacy or exact value if known 7;
    Upper bound on strong determinacy = 7
Is ordinary determinacy known?  (Y or N) N
Now please type in the unfolding :  use array elements s(1), ..., s(n) for the
state variables and c(1), ..., c(m) for the control variables, your own notation
```

```
if you have altered the defaults, or a predefined function.
Initial unfolding ?  :  x**3 + x*y**3 + a*y**5 + c*x ;
    Initial unfolding is x^3 + x * y^3 + x * c + y^5*a
    Is this correct?  (Y or N) Y
Is the unfolding in Taylor polynomial form?  (Y or N) Y
Do you want to see the steps in the right equivalence ?  (Y or N) N
Do you want to normalise the singularity coefficients ?  (Y or N) Y
    Diagonalising quadratic terms of unfolding
    rank = 0      corank = 2
    Calculating the codimension and determinacy of singularity
    codimension = 7      determinacy = 4
Catastrophe is of type E7
The initial unfolding has been successfully reduced to normal form.  The initial
singularity is stored in unfi, and in unf0 in Taylor series form.  The reduced
singularity is stored in unf, in Taylor series form.  If the singularity
contains unfolding terms, then a symbolic expression for the singularity may be
found in unfolding, and coefficients of the unfolding basis may be found in the
array elements u(j) :  j = 0, ..., codimension-1.  Coefficients of the
singularity monomials are stored in the array cf(j):  j = 1, ...  .
2:  unfolding;
    x^3*cf1 + x * y^3*cf2 + x * y * u4 + x * u2 + y^4*u6 + y^3*u5 + y^2*u3 + y * u1 + u0
3:  u(0);
    1/9a^3*c^2
4:  bye;
```

---

4. *THE RECOGNITION PROBLEM AND GRÖBNER BASES*

In order to reduce an unfolding to normal form it is necessary to know in advance to which equivalence class the unfolding belongs. At the present time all (complex) singularities having 2 or fewer modal parameters have been completely classified (by Arnol'd, 1974). For the simple singularities (modality = 0) and the exceptional singularities of modality ≤ 2 a set of procedures has been written in REDUCE 3.3 (approximately 240 lines of code) which solves this recognition problem. A call to a single procedure – CLASSIFY – is all that is required for the recognition package. The procedure requires three arguments: (i) the singularity in taylor series form, truncated at some appropriate degree, (ii) a list of state variables, and (iii) an integer greater than the determinacy of the singularity (see below). For example, the input:

```
CLASSIFY( a*x*x*y + y**5,{x,y},8);
```

will identify the (D_6) singularity $x^2y + y^5$ and write to the standard output:

```
singularity type :  d6 = x^2 * y + y^5
Boardman symbol :  {1,2,1,1,1}
local determinacy :  4
strong determinacy :  5
determinacy :  5
multiplicity :  6
modality :  0
```

The package is based on ideas closely related to the Buchberger algorithm (Buchberger, 1985; see also these proceedings). The similarity arises because both catastrophe theory and polynomial ideal theory pose problems concerning the zeros of equations – for catastrophe theory, the location of stationary points of a function, and for polynomial ideal theory, the common roots of sets of polynomial equations. However, catastrophe theory is a *local* theory concerning (local stability properties of) the stationary points of functions near to the origin (by assumption) of the state space. In contrast, the application of the theory of Gröbner bases is applicable to finding *all* common roots of a set of equations, which is a *global* problem. The restriction to local analysis in catastrophe theory leads naturally to a simple variation of the Buchberger algorithm. Before describing this modification, we first introduce some terminology and results from catastrophe theory. For simplicity we work with complex singularities (real singularities have a more refined classification scheme) and we assume that the singularites are of zero rank, and hence have no Morse terms.

Let $s \in \mathbf{C}^n$, and let $\phi : \mathbf{C}^n \to \mathbf{C}$ be a singularity with critical point at the origin of s. This implies that at $s = 0$,

$$\partial\phi(s)/\partial s_\alpha = 0, \quad \alpha = 1, \ldots, n,$$

and

$$\det\ \partial^2\phi(s)/\partial s_\alpha \partial s_\beta \ = 0.$$

Let $\langle\partial\phi\rangle$ denote the Jacobian ideal generated by the set of first order partial derivatives of $\phi(s)$, and let $J^k\phi(s)$ denote the k-jet of $\phi(s)$, i.e. the (multivariate) taylor expansion of the singularity up to and including terms of degree k in s. Finally let m denote the (unique) maximal ideal (generated by s) of the ring of germs of C^∞ functions $\mathbf{C}[s]$, and define the *multiplicity* of the singularity $\phi(s)$ to be $\mu \equiv \dim(m/\langle\partial\phi\rangle)$. Then the following are equivalent (Poston and Stewart, 1978):

(i) $s = 0$ is an isolated critical point of the singularity $\phi(s)$;
(ii) $\mu < \infty$;
(iii) there exists a smallest finite integer k (called the determinacy of ϕ) such that $\phi(s) \sim \phi(s) + h \sim J^k\phi(s)$ for all $h \in m^{k+1}$.

The importance of (iii) is that for singularities of finite determinacy calculations may be performed on finite taylor polynomials (jets), which in general is a great simplification. The number of coincident roots at $s = 0$, into which the singularity breaks up under versal deformation, is equal to the multiplicity μ.

Assume that a singularity $\phi(s)$ has finite determinacy k. Define a set of quotient spaces

$$V_i\phi \equiv J^i(m^i/\langle\partial\phi\rangle) \quad i = 0,\ldots,k,$$

and denote the dimension of each space by $\mathrm{cod}_i\phi = \dim(V_i\phi)$. Then the *Boardman symbol* of ϕ is the ordered set of integers $\{\mathrm{cod}_0\phi, \ldots, \mathrm{cod}_k\phi\}$, and it is a topological invariant under diffeomorphism (Gibson, 1979).

The Boardman symbol is unique for each of the low modality singularities classified by Arnol'd (we do not know if this is true in general) and calculating it is the basis of our recognition package. It could be calculated by doing Gaussian elimination on a suitable matrix. However, a more efficient method is to employ ideas from Gröbner basis theory in order to implement Siersma's trick (Poston and Stewart, 1978). Thus we impose a total ordering on the monomials generated by the (specified) coordinates s, which for our purpose we choose to be a *total degree ordering* denoted by $<_T$ (we have not yet explored the possibility of using other orderings). Then for $f \in \mathbf{C}[s]$ we define

$MinMon(f) \equiv$ the *minimum* monomial (with respect to $<_T$) in f with non-zero coefficient;
$MinCoef(f) \equiv$ the coefficient of $MinMon(f)$.

Then making the replacements

$$MinMon \leftarrow \textit{LeadingPowerProduct},$$
$$MinCoef \leftarrow \textit{LeadingCoefficient},$$

the notion of the *S-Polynomial* of two polynomials may be defined as in Gröbner basis theory (Buchberger, 1985). From the set of k-jets of the first order partial derivatives of a singularity $\phi(s)$, which has determinacy k, the Buchberger Algorithm may be recast using the above replacements to derive a 'Gröbner basis' for the singularity. This 'modified Buchberger Algorithm' will terminate if at each stage all terms which are of degree $k+1$ or higher in s, generated by the reduction process, are set to zero. This truncation of high-order terms is justified by the finite determinacy k of the singularity. From the set of minimum monomials of the elements of this 'singularity Gröbner basis' the Boardman symbol of the singularity may be readily evaluated (Cowell 1989).

The correspondence of (local, minimum polynomial) $\leftrightarrow$ (global, leading polynomial) gives rise to some similarities between the results of catastrophe theory and Gröbner basis theory. For example, if $F = \{f_1, \ldots, f_m\}$ is a finite set of polynomials in the finite set of variables x, and G is a Gröbner basis derived from F, then the set of coupled equations $f_i = 0,\ i = 1, \ldots, m$ has finitely many solutions if and only if for each x_i in x a monomial of the form $x_i^{k_i}$ appears in the leading power products of G. The analogous result for catastrophe theory, using the 'modified Buchberger Algorithm' but *without* truncation is as follows: If G is a singularity Gröbner basis obtained from the singularity $\phi(s)$ with critical point at $s = 0$, then ϕ has finite determinacy (and hence an isolated critical point at the origin) if and only if for each s_i a monomial of the form $s_i^{k_i}$ occurs among the minimum monomials of the elements of G (Cowell, 1989). We expect there to be many other analogies between the two fields.

5. *FUTURE DEVELOPMENTS FOR* CATFACT

There are several ways of extending the CATFACT package from its present state. The reduction of modal singularities to normal form is being pursued at present. In view of the large number of low-modal singularities and the lack of an efficient algorithm for calculating the modality of a singularity, and partly because of the differences between REDUCE 3.2 and REDUCE 3.3, the most efficient way to do this will probably be to build upon the recognition package described in §4. The reduction process would then consist of the following three steps:

(1) classify the unfolding;

(2) apply the splitting lemma to remove inessential variables;

(3) transform the remaining core unfolding to normal form.

Also under development is a package which transforms a user specified unfolding *from* its (known) normal form. This requires constructing the unfolding functions $u(c)$ *together with* the diffeomorphism $s = s(s'; c)$ which is inverse to the mapping $s \rightarrow s'(s; c)$ that the CATFACT package constructs. (Note that CATFACT gives the unfolding functions $u(c)$ *automatically* by effecting the reduction *to* normal form.) At the present time this package has been implemented in both REDUCE 3.2 and REDUCE 3.3 for unfoldings of arbitrary rank cuspoid singularities by Colvin (1988), who has improved and extended algorithms developed by Millington (1985) (see also Millington and Wright, 1985).

A future extension of CATFACT, not being pursued at present, may also include the development of normal form recognition and reduction algorithms for equivariant singularities (singularities possessing symmetries) and for boundary catastrophes.

REFERENCES

Arnold, V. I., 1974 'Critical points of smooth functions and their normal forms', *Usp. Mat. Nauk* **29** pp 11-49. (Translated as *Russ. Math. Surveys* **29** pp 10-50).

Bruce, J. W., du Plessis, A. A. and Wall, C. T. C., 1987 'Unipotency and Determinacy', *Invent. Math.* **88** pp 521-554.

Buchberger, B., 1985 'Gröbner Bases : An algorithmic method in polynomial ideal theory', in *Multidimensional systems theory: progress, directions, and open problems in multidimensional systems* (Edited by N. K. Bose) Dordrecht, Holland: D. Reidel.

Colvin, A. P., 1988 *Private communication.*

Cowell, R. G., 1989 'Application of ordered standard bases to catastrophe theory' (Submitted to *Proc. LMS.*)

Cowell, R. G. and Wright, F. J., 1989a 'Truncation criteria and algorithm for the reduction to normal form of catastrophe unfoldings I: Singularities with zero rank' (Submitted to *Phil. Trans. R. Soc. Lond.*)

Cowell, R. G. and Wright, F. J., 1989b 'Truncation criteria and algorithm for the reduction to normal form of catastrophe unfoldings II: Singularities with non-zero rank' (Submitted to *Phil. Trans. R. Soc. Lond.*)

Gibson, C. G., 1979 *Singular points of smooth mappings*, London: Pitman.

Millington, K., 1985 'Using computer algebra to determine equivalences in catastrophe theory' (Ph.D. Thesis, University of London).

Millington, K. and Wright, F. J., 1985 'Algebraic computations in elementary catastrophe theory,' *EUROCAL'85 Lecture Notes in Computer Science* **204** (Berlin, Heidelberg: Springer) pp 116-125.

Poston, T. and Stewart, I. N., 1978 *Catastrophe Theory and its Applications*, London: Pitman.

Stewart, I. N., 1981 'Applications of catastrophe theory to the physical sciences', *Physica* **2D**, pp 245-305.

Stewart, I. N., 1982 'Catastrophe theory in Physics', *Rep. Prog. Phys.* **45** pp 185-221.

Thom, R., 1972 *Stabilité Structurelle et Morphogénèse*. Reading, Mass.: Benjamin. (English translation by D. H. Fowler, 1975, *Stuctural Stability and Morphogenesis*. Reading, Mass.: Benjamin.)

Wright, F. J. and Cowell, R. G., 1987 'Computer algebraic tools for applications of catastrophe theory', in *The Physics of Structure Formation : Theory and Simulation*. Eds. W. Güttinger and G. Dangelmayr. Berlin: Springer, pp 402-415.

Wright, F. J. and Dangelmayr, G., 1985 'Explicit iterative algorithms to reduce a univariate catastrophe to normal form', *Computing* **35** pp 73-83.

Zeeman, E. C., 1977 *Catastrophe Theory*. Reading, Mass.: Addison-Wesley.

COMPUTER ALGEBRA APPLICATION FOR INVESTIGATING INTEGRABILITY OF NONLINEAR EVOLUTION SYSTEMS

V.P.Gerdt
Laboratory of Computing Techniques and Automation
Joint Institute for Nuclear Research
Head Post Office, P.O. Box 79, Moscow, USSR

A.B. Shabat, S.I. Svinolupov
Bashkir Branch of the USSR Academy of Sciences
Ufa, USSR

A.Yu. Zharkov
Saratov State University
Astrakhanskaya 83, Saratov, USSR

At present an intensive work on classification of integrable non-linear partial differential equations with two independent variables is carried out. In a number of the cases ([1],[5],[9]) the formulation of effective criteria of integrability has been achieved and the complete lists of integrable systems have been obtained. For example, in [4],[5],[9] the complete list of integrable systems of the Schrödinger type

$$u_t=u_{xx}+f(u,v,u_x,v_x),\quad v_t=-v_{xx}+g(u,v,u_x,v_x) \qquad (1)$$

is obtained.

In the frames of the symmetry approach (see [6]-[9]) the definition of integrability is based on internal properties of the equations and the conditions used for classification are the necessary conditions for existence of higher-order symmetries and conservation laws. The general derivation scheme for this conditions (see [8],[9]) covers the systems of the form

$$u_t=\Phi(x,u,\dots u_N)=\Lambda u_N+F(x,u,\dots u_{N-1}),\quad N\geq 2 \qquad (2)$$

where

$$u=u(t,x),\quad u_k=\partial^k u/\partial x^k,\quad u=(u^1,\dots u^M),\quad F=(F^1,\dots F^M),$$
$$\Lambda=diag(\lambda_1,\dots \lambda_M),\quad \lambda_i\in C,\quad \lambda_i\neq\lambda_j\quad (i\neq j). \qquad (3)$$

Evolution equations integrable by inverse spectral transform and linearizable like the Burgers equation satisfy the conditions of integrability arising in the symmetry approach.

There are two types of problems solvable by means of the symmetry approach: 1) properly classification problems of obtaining the complete list of the fixed form systems and describing the most general transforms connecting these systems; 2) for a given concrete system testing the conditions of integrability and computing symmetries and conservation law densities.

In the present paper we discuss the problems of the second type. Note that such problems remain actual even after the complete lists of systems under consideration have been already obtained. For example, since the list of integrable systems (1) presented in [4],[5],[9] is too large, it's more convenient for a given concrete system to check us the integrability conditions directly rather than to identify it with that from the list.

The procedure of derivation and checking the necessary condition of integrability demands tedious algebraic computations. To carry them out automatically it's worth while to use computer algebra ([10]).

In the case of scalar evolution equations the algorithms for checking up the integrability conditions for a given equation is already implemented in computer algebra systems REDUCE ([11]) and FORMAC ([12]). In the present paper we suggest the algorithm for checking the necessary conditions of integrability and its computer implementations for evolution systems (2)-(3) in FORMAC. In addition, our FORMAC program allows one to find symmetries and canonical conservation law densities.

2. Let us remind the basic concepts of the symmetry approach and the general derivation scheme for the necessary integrability conditions.

The <u>symmetry</u> (generator of infinitesimal symmetry) of evolution system (2) is called the vector function $f=(f^1,\dots,f^M)$ of a finite number of dynamical variables from the infinite set $x,u,u_1,u_2,\dots$ such that

$$\frac{df}{dt} = \Phi_*(f), \tag{4}$$

where Φ_* is a matrix differential operator

$$\Phi_* = \Phi_u + \Phi_{u_1}\mathbb{D} + \dots + \Phi_{u_N}\mathbb{D}^N. \tag{5}$$

Φ_{u_i} - Jacobi matrix, that is

$$\left[\Phi_{u_i}\right]_{kj} = \frac{\partial \Phi^k}{\partial u_i^j}$$

The operators d/dt and $\mathbb{D}=d/dx$ in (4)-(5) are the total differentiation operators with respect to variables t and x correspondingly, both acting on the functions of a finite number of dynamical variables:

$$\frac{d}{dt} = \sum_{i=1}^{M} \sum_{j=0}^{\infty} \mathbb{D}^j(\Phi^i) \frac{\partial}{\partial u_j^i} ,$$
$$\mathbb{D} = \frac{\partial}{\partial x} + \sum_{i=1}^{M} \sum_{j=0}^{\infty} u_{j+1}^i \frac{\partial}{\partial u_j^i} . \tag{6}$$

Equation (4) defining symmetries means that equation (2) and equation

$$u_\tau = f(x,u,u_1,\dots)$$

are compartible or, in other words, this means the invariance of equation (2) under infinetisimal transformations $\underline{t}=t, \underline{x}=x, \underline{u}=u+\tau f(x,u,u_1,\dots)$. The linearization operator * defined by the formula

$$f_*(v) = \frac{\partial}{\partial\varepsilon} f(x,u+\varepsilon v,\mathbb{D}(u+\varepsilon v),\dots)\Big|_{\varepsilon=0}$$

transforms the equation (4) to the following operator relation

$$L_t + [L,\Phi_*] = (\Phi_*)_\tau \tag{7}$$

where $L = f_* = f_u + f_{u_1}\mathbb{D} + \dots$ (compare with (5)).

Let us define the <u>n-th order formal symmetry</u> of the system (2) as any differential operator of the n-th order

$$L = \sum_{k=0}^{n} A_k \mathbb{D}^k, \quad deg(L)=n \tag{8}$$

with matrix coefficients A_k which depend on a finite number of dynamical variables and such that

$$L_t - [\Phi_*,L] = Q \tag{9}$$

where Q is a differential operator of the form

$$Q = \sum_{k<N} Q_k \mathbb{D}^k.$$

Substitution of (8) in (9) leads to a chain of equations arising after equating the coefficients of $\mathbb{D}^j$, j=N+n,...,N+1,N. So, putting the coefficient of $\mathbb{D}^{N+n}$ equal to zero one gets

$$[\Lambda,A_n] = 0. \tag{10}$$

Then for j=N+n-1 one can obtain

$$[\Lambda,A_{n-1}] + N\cdot\Lambda\cdot\mathbb{D}(A_n) + [\Phi_{u_{N-1}},A_n] = 0. \tag{11}$$

Next equations have the following general form

$$[\Lambda,A_{j-N}] + N\cdot\Lambda\cdot\mathbb{D}(A_{j-N+1}) + [\Phi_{u_{N-1}},A_{j-N+1}] + B_j = 0, \quad j\ge N, \tag{12}$$

where B_j are matrices with components expressed in terms of the differential operator (5) coefficients and the components of A_i where $i>j-N+1$. One can find from equation (12) non-diagonal part of the matrix A_{j-N} and the diagonal part of the matrix A_{j-N+1}.From equations (10)-(11) it follows that

$$A_n = diag(\mu_1,\mu_2,\dots\mu_M), \quad \mu_i\in C \tag{13}$$

The formal symmetry (8) is called <u>non-degenerated</u> if

$$det(A_n) = \prod_{k=1}^{M} \mu_k \neq 0.$$

Evidently the formal symmetry is defined up to addition an arbitrary diagonal matrix depending on a finite number of the dynamical variables (this arbitrariness can be elimineted with normalization $diagA_0=0$).

The conditions of existence the non-generated formal symmetry being the conditions of solvability of eqs. (10)-(12) in terms of matrix - functions of dynamical variables are the criteria for constructing the lists of integrable equations. For instance, the complete list of the secondintegrable scalar equations of the general form $u_t=\Phi(x,u,u_1,u_2)$ presented in 3] corresponds to the choice n=5. For the equations of the form (2) with M=1, N=3 considered in paper (2) it was supposed that n=9.

It's easy to verify that the existence of the n_1-order symmetry (the order of the symmetry is called $deg(f_*)$) with $n_1\geq n > N$ leads to the existence of the n-th order formal symmetry (compare (7) and (9)). Thus the conditions of the existence of the n-th order symmetries are the necessary conditions of the existence of the n_1-th order symmetries, where $n_1\geq n$. One can prove ([4],[8],[9]) that from the existence of the pair of the local high-order conservation laws it follows the existence of the formal symmetry.

It should be reminded that the local conservation law of system (2) is given by the pair of scalar functions (ρ,σ) of the dynamical variables such that

$$\frac{d\rho}{dt} = \mathbb{D}(\sigma) \tag{14}$$

The function ρ is the density of the conservation law (14) and the order of the conservation law is defined as a degree of the following polynomial in $\mathbb{D}$ (see [4],[9]):

$$R = (\frac{\delta\rho}{\delta u})_* = \sum_{i\geq 0} R_i \mathbb{D}^i, \qquad \frac{\delta}{\delta u} = \sum_{k=0}^{\infty} (-1)^k \mathbb{D}^k \frac{\partial}{\partial u_k} . \tag{15}$$

Using the algorithms of the manipulation with power series of the form

$$L = \sum_{k\geq m} A_k \mathbb{D}^k = A_m \mathbb{D}^m + \ldots + A_0 + A_{-1}\mathbb{D}^{-1} + \ldots, \tag{16}$$

with negative powers of the symbol $\mathbb{D}$ it is possible to generalize the above definition of the formal symmetry and to denote by the formal symmetry of degree m and order n every formal series (16) with matrix coefficients depending on dynamical variables and satisfying the relation

$$deg(L_t - [\Phi_*, L]) < m+N-n. \tag{17}$$

Multiplication of formal series (16) is defined by the formula

$$a\, \mathbb{D}^l \cdot b\, \mathbb{D}^k = a \sum_{i=0}^{\infty} \binom{l}{i} \mathbb{D}^i(b) \cdot \mathbb{D}^{l+k-i},$$
$$\binom{l}{i} = \frac{l(l-1)\ldots(l-i+1)}{1\cdot 2\cdot \ldots i} \tag{18}$$

generalizing the well-known differential operator multiplication rule.

One of the principal propositions of the theory of formal symmetries is the following

Theorem 1 ([4],[8],[9]). The conditions of the n-th order formal symmetry existence don't depend on the choice of the degree m of the formal symmetry (16), and on the choice of integration constants arising in the chain of eqs. (10)-(12) for the coefficients of the formal symmetry.□

Note that the system of equations for the first n coefficients of the m-th order formal symmetry obtained by equating the coefficients at $\mathbb{D}^j$, j=m+N,...m+N-n+1 has the same form as the chain of eqs. (10)-(12) for the formal symmetry. In particular, for all m=0,±1,±2,... the leading coefficient of the formal symmetry (16) is a constant diagonal matrix, that's

$$A_m = diag(\mu_1, \ldots \mu_M), \quad \mu_i \in C. \tag{19}$$

One can formulate the existence conditions of the formal symmetry of the order n>N in terms of the lower-order formal symmetry coefficients as it's given below.

Theorem 2 ([4],[8],[9]). Let the non-generated formal symmetry of the order n=N+i, i≥0 exists. Then the existence of the n-th order formal symmetry, where n=N+i+1 is equivalent to the following conditions (see (14))

$$\frac{d}{dt}(R(i,j)) \in \mathrm{Im}\mathbb{D}, \qquad j=1,2,\ldots M \tag{20}$$

where

$$R(i,j) = \begin{cases} \dfrac{\partial F}{\partial u^j_{N-1}} & , i=0 \\[2ex] \dfrac{\partial}{\partial \mu_j}\, trace(res\ L) & , i>0 \end{cases} \tag{21}$$

L is a formal symmetry of the order $i+2$ and the degree i with the leading coefficient (19) depending on M arbitrary parameters $\mu_1, \mu_2, \ldots \mu_M$ and $res\ L \equiv A_{-1}$ (see (16)).□

The condition $dR/dt \in \mathrm{Im}\mathbb{D}$ in (20) means the existence of the function σ depending on the dynamical variables $x, u, u_1, \ldots$ such that the pair (R,σ) specifies the conservation law (14) for the system (2). Formula (21) determines the algorithm for constructing the local conservation law densities from the infinite series which we shall call the canonical series. For all known examples of equations integrable by inverse spectral transform this canonical seris coincides with the series of local conservation laws which can be constructed using the scattering matrix in frames of the inverse problem method ([13]).

3. Algorithm for checking the integrability conditions for the system (2) (algorithm (I)) is the following. First one tests conditions (20) for $i=0$, $j=1,\dots M$ which are equivalent (theorem 2) to existence of the N+1- order formal symmetry. At the second step one constructs the formal symmetry. At the next step one constructs the formal symmetry of the order 1 and the degree 3 and tests M conditions (20) for $i=1$, $j=1,..M$. The fulfilment of these conditions guarantees the existence of the N+2- order formal symmetry and so on. The elements of matrix coefficients A_k for the m-th degree formal symmetry (16) can be found from the following recurrence relations (see (11)-(13),(17)-(19)):

$$A_k = 0, \quad k>m,$$

$$[A_m]_{ij} = \begin{cases} 0, & i\neq j \\ \mu_i, & i=j \end{cases}$$

$$[A_k]_{ij} = \begin{cases} \dfrac{1}{\lambda_i-\lambda_j}\left\{\left[\dfrac{d}{dt}([A_{N+K}]_{ij}) - [C_{N+k}]_{ij}\right|_{A_k=0}\right\}, & i\neq j \\ \dfrac{1}{N\lambda_i}\,\mathbb{D}^{-1}\left\{\left[\dfrac{d}{dt}([A_{N+k-1}]_{ii} - [C_{N+k-1}]_{ii}\right|_{diag(A_k)=0}\right\} + \gamma_i^k, & i=j \end{cases} \tag{22}$$

where $\mu_i, \gamma_i^k \in C$ are arbitrary constants, C_i are the coefficients of the commutator

$$[\Phi_*, L] = \sum C_i \mathbb{D}^i$$

The main computational difficulty of the above algorithm is the inversion of the operator $\mathbb{D}$ which is necessary for constructing the diagonal elements of A_k.This problem is reduced to solving the following equation

$$\mathbb{D}(Q) = S, \tag{23}$$

where Q and S are the scalar functions of dynamical variables $x, u, u_1, \dots$ and $\mathbb{D}$ is the total differentiation operator with respect to x. Note that the equation (23) can be solved not for any right hand side S. Its solvability is equivalent to the equality $\delta S/\delta u = 0$ (see [14]).

The algorithm for the operator $\mathbb{D}$ inversion (algorithm (II)) is given below. It allows to find the function Q as well as the set of relations for S which must be satisfied. The algorithm is based on the following conditions which must be hold for any function $S(x,u,u_1\dots u_k) \in \mathrm{Im}\mathbb{D}$:

$$\frac{\partial^2 S}{\partial u_k^i \partial u_k^j} = 0, \qquad \frac{\partial^2 S}{\partial u_k^i \partial u_{k-1}^j} - \frac{\partial^2 S}{\partial u_k^j \partial u_{k-1}^i} = 0, \ k>0$$
$$\frac{\partial S}{\partial u^i} = 0, \quad k=0, \tag{24}$$

where $i, j = 1,2,\dots M$.

If conditions (24) are satisfied, then S can be written as:

$$S = \mathbb{D}(q(x,u,\dots u_{k-1}) + \bar{S}(x,u,\dots u_{k-1}). \tag{25}$$

Determining q from (25), we find the solution of (23) is represented in the form $Q = q + \bar{Q}$, where $\bar{Q}$ satisfies the equation $\mathrm{D}(\bar{Q}) = \bar{S}$. Therefore the condition $S \in \mathrm{Im}\mathbb{D}$ is reduced to the condition $\bar{S} \in \mathrm{Im}\mathbb{D}$, where the order of $\bar{S}$ is lower than the order of S.

To find the function q from equation (25) it is necessary to compute indefinite integrals of the form

$$\int \frac{\partial S}{\partial u_k^i} du_k^i.$$

Our program allows to compute indefinite integrals $\int g(w)dw$ for the following class of integrands:

$$g(w) = \sum_j [P_j(w)e^{\lambda_j w} + (\alpha_j w + \beta_j)^{\gamma_j}], \tag{26}$$

where $\alpha_j, \beta_j, \gamma_j, \lambda_j$ are constants, $P_j(w)$ are polynomials in w. The integration constants are put to zero. The formal description of the above algorithm is the following

input: $S(x,u,\dots u_k)$, k=*ord* S
output: $Q(x,u,\dots u_{k-1})$, Z
$Q := 0$, $Z := 0$
for m:=k *to* 1 *step* -1 *do*
  for i:=1 *to* M *do*
    for j:=i *to* M *do*
    $$Y := \iint \frac{\partial^2 S}{\partial u_m^i \partial u_m^j} du_m^i du_m^j$$
    $S := S - Y$
    $Z := Z + Y$

for $j:=1$ *to* $i-1$ *do*

$$Y:=\iint \frac{\partial^2 S}{\partial u^i_m \partial u^i_{m-1}} du^i_m du^j_{m-1}$$

$$S:=S-Y$$

$$Z:=Z+Y$$

$$Q:=Q+\int \frac{\partial S}{\partial u^i_m} du^i_{m-1} \tag{27}$$

$$S:=S-\mathbb{D}\left(\int \frac{\partial S}{\partial u^i_m} du^i_{m-1}\right)$$

for $i:=1$ *to* M *do*

$$Y:=\int \frac{\partial S}{\partial u^i} du^i$$

$$S:=S-Y$$

$$Z:=Z+Y$$

$$Q:=Q+\int S dx$$

Applying the algorithm (II) we have $S=\mathbb{D}(Q)+Z$, where Z is "non-integrable" expression. The equality $Z=0$ is equivalent to $S\in\mathrm{Im}\,\mathbb{D}$.

Example. Let M=2, $u=(v,w)$ and

$$S = vw_2 + \alpha v_1 w_1 + \beta v_1^2, \qquad \alpha,\beta\in C$$

Applying the algorithm (II), we get

$$Q = vw_1, \quad Z = (\alpha-1)v_1 w_1 + \beta v_1^2.$$

Thus $S\in\mathrm{Im}\,\mathbb{D}$ is equivalent to $\alpha=1$, $\beta=0$.

The algorithm for computing the n-order symmetry (algorithm (III)) is based on formulae (7)-(12) and includes three steps.

At the first step one computes the n-th order formal symmetry of the form (8) applying the algorithm (I). The existence of the formal symmetry is ensured by the corresponding integrability conditions (theorems 1,2). Otherwise the algorithm informs about the absence of symmetry and terminates. Remind that recurrence relations (22) lead to the formal symmetry depending on parameters $\mu_1,\ldots\mu_M$ and integration constants γ_i^k.

The symmetry of the order n defined by (4) exists only if the equality $L=f_*$ holds with the n-th order formal symmetry L already found. The equation $L=f_*$ is equivalent to the following system of equations for the formal symmetry coefficients

$$\frac{\partial}{\partial u^m_l}[A_k]_{ij} - \frac{\partial}{\partial u^j_k}[A_l]_{im} = 0, \qquad \begin{matrix} k,l=0,1,\ldots n \\ i,j,m=1,\ldots M \end{matrix} \tag{28}$$

At the second step one verifies the solvability of this system to obtain concrete values of μ_i, γ_i^k. Note that conditions (28) are equivalent to the existence of the function $H=H(x,u,\ldots u_n)$, $H=(H^1,\ldots H^M)$ satisfying the equation

$$\mathbb{D}(H) - \frac{\partial}{\partial x}(H) - (\frac{\partial H^1}{\partial u^1} u_1^1, \ldots \frac{\partial H^M}{\partial u^M} u_1^M) = L(u_1) \tag{29}$$

If H is a solution of the equation (29),then the symmetry f can be represented in the form

$$f = H + h(x,u), \quad h = (h^1,\ldots h^M), \quad h^l = h^l(x,u^l). \tag{30}$$

At the third step by substitution of (30) in (4) we derive the overdetermined system of equations for $h(x,u)$. To test the conditions (28) and solve the equation (29) it is sufficient to modify the algorithm (II) slightly. Equation (29) is the set of M scalar equations

$$\hat{\mathbb{D}}(Q) = S, \quad \hat{\mathbb{D}}_l = (\mathbb{D} - \frac{\partial}{\partial x} - u_1^l \frac{\partial}{\partial u^l}), \quad l=1,\ldots M \tag{31}$$

To modify the algorithm (II) for the equation (31) one has to replace $\mathbb{D}$ by $\hat{\mathbb{D}}_l$ in (27) and to modify conditions (31) for k = $ord(S)$ =0,1 which become the following (compare with (24))

$$\frac{\partial S}{\partial u_1^l} = 0, \quad \frac{\partial^2 S}{\partial u_1^i \partial u_1^j} = 0, \quad \frac{\partial^2 S}{\partial u_1^j \partial u^i} - \frac{\partial^2 S}{\partial u_1^j \partial u^j} = 0,$$

$$i,j = 1,\ldots M, \quad i\neq l, j\neq l$$

for k=1 and $\partial S/\partial u^i=0$, $i=1,\ldots M$, $\partial S/\partial x=0$ for k=0. Note that the general solution of the equation (31) containts an additive arbitrary function $g(x,u^l)$.

4. The above algorithms are implemented in the frame of computer algebra system FORMAC ([15]). The choice of the FORMAC is caused by its high performance, effective tools for expression analysis and the possibility to extend the system by means of the PL/1 language. Our program includes two basic procedures (CONDS and SYM) and about 20 auxiliary procedures. The input of the program includes M - the number of equations in the system (2); N - the order of the system (2); FF(I) (I=1,...M) - the scalar components of (2) right-hand side. The variables u_j^i in the right-hand side of (2) are coded as U(I,J). It is sufficient to use procedures CONDS and SUM to check up the integrability of a given system (2) and to compute symmetries and canonical conservation law densities.

Procedure CONDS implements the algorithm (I). It allows to test the conditions of integrability (that is the existence of the formal symmetry) and to compute the canonical coservation law densities (21). The call for CONDS has the form CONDS(I,J,K) where I,J,K are integers such that I≥0, J≥I, K=0,1. For an input

I=0, J=n-N-1, K=1

CONDS tests the conditions (20) of the n-th order formal symmetry existence for n>N. If these conditions are not satisfied, the following messages

ZERO = < expression >

are printed out. For existence of the formal symmetry the right-hand sides of these messages must be put equal to zero. If it leads to contradiction, then the evolution system under investigation has not any formal symmetry of the order greater or equal to n. Therefore it is not integrable.

For integrable systems (2) CONDS computes and prints out the canonical conservation laws densities (21). For an input I,J with I≤J and K=0 the canonical densities $R(i,j)$ for

i=I,I+1,...J, j=1,...M

are computed and printed out (for K=0 the corresponding conditions (20) are not tested).

Procedure SYM implements the algorithm (III). It allows to compute symmetries of the form (4) for the system (2). The procedure call is SYM(N) where N is a symmetry order. SYM computes and prints out the function H satisfying the equation (29). Moreover the system of equations for constants μ_1, γ_i^k (coded as MU(I),G(I,K)) and the system of equations for $h(x,u)$ (its components are coded as HO(I).(X,U(I,0)) are printed out as

ZERO = < expression >

which right-hand sides must be equal to zero.

Note that the current version of our program should be applied only to the systems (2) with right-hand sides being functions of u_i^j from the class (26). To extend the class of appropriate systems it is sufficient to modify the procedure INT implementing the indefinite integration.

5. Here we give some examples of the program application.

1) Let us consider the following nonlinear Shrödinger-like equation

$$u_t = u_4 + u^2 v, \quad -v_t = v_4 + v^2 u \tag{32}$$

Using CONDS(0,6,1), we obtain that system (32) has the 10-th order formal symmetries, therefore, it is not integrable.

2)The following system was obtained in [5], where the integrable systems (1) have been classified:

$$u_t = u_2 + v_1^2 + \frac{\partial z}{\partial v}, \quad -v_t = v_2 + u_1^2 + \frac{\partial z}{\partial u}, \tag{33}$$

where

$$z = \beta_1 exp(u+v) + \beta_2 exp(\bar{\lambda}u+\lambda v) + \beta_3 exp(\lambda u+\bar{\lambda}v),$$

$$\lambda = exp(2\pi i/3), \qquad \bar{\lambda} = \lambda^2,$$

β_i are arbitrary constants.

The system (33) is known to have the formal symmetry of the order n=6. By means of the program developed we found that (33) has the 4-th order symmetry $f=(f^1,f^2)$, where

$$f^1 = u_4 + 2v_1v_3 - 2v_1u_1u_2 + v_2^2 - \frac{4}{3}v_1^3u_1 + \frac{1}{3}u_1^4 +$$

$$z(2u_2+ v_1^2) + \frac{\partial z}{\partial u}(v_2+ u_1^2) + \frac{1}{2}\frac{\partial z}{\partial v}z ,$$

$$f^2 = -v_4 - 2u_1u_3 + 2u_1v_1v_2 - u_2^2 + \frac{4}{3}u_1^3v_1 - \frac{1}{3}v_1^4 -$$

$$z(2v_2+ u_1^2) - \frac{\partial z}{\partial v}(u_2+ v_1^2) - \frac{1}{2}\frac{\partial z}{\partial u}z .$$

3)In the classifying integrable third-order systems of two equations the following systems were obtained

$$u_t = u_3 + uu_1, \quad v_t = 4v_3 + uv_1 + \frac{1}{2}vu_1 \tag{34}$$

and

$$u_t = 5u_3 + \frac{5}{2}uu_1, \quad v_t = 8v_3 + \frac{3}{2}uv_1 + \frac{1}{4}vu_1 \tag{35}$$

which were known to have the 9-th order formal symmetries only. Using the program developed, we tested the existence of the 11-th order formal symmetries. It is turned out that the system (34) posseses and system (35) doesn't posses such a formal symmetry. Moreover, we computed the 5-th order symmetry (4) for the system (34) which is the form $f=(f^1,f^2)$, where

$$f^1 = u_5 + \frac{5}{3}uu_3 + \frac{10}{3}u_1u_2 + \frac{5}{6}u^2u_1 ,$$

$$f^2 = 16v_5 + \frac{20}{3}uv_3 + \frac{5}{2}vu_3 + 10u_1v_2 + \frac{25}{3}v_1u_2 + \frac{5}{6}u^2v_1 + \frac{5}{6}vuu_1 .$$

All above examples take from 5 to 20 minutes of ES-1061 running time and about 300 K memory.

REFERENCES

1. Zhiber A.V.,Shabat A.B.(1979).Klein-Gordon equation with nontrivial group.*DAN SSSR* 247,1103 (in Russian).
2. Svinolupov S.I.,Sokolov V.V.(1982).On the evolution equations with nontrivial conservation laws.*Funct.Anal.*16,86 (in Russian).
3. Svinolupov S.I.(1985).Second order evolution equations possesing metries.*Usp.Mat.Nauk* 40,263 (in Russian).
4. Mikhailov A.V.,Shabat A.B.(1985).Integrability conditions for system of two equations of form $u_t=A(u)u_{xx}+F(u,u_x)$ I.*Theor.&Math.*Phys. 62,163 (in Russian).
   Mikhailov A.V.,Shabat A.B.(1986).Integrability conditions for system of two equations of form $u_t=A(u)u_{xx}+F(u,u_x)$ II.*Theor.*& Math. *Phys.*66,47 (in Russian).
5. Shabat A.B.,Yamilov R.I.(1985).On complete list of integrable equations of form: $iu_t=u_{xx}+f(u,v,u_x,v_x)$, $-iv_t=v_{xx}+g(u,v,u_x,v_x)$.Preprint BF AN SSSR,Ufa (in Russian).
6. Ibragimov N.H.,Shabat A.B.(1984).On infinite Lie-Bäcklund algebras. *Funct.Anal.*14,79 (in Russian).
7. Fokas A.S.(1987).Symmetries and integrability.*Stud.Appl.Math.*77,253.
8. Sokolov V.V.,Shabat A.B.(1984).Classification of integrable evolution equations.*Math.Phys.Rev.*4,221,New York.
9. Mikhailov A.V.,Shabat A.B.,Yamilov R.I.(1987).Symmetry approach to classification of nonlinear equations. Complete list of integrable systems.*Usp.Mah.Nauk* 42,3 (in Russian).
10. Buchberger B.,Collins G.E.,Loos R. (eds.) (1983). Computer algebra: symbolic and algebraic computation,2nd edition,Vienna:Springer-Verlag.
11. Zharkov A.Yu.,Shvachka A.B.(1983).REDUCE-2 program for investigating integrability of nonlinear evolution equations.Preprint JINR,R11-83-914,Dubna (in Russian).
12. Gerdt V.P.,Shvachka A.B.,Zharkov A.Yu.(1985).FORMINT- a program for classification of integrable nonlinear evolution equations.*Comp. Phys.Comm.* 34,303.
    Gerdt V.P.,Shvachka A.B.,Zharkov A.Yu.(1985).Computer algebra application for classification of integrable nonlinear evolution equations.*J.Symb.Comp.*1,101.
13. Zakharov V.E.,Manakov S.V.,Novikov S.P.,Pitaevsky L.P.(1980).Theory of solitons.The inverse problem method,Moscow,Nauka (in Russian).
14. Gelfand I.M.,Manin Yu.I.,Shubin M.A.(1976). Pouasson brackets and nel of variational derivative in formal variational calculus. *Funct. Anal.*10,30 (in Russian).
15. Bahr K.A.(1973).FORMAC 73 user's manual,Darmstadt: GMD/IFV.

COMPUTER CLASSIFICATION OF INTEGRABLE SEVENTH ORDER MKdV - LIKE EQUATIONS

V.P. Gerdt
Laboratory of Computing Techniques and Automation
Joint Institute for Nuclear Research
Head Post Office, P.O. Box 79, 101000 Moscow, USSR

A.Yu. Zharkov
Saratov State University
410601 Astrakhanskaya 83, Saratov, USSR

Extended abstract

At present paper we show that the most tedious steps of classification of integrable nonlinear evolution equations

$$U_t=F(x,U,U_1,..U_N,\lambda_1,\lambda_2,..\lambda_M),\ U_i=\mathbb{D}^i(U),\ \mathbb{D}=d/dx,\ \lambda_j\in C \tag{1}$$

where F is known up to arbitrary parameters λ_j can be completely automated by means of computer algebra. The classification problem consists in obtaining a complete list of integrable equations (1) for some fixed N and to describe the most general transforms connecting these systems. As it was shown in [1,2] the necessary integrability conditions of eqs.(1) with arbitrary F may be written as an infinite series of local conservation laws

$$\frac{d\rho_k}{dt}=\frac{d\sigma_k}{dx}\leftrightarrow\frac{d\rho_k}{dt}\in \mathrm{Im}\mathbb{D},\quad k=1,2,3,...,\tag{2}$$

where the densities ρ_k can be recurrently expressed in terms of F. The algorithms for computing the densities ρ_k, testing conditions (2) and derivation the overdetermined systems of equations in F from the condi-tions (2) are already implemented in the program FORMINT [3] on the basis of the computer algebra system FORMAC. The next step is to solve the overdetermined system in F equivalent to several first conditions (2) which reduces to the system of polinomial equations in λ_j in the case of equation (1). We suggest the effective algorithm for solving the above problem which can be easily implemented. The main idea of the algorithm is to consider all the possible cases when the some of relevant conservation laws (2) are trivial ($\rho_k\in\mathrm{Im}\mathbb{D}$) and the others are non-trivial ($\rho_k\notin\mathrm{Im}\mathbb{D}$). For each such case the source polynomial system is a strictly simplified one and can be solved succesively by computer using general approach based on Groebner basis computation [4].

Our algorithm is applied for classification of integrable 7th order MKdV - like equations of the following form

$$U_t=\mathbb{D}\frac{\delta}{\delta U}\left(-\frac{1}{2}U_3^2+\lambda_1 U_1 U_2^2+\lambda_2 U^2 U_2^2+\lambda_3 U_1^4+\lambda_4 U^2 U_1^3+\lambda_5 U^4 U_1^2+\lambda_6 U^8\right),$$

where

$$\frac{\delta}{\delta U}=\sum^{\infty}(-1)^i \mathbb{D}^i \frac{\partial}{\partial U}$$

Using the FORMINT we derived the system of 9 polynomial equations of degree 3 in 6 parameters λ_j, equivalent to conditions (2) with k=1,3, 5 (for k=2,4 conditions (2) hold without restrictions on λ_j). Applying the above algorithm, we obtain that the system is nontrivially solvable in two cases

1) $\rho_1 \notin \mathrm{Im}\mathbb{D}$, $\rho_3 \notin \mathrm{Im}\mathbb{D}$, $\rho_5 \notin \mathrm{Im}\mathbb{D}$: $\lambda_1=0, \lambda_2=7, \lambda_3=21, \lambda_4=0, \lambda_5=35/2, \lambda_6=-35/2$
2) $\rho_1 \notin \mathrm{Im}\mathbb{D}$, $\rho_3 \in \mathrm{Im}\mathbb{D}$, $\rho_5 \notin \mathrm{Im}\mathbb{D}$: $\lambda_1=7, \lambda_2=-7, \lambda_3=-14, \lambda_4=-14/3, \lambda_5=14, \lambda_6=-28/3$.

This means that all integrable cases for equation (3) are exhausted by the symmetries of well-known lower order integrable equations.

REFERENCES

1. Sokolov V.V., Shabat A.B. (1984). Classification of integrable evolution equations. *Math. Phys. Rev.*, 4, 221, New York.
2. Mikhailov A.V., Shabat A.B., Yamilov R.I. (1987). Symmetry approach to classification of nonlinear equations. Complete list of integrable systems. *Usp. Mat. Nauk*, 42, 3 (in Russian).
3. Gerdt V.P., Shvachka A.B., Zharkov A.Yu. (1985). FORMINT - a program for classification of integrable nonlinear evolution equations. *Comp. Phys. Comm.*, 34, 303.
   Gerdt V.P., Shvachka A.B., Zharkov A.Yu. (1985). Computer algebra application for classification of integrable nonlinear evolution equations. *J. Symb. Comp.*, 1, 101.
4. Buchberger B. (1985). Groebner basis: a method in symbolic mathematics. In: Progress, directions and open problems in multidimensional system theory (ed. Bose N.K.), Dorbrecht, Reidel, p.184.

Symbolic Computation and the Finite Element Method

John Fitch
School of Mathematical Sciences, University of Bath
Richard Hall
South West Universities Regional Computing Centre, Bath

Introduction

A number of years ago Norman (1972) created a system for the numerical solution of ordinary differential equations in which the user specified his problem in a high level way, and the system, called TAYLOR, generated a suite of FORTRAN functions which could initialise the equations and give the value at any point. Here we give an outline of the FESTER package which is based on the same principle of high level input and the production of a runnable FORTRAN program as the output, but the aim is to use the finite element method as the foundation of the generated suite, and instead of ordinary differential equations we are interested in three dimensional elliptic partial differential equations. The user specifies the region of interest, the values of the function on the boundary, and the differential equation of to be investigated, all in a natural style, and then the software package manages all the organisation of elements, assembly of equations and generation of code (Hall 1981). In this respect FESTER is like MODULEF (Laug and Vidrascu, 1985), in attempting to provide a complete finite elements system.

The totality of FESTER contains a number of interesting sub-problems, such as the division of the region and its surface into elements, generation of suitable trial functions, and the presentation of the answer. This extended abstract introduces the sub-problems and indicates the solutions adopted, and assesses the extend to which this work provides a basic work plan for future work on partial differential equations.

Steps to a Solution

In order to construct a totally automatic solution we need to be able to specify the problem; we choose to do this as a partial differential equation and a three dimensional region with boundary conditions. The user can add to this basic information additional information, such as the degree of the trial function on each element, or a density function which indicates the parts of the region in which there is a need for more elements. Figures 1 and 2 show samples of what the user is required to provide as input. In the current version of FESTER the region is determined by planes, which is a restriction, although this restriction is not intrinsic to the internal workings.

The first calculation that is required from the input is to construct the elements which cover the region. It was decided to use only tetrahedral elements, but we cannot use a regular assembly because of the shape of the region, and the user supplied density function. The approach is first to divide the surface into triangles, using a revised version of the SHELL program of Gill (1972). This requires as input that the surface be described by a set of Coon patches (1967). Based on the triangles tetrahedra are grown into the region, attempting to keep the elements as near regular as can be achieved with the constraints of the density function and the shape. This is repeated until the whole region is covered by independent elements. This yields a controlled and guaranteed division of the region into elements.

Following the lead of TAYLOR we specify the equation in a natural notation. The system must convert this equation into a minimisation principle. While it would be possible to utilise a table of known equations, FESTER uses a REDUCE program to determine a particular function to be minimised, by doing the inverse of the usual variational principle. The methods works well for linear partial differential equations, and can be used in a class of non-linear ones.

The trial function on each element is constructed to a specified degree. Theoretical work by Hall (Fitch and Hall, 1988) has shown that for polynomial functions $C^{(1)}$ continuity is impractical, requiring an eleventh degree polynomial with 364 modal parameter. For this reason we use $C^{(0)}$ functions tailored to the actual element.

The equations are assembled, and an efficient FORTRAN program is generated using REDUCE, using specialised techniques for integrating over the elements. This output stage would certainly benefit from the more recent FORTRAN generation techniques for GENTRAN (Gates 1986). Ultimately the solution method is linear equations, the techniques for which are well known. We use at present routines from the NAG library, but there will ultimately be a need for sparse matrix methods.

Conclusions

The FESTER system uses an amalgamation of sections of BCPL, REDUCE and FORTRAN for the region definition, construction of minimisation principle and trial functions, and eventual solution. They produce a harmonious whole which can solve a class of partial differential equations with a minimum of assistance.

References

Coons, S. A. (1967) *Surfaces for computer-aided design of space forms*, MAC–TR–41, Project MAC.
Fitch, J. P. and Hall, R. G. (1988) *The Minimum Degree of a Three Dimensional $C^{(1)}$ Trail Function*, in preparation.
Gates, B. L. (1986) *GENTRAN*, Proc. SYMSAC '86, Waterloo, Ontario.
Gill, J. I. (1972) *Computer-aided design of shell structures using the finite element method*, Ph.D. Thesis, University of Cambridge.
Hall, R. G. (1981) *Symbolic Computation and the Finite Element Method*, Ph.D. Thesis, University of Cambridge.
Laug, P. and Vidrascu, M. (1985) *The MODULEF Finite Element Library*, Proc. IFIP WG2.5 Working Conference 4, Sophia-Antipolis, France.
Norman, A. C. (1972) *A System for the Solution of Initial and Two-point Boundary Value problems*, Proc. ACM 25th Anniversary Conference 2 pp826–834.

```
COORDINATES CARTESIAN X,Y,Z
DOMAIN
0 <= X <= 2,
Y >= 0,
Y-X <= 1,
Y+X <= 3,
0 <= Z <= 1
EQUATION d2U/d2X + d2U/d2Y + d2U/d2Z = 0
CONDITIONS
ON X=0, U = 1 + 2*Y + 3*Z
ON X=2, U = 3 + 2*Y + 3*Z
ON Y=0, U = 1 + X + 3*Z
ON Y=1+X, U = 3 * (1 + X + Z)
ON Y=3-X, U = 4 + Y + 3*Z
ON Z=0, U = 1 + X + 2*Y
ON Z=1, U = 4 + X + 2*Y
DENSITY 2/3
METHOD QUADRATIC
```

Figure 1: Pentagonal Prism Region for Simple Function

```
COORDINATES CARTESIAN X,Y,Z
DOMAIN 0 <= X <= 1, 1 <= Y <= 2, 0 <= Z <= 1
EQUATION
X*Y**2*Z*d4U/d2Xd2Y + 2*X*Y*Z*d3U/d2XdY + y**2*Z*d3U/dXd2Y +
2*Y*Z*d2U/dXdY + Y**3*Z*d4U/d2Yd2Z + Y**3*d3U/d2YdZ +
3*Y**2*Z*d3U/dYd2Z + 3*Y**2*d2U/dYdZ + X*Y*d2U/d2X + Y*dU/dX +
Y*Z*d2U/d2Z + Y*dU/dZ - 12*X - Y - 1 = 0
CONDITIONS
ON X=0, U = Z + (1 + Z)/Y
ON X=1, U = Z + (5 + Z)/Y
ON Y=1, U = 1 + X + 3*X**2 + 2*Z
ON Y=2, U = (1 + X + 3*X**2 + 3*Z)/2
ON Z=0, U = (1 + X + 3*X**2)/Y
ON Z=1, U = (2 + X + 3*X**2 + Y)/Y
DENSITY 1/3
METHOD LINEAR
```

Figure 2: Input for a Complex Equation

APPLICATION OF LIE GROUP AND COMPUTER ALGEBRA TO NONLINER MECHANICS.

Klimov D.M., Rudenko V.M, Zhuravlev V.F.
Institute for Problem of Mechanics, Academy of Sciences of USSR, Prospect Vernadskogo 101, 117526 Moscow, USSR.

On the basis of Lie group analysis a method has been proposed, which is applied here to nonliner mechanical systems.

Hori [1] and other authors used such a method only for autonomous mechanical systems. In this paper we shall describe an Krylov-Bogoliubov asymptotic method, based on certain properties of Lie groups and implemented on the basis of the computer algebra system REDUCE. This method based on the properties of Lie groups, is very effecient because: 1) in the algorithm asymptotic series are not used; 2) the formulae for change of variables are easily calculable; 3) the procedure of solving this problem is of a recurrent form. The method is described for mechanical systems with a single frequency. For a resonant case we shall follow papers [2,4].

Consider a mechanical system governed by equations

$$\frac{dx}{dt} = X(x,y,e), \quad \frac{dy}{dt} = Y(x,y,e), \qquad (x^1 \in R, \quad y^n \in R, \quad e<<1) \qquad (1)$$

where x is a scalar, y is a vector, e is a small parameter.

Equation (1) will be transformed into a new set, which is suitable for later analysis.

The system (1) generates a Lie group with the operator

$$A = X(x,y,e) \frac{\partial}{\partial x} + Y(x,y,e) \frac{\partial}{\partial y} \qquad (2)$$

Therefore transformation of system (1) shall be chosen in the same space as the system (1)

$$dx/d\varkappa = M(x,y,e) , \qquad dy/d\varkappa = N(x,y,e) \qquad (3)$$

Here $\varkappa$ is the group's parameter, and U is the operator

$$U = M(x,y,e)\frac{\partial}{\partial x} + N(x,y,e)\frac{\partial}{\partial y} \tag{4}$$

In this case the object (2) and object (4) are linear. Let us find the change of a variable $(x,y) \Rightarrow (p,q)$:

$$p = f_1(x,y,\varkappa,e) , \qquad q = f_2(x,y,\varkappa,e) ,$$

such that p and q are the solution of equation (3) and this relations

$$f_1(x,y,0,e) = 0, \quad f_2(x,y,0,e) = 0 .$$

This change of variable can be obtained as Lie series.

$$p = x + \varkappa U x + \frac{1}{2!} \varkappa^2 U^2 x + \dots = E^{\varkappa U} x$$

$$q = y + \varkappa U y + \frac{1}{2!} \varkappa^2 U^2 y + \dots = E^{\varkappa U} y \tag{5}$$

$$x = p - \varkappa U p + \dots = E^{-\varkappa U} p, \quad y = q - \varkappa U q + \dots = E^{-\varkappa U} q$$

On the other hand U should be expressed in terms of x, y, if an inverse transformation is needed.

If G is a group symmetries for e = 0 (G is the operator of the group) such that $[A , B]|_{e=0} = 0$, then it is necassary to find an operator U, which transforms the operator A into the operator B, where B generates the equation set of smaller order.

The operator [ ] is linear and is calculated according to the rule :

$$[A , G] = A G - G A$$

System (1) written by means of canonical coordinates of the group G as :

$$dx / dt = 1 + e X(x,y,e) , \quad dy / dt = e Y(x,y,e) \tag{6}$$

therefore

$$A = \frac{\partial}{\partial x} + e \left(X(x,y,e) \frac{\partial}{\partial x} + Y(x,y,e) \frac{\partial}{\partial y} \right)$$

If A is the operator of the original mechanical system, U is the operator of the change of a variables and B is the operator of transformed system, then these operators are related by Cauchy problem for an equation :

$$dB / d\varkappa = [B , U]; \qquad B \Big|_{\varkappa = 0} = A.$$

The solution of this problem may be represented by the following series :

$$B = A + \varkappa [A , U] + \frac{1}{2!} \varkappa^2 [[A, U], U] + \ldots \tag{7}$$

If B is not dependent on p and $\varkappa = e$ then we shall find an

equation for the operator U. This problem is solved asymptotically in the form

$$A_k = A + O(e^k) , \; B_k = B + O(e^k) , \; U_k = U + O(e^k)$$

where A_k , B_k , U_k are operators with deviations from the exact ones being of the order less than e^k.

From the formula (7) it follows that

$$B_o = A_o , \; B_1 = A_1 + e [A_o , U_o] , \ldots \tag{8}$$

$$\ldots , B_n = A_n + \sum_{k=1}^{n} \frac{e^k}{k\,!} \left[\cdot\cdot \left[\left[A_{n-k}, U_{n-k} \right] \cdot\cdot \right] , U_{n-k} \right]$$

The expression for B with k = 1 yields

$$\left[A_{n-1}, U_{n-1} \right] = \left[A_{n-1} - A_o, U_{n-1} \right] + \left[A_o, U_{n-1} \right] \tag{9}$$

Since $A_{n-1} - A_o \approx e$ then $\left[A_{n-1}, U_{n-1} \right] =$

$$= \left[A_{n-1} - A_o, U_{n-1} \right] = \left[A_{n-1} - A_o, U_{n-2} \right] + \left[A_o, U_{n-1} \right] \Rightarrow$$

$$\Rightarrow B_n = e \left[A_o, U_{n-1} \right] + L_n$$

where L_n is an operator, which depends only on U_1, U_2, U_3, ... , U_{n-1}, such that

$$L_n = A_n + e\left[A_{n-1} - A_o, U_{n-2}\right] + \sum_{k=2}^{n} \frac{e^k}{k!}\left[\cdots\left[\left[A_{n-k}, U_{n-k}\right], \cdots\right], U_{n-k}\right] \qquad (10)$$

The relation (8) may be rewritten as

$$B_o = A_o\,,\; B_1 = A_1 + e\,\partial U_o/\partial p\,,\; B_2 = A_2 + e\,\partial U_1/\partial p + e\left[A_1 - A_o, U_o\right] + 1/2\, e^2\left[\left[A_o, U_o\right], U_o\right], \dots$$

$$B_n = e\;\partial U_{n-1}/\partial p + L_n \qquad (11)$$

Indeed let B_1 be chosen in the form :

$$B_1 = \lim_{h \to 0} \frac{1}{h}\int_0^h A_1\, dp = \langle A_1 \rangle$$

then U_o is defind as an integral :

$$U_o = -\frac{1}{e}\int (A_1 - \langle A_1 \rangle)\, dp \equiv -\frac{1}{e}\int \bar{A}_1\, dp$$

where $\langle A_1 \rangle$ denotes averaged value of A_1 and $\bar{A}_1$ denotes a complement to A_1.

In the second approximation

$$U_1 = -\frac{1}{e}\int \left(\bar{A}_2 + e\,\overline{\left[A_1 - A_o, U_o\right]} + 1/2\, e^2\,\overline{\left[\left[A_o, U_o\right], U_o\right]}\right) dp$$

and so on.

The general solution of n-th approximation is

$$B_n = \langle L_n \rangle\,, \qquad U_{n-1} = -\frac{1}{e}\int \overline{L_n}\, dp \qquad (12)$$

Since B_n is independent on p and $B_o = A_o = \partial/\partial p$

$$B_n = \frac{\partial}{\partial p} + e\left[P_n(q,e)\frac{\partial}{\partial p} + Q_n(q,e)\frac{\partial}{\partial q}\right]$$

and the relation (6) becomes

$$dp / dt = 1 + e\, P_n(q,e), \qquad dq / dt = e\, Q_n(q,e) \tag{13}$$

The solutions of equations (13) provide a following transformation of variables from (p , q) to (x , y)

$$x = E^{-e\, U_{n-2}}\, p, \qquad y = E^{-e\, U_{n-2}}\, q \tag{14}$$

So, the procedure is introduced as a recursive method with the formulae (10), (12). The expression for B_n defines the right hand side of the equations (13), whereas U_{n-2} defines the change of a variables from (x, y) to (p, q), with system (6) independent on p.

For example, consider the problem of the resonance in a non -linear gyroscopic system balanced in a gimbal with a periodical perturbation. The equations of this system are

$$\frac{d}{dt^{\wedge}} \left(A_o \frac{d\alpha}{dt^{\wedge}} + H \sin \beta \right) = e^2 J \cos(W_1 t^{\wedge} + \delta)$$

$$B_o \frac{d^2 \beta}{dt^{\wedge}} 2 - H \frac{d\alpha}{dt^{\wedge}} \cos \beta = e^2 \left(G \sin W_1 t^{\wedge} - h \frac{d\beta}{dt^{\wedge}} \right) \tag{15}$$

$$H = C \left(\frac{dq}{dt^{\wedge}} + \frac{d\alpha}{dt^{\wedge}} \sin \beta \right) = \text{Const.}$$

where e is a small parameter ; δ - is phase; A_o is the sum of moments of inertia of the rotor, flywheel casing and the external ring ; B_o is a moment of inertia of the flyweel casing and rotor about rotor axis of symmetry; H is a kinetic moment of inertia of the gyroscope; α, β, γ are Krylov angles.

Using the integral of the first of the eguations (15) yields

$$\beta'' - (1 - \sin \beta) \cos \beta = e^2 F(\beta , \beta', W t)$$
$$F(\beta, \beta', W t) = g \sin W t - k \beta' + j \cos \beta \sin(W t + \delta) \tag{16}$$

where

$$t = \frac{H\,\hat{t}}{(A_o B_o)^{1/2}} = n\,\hat{t}; \quad g = \frac{A_o G}{H}, \quad k = \frac{h}{H}(A_o/B_o)^{1/2}$$

$$j = J/(W_1 H); \quad W = (A_o B_o)^{1/2}/H$$

$$l = \frac{A_o}{H}\frac{d\alpha}{dt} + \sin\beta \qquad \text{for} \quad \sin(W_1 \hat{t}_o + \delta) = 0 \quad \text{and } \hat{t} = 0$$

The relation (16) are nondimensional and F is periodic function of W t with period 2π.

For e = 0 the integral of energy exists

$$(\beta')^2 + (l - \sin\beta)^2 = a^2$$

Using this relation the solution may be obtained in terms of elliptical functions.

$$\beta = \arcsin\left[\frac{1 + l + a}{1 - \dfrac{2a}{1 + l + a}\,Sn^2(K s/\pi)} - 1\right] \tag{17}$$

where Sn is a elliptic sine; K is an elliptic integral of the first kind.

The solution is the periodic function of W t with period 2π. The parameter s is related to t as

$$s = \omega(a)(t - t_o + I_o/m), \quad \omega(a) = \pi\, m/K(k) = \frac{\pi\,((1+a)^2 - l^2)^{1/2}}{2\,K(k)}$$

$$I_o = \int_0^{v_o} \left[(1-v^2)(1-k^2v^2)\right]^{-1/2} dv, \qquad k^2 = \frac{4a}{(1+a)^2 - l^2}$$

Consider now equations (16) with $e \neq 0$. The system of differential equations (16) yields :

$$a' = \frac{e^2 Q_s}{D} F(Q, \omega(a) Q_s', W\,t), \quad s' = \omega(a) - \frac{e^2 Q_s}{D} F(Q, \omega(a) Q_s', W\,t)$$

$$D = \omega(a)\,(Q_s'\,Q_{sa}'' - Q_a'\,Q_{ss}'') + \omega_a'(a)\,Q_s'^2 = a/\omega(a)$$

and the variables a, s are defined as

$$b = Q(a,s) \quad , \qquad b' = Q'_s(a,s)\ \omega(a) \tag{18}$$

In resonant case the oscilation frequency $\omega(a)$ is equal to (W/p). Introducing a new variable R

$$R = p\,s - W\,t$$

the equations (16) are transformed to

$$\begin{aligned} R' &= p\,\omega(a) - W + e^2 p\, X\,(\,s,\ a,\ p\,s - R\,) \\ a' &= \qquad\qquad\quad e^2\ \ Y\,(\,s,\ a,\ p\,s - R\,) \\ s' &= W\,/\,p \end{aligned} \tag{19}$$

If e=0, then the first of (19) yields an expression for the amplitude of oscilations. This equation defines a backbone curve

$$p\,\omega(a\hat{}) - W = 0$$

To study motion in the vicinty of the stationary one a new variable z is introduced as

$$a = a\hat{}(W) + e\,z$$

Then the equations (19) yield

$$\begin{aligned} R' &= p\,\omega(a) - W - e^2 p\, Q'_a \omega(a)\ F\,/a \\ z' &= \qquad e\ Q_s \omega(a)\ F\,/a \\ s' &= W\,/\,p \end{aligned} \tag{20}$$

The transformation of variables $s,R,z \Rightarrow u,r,w$ is now sought for which corresponds to the above procedure. The operator is

$$A = \frac{W}{p}\frac{\partial}{\partial u} + p\, e\, (\omega' w + \omega'' \frac{e\, w^2}{2!} - e\, Q'_a \frac{\omega F}{a}) \frac{\partial}{\partial r} + e\, Q_s \frac{\omega F}{a}\frac{\partial}{\partial w}$$

therefore

$$B_o = A_o = \frac{W}{p}\frac{\partial}{\partial u}; \quad B_1 = \frac{W}{p}\frac{\partial}{\partial u} + (p\,\omega - W)\frac{\partial}{\partial r} + \langle e\, Q'_s \frac{F\,\omega}{a} \rangle \frac{\partial}{\partial w}$$

$$U_o = -\frac{p}{W}\int \overline{Q'_s \frac{\omega F}{a}}\, du \frac{\partial}{\partial w} = -\frac{\omega\, p}{W\, a}\int \overline{Q'_s F}\, du \frac{\partial}{\partial w} = -\frac{1}{a}\int \overline{Q'_s F}\, du \frac{\partial}{\partial w}$$

The equations which correspond to this approximation are

$$u' = W/p$$

$$r' = p\,\omega - W$$

$$w' = e\,\omega \langle Q'_a F \rangle / a$$

In a second approximation we have

$$B_1 = \frac{W}{p}\frac{\partial}{\partial u} + (p\,\omega - W - e^2 \frac{\omega}{a}\langle Q'_a F \rangle)\frac{\partial}{\partial r} + \langle e\, Q'_s \frac{F\,\omega}{a} \rangle \frac{\partial}{\partial w}$$

The equations of second approximation are correspondingly

$$u' = W/p$$

$$r' = p\,\omega - W - e^2\omega \langle Q'_a F \rangle / a \tag{21}$$

$$w' = e\,\omega \langle Q'_s F \rangle / a$$

The change of variables which transforms (20) into (21) is

$$s = u, \qquad R = r, \qquad z = w - e/a \int \overline{F\, Q'_s}\, du$$

After applying the procedure of averaging we obtain [3] from (21)

$$r' = p\,\omega - W - e^2\omega/a \left[A_{3,p}\sin r + A_{4,p}\sin(r-\delta) \right]$$

$$w' = e\,\omega \left[A_{1,p}\cos r - \xi J(a) - A_{2,p}\cos(r-\delta) \right]$$

$$u' = W/p$$

where $$A_{1,p} = g \sin \frac{p\,S}{K}\ \mathrm{sch}\ \frac{p\,K'}{K}, \quad S = F(j,k)$$

$$A_{2,p} = \frac{p\,f\left[(1+a)^2 - 1^2 \right]^{1/2}}{2\ K} \sin \frac{p\,S}{K}\ \mathrm{csch}\ \frac{p\,K'}{K}$$

$$j = \arcsin\left[(1+a - 1)/2 \right]^{1/2}, \quad K' = K(k')$$

$$J(a) = \frac{1}{\pi} \int_{1-a}^{1+a} \left[(u-u_1)(u_2-u) \,/\, (1-u)^2 \right]^{1/2} du$$

Another example of the application of Lie groups is formulation of method of numerical integration. Solving a Cauchy problem for a certain mechanical system with the use of computer algebra system REDUCE we obtain the solution as a power series of time. The length of the series and desired accuracy of the results depend on the number of integration steps and the number of terms of power series. This method provides economy of computer time and guarantees the accuracy of the solution.

For specific examples method of the group may sometimes provide an exact solution. For example consider the system of equations, for which the dependence on p should be excluded :

$$dx/dt = 1\ ; \qquad dy/dt = -\,e\,y^3\cos^2 x$$

(this system may be integrated in a closed from by separation of variables). The operator of this system is

$$A = \frac{\partial}{\partial p} - e\,q^3\cos^2 p\,\frac{\partial}{\partial q}$$

The application of the above general procedure yields

$$B_1 = \frac{\partial}{\partial p} - \frac{1}{2} e q^3 \frac{\partial}{\partial q}, \quad U_0 = (\frac{q^3}{2} \quad \cos 2p \; \partial p \;) = \frac{1}{4} q^3 \sin 2p \frac{\partial}{\partial q}$$

$$B_2 = B_1, \qquad U_1 = U_0, \qquad \ldots\ldots$$

Thus the first approximation is seen to coincide with the second one. It is seen from (10), that such a coincidence is found for every approximation. The transformed equations are

$$dp \,/\, dt = 1 \quad , \qquad dq \,/\, dt = + e q^3$$

with the corresponding transformation being

$$x = p \,, \quad y = q \,(\, 1 + 1/2 \; e \; q^2 \sin 2p \,)^{1/2}$$

These examples clearly illustrate the efficiency of Lie qroup approach in mechanics, particularly in cojunction with the use of computer algebra. The convenience of Lie groups for symbolic calculations is due to conciseness of algorithms and due to possibilities for recurrent applications of formulae.

References

1. Hori G.I. Theory of general perturbation with unspecified cononical variables. - Publ. Astron. Soc. Japan, 1966, v. 18, n. 4, p. 287-296.
2. Zhuravlev V.F. About the application of single Liegroups to a problem asymptotical integreable the equations of mechanics. Jour. App. Math. Meth., USSR, v. 50, n. 3, 1986, p. 344-352, (in Russian).
3. Filippov V.A., Klimov D.M. The resonance in the nonlinear gyroscopic system. Jour. Soviet Academy Science News Mech. Rigid Body, n. 6, 1970, p. 42-54.(in Russian).
4. Zhuravlev V.F. Method of Lie groups in the problem of a motion selection for nonlinear mechanics. Jour. App. Math. Meth., USSR, v. 47, n. 4, 1983, p. 559-565.(in Russian).

HIERARCHICAL SYMBOLIC COMPUTATIONS IN THE ANALYSIS OF LARGE-SCALE DYNAMICAL SYSTEMS

Jerzy Paczyński
Institute of Automatic Control
Warsaw University of Technology
ul.Nowowiejska 15/19, 00-665 Warszawa, Poland

ostract

The paper presents a case study of an application of symbolic omputation in a new area - in the qualitative analysis of large scale ynamical systems. The main distinctive feature is that the underlying lgebra is not usual but that of block-diagrams. Identifiers (denoting ynamical systems) have two properties: linear/nonlinear and ne/multi-dimensional. Properties of arguments affects the properties of ne multiplication. It is commutative only for the zero, identity and for inear one-dimensional operators, for a nonlinear multiplier it is only ight- distributive over addition. Simplification must be thought of in erms of the clarity of block diagrams, what contradicts sometimes the sual algebraic notion of simplicity. The paper presents briefly the ackground and the way to a specialized symbolic algebra package.

The concept of hierarchical calculations is introduced. The idea is nat the same object may be represented by several values at different evels of hierarchy of specifications. Eg. some calculations can be done on atrices denoted by their identifiers, some on the same matrices with xplicit block elements denoted by another identifiers and some on matrices ith explicit elements. Calculations may "wander" from one level of ecification to another in a complicated pattern. But matrices are stored n a multivalued form and any calculation implies automatic recalculation : other involved levels - may be in completely different computational omains. This is easier to use than direct substitutions done explicitely the user. In a specialized system, as presented here, this notion is elatively easy to implement. It is believed, however, that it would be a welcomed programming convenience in general purpose systems. Some implementation requirements are discussed for that case.

1. Introduction

The paper presents a case study of an application of symbolic computation in a new area - in the qualitative analysis of large-scale dynamical systems. The necessery background is introduced in Chapter 2. Peculiarities of the problem are presented in Chapter 3. Identifiers denote here *dynamical operators*, their algebra has some special properties. Even the notion of simplicity is slightly different from that used in "conventional" algebra systems. But the most general contribution of the paper is the introduction of the concept of *hierarchical calculations*.

Due to obvious limitations of human brain, many ideas in science and engineering are introduced in a hierarchical way. The most obvious example are matrices. In general theory of matrices they are represented by identifiers. But they may have special structure or values - so the notion of block matrices is introduced, perhaps recursively. At last there is the level of matrices with explicit elements. If the order of introduction is strictly "top-down", then usual substitutions would be sufficient in corresponding symbolic computations. But matrices are used in applications in many "mixed" ways. To computerize such calculations, a matrix should be stored in multivalued form - with values at all needed levels. Any calculation done at some "level" should imply automatic recalculation of other involved levels. The author believes, that this notion is applicable not only in dedicated systems, but should also be implemented in general purpose systems as a programming convenience. Some considerations about such implementation are presented in Chapter 4.

Drouffe(1985) distinguishes between the two levels of usage of computer algebra:

- "on a small scale", when "it can be compared to a pocket calculator in numerical calculations",
- "as an intelligent tool", when "the program has to perform an analysis of the situation and to take some decisions on the strategy".

It seems that this classification is not directly connected with the "complication level" of computations, "calculator" computations can be very complex while "intelligent" - may not.It is believed that the presented aper belongs to the second category.

. An inspiring example

One of the possible ways of describing a *dynamical system* is the *nput-output* formalism. A dynamical system is represented as an operator, hich transforms a function of time - u, called *input*, into another unction of time - y, called *output*. Operators are denoted by capital etters, so say

$$y=G*u \tag{1}$$

) formulate the problem one must define spaces of inputs and outputs. In ne following only intuitive ideas are presented, without mathematical etails. Detailed presentation and vast references can be found eg. in esoer and Vidyasagar(1975) or Mees(1981). Inputs and outputs can be scalar one-dimensional) or many-dimensional functions, time can be continuous or .screte. More complex dynamical systems can be represented as a nterconnection of simpler subsystems. ***Block diagrams*** are often used as .ternative representation of algebraic expressions. Elementary uivalences are presented in Table 1.

Operators may have different properties. Multiplication is commutative ly for the operators I and O and linear scalar operators. The most portant property is that multiplication (for nonlinear C and D) is only ght-distributive over addition:

$$\begin{aligned} (A+B)*C &= A*C+B*C, \\ D*(A+B) &\neq D*A+D*B. \end{aligned} \tag{2}$$

One of the most important properties of a dynamical system is at of ***stability***. There are some slightly different definitions stability in input-output problems, here it will be assumed

$$\exists\ c>0\quad \forall\ u\in U \qquad \|y\| < c\ \|u\|$$

When the system operator transforms zero into zero it is just the undedness). The other definitions are similar.

When considering various examples of both engeneering objects and ysical, biological and other phenomena in terms of dynamical systems thodology, it turns out that very often they contain *feedback* and it is ually the underlying principle of their work. Feedback, broadly speaking,

Table 1

Expression	Block diagram	Remarks
I		Identity operator
0		Zero operator (no signal flow)
A		Operator A
A*B		Serial connection (note the order of blocks!)
A±B	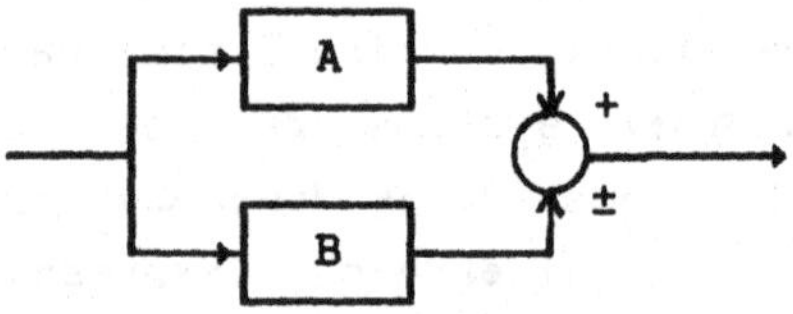	Parallel connection (note the summing junction)
$A*(I-B*A)^{-1}$	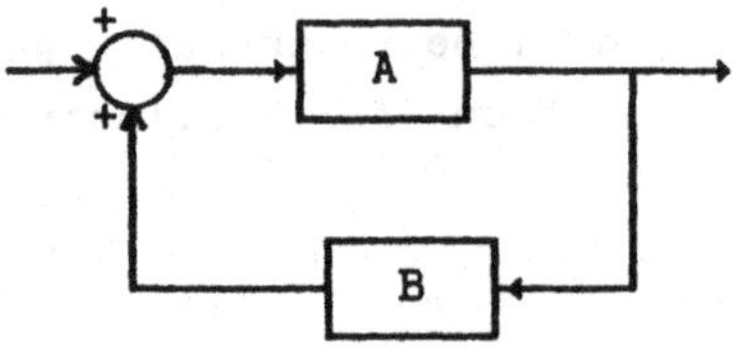	Feedback connection (note the signs!)

affects a system in one of two opposing ways : depending on circumstances it is either *stabilizing* or *destabilizing* – either degenerative or regenerative.

In order to analyze properties of the system, more information about it must be possessed. So the next level of specification of the system operator G (or the operator equation y=G*u) is the following presentation, explicitely showing the feedback loop :

$$u=(I+H)*e, \qquad y=H*e \tag{3}$$

$$u=\begin{bmatrix}u_1\\u_2\end{bmatrix}, \qquad e=\begin{bmatrix}e_1\\e_2\end{bmatrix}, \qquad y=\begin{bmatrix}y_1\\y_2\end{bmatrix}, \qquad H=\begin{bmatrix}0 & H_2\\H_1 & 0\end{bmatrix} \tag{4}$$

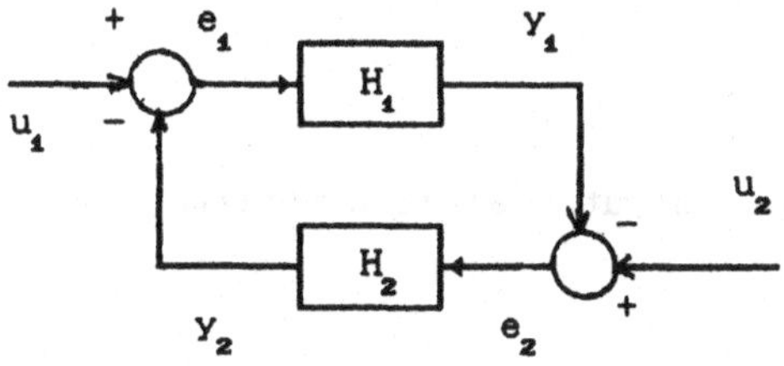

The circle denotes a *summing junction* - a linear operator. In further onsiderations it is sufficient to consider only the dependence of the ror *e* on the input *u*. The equation (3) is of implicit type, and we are in act interested in the invertibility of the operator (I+H). Simple anipulations lead to the sufficient condition

$$\|H\| < 1,$$

own (in many variants) as the small-gain theorem. However, this condition very conservative, very unlikely to be satisfied by a real feedback stem, given that high loop gain is usually required for good control.

Let for a moment u_1, u_2, e_1, e_2 be scalar functions. There is a chnique known as *loop-transformations*. The aim is to produce an quivalent" system whose operators have smaller norms. Some recipes are own how to transform the system depending on the properties of the iginal operators H_1 and H_2. Application of the small gain theoremto the ansformed systems yields different theorems on stability of feedback stems.

The theory was generalized for the multidimensional case. In the most neral formulation only the definitions of spaces are different. For actical applications, however, more important is the case, when a ltidimensional system is treated as an interconnection of smaller bsystems. H_1 and H_2 in (4) can then be expressed as matrix operators. The vestigator has usually some freedom in establishing partition between em. Usually "more important" operators are grouped together as a diagonal matrix operator H_1 and the operator H_2 expresses *interconnections*

The practical application (in a traditional way) of this theory is handicapped by very large amount of analytical calculations (only part of them is shown in the paper). In any real case the analysis should be repeated many times but it is unprobable that anybody would do so without ample time and the strongest motivation. Such was the genesis of author's interest in symbolic computations.

3. A way to the specialized computer algebra system

The first step towards the computerization of stability analysis was putting the system transformations into an algebraic theory. The first ideas were published in Paczyński(1978). It turned out, that there is whole family of eligible transformations, characterized (in two levels) by its type and specific values of matrix elements. However, the hand calculations were very laborious. It is perhaps the reason that traditional transformations form only a small percentage of eligible ones. A generalization of system transformations for large scale systems was presented in Paczyński(1980). Applications problems are here even sharper. (After completing the described software many examples from literature were verified. Typically it turned out that the analyzed objects were carefully "prepared" - usually as to the number and positions of nonzero matrix elements. Any perturbation of them caused such growth of the number of nonzero elements in transformed systems, that they would rather illustrate the nonapplicability of the classical approach.)

In the result the author turned to computer algebra and, as no system was available, started to develop an own system. The problem turned out to be very specific. To simplify the implementation, properties of the multiplying operator were assumed argument-independent, taken for the "worst" case. Secondly, it turned out that simplification must be understood in terms of the clarity of block diagrams, what is sometimes in conflict with the usual algebraic simplification. This point will be illustrated by an example.

One of the possible transformations, defined at the level of specification as in (3), consists of the following manipulations with some operator H_o of "desired" properties:

i) $u = (I+H-H_o+H_o)*e = (I+H_o)*e+(H-H_o)*e$;

ii) $(I+H_o)^{-1}*u = [I+(I+H_o)^{-1}*(H-H_o)]*e$;

iii) $u' = (I+H_o)^{-1}*u$, $H' = (I+H_o)*(H-H_o)$

ne of the possible specifications as in the equation (4) with

$$H_o = \begin{bmatrix} 0 & 0 \\ -H_{z1} & 0 \end{bmatrix}$$

esults in

$$H' = \begin{bmatrix} 0 & H_2 \\ -H_1+H_{z1} & H_{z1}*H_2 \end{bmatrix}$$

owever, from the block diagram point of view the following equivallent epresentation should be preferred

$$H' = \begin{bmatrix} 0 & H_2*(I+H_{z1}*H_2)^{-1} \\ -H_1+H_{z1} & 0 \end{bmatrix}$$

Then the need of recognizers of some syntax patterns was noticed. A rect way of finding valid transformations consists of calculation of all ossible results and their final inspection. The number of results grows, owever, in the combinatorial way. The system theory asserts that pearance of certain subexpressions results in too loose norm bounds. On e recognition of such subexpression, the corresponding calculations can abandoned. Finally, the notion of hierarchical computing grew up. In the stematic investigations of stability of dynamic systems, the calculation ok a simple form of some iteration loops, the inner was the loop - the wer was the level of object specification. It is unreasonable to repeat e whole calculations for the innermost case. Instead the symbolic values objects are substituted into the results given by the next higher level computations. During an interactive session on analysis of a dynamical stem the situation is different - the hierarchical computing plays more gnificant role, as the "levels" are changed according to current user's shes, in the unpredictable manner.

The system itself was written in PASCAL. In the present version it nsists of several packages of procedures for computations for pressions, matrices (in dense and scarce forms) and block matrices. The er can create a specialized vvariant, like that for investigations of namical systems, by defining his own variables, specialized procedures d a part of the main program. The system works rather fast but is not very user friendly. It has been transferred from the original main - fram computer to the IBM-AT and is available for demonstration. A new interactive version of the system (with some other enhancements) is unde preparation.

4. Hierarchical computations in interactive environment

It is believed, that hierarchical computation would be a welcome programming convenience in general purpose systems too. Some implementatio requitements are summarized below.

i) The same mathematical objects may be represented simultaneuosly b several symbolic objects at different levels of hierarchy.

ii) The user may wish to introduce new levels during computations. Stif "top-down" or "bottom-up" orders are too conservative, the use should be able to "zigzag" freely between different levels.

iii) Symbolic objects from different levels may have different properties it may happen that different algebra systems must be used.

iv) Introduction of a new level consists of some description of object and their algebraic properties. This process should be especiall user-friendly. Unrealistic user demands must be foreseen.

v) Recognizers of user-defined lexical patterns should be available Their use can provide more intelligent control structures.

vi) Different objects may be specified to different depth. Th computation strategy must then be rather flexible. Computation should be performed automatically at all levels, when th corresponding expressions are completely specified. Expression specified only partially must be left without evaluation, but som trace of them should be stored. Eventual future specification o missing variables should cause an automatic evaluation o "uncompleted" expressions in order to keep the whole "value system consistent.

In a dedicated software the number of levels and their properties may b fixed in advance. This makes the implementation much simpler, yet wholl satisfactory for the user.

. Conclusions

The presented system works in fact with the algebra of block diagrams nd not with the "usual" one. It was written from the beginning primarly rom the lack of access to any ready system. However, it seems doubtful hether the adaptation of a general purpose system would be possible with easonable efforts.

On the other hand, surely there are many areas with untypical lgebraic properties. The application of symbolic computations will spread o them only if there will be general purpose systems, which can be tailored" to user s needs by an experienced user itself - not by the ystem designer only.

. References

soer C.A., Vidyasagar M. Feedback Systems: Input - Output Properties. Academic Press. 1975.

rouffe J. -M. Computer Algebra as a Research Tool in Physics. In: EUROCAL'85. Vol.1. Springer, LNCS 203. 1985. 58-67.

es A.I. Dynamics of Feedback Systems. Wiley. 1981.

aczyński J. Norm Positive-Definiteness in the Stability Theory of Feedback Systems. Systems Science. Vol.4. No.3. 241-249. 1978.

aczyński J. Input-Output Stability of Multidimensional Systems. Proc.VIII Conf.Aut.Contr.,Szczecin,Vol.I. 236-244.1980.

SCHOONSCHIP FOR COMPUTING OF GRAVITINO INTERACTION CROSS SECTIONS IN N=2 SUPERGRAVITY

N.I. Gurin
Institute of Physics, BSSR Academy of Sciences,
220602, Minsk, USSR

A calculation of exact analytical expressions for cross sections of spin-3/2 particles interaction [1,2] is a difficult task which can be realized in practice only by using the modern systems of computer algebra. But it is easy to overload any system without sufficient optimization of computational algorithm. We have used the system "SCHOONSCHIP" [3] as a most suitable for this kind of computation.

In N=2 supergravity the matrix element of gravitino-photon interaction [4] in the first order of the pertubations theory is

$$\mathcal{M} = \bar{u}_2 Q u_1 = \bar{u}_2 \left(\Gamma_2 \mathcal{D}_1 \Gamma_1 + \Gamma_1 \mathcal{D}_2 \Gamma_2 \right) u_1 , \qquad (1)$$

where $u_{1,2}$ are wave functions of gravitino, $\Gamma_{1,2} = \Gamma(a_{1,2})$ are vertices of gravitino electromagnetic interaction ($\Gamma_{\mu\nu}(a) = i g \varepsilon_{\mu\nu\lambda\sigma} \gamma_5 \gamma_\lambda a_\sigma$, $a_{1,2}$ are 4-vectors of photon polarization, g is a gauge charge), $\mathcal{D}_{1,2} = \mathcal{D}(f_{1,2})$ are the gravitino propagators ($\mathcal{D}_{\mu\nu}(f) = (6 \varkappa (1-\hat{f}) \gamma_\nu \gamma_\mu - (2 f_\mu \hat{f} + (1 - 6\varkappa + \hat{f}) \gamma_\mu)(2i f_\nu + \gamma_\nu))/(3\varkappa^2)$, $\varkappa = f^2 + m^2$, $\hat{f} = f_\mu \gamma_\mu$, γ_μ are Dirac matrices, $m f_1 = p_1 + k_1$, $m f_2 = p_1 - k_2$, m is mass of gravitino, $p_{1,2}$ and $k_{1,2}$ are 4-pulses of gravitino and photon).

By computing the cross sections of particles interaction one of the main problems is to obtain an exact expression for $|\mathcal{M}|^2$. For optimization of the program we present $|\mathcal{M}|^2$ as

$$|\mathcal{M}|^2 = g^2 \sum_{i=1}^{2} \sum_{j=1}^{2} \mathcal{M}_i \mathcal{M}_j , \quad \mathcal{M}_i = \bar{u}_{2\mu} \varepsilon_{\mu\nu\lambda a_2} \gamma_5 \gamma_\lambda \mathcal{D}_{i\nu\rho} \varepsilon_{\rho\sigma\alpha a_1} \gamma_5 \gamma_\alpha u_{1\sigma} . \quad (2)$$

For summation over photon polarizations we used covariant exchange $a_\mu a_\nu \to \delta_{\mu\nu}$ [5,6] which is realized by declaring vectors $a_{1,2}$ as indices with a subsequent summation over them in the program text.

It is possible to compute the expression $\mathcal{M}_1 \mathcal{M}_1^*$ and $\mathcal{M}_2 \mathcal{M}_2^*$ very economicaly by preliminary reduction of the product of antisymmetric tensors $\varepsilon_{\mu\nu\lambda\sigma} \ldots$ into the sum of products of $\delta_{\mu\nu} \delta_{\lambda\sigma \ldots}$-functions and then to summarize succesively all quantities in $|\mathcal{M}|^2$ (2) over repeated indices. As a result the computation of sum $\mathcal{M}_1 \mathcal{M}_1^* + \mathcal{M}_2 \mathcal{M}_2^*$ demands about 5 minutes of CPU-times and 250 kbytes operational memory region on the EC-1060 computer.

But for computing the expressions $\mathcal{M}_1 \mathcal{M}_2^*$ and $\mathcal{M}_2 \mathcal{M}_1^*$ this procedure is insufficient. First of all it is necessary to express the operator Q in (1) in the most universal form [2,5] :

$$Q = \xi + \gamma_5 \eta + \hat{k} + \gamma_5 \hat{d} + \mathcal{I}\ , \quad \mathcal{I} = \frac{1}{2} \alpha_{\mu\nu} \gamma_\mu \gamma_\nu\ , \quad \alpha_{\mu\nu} = -\alpha_{\nu\mu} \tag{3}$$

and then to use the conditions which are put on the wave functions of free gravitino. It takes about 2 hours to compute each of the expressions $\mathcal{M}_1 \mathcal{M}_2^*$ and $\mathcal{M}_2 \mathcal{M}_1^*$ [7] .

The program gives the differential cross section of process (1) as a double sum on exponents of invariant variables $s = f_1^2$ and $t = f_2^2$ [2]

$$\frac{d\sigma}{dt} = \frac{d\sigma}{dt}(s,t) = \sigma_0 \sum_{i=0}^{8} \sum_{j=0}^{8} A_{ij}\, s^i t^j \ , \tag{4}$$

where A_{ij} is a number matrix of polynomial coefficients, σ_0 is a coefficient of proportionality. By a simple exchange of variables in the program it is possible to receive the cross sections of a number of processes of interaction gravitino with photons: compton-effect, two photon creation of a gravitino pair, annihilation of the pair in two photons. The computation of total cross sections for these processes is realized by preliminary organized blocks of integration on scattering angles of interacting particles.

References

1. Gurin N.I. Preprint/Institute of Physics AN BSSR,Minsk,N356,1984.
2. Gurin N.I. Covariant methods in theoretical physics.-Minsk,1986, p.102-108 (in Russian).
3. Strubbe H.S. Comp.Phys.Comm.,1974,v.8.p.1-30.
4. Freedman D., Das A. Nucl.Phys.,1977,B120,p.221-230.
5. Fedorov F.I. Lorentz Group.-Moscow,1979,Nauka (in Russian).
6. Gurin N.I., Lashkevich V.I., Fedorov F.I. Docl.AN BSSR,1983,v.27, N8,p.690-693 (in Russian).
7. Skomorokhov A.G., Gurin N.I., Zirkov L.F. Preprint/Institute of Physics AN BSSR,N380,1985.

Creation of efficient symbolic-numeric interface.

N.N.Vasiliev

Institute for Theoretical Astronomy, 10, Kutuzov Quay, Leningrad, 191187, USSR

With increasing development of symbolic manipulation the problem of constructing an efficient numerical-analytycal interface is of particular importance. Such problem has two aspects: 1) communication of Computer Algebra System (CAS) with the front-end software environment involving particularly the use of the internal package of numerical algorithms in combination with symbolic manipulation, and 2) the use of the results obtained by CAS for the numerical applications. It is assumed herewith that these results are not subjected to symbolic transformations anymore and serve only for evaluation of corresponding analytical expressions in substituting numerical values of occuring variables.

The first question is solved quite satisfactorily in the majority of CAS which contain necessary facilities for their extension by user.

The second question is more complicated. Its solution involves high efficient algorithms of substituting numerical values into symbolic expressions as well as special programming facilities for evaluating expressions outside CAS.

All more or less advanced CAS include as a rule the internal facilities to substitute numerical values into symbolic expressions. But such facilities usually are not too effective since the CAS object internal representation is oriented just to symbolic (not-numerical!) manipulation. For this reason the internal facilities schould be applied only for occasional, not too often evaluation of expressions.

For realization of efficient symbolic-numeric-interface the specialized system SPRINT has been developed. SPRINT is the system for synthesis of programs for evaluation of polynomial-trigonometric expressions. It isn't connected with any particular CAS. But it can accept expressions, obtained by almost any CAS. In calling SPRINT it remains only to give information about the employed representation of polynomial-trigonometric expressions. The first implementation of SPRINT was described in [1]. There also were described the algorithms for values substitution in sparse polynomial-trigonometric expressions. These algorithms are very fast. Their complexity is near to theoretical low bound for algorithms without adaptation of coefficients.

The second version of SPRINT has means for joint evaluation of polynomial-trigonometric expression sets. The algorithms for such joint evaluations were described in [2].

SPRINT system had been used at the Institute for Theoretical Astronomy for solving some practical problems. Namely, the programs for evaluation of astronomical ephemerides from analytical theories of motion were synthsised by SPRINT. It should be pointed out that the expressions in these theories have thousands of terms. The programs for evaluation of coordinates of the satellites of Venus, Mercury, Mars and Jupiter were synthesised by SPRINT. Application of synthesised programs for the joint evaluaton of expression sets in all of these cases reduced computer time by 10-20 times as compared with the standard algorithms.

References

1. N.N.Vasiliev. Synthesis programs system for evaluation of polynomial-trigonometric expressions. In: Proc. Int. Conf. on Systems and Techniques of Analytical Computing and Their Appl. in Theor. Phys., JINR, Dubna 1982.

2. N.N.Vasiliev. Joint evaluations of sparse polynomials and the synthesis of evaluating programs. In: Proc. Conf. on Computer Algebra and its Appl. in Theor. Phys., JINR, Dubna 1985.

Automatic Generation of FORTRAN-Coded Jacobians and Hessians

P. van den Heuvel, J.A. van Hulzen, V.V. Goldman

University of Twente, Department of Computer Science
P.O. Box 217, 7500 AE Enschede, The Netherlands

ABSTRACT

A package for automatically generating efficient (FORTRAN) code for calculating Jacobians and Hessians is presented. It is meant as an extension to the symbolic-numeric interface of REDUCE. GENTRAN and the code optimizer prove to be a powerful combination. GENTRAN, a code GENeration and TRANslation package, is both used for translating Lisp code into FORTRAN and for *template processing*. The code optimizer is used for locally optimizing the generated code.

1. Introduction

Computer algebra systems (cas's) are not primarily designed for performing numeric calculations. However, a need exists for using the algebraic facilities of cas's to generate code for languages like FORTRAN, as recently indicated by Hearn [4] and Van Hulzen [10].

In attempting to avoid redundancy in the generated code, a *code optimizer* has been developed for REDUCE. It considerably reduces the arithmetic complexity of sets of expressions to be translated. Another recent development has been the implementation of GENTRAN [2], a code GENeration and TRANslation package. In addition to being able to translate REDUCE expressions and assignment statements into FORTRAN, RATFOR or C, it has extended facilities for generating control structures, subprogram definitions and type declarations. An additional feature contained in GENTRAN, is its so-called *template processor*. It allows the user to offer the system a partially completed program, the *template*. It is a mixture of *active parts* to be processed by GENTRAN, and *passive parts* which are throughputted literally.

So far, the symbolic-numeric interface in REDUCE only consists of tools for transforming REDUCE input into FORTRAN. However, a possibility for generating complete FORTRAN programs for solving numerical problems is more interesting. We discuss methods for automatically generating (FORTRAN) code for Jacobians and Hessians, making a combined use of REDUCE, GENTRAN and the code optimizer. A general class of problems where this might be useful is the numeric solution of a set of non-linear equations or the minimization of a multivariate function. Assume for instance the need to solve a system of d non-linear equations in d indeterminates, denoted by:

$$\mathbf{f}(\mathbf{x}) = \mathbf{0}. \tag{1}$$

Application of the generalized Newton-Raphson method requires the construction of the iteration function:

$$\mathbf{x}^k = \mathbf{x}^{k-1} - J^{-1}(\mathbf{x}^{k-1})\mathbf{f}(\mathbf{x}^{k-1}) \ , \tag{2}$$

starting with an initial estimate $\mathbf{x}^0$. $J(\mathbf{x}^k)$ is the *Jacobian* evaluated at point $\mathbf{x}^k$. An element of this matrix is defined as: $J_{ij} = \frac{\partial f_i}{\partial x_j}$. A slight extension of such facilities is the computation of the Hessian $H(\mathbf{x})$, where $H_{ij} = \frac{\partial^2 g}{\partial x_i \partial x_j}$, which is used in the computation of the extrema of the given function $g(\mathbf{x})$.

Some numerical FORTRAN packages for solving such problems require the user to provide subroutines for calculating $\mathbf{f}(\mathbf{x})$ and $J(\mathbf{x})$ (or for $g(\mathbf{x})$, its first derivatives $\mathbf{Dg}(\mathbf{x})$ and $H(\mathbf{x})$). In the case of large sets of equations, writing these routines by hand is tedious, error prone and thus sometimes almost impossible. The number of equations can quickly increase upto hundreds, resulting in the calculation of *thousands* of derivatives!

We describe a program that, given a set of equations as in (1), automatically generates a FORTRAN subroutine*, that efficiently calculates $J(\mathbf{x})$ and $\mathbf{f}(\mathbf{x})$, for a given value of $\mathbf{x}$. In sections 2 to 5 we discuss the analysis of the problem , a functional description of the code to be generated and of the generating program and the implementation of this program, respectively. In section 6 we consider program optimization. Finally some conclusions are drawn in section 7.

2. Analysis of the problem

In practice it often happens that variable names are subscripted. This implies that each equation represents a *class* of x equations, and each variable name a *class* of y indeterminates, where x and y depend on the *index ranges*, i.e. the values the indices may take. Since the set of equations is independent and solvable, each equation class is also an *equivalence class*. Hence, all equation classes together define a *partitioning* of the total set of equations. Similarly, all classes of indeterminates form a partitioning of the total set of indeterminates, since each indeterminate has a unique *name*. For some combination of indices, some indeterminates may have constant values. These cases are called *constraints* and are not indeterminates. A particular instance of an equation class which contains such a constraint is called a *degeneration*.

Example 1.

Consider the set of equations formed by:

$$E_1 = \{\, x_{pqr}^2+3y_{pq}+z_r = 0 \mid 1\le p\le 2\,,\ 1\le q\le 3\,,\ 1\le r\le 4 \,\}$$

$$E_2 = \{\, z_2\, y_{p+1,\,q+1}^3+1 = 0 \mid 0\le p\le 1\,,\ 0\le q\le 2 \,\},$$

with the constraints: $x_{111}=1, y_{23}=2, z_1=0, z_4=0$. We will use it as a case study throughout this paper. Note that the number of indices is *not* restricted to one. The total number of equations is 30, with an equal number of indeterminates. The example shows that one expression can have more than one degeneration: among others, E_1 will degenerate for $p=1, q=1, r=1$ to $3y_{11}+z_1=0$, and for $r=1,4$ to $x_{pqr}^2+3y_{pq}=0$. □

Generally there are r equation classes $E_1, E_2, \ldots, E_r$, n indices $i_1, i_2, \ldots, i_n$, and m indeterminates $(x_1)_{i_1,i_2,\ldots,i_n}$, $(x_2)_{i_1,i_2,\ldots,i_n}, \ldots,$ $(x_m)_{i_1,i_2,\ldots,i_n}$†. An index i_j occuring in an equation class E_k has a corresponding *range* within it, denoted by $a_{kj}..\,b_{kj}$. When an index does *not* occur within an equation class, its lower and upper bounds are equal by definition. Hence, the total number, d, of equations is:

$$d = \sum_{i=1}^{r}\prod_{j=1}^{n}(b_{ij}-a_{ij}+1) \tag{3}$$

In general there are m different indeterminate *names* x_k, each with a fixed number of indices. Each index i_j of an indeterminate has a certain range, denoted by $l_{kj}..u_{kj}$, independent of its ranges within the equation classes. When an index does not belong to an indeterminate, its lower and upper bounds are equal by definition.

A *constraint* is always of form: $(x_j)_{i_1 i_2 \cdots i_k} = c$. The indices in a constraint variable are supposed to always have either their *lowest* possible values or their *highest* possible values for that variable. Thus, the indices $i_1, i_2, \ldots, i_k$ of x_j are equal to either $l_{j1}, l_{j2}, \ldots, l_{jk}$ or $u_{j1}, u_{j2}, \ldots, u_{jk}$ only. This assumption is not restrictive.

*) The choice of the target language is not significant. Our main interest is the analysis of the problem and the algorithms for obtaining the desired results.

†) Only for convenience we simply assume that all variables have n indices. In reality, even *non*-subscripted variables can occur.

3. A FORTRAN Subroutine for Calculating f and J

We now describe in some detail FDF, a FORTRAN routine for calculating **f** and J.

3.1. Data Structures used in FDF

Concrete numerical library implementations of Newton's iteration formula (2) will usually work with data structures which are arrays of length d and matrices of dimensions $d{\times}d$. Let **X** and **F** be arrays of length d to store the values of the indeterminates **x** and the values of **f(x)**, respectively, and **J** be a $d{\times}d$ matrix to store the Jacobian J. The use of these data structures requires functions to map the equation classes onto **F** and the rows of **J**, i.e. a *row-mapping* function, and to map the (indexed) indeterminates onto **X** and the columns of **J**, i.e. a *column-mapping* function.

The order in which the user defines the equation classes constitutes the global ordering for the row-mapping. Within an equation class E_j, the equation ordering is determined by the actual values of the indices $i_1,i_2,..,i_n$ occuring in E_j. In section 4 it is shown how the index-ordering has to be defined in the generating program. Our row-mapping function is defined as follows.

$$\mathrm{Rmap}(E_j,i_1,i_2,..,i_n) = \mathrm{Rbase}(E_j) + \mathrm{Roffset}(E_j,i_1,i_2,..,i_n) + 1\ ,$$

with:

$$\mathrm{Rbase}(E_1) = 0\ ,$$

$$\mathrm{Rbase}(E_j) = \mathrm{Rbase}(E_{j-1}) + \prod_{i=1}^{n}(b_{j-1,i}-a_{j-1,i}+1)\ ,\quad j>1\ ,$$

$$\mathrm{Roffset}(E_j,i_1,i_2,..,i_n) = \mathrm{R}(j,(i_1 i_2 \cdots i_n))$$

with:

$$\mathrm{R}(j,()) = 0\ ,$$
$$\mathrm{R}(j,(i_1 i_2 \cdots i_{k-1} i_k)) = \mathrm{R}(j,(i_1 i_2 \cdots i_{k-1})) \times (b_{jk}-a_{jk}+1) + (i_k-a_{jk})$$

The column-mapping function $\mathrm{Cmap}((x_j)_{i_1 i_2 .. i_n})$ is analogously defined, assuming we deal carefully with the possible occurrence of constraints.

Example 2.

The figure below schematically shows how the equations and indeterminates of example 1 are mapped onto the Jacobian.

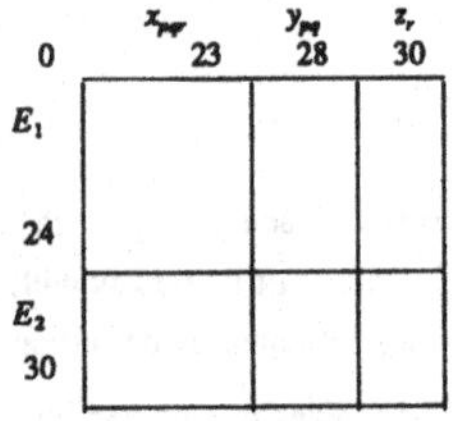

We find for instance: $\mathrm{Rmap}(E_1,p,q,r) = 12p+4q+r-16$, and $\mathrm{Cmap}(y_{pq}) = 3p+q+20$, implying that $\dfrac{\partial(E_1,p{=}2,q{=}1,r{=}3)}{\partial y_{11}}$ is mapped onto $J(15,24)$. □

3.2. General Structure of FDF

FDF is used to calculate **F** and **J**, given **X**. The occurrence of indexed variables in the set of equations leads to *nested for-loops* in the body of FDF. The jth loop corresponds to the index i_j and runs through its whole range, i.e. from $\min_{1\le i\le r}(a_{ij})$ to $\max_{1\le i\le r}(b_{ij})$. Before entering these loops, all elements in **J** have to be initialized to 0. Within these loops parts of **F** and **J** are calculated, depending on the actual values of the indices. All code corresponding to an equation class E_i is called the *function block for* E_i. Whether a function block has to be entered or not, depends on the actual values of the indices and is tested in an **if**-statement. Within a function block, the evaluation of the degenerated cases versus the 'normal' case, is also controlled by **if**-statements. In appendix A an ad hoc implementation for the set of equations of example 1 is listed. The equations and indeterminates have already been mapped. Both the first **if**-statement and the one at label L_7 are used to control the entry of the function blocks for E_1 and E_2, respectively. Note that in the case of E_1, among others, the equation degenerates *differently* for $p=1, q=1, r=1$ and $r=1$. Testing for these cases in the opposite order, i.e. first for $i=1$ and then for $p=1, q=1, r=1$, would result in erroneous code: the latter case would *never* be tested. Special care in the implementation of the generating program should be taken here.

3.3. Improving FDF

The previously mentioned ad hoc solution for the body of FDF can be highly improved. We can distinguish between making the *passive* part of the code as efficient as possible and reducing the arithmetic complexity.
Because of its specific nature, the latter improvement will be discussed seperately in section 6. *Passive* code, occuring in the *template*, consists of the control structures. In the following sections we will dicuss some possible improvements of the passive code.

3.3.1. Moving Function Blocks outside Loops

Generally not all equation classes depend on all indices. For instance, equation class E_2 of example 1 does not contain index r. By definition, r's lower and upper bound in E_2 are equal, so the function block corresponding to E_2 will only be entered for one value of r. However, for all its remaining values, r still has to be tested each time before passing on to the next function block, as illustrated by the condition in the **if**-statement at label L_7 in appendix A. The number of these 'never succeeding' tests can quickly increase when more indices are missing in an equation class. Thus, placing function blocks outside loops whenever possible improves the code. In the routine of appendix A, the function block corresponding to E_2 can be placed between the second and third loop, thus saving 36 tests on $1 \le r \le 1$.

3.3.2. Changing the Loop Order

Consider again the body of FDF in appendix A. Suppose the loop order would have been *prq* instead of *pqr**. Clearly, this does not change the result of an FDF application. However, the *prq*-order does not allow to move E_2 outside the r-loop. This implies that the loop order influences possibilities for moving blocks, i.e. for improving the code's efficiency. The generating program has to determine an intuitive *optimal* loop order. Such a loop order allows to remove the highest possible number of function blocks from the most deeply nested loops. Suppose $i_{p_1}, i_{p_2}, \ldots, i_{p_n}$ is determined to be an optimal loop order, where $(p_1, p_2, \ldots, p_n) = \mathrm{PERM}(1,2,\ldots,n)$. The general structure of the body of FDF can then be given by

> Prog(0)
> **L_s continue**

where, for $0 \le j < n$ holds:

> Prog(j) =
>
> ACTIVE(j)
> **do** L_s $i_{p_{j+1}} = \min\limits_{1 \le i \le r}(a_{ip_{j+1}})\, ,\ \max\limits_{1 \le i \le r}(b_{ip_{j+1}})$

---

*) The loop order is expressed in terms of the indices corresponding to the loops. Starting at the outermost loop, they are listed in the order in which they are nested.

Prog($j+1$ **)** ,

and for $j=n$:

Prog(n) = ACTIVE(n)

where

ACTIVE(j) is defined as

all code belonging to the function blocks contained between the jth and $(j+1)^{st}$ loops, i.e. all function blocks that contain the index i_{p_j} but not $i_{p_{j+1}}..i_{p_n}$.

This simple description will prove to be very useful for the implementation of the generating program.

3.3.3. Introduction of Boolean Variables

Immediately after entering the kth loop, all boolean (sub-) expressions occuring in the kth up to the nth loop, containing i_{p_k} and governing the actual computations inside these loops, can already be evaluated. The results then have to be temporarily assigned to *boolean variables*, thus allowing later use wherever needed. This is an improvement over the normal situation where (sub-) expressions which depend on i_{p_k} and occur in the $(k+1)^{st}$ up to the nth loop are re-evaluated many times.

3.3.4. Combining Degenerations

In some cases the validity of different constraints leads to identical degenerations. This happens for instance with E_2 in example 1, for $r=1$ and $r=4$. Instead of separately testing for these constraints, this can be done only *once* by performing these tests in one if-statement. Special care must be taken that after combining the blocks the resulting boolean expression neither implies a previous one within the function block, nor is implied by one succeeding it. Appendix C contains the improved version of the body of FDF listed in appendix A. Note that all indices whose value is known beforehand are already replaced by that value.

4. Functional Description of GENFDF

The problem analysis and the description of FDF have provided sufficient insight for designing a generating program for FDF, hereafter called *GENFDF*. The advantages of such a system are more evident now: besides eliminating the tedious job of writing the numeric code, the just described code improvements are also hard to correctly incorporate by hand. In this section we will give GENFDF's functional description.

GENFDF accepts a complete description of a set of (non-linear) equations and generates a corresponding FORTRAN subroutine FDF, as described in the previous sections. The program runs under REDUCE and input can be given in a comfortable, REDUCE compatible syntax. The use of these commands is illustrated by example 3.

Example 3.

The command sequence for specifying the set of equations of example 1 is shown below. Output will be sent the file FOO and the terminal. The resulting routine is shown in appendix B.

```
eqvars X,Y ;

eqindices P,Q,R ;

puteq X(P,Q,R)^2+3*Y(P,Q)+Z(R)
  index P,Q,R from 1,1,1 upto 2,3,4 ;

puteq Z(2)*Y(P+1,Q+1)^3+1
  index P,Q from 0,0 upto 1,2 ;
```

constraints

X(1,1,1) = 1, Y(2,3) = 2, Z(1) = 0, Z(4) = 0 ;

genfdf FOO, NIL ; □

The **eqvars**-command is used to specify the indeterminate names. The order in which this is done, constitutes the global ordering of the mapping of the indeterminates onto the rows of **X** and the columns of **J**. The command can be given more than once.

The **eqindices**-command is used to specify the indices occuring in the set of equation classes. No other indices may be used in the specification of the equation classes. The order in which the indices are specified, partially constitutes the ordering of both the row and column mapping functions. The command can be given more than once, but only the last one will be remembered.

The optional **cleareq**-command tells the program that a new set of equations is about to be specified. It is meant for cases in which the user has made mistakes while entering the equations. Of course, this command is only helpful when using the program interactively. Normally, the input will be read from a file.

The **puteq**-command is used to add a new equation class to the total set of equations. An equation may contain indexed indeterminates, in which case the indices must be fully specified. Indexed indeterminates are denoted in REDUCE as *operators*. The equation classes have to be entered one at the time, but not necessarily immediately after one another. The order in which the user enters the equation classes, constitutes the global ordering in which they are mapped onto the rows of **F** and **J**.

The **constraints**-command is used to specify all constraints. It can be repeated more than once, but only the last command will be remembered.

The **genfdf**-command initiates the code generation. Output is sent to the specified output files. NIL specifies the standard output, usually the terminal. The command knows a **split**-option, allowing for generating seperate routines for calculating **F** and **J**. After completion, an automatic **cleareq** is executed, so that a new set of equations can be entered.

Example 4.

Suppose we wish to generate a routine for calculating the first derivatives (**Dg**) and the Hessian matrix H of the function $g(x,y) = 3x^2+2xy+y^2$. After entering:

eqvars X,Y ;

genfdfd2f 3*X^2+2*X*Y+Y^2
out FOO ;

the required routine can be found on file FOO. X will be mapped onto X(1) and Y onto X(2). □

The **genfdfd2f**-variant is used for Hessian computation. Output is send to the specified files. It also has a **split**-option, allowing to provide seperate routines for calculating **Dg** and H. In all of the above cases, error messages are given whenever necessary. A complete user manual for GENFDF can be found in [8].

5. Implementing GENFDF

The program runs through three different phases, which will now be explained.

5.1. The Input Phase

The parser for the input is fully syntax driven, and accepts the syntax described in the previous section. A complete description of the methods and tools used can be found in [7]. All input is stored in internal data structures. The design of these data structures allows for convenient and efficient processing.

5.2. The Preparation Phase.

During this phase all remaining data needed for code generation is calculated, i.e. a map of the equation classes onto F and the rows of J and of the indeterminates onto X and the columns of J, Calculation of all degeneration cases and partial derivatives, Determination of the optimal loop order, and a partitioning of the set of equations over the ACTIVE-blocks.

5.3. The Code Generation Phase

During this final phase the code of the FDF routine is produced. Since most *active* code, i.e. its arithmetic part, has already been generated during the preparation phase, *this* phase is used to generate the *passive* part of FDF, i.e. its control structures and boolean assignments. It merely consists of a 1:1 mapping of the internally stored information to a syntactically correct Standard Lisp [5] S-expression, semantically equivalent to the FDF routine in FORTRAN. The motivation for choosing this strategy is twofold : Since no difference exists between *data* and *instruction* in Lisp, generating code is as convenient as constructing any other data object and the GENTRAN package is specially designed for translating Lisp to FORTRAN.
We use GENTRAN just under its top level implementation. In top level use, GENTRAN translates Rlisp code into FORTRAN, RATFOR or C. Actually, the Rlisp code to be translated is first transformed by the Rlisp parser Xread into an equivalent prefix form. This prefix form is subsequently passed through to the lower level Gentran function. Thus, the idea is to generate FDF in Lisp prefix form, which is subsequently handed over to the function Gentran.

6. Optimizing FDF's Active Part

As mentioned, the program code can be divided into an *active* and a *passive* part. The passive part, or *passive code*, consists of declarations and control structures, such as **if**- and **for**-constructs. It is passive in the sense that the code is not used for calculations or other time consuming actions. It merely describes the logical order in which these actions have to take place. Active code consists of assignment statements, function calls and expressions. In FDF the active part mainly consists of arithmetic code, especially when generated for a *large* set of equations. Normally active code claims most of the CPU time*.
The arithmetic complexity of this active code can be reduced, using both the code optimizer and GENTRAN's *template processor*, as explained below.

6.1. The Code Optimizer

In order to reduce the arithmetic complexity of straight line code, a *code optimizer* has been developed [9]. It performs a heuristic search for *common sub-expressions* (*cse's*) in the set of input expressions, and assigns them to unique identifiers. All occurences of the cse's are then replaced by identifers.

Example 5.

Consider the following set of (simple) equations:

$$q_1 = a + b + cd + (g + f)\,(a + bcd)$$
$$q_2 = a + f + cd(g + f)(a + bcd)$$
$$q_3 = a + gcd(g + cd)(a + b)$$

They define 14 multiplications and 12 additions. After optimizing these equations, the resulting code is:

$$p_1 = cd$$
$$s_1 = a + b$$
$$p_2 = bp_1$$

*) Observe an interesting contradiction here. Although active code consumes most of the CPU time, it is the control structure, i.e. passive code, that determine the program's complexity!

$s_2 = f + g$
$s_3 = a + p_2$
$p_3 = s_2 s_3$
$q_1 = s_1 + p_1 + p_3$
$q_2 = a + f + p_1 p_3$
$q_3 = a + g p_1 (g + p_1) s_1$

Only 7 multiplications and 9 additions remain. □

Besides finding cse's, the code optimizer also performs optimizations such as removal of all exponentiations by decomposing them into minimal sequences of multiplications and factoring out all local factors in case this proves profitable.
The results are often quite dramatic, which is shown by experiments done by Wang e.a. [11].

6.2. Local Optimization

When considering optimization only as an attempt to reduce the arithmetic complexity of a (set of) expression(s) by using the code optimizer, we can distinguish two kinds of code optimization: global and local. Global optimization is meant to reduce the arithmetic complexity of programs as a *whole*. Local optimization restricts itself to the *basic blocks* within a program, i.e. to sequences of consecutive statements in which flow of control enters at the beginning and leaves at the end without halt or possibility of branching except at the end [1]. A basic block may thus contain only a sequence of assignment statements. In this context, we also assume that all right hand sides contain arithmetic expressions. Thus, the code optimizer is perfectly suited for local optimization.

6.3. Using GENTRAN's Template Processor for Local Optimization

Besides translating and generating FORTRAN, RATFOR and C code from Lisp code, GENTRAN is also capable of processing *template files*. These files contain fixed parts of code, the *passive parts*, and *active* parts, which have to be evaluated by REDUCE. Its purpose is to offer the user the possibility of generating code within an already existing part of a program.
The template processor is invoked by the following command:

gentranin F1,F2,...,Fn [**out** G1,G2,...,Gm] ;

F1 F2,...,Fn are template files which are read in sequence, processed and concatenated. When **out** is used, output is sent to the files G1,G2,...,Gm, otherwise to the current output file, usually the terminal.
All active parts in the template files start with the character sequence **;begin;** and end with **;end;**. Passive parts of the template files, i.e. the template itself, are simply copied to the output. Active parts may contain any number of REDUCE expressions, statements and commands. They are given to REDUCE for evaluation in algebraic mode, before being sent to the output.
Active parts will most likely contain calls to GENTRAN to generate code. This means that the result of processing a template file will be the original template file with all active parts replaced by active code.
GENTRAN recognizes an important flag: GENTRANOPT. With **on** GENTRANOPT, GENTRAN will isolate each group of consecutive assignment statements in its input and have it processed by the code optimizer prior to translation. Together with GENTRAN's template processor, this option can be used quite effectively for locally optimizing programs. First, each basic block *BB* has to be traced and transformed into:

;begin;
  gentran
  ≪ *BB'* ≫
;end;

where *BB'* is the REDUCE equivalent of *BB*. We have thus obtained a template file, which can subsequently be processed by the template processor. The result is a locally optimized version of the original program.

6.4. Generating FDF as a Template File

Confusing as it may seem, GENTRAN can also be used for *producing* template files. These template files can then be processed by GENTRAN's template processor. This technique can be used for generating a locally optimizing version of FDF as follows.

After GENFDF passed its preparation phase (see sections 4), all information about FDF's structure and contents is available. Consequently, basic blocks can easily be localized: they consist of all groups of assignment statements defining elements of F and J contained in the function blocks.

Instead of directly generating the code for FDF, a template file FDF' is generated first. This file is almost identical to FDF, except that all basic blocks are now active parts for the template processor. The remaining part of FDF, i.e. its control structures and boolean assignment statements, is to be considered as the template. The template file FDF' can subsequently be processed by GENTRAN's template processor.

Example 6.

Consider the second basic block in the code listed in appendix B.

```
F(3*P+Q+25) = X(3*P+Q+24)**3*X(29)+1
J(3*P+Q+25,3*P+Q+24) = 3*X(29)*X(3*P+Q+24)**2
J(3*P+Q+25,29) = X(3*P+Q+24)**3
```

The corresponding active block in the template file FDF' is:

```
;begin;
  gentran
  <<
    F(3*P+Q+25) ::=: X(3*P+Q+24)**3*X(29)+1 ;
    J(3*P+Q+25,3*P+Q+24) ::=: Ddf(X(3*P+Q+24)**3*X(29)+1,X(3*P+Q+24)) ;
    J(3*P+Q+25,29) ::=: Ddf(X(3*P+Q+24)**3*X(29)+1,X(29)) ;
  >>
;end;
```

The special ::=: assignment operator, is recognized by GENTRAN and causes both right- and left-hand side of the assignment statement to be evaluated prior to translation. The operator Ddf is similar to the REDUCE differentiation operator DF, but allows differentiation with respect to *non-atoms*. After processing the template file with on GENTRANOPT, the resulting code from the active block above is:

```
IT12 = 3*P+Q
IT0 = IT12+25
IT2 = IT12+24
T15 = X(IT2)*X(IT2)
T14 = T15*X(IT2)
F(IT0) = T14*X(29)+1
J(IT0,IT2) = 3*X(29)*T15
J(IT0,29) = T14
```

All temporary variables of INTEGER type are prefixed by the character I. This is an additional feature provided by the GENFDF program, and is not yet contained in GENTRAN or the code optimizer. □

Local optimizations can dramatically reduce the arithmetic complexity : We have used GENFDF for generating code for a large set of equations, both with and without local optimization. The difference in arithmetic complexity between the two versions is given in table 1.

Table 1. *Number of operations used in the optimized and non-optimized versions of FDF.*

operator	*# occuring in non-optimized FDF*	*# occuring in optimized FDF*
+/-	1131	124
×	711	168
↑	288	0
others	145	33

7. Conclusions

One motivation for developing the facilities we just discussed was to increase our insight in the various aspects of the symbolic-numeric interface. GENTRAN and the code optimizer have proven to be indispensable tools in this area. GENTRAN's usefulness has manifested itself in two ways. First of all, its code translation facilities have made it possible to obtain the desired FORTRAN code with great ease. Secondly, GENTRAN's template processing facilities have provided a general tool for locally optimizing program code. Future research should investigate the possibilities of using this technique for global optimization. For this purpose an extension of the code optimizer will have to make use of data flow and dependency analysis information.

We conclude that GENTRAN and the code optimizer can be effectively paired up to fill the gap between REDUCE's algebraic facilities and the numeric power of other languages. It is worth emphasizing that code, generated by GENFDF, is directly usable in combination with NAG-library [6] routines.

8. References

[1] Aho A.V., R. Sethi, J.D. Ullman, *Compilers Principles, Techniques, and Tools.* Addison-Wesley, Reading, Mass., 1986.

[2] Gates, B.L., "An Automatic Code Generation Facility for REDUCE". *SYMSAC Proceedings '86*, B.W. Char (ed), Waterloo, July 21-23, 1986.

[3] Hearn, A.C. (ed), *REDUCE User's Manual, Version 3.2.* The Rand Corporation, Santa Monica, California. April 1984.

[4] Hearn, A.C., "Optimal Evaluation of Algebraic Expressions". *Proceedings AAECC-3*, J. Calmet (ed), Springer LNCS Series **229**, pp392-403, Springer, Heidelberg. 1986.

[5] Marti, J.B., A.C. Hearn, M.L. Griss, C. Griss, *Standard Lisp Report.* University of Utah, August 1978.

[6] *NAG FORTRAN Library Manual Mark 11.* Numerical Algorithms Group, Oxford, 1984.

[7] van den Heuvel, P., "Adding Statements to REDUCE". *ACM SIGSAM Bulletin* **20,1/2**, pp 8-14, 1986.

[8] van den Heuvel, P., *Aspects of Program Generation related to Automatic Differentiation.* Master's Thesis, University of Twente, Department of Computer Science, Enschede, The Netherlands, December 1986.

[9] van Hulzen, J.A., "Code Optimization Of Multivariate Schemes: A Pragmatic Approach", *Proceedings EUROCAL '83*, J.A. van Hulzen (ed.), Springer LNCS-series **162**, pp 286-300, Springer, Heidelberg, 1983.

[10] van Hulzen, J.A., *Program Generation Aspects of the Symbolic-Numeric Interface.* Memorandum **INF-85-25**, Twente University of Technology, Department of Computer Science, Enschede, The Netherlands, November 1985.

[11] Wang, P.S., T.Y.P. Chang, J.A. van Hulzen, "Code Generation and Optimization for Finite Element Analysis", *EUROCAM '84 Proceedings*, J.P. Fitch (ed.). Springer LNCS series **174**, pp. 237-247. Springer, July 1984.

Appendix A.

```
          DO L P=0,2
             DO L Q=0,3
                DO L R=1,4
                IF NOT (0<=P<=1 AND 0<=Q<=2 AND 1<=R<=4) GOTO L7
                   IF NOT (P=1 AND Q=1 AND R=1) GOTO L1
                      F(12*P+4*Q+R-16)=1+3*X(3*P+Q+20)
                      J(12*P+4*Q+R-16,3*P+Q+20)=3
                      GOTO L7
L1                 IF NOT (P=2 AND Q=3 AND R=1) GOTO L2
                      F(12*P+4*Q+R-16)=X(12*P+4*Q+R-17)**2+6
                      J(12*P+4*Q+R-16,12*P+4*Q+R-17)=2*X(12*P+4*Q+R-17)
                      GOTO L7
L2                 IF NOT (P=2 AND Q=3 AND R=4) GOTO L3
                      F(12*P+4*Q+R-16)=X(12*P+4*Q+R-17)**2+6
                      J(12*P+4*Q+R-16,12*P+4*Q+R-17)=2*X(12*P+4*Q+R-17)
                      GOTO L7
L3                 IF NOT (P=2 AND Q=3) GOTO L4
                      F(12*P+4*Q+R-16)=X(12*P+4*Q+R-17)**2+X(R+27)+6
                      J(12*P+4*Q+R-16,R+27)=1
                      J(12*P+4*Q+R-16,12*P+4*Q+R-17)=2*X(12*P+4*Q+R-17)
                      GOTO L7
L4                 IF NOT (R=1) GOTO L5
                      F(12*P+4*Q+R-16)=X(12*P+4*Q+R-17)**2+3*X(3*P+Q+20)
                      J(12*P+4*Q+R-16,3*P+Q+20)=3
                      J(12*P+4*Q+R-16,12*P+4*Q+R-17)=2*X(12*P+4*Q+R-17)
                      GOTO L7
L5                 IF NOT (R=4) GOTO L6
                      F(12*P+4*Q+R-16)=X(12*P+4*Q+R-17)**2+3*X(3*P+Q+20)
                      J(12*P+4*Q+R-16,3*P+Q+20)=3
                      J(12*P+4*Q+R-16,12*P+4*Q+R-17)=2*X(12*P+4*Q+R-17)
                      GOTO L7
L6                   F(12*P+4*Q+R-16)=X(12*P+4*Q+R-17)**2+3*X(3*P+Q+20)+X(R+27)
                     J(12*P+4*Q+R-16,12*P+4*Q+R-17)=2*X(12*P+4*Q+R-17)
                     J(12*P+4*Q+R-16,3*P+Q+20)=3
                     J(12*P+4*Q+R-16,R+27)=1
L7              IF NOT (0<=P<=1 AND 0<=Q<=2 AND 1<=R<=1) GOTO L
                   IF NOT (P=1 AND Q=2) GOTO L8
                      F(3*P+Q+25)=8*X(29)+1
                      J(3*P+Q+25,29)=8
                      GOTO L
L8                 F(3*P+Q+25)=X(3*P+Q+24)**3*X(29)+1
                   J(3*P+Q+25,3*P+Q+24)=3*X(3*P+Q+24)**2*X(29)
                   J(3*P+Q+25,29)=X(3*P+Q+24)**3
L  CONTINUE
```

Appendix B.

```
      LOGICAL B1S2,B1S7,B1S1,B1S5,B2S2,B2S7,B2S1,B2S5,B2S3,B3S6,
     .B3S4,B3S3
      INTEGER P,Q,R
      REAL J(30,30),F(30),X(30)
      NDIM = 30
      DO 25018 I1=1,30
          DO 25019 I2=1,30
              J(I1,I2)=0
25019     CONTINUE
25018 CONTINUE
      DO 25020 P=0,2
          B1S2=P.LE.1
          B1S7=P.EQ.1
          B1S1=1.LE.P
          B1S5=P.EQ.2
```

```
           DO 25021 Q=0,3
              B2S2=B1S2.AND.Q.LE.2
              B2S7=B1S7.AND.Q.EQ.2
              B2S1=B1S1.AND.1.LE.Q
              B2S5=B1S5.AND.Q.EQ.3
              B2S3=B1S7.AND.Q.EQ.1
              IF (.NOT.B2S2) GOTO 25022
                 IF (.NOT.B2S7) GOTO 25023
                    F(30)=8*X(29)+1
                    J(30,29)=8
                    GOTO 25024
25023            CONTINUE
                    F(3*P+Q+25)=X(3*P+Q+24)**3*X(29)+1
                    J(3*P+Q+25,3*P+Q+24)=3*X(3*P+Q+24)**2*X(29)
                    J(3*P+Q+25,29)=X(3*P+Q+24)**3
25024            CONTINUE
25022         CONTINUE
              DO 25025 R=1,4
                 B3S6=R.EQ.4.OR.R.EQ.1
                 B3S4=B2S5.AND.B3S6
                 B3S3=B2S3.AND.R.EQ.1
                 IF (.NOT.B2S1) GOTO 25026
                    IF (.NOT.B3S3) GOTO 25027
                      F(1)=3*X(24)+1
                      J(1,24)=3
                       GOTO 25028
25027               CONTINUE
                       IF (.NOT.B3S4) GOTO 25029
                      F(R+20)=X(R+19)**2+6
                      J(R+20,R+19)=2*X(R+19)
                          GOTO 25030
25029                  CONTINUE
                          IF (.NOT.B2S5) GOTO 25031
                      F(R+20)=X(R+27)+X(R+19)**2+6
                      J(R+20,R+27)=1
                      J(R+20,R+19)=2*X(R+19)
                             GOTO 25032
25031                     CONTINUE
                                IF (.NOT.B3S6) GOTO 25033
                      F(12*P+4*Q+R-16)=X(12*P+4*Q+R-17)**2+3*X
     .                 (3*P+Q+20)
                      J(12*P+4*Q+R-16,3*P+Q+20)=3
                      J(12*P+4*Q+R-16,12*P+4*Q+R-17)=2*X(12*P+
     .                 4*Q+R-17)
                                   GOTO 25034
25033                          CONTINUE
                      F(12*P+4*Q+R-16)=X(12*P+4*Q+R-17)**2+3*X
     .                 (3*P+Q+20)+X(R+27)
                      J(12*P+4*Q+R-16,12*P+4*Q+R-17)=2*X(12*P+
     .                 4*Q+R-17)
                      J(12*P+4*Q+R-16,3*P+Q+20)=3
                      J(12*P+4*Q+R-16,R+27)=1
25034                          CONTINUE
25032                     CONTINUE
25030                  CONTINUE
25028               CONTINUE
25026            CONTINUE
25025         CONTINUE
25021      CONTINUE
25020 CONTINUE
      RETURN
      END
```

LAPLACE TRANSFORMATIONS IN REDUCE 3

Christomir Kazasov
Laboratory for Informatics with Computing Centre
Sofia University "Klimemt Okhridski"
boul. Anton Ivanov 5, Sofia 1126, Bulgaria

The leading principles, I have decided to follow in implementation of Laplace Transformations (LTs) in REDUCE 3, are: performance of the LTs without explicit use of the integration operator applying the rules and table of Laplace-transform pairs; reduction of the table of pairs to a minimal size without restriction of the class of most useful functions, owing to the more intensive work of the rules, and ensuring, thereby, an easy inclusion of new special functions, if desired; and execution of an optimal checking of expressions' correctness.

The both transformations are given by the formulae:

$$F(p)=L[f(t)]=\int_0^\infty e^{-pt}f(t)dt; \quad f(t)=L^{-1}[F(p)]=\frac{1}{2\Pi i}\int_{s-i\infty}^{s+i\infty} e^{pt}F(p)dp; \quad (1)$$

where the function f(t) is called object function, the real variable t - object variable, the complex function F(p) - image function, and the complex variable $p = s+i\omega$ - parameter of the Laplace transform.

The rules I have implemented are the following:

i) the linearity rules for both LTs;

ii) the translation rules, namely:

$$L[e^{\lambda t}f(t)] = F(p-\lambda); \quad L^{-1}[e^{-p\tau}F(p)] = f(t-\tau), \text{ for } \tau>0 ; \quad (2)$$

iii) *one generalization* of the translation of the object function and change of scale rules, for the Direct Laplace Transformation (DLT) only:

$$L[f(kt-\tau)]= e^{-p\tau/k}\, L[f(t)], \text{ for } k>0,\ \tau\geq 0 ; \quad (3)$$

iiii) *a modification* of the rule for differentiation of the image function, but which is much useful for *the DLT only*:

$$L[t^n f(t)] = (-1)^n \frac{d^nF(p)}{dp^n} ; \quad (4)$$

iiiii) the rule for integration of the object function, for DLT only.

Thanks to the precisely performed analysis of everyone part of the expressions, which have been done *in symbolic mode* only, and some tricks, too, each term of the *object function* in performing of the DLT (provided the time and memory are unlimited), can have the form:

$$f(t) = Ct^n e^{\lambda t} \sin^{m_1}(t_1) \cos^{m_2}(t_2) \sinh^{m_3}(t_3) \cosh^{m_4}(t_4)\, f_0(t_0) \; ; \qquad (5)$$

where $n, m_i > 0$, integer, C is a constant, and *either*:

1) $t_i = k_i t$, $k_i \geq 0$, for $i=1 \div 4$, and then $f_0(t_0)$ may be unit-step function $1(t_0)$, or unit-impulse function $\delta(t_0)$, or t_0^{α}, for $\alpha > -1$, non-integer (in which case gamma function $\Gamma(\alpha)$ is involved, and if α is semi-integer, then $\Gamma(\alpha)$ is expressed by powers of p), *all this - for an argument* $t_0 = k_0 t - \tau_0$, for $k_0, \tau_0 \geq 0$, or $f_0(t_0)$ may be a *new function*, for which we had already given a LET rules for DLT; *or*:

2) $t_i = k_i t - \tau_i$, $k_i, \tau_i \geq 0$, for all i, and then $f_0(t_0) = 1^{m_0}(k_0 t - \tau_0)$.

In case 1) we could have an integral from 0 to t of the term (5), too.

The admissible form for each term of the *image function* in performing of the Inverse Laplace Transformation (ILT) is, respectively:

$$F(p) = Ce^{-p\tau} \frac{\Phi(p)}{\Psi(p)}; \; or \; F(p) = Ce^{-p\tau}(p-\lambda)^{-\alpha}; \; or \; F(p) = Ce^{-p\tau}\Gamma(\alpha); \qquad (6)$$

where $\tau \geq 0$, C is a constant, $\alpha > 0$, non-integer, and $\Phi(p)$ and $\Psi(p)$ are polynomials of p, *no matter of what degree*. In the most often used *ratio-case* (the most left formula above), if $\Psi(p)$ has a higher degree than $\Phi(p)$ then the *residue theorem* is applied, provided the zeros of $\Psi(p)$ might be found (not always, but in many cases); otherwise the polynomials are first divided, and for polynomial-quotient the Laplace-transform pairs for δ function and its derivatives are applicable.

The realized program has been widely tested in performing of the LTs for commonly used in the literature expressions and for solving linear differential equations with constant coefficients or set of them.

To conclude, I would note that one restricted variant of the method exposed was realized for REDUCE 2 in 1982 by the author under the supervising of V. Tomov and M. Spiridonova, of the Mathematical Institute at the Bulgarian Academy of Sciences.

References

1. Kolobov A.M., Selected chapters of mathematics. Fourier series and integrals. Operational calculus, Minsk '62 (in Russian).
2. Martinenko V.S., Operational calculus, Kiev Univ. '68 (in Russian).
3. Korn A.G., Korn T.M., Mathematical handbook, McGraw Hill '68.
4. Spiridonova M., Kazasov Ch., Computer-aided symbolic Laplace Transformations, Automated control systems, vol. 3 '83 (in Bulgarian).

Reduce 3.2
on
iAPX86/286-based personal computers

Tsuyoshi Yamamoto and Yoshinao Aoki
Department of Information Engineering,
Hokkaido University, Sapporo, Japan 060
Email:yamamoto%huie.hokudai.junet@japan.cs.net

1. Introduction

In this paper, we report a full implementation and performance evaluation of REDUCE3.2 [1] for the standard personal computers that use iAPX-86/286 CPU. To date, number of algebraic systems have been implemented on personal computers and, generally speaking, such systems are designed by two strategies. One approach is to design special algebraic system that is small enough for personal machines and the MuMATH is outstanding system of this approach. However, this approach has problems of syntactical incompatibility to de facto *standard* system on larger machines and missing features by less computing power. Another approach is to port existing algebraic system to high-end personal computers that use 32 bit architecture. This approach will result in good compatibility and enough performance while such machine is not so personal as so called *The Personal Computer*. Our approach is little different from them. Our goal is to realize fully loaded standard algebraic system on cheap hardware. We chose two standards, REDUCE3.2 as the standard of algebraic system and IBM-PC like personal computer that use iAPX-86/286 and MS-DOS as the standard of personal computer. Because of its 16 bit addressing architecture, it is hard to port REDUCE on iAPX-86 using standard porting procedure that is used to implement it on 68000 based machines. We designed special lisp kernel or AMI-Lisp on which fully loaded REDUCE3.2 runs. The architectural design of lisp kernel and performance analysis of REDUCE3.2 on personal computers will be shown.

2. The design of AMI-Lisp kernel

As well known, REDUCE is entirely written using RLISP syntax sugar and it is semantically equivalent to the STANDARD LISP [2] . What we need to realize personal REDUCE environment is to make a STANDARD LISP compatible lisp kernel that can load all REDUCE source code in it. Simple analysis of REDUCE source code shows us that it contains over 4000 symbols and 2000 functions. Of course, enough free space must be left after loading REDUCE kernel for user program. As the nature of iAPX-86 architecture, 16 bit pointer representation of lisp object is natural and it means that only 32K lisp objects can be exist at a time. It seems to be enough to load REDUCE, however such direct implementation will result in less free space and poor performance.

We designed a new lisp kernel for iAPX-86 to run huge application program. By this kernel architecture, number of lisp objects is limited to 32K while compiled code space can grow up to iAPX-86's physical limit of 1M byte. One of unique feature of this kernel is semiautomatic hiding of internal symbols on module compilation. As a nature of large lisp program, most functions are referenced only from inside of the module and such functions need not to be defined as external function associated with symbols. If it is possible to distinguish a function is external or not, we can reduce number of symbols. However, REDUCE source program doesn't have such declaration in it, so that we need a rule to pick up all external functions. A function should be external if

1. **It is documented as a function**
2. **Its name symbol is stored in global variable by load-time evaluation**
3. **Its name symbol has non NIL property-list or it is stored in property-list of other symbols.**

Apparently, these rules are not sufficient and it is required to declare some symbols as external explicitly, however, the number of such symbols was very small. As the result of hiding internal symbols, bare-REDUCE uses only 1142 symbols including all AMI-Lisp specific symbols and more than 23K free cells are left for user program.

3. Implementation and evaluation of REDUCE3.2 on AMI-Lisp

Implementation of REDUCE3.2 on AMI-Lisp has been done in number of steps as follows,

1. **Translate all RLisp level REDUCE source programs into S-expression using host machine.**
2. **Compile resident modules(RLISP,ALG1,ALG2 etc) into relocatable object file(OBJ file).**
3. **Link AMI-Lisp kernel and REDUCE resident modules and generate bare-REDUCE(EXE file)**
4. **Compile other REDUCE modules and make dynamic load module (FASL file)**

Because of small data space of iAPX-86, self translation of huge RLISP source program couldn't complete and we used cross environment to do it. To make runtime image small, we employed two step generation, that is, making bare-REDUCE that is tightly linked image of lisp kernel and REDUCE resident modules that include RLISP,ALG1,ALG2,MATR and HEPHYS modules. All other modules are compiled with additional declaration of external functions and the results will be loaded into bare-REDUCE on demand. Table 1 shows the memory size of each modules and it is clear that more than 640KB of free space, that is the physical limit of most personal computers, is required to load all modules at a time. It is one of limitation of this implementation, however, it is enough in most application if it can be configured by selected modules. Table 2 shows the performance data measured on typical iAPX-286 based PC *(Toshiba T-3100,iAPX-286 8MHz)*.

4. Conclusion

An implementation of *de facto* standard algebraic system into personal computers and its performance measurements were shown. Of course, this implementation has major limitation of small data space, however its algebraic processing capability is very close to 32 bit machine's one. We hope that our implementation could boost the population of algebraic system user in many fields where the cost of such system was too expensive.

Table 1: Memory size of modules

Module name	Size(KB)
bare-REDUCE3.2	459.2
FACTOR	94.5
INT	44.1
BFLOAT	24.1
SOLVE	12.5
CEDIT	10.8

Table 2: Timings

Problem	Time(Sec)
Reduce.tst(standard benchmark)	49.1
Inverse of matrix of rank 5	38.3
(X1+X2+......+X10)**5$	6.8
(X1+X2+......+X5)**12$	25.7
Factorization (Wang test case 5)	24.2
Factorization (Wang test case 7)	3.7

References

(1) Hearn,A.C.:REDUCE User's Manual, Version 3.2, The Rand Corporation,(1985)
(2) Marti,J.B. et.al.:Standard Lisp Report, SIGPLAN Notices 14,10,pp.48-68(1979)
(3) Yamamoto,T et.al.:A Compiler Based Lisp System and REDUCE3.2 for the MS-DOS, WGSYM-37-5, Information Processing Society of Japan,(1986) *In Japanese*

SOME EXTENSIONS AND APPLICATIONS OF REDUCE SYSTEM

M. Spiridonova
Institute of Mathematics with Computer Center
Bulgarian Academy of Sciences
Acad. G.Bontchev str., bl.8, 1113 Sofia, Bulgaria

Some results on extension and application of REDUCE system are briefly described.

1. Laplace transformation and inverse Laplace transformation.

This implementation (in symbolic mode) may be considered as a subsystem or as an extension of REDUCE [4]. It allows to use two new commands named LPS (for Laplace transformation) and RLS (for inverse Laplace transformation) with the following syntax:

LPS (<expr> , <varorig> , <varim>);

RLS (<expr> , <varim> , <varorig>);

where <expr> is the expression to be transformed, <varorig> is the variable of the original expression, <varim> is the variable of the Laplace transform.

In LPS <expr> may be: sin(t), cos(t), 1(t), t^n (n>0, integer), e^{at}, a^{kt} (a>0), δ-function; any of them multiplied by a number; sums, products, derivatives and integrals of these functions.

In RLS <expr> may be a polynomial, a rational function or any of them multiplied by $e^{-a\langle varim\rangle}$ (a>0).

An improved version of this implementation using the capabilities of REDUCE 3 is under development.

Some applications of REDUCE were made for solving problems in mechanics. For example, REDUCE extended by Laplace transformation and inverse Laplace transformation was applied for solving the finite elements equations for dynamic loading [1].

2. Building, maintenance and usage of symbolic transformations bases in REDUCE programs.

This extension of REDUCE (implemented in symbolic mode as well) allows more mathematical knowledge to be used in REDUCE programs.

Language tools for building, maintenance and usage of bases of symbolic transformations are implemented ([2], [3]). Such bases can consist of sets of formulae, represented as substitution rules. A set of rules can be divided into several groups and for different applications different groups can be used. REDUCE procedures can be used to represent procedural knowledge. A base can include only substitution rules, only procedures, or substitution rules plus procedures.

Several commands with simple syntax allow building of a base of the described type, using it (the whole content or a part of it only), suppressing the use of a base, adding or deletion of rules and procedures, backtracking and receiving information about the content of a base.

An important feature of the described tools is that the planning and controlling of the manipulations are done by the user.

3. Some applications of REDUCE for solving problems in mathematical geodesy (in algebraic mode) including manipulations on power series of one and two variables, solution of spherical triangles using the described tools for dealing with bases of symbolic transformations, etc. are under development.

The works reported here are performed under the general management of Dr. Valentin Tomov.

REFERENCES

1. Brankov G., M. Spiridonova, K. Ishtev, Ph. Philipov: Modelling of Building Structures Behaviour under Seismic Loading with Use an Algebraic Manipulation System. Proc. of the 7-th Europ. Conf. on Earthquake Engin., Athens, 1982.

2. Spiridonova M.: Organization and Application of Bases of Symbolic Transformations Using REDUCE System. Intern. Conf. on Computer Algebra and its Applications in Theoretical Physics, Dubna, 1985 (in Russian).

3. Spiridonova M., M. Djambazova: A Knowledge Based Extension of REDUCE 2 System. Proc. of the Intern. Conf. on Artificial Intelligence - AIMSA'84, Varna, 1984.

4. Spiridonova M., Ch. Kazassov: Computer - Aided Symbolic Laplace Transformation. ASU, Sofia, 1983, No 3 (in Bulgarian).

INFINITE STRUCTURES IN SCRATCHPAD II

William H. Burge
Stephen M. Watt

IBM Thomas J. Watson Research Center
Box 218, Yorktown Heights, NY 10598 USA

Abstract

An *infinite structure* is a data structure which cannot be fully constructed in any fixed amount of space. Several varieties of infinite structures are currently supported in Scratchpad II: infinite sequences, radix expansions, power series and continued fractions. Two basic methods are employed to represent infinite structures: self-referential data structures and lazy evaluation. These may be employed either separately or in conjunction.

This paper presents recently developed facilities in Scratchpad II for manipulating infinite structures. General techniques for manipulating infinite structures are covered, as well as the higher level manipulations on the various types of mathematical objects represented by infinite structures.

1. Introduction

An *infinite structure* is a data structure which cannot be fully constructed in any bounded amount of space. Examples are a list of the digits of π or an explicit representation of $n^3: Z \to Z$ as a set of ordered pairs. On the other hand a *finite structure* is the opposite: it is a data structure which can be be fully constructed in finite storage.

It is common practice to use different data structures to represent the same value, according to the operations which are expected to be performed. For example, a collection of strings may be stored in any one of a linked list, a binary search tree or a hash table. Normally, these data structures will have differing storage requirements.

A value which has a representation as an infinite structure might also be represented by other infinite or finite structures. For example, the Fibonacci numbers can be given either as an infinite sequence of integers or as a finite recurrence with initial values. Likewise, $\sin(x)$ can be represented by infinite structures (a power series, a continued fraction, an infinite product) or by finite structures (a differential equation with initial value, an expression tree).

Scratchpad II has a number of facilities for creating and manipulating infinite structures. Two principal techniques are employed: (i) folding recursive data structures to be self-referential, and (ii) lazy evaluation.

This paper describes how these techniques are used to create and manipulate infinite structures in Scratchpad II.

We begin in section 2 by summarizing the general techniques for manipulating infinite structures. In section 3 the use of these techniques in creating *streams*, a low-level data type for representing infinite sequences is shown. In sections 4, 5 and 6 we give a number of examples of mathematical objects in terms of infinite structures. Section 4 discusses radix expansion of numbers. Section 5 describes the implementation of power series in terms of streams. Section 6 shows infinite continued fractions in normal form. Finally, in section 7, future developments are discussed.

2. Techniques for Infinite Structures

2.1. Self-Referential Data Structures

Structures for composite data are often defined recursively. Two common examples are linked lists and binary trees. The tail of a linked list has the same type as the whole list, and both branches of a binary tree have the same type as the whole tree.

Programs which traverse recursively defined data structures often have a logical structure which mi rs the data structure. Such programs have only a al view of the data. On recursive calls, they conler their argument as being the whole data structre and positional information relating to the iginal structure is lost.

it is possible to update parts of a data structure g. record field assignment), then a self-referential cled object may be created. This can lead to bugs programs which are written to have only a local w of the structure and which are expected to mpletely traverse a finite structure.

the other hand, self-referential data structures are al for representing infinite structures which have me form of periodicity. Programs written to trave a logically infinite structure can use an equivnt self-referential finite structure so long as they not modify it as they proceed. Such programs are itten to traverse only a finite portion of the loglly infinite structure or are intentionally nonminating.

an example, consider power series represented as infinite list of coefficients, with the ith element be- the coefficient to the term of degree i. Then the ies for $1/(x+1)$ is a cycle of two elements: , − 1] and the series for a polynomial has a one ment cycle of zeros at the end.

2. Lazy Evaluation

is common practice to use procedural abstraction programming languages to represent mappings fiely. This allows a possibly infinite set of ordered irs of the mapping to be represented by an equivnt structure: the program. Then when points from mapping are needed they are computed on demnd by invoking the program on elements of the apping's domain.

is is the fundamental idea of lazy evaluation: nstruct a value when it is to be used rather than en it is defined. For extended or aggregate data uctures, construct the parts when they are reired.

construct an object as it is used, each unconucted portion is represented as a program with me state information. The program part may be explicit (e.g. a pointer to a function) or implicit (a fixed interpreter) and the state information, if any, may be stored in global variables, in the data structure, or in a closed environment of the program part.

2.3. Fixed Point Operations

A fixed point x of a map f is a point in the domain of f such that $x = f(x)$. Given a function which operates on a recursively defined data type, it is often possible to compute a useful fixed point. A powerful method for manipulating infinite structures is to compute the fixed point of structure transforming functions. As well as providing a functional mechanism for constructing self-referential structures, a combination of lazy evaluation and self-reference may be achieved.

Consider a recursively defined data type T and the class of functions mapping $T \rightarrow T$. Certain functions in this class have trivial fixed points: the identity and constant valued functions. Some functions in the class may have no fixed point. The fact that negation has no fixed point leads to the Russell paradox. Other functions may have a fixed point which it is impossible to compute effectively.

Let us restrict our attention to functions which do not perform operations on their argument but rather just include it in a new structure which is returned as the value. Then we may always compute a fixed point as follows:

```
fixedPoint(f) ==
    arg := generateUnique()
    ret := f(arg)
    if arg = ret then
        -- f is the identity
        return arb. element from the domain of f
    else
        ret := subs(arg = ret in ret)
        return ret
```

Here `generateUnique` is a function which returns a unique system-wide value. Since f does not perform any operations on its argument, it is safe to pass it this generated unique value, which strictly speaking does not lie in its domain.

From the definition of `fixedPoint` above we see that, for functions in our restricted class, the set of fixed points will be one of

- a single constant (for functions which ignore their argument),
- the entire domain (for the identity function), or
- a single infinite structure

As an example, an infinite repeating list of values can be obtained as follows:

```
cons1234(l) == cons(1,cons(2,cons(3,cons(4,l))))
repeating1234 := fixedPoint cons1234
```

To combine lazy evaluation and self-reference, one builds a lazily evaluated object in which the state information for the unconstructed part contains references back to the whole object. It is convenient to use a fixed point calculation to build such objects. Examples of this are given in sections 3.4.

3. Streams

3.1. Language Support

Streams have been in Scratchpad II for some time. They have recently been reorganized into a domain[1] similar to lists, but with the difference that they might be infinite.

The language provides special syntax for creating and iterating over lists and streams. This syntax is merely a convenience – it is ultimately translated into calls to operations on a list or stream type, as appropriate. Some of the stream functions are `cons`, `null`, `first`, `rest` take, drop and `elt` (similar to the list functions), together with functions for creating finite and infinite streams.

We begin this section on streams by illustrating with some examples of the language support.

```
a := [1..]

   (1)  [1,2,3,4,5,6,7,8,9,10,...]
```

By default, 10 elements are computed for display.

```
b := [i+1 for i in a]

   (2)  [2,3,4,5,6,7,8,9,10,11,...]
```

Select the 20th element (0-based indexing):

```
b.20

   (3)  22
```

At this point, the first 21 elements of `b` have been evaluated (as well as the first 21 elements of `a`).

```
b

   (4)  [2,3,4,5,6,7,8,9,10,11,12,13,14,15,
         16,17,18,19,20,21,22,...]
```

The stream of odd integers can be created using filter.

```
[i for i in a | oddp i]

   (5)  [1,3,5,7,9,11,13,15,17,19,...]
```

It is possible to combine multiple streams using parallel for iterators.

```
[[i,j] for i in a for j in b]

   (6)  [[1,2], [2,3], [3,4], [4,5], [5,6], [6,7],
         [7,8], [8,9], [9,10], [10,11], ...]
```

`append(x,y)` is the concatenation of streams `x` and `y`.

```
append([i for i in a while i<7],a)

   (7)  [1,2,3,4,5,6,1,2,3,4,5,6,7,8,9,10,11,12,13,
         14,15,16,17,18,19,20,21,...]
```

The sum of a finite stream of integers:

```
reduce(0, _+$I, take(a,10))

   (8)  55
```

A stream of partial sums:

```
scan(0, _+$I, a)

   (9)  [1,3,6,10,15,21,28,36,45,55,...]
```

1 Here the word "domain" is Scratchpad II terminology for "abstract data type".

2. *Lazy Evaluation in Streams*

stream is represented by a list whose last element a function that contains the wherewithal to create e rest of the list from that point, should it ever be quired. The function takes no arguments and ıen invoked returns a stream of the appropriate pe. The stream that it returns is used as the re- ainder of the original stream, and the value it re- rns becomes the first member of the stream. The nction is paired with an environment which con- ins whatever state information is required to ex- nd the stream.

nce functions that extract elements from a stream ust test whether it is null before proceeding, the cessity of extending the stream can be determined the function which tests for a null stream. It is e function `null` which does this. If `null` determines at more of the stream need be computed, then the rticular stream-extending function is invoked. The eam is updated in place to obviate re-evaluation.

ratchpad II is an abstract data type language so e representation of streams is invisible outside of e code which defines the data type. To allow pro- ams to create infinite streams using lazy evaluation e stream data type exports the primitive operation `lay`.

`lay` takes a nullary stream-valued function and re- rns a stream of the same type. I.e.

$$delay\colon (() \rightarrow Stream\ T) \rightarrow Stream\ T$$

nis, by itself, is not very exciting. All that has ppened is that the stream valued function has been ved away somewhere, to be evaluated when ele- ents are required from the stream. The power of `lay` comes from its use in the recursive definition functions.

 illustrate, consider the definition of the `scan` nction. This function takes as parameters, an ini- l value b, a binary function h, and a stream x. The lue returned is the stream

$$(b, x_0), h(h(b, x_0), x_1), h(h(h(b, x_0), x_1), x_2), \ \ldots\ ]$$

an may be defined in terms of `delay` as follows:

```
scan(b,h,x) ==
    null x => nil()
    delay
       c := h(b, frst x)
       cons(c,scan(c,h,rst x))
```

The last two lines form the body of a nullary func- tion to which `delay` is applied.

`delay` also forms the basis of a style of programming for the creation of infinite streams. The style is to use intentionally recursive functions with no base case. For example, the following function returns the stream of Fibonnaci numbers when invoked on (1,1).

```
fib(a0, a1) ==
    cons(a0, delay fib(a1, a0 + a1))
```

3.3. Self-Referential Streams

A second way to create infinite streams is through the use of self-reference. The simplest way to do this is with the function `repeating`. This function takes a list of elements and produces a stream which re- peats them indefinitely.

```
repeating [1, 2, 3]
```

(1) $[\overline{1,2,3}]$

Although this could be implemented using lazy eval- uation, it is more efficient to represent this stream as a list with the third tail pointing back to the be- ginning.

Other self-referential streams can be created using fixed point operations. The simplest are again re- peating streams. A subtler form of self-reference can be achieved by computing a fixed point in which the state information paired with the function contains a pointer back into the stream itself. Examples of both sorts are given in the following section.

3.4. Fixed Points of Stream Transforming Functions

A fixed point finding operation is provided which operates on a stream transforming function and finds its fixed point, a stream.

```
a:=integers 1
```

(2) [1,2,3,4,5,6,7,8,9,10,...]

The function below prefixes a 1 to an integer stream.

```
f1(x: ST I): ST I == cons(1,x)
f1 a
```

(4) [1,1,2,3,4,5,6,7,8,9,10,...]

and the fixed point of f is an infinite stream of 1's

```
b := fixedPoint f
```

(5) $[\overline{1}]$

Similarly

```
f2(x: ST I): ST I == append([1,2,3,4,5,6], x)
fixedPoint f2
```

(8) $[1,\overline{2,3,4,5,6,1}]$

Here is another way to define the Fibonacci number stream. The plus operation takes two streams and adds them pair-wise.

```
f3(fib: ST I): ST I == cons(1,fib+cons(0,fib))
f3 b
```

(10) [1,1,2,2,2,2,2,2,2,2,...]

```
fixedPoint f3
```

(11) [1,1,2,3,5,8,13,21,34,55,...]

The stream of Catalan numbers:

```
f4(cat: ST I): ST I == cons(1,cat*cat)
fixedPoint f4
```

(14) [1,1,2,5,14,42,132,429,1430,4862,...]

The function `integ` integrates a stream viewed as the coefficients of a power series.

```
integ b
```

(15) $[1,\frac{1}{2},\frac{1}{3},\frac{1}{4},\frac{1}{5},\frac{1}{6},\frac{1}{7},\frac{1}{8},\frac{1}{9},\frac{1}{10},...]$

Here we compute the fixed point of the function that integrates a stream, and adds the constant tern 1.

```
g(e: ST RN -> ST RN) == cons(1,integ e)
fixedPoint g
```

(18) $[1,1,\frac{1}{2},\frac{1}{6},\frac{1}{24},\frac{1}{120},\frac{1}{720},\frac{1}{5040},\frac{1}{40320},\frac{1}{362880},...]$

It is also possible to find the fixed point of a functio that transforms a pair of streams to a pair o streams.

```
k(tr: L ST I): L ST I == [cons(0,tr.1),1/(1-tr.0)]
k([cons(0,b),b])
```

(20) $[[0,\overline{1}],[1,1,2,4,8,16,32,64,128,256,...]]$

The fixed point of k is two mutually recursiv streams. Computing this provides another way t obtain the stream of Catalan numbers.

```
fixedPoint(k, 2)
```

(21)
[[0,1,1,2,5,14,42,132,429,1430,...],
[1,1,2,5,14,42,132,429,1430,4862,...]]

4. Radix Expansions

In Scratchpad II it is possible to evaluate certai numeric types to decimal expansions or radix expar sions in other bases. The simplest of these is the ex pansion of rational numbers. Here we give som examples.

First we define a couple of functions for coercion.

```
decimal r    == r::DecimalExpansion
radix(n, r) == r::RadixExpansion(n)
```

All rational values have repeating decimal expar sions

```
decimal(22/7)
```

(49) $3.\overline{142857}$

The arithmetic of decimal expansions is exact.

```
% + decimal(6/7)
```

(50) 4

[Th]e periods can be short or long:

```
[decimal(1/i) for i in 350..353]
```

(51)

$[0.00\overline{285714},\ 0.\overline{002849},\ 0.00284\overline{09},$

$0.\overline{00283286118980169971671388101983}]$

```
decimal(1/2049)
```

```
(52)
0.
  OVERBAR
     000488042947779404587603709126403123474865788189
        3606637384089799902391410444119082479258174719
        3753050268423621278672523182040019521717911176
        1835041483650561249389946315275744265495363591
        9960956564177647632991703269887750122010736944
        8511469009272816007808687164470473401659346022
        4499755978526110297706198145436798438262567105
        90531966813079551
```

[Sc]ratchpad II can do radix expansions in other [ba]ses.

```
[radix(i, 5/24) for i in 2..10]
```

(53)

$[0.001\overline{10}\text{:RADIX } 2,\ 0.0\overline{12}\text{:RADIX } 3,\ 0.03\overline{1}\text{:RADIX } 4,$

$0.\overline{10}\text{:RADIX } 5,\ 0.113\text{:RADIX } 6,\ 0.\overline{13}\text{:RADIX } 7,\ 0.1\overline{52}\text{:RADIX } 8,$

$0.1\overline{7}\text{:RADIX } 9,\ 0.208\overline{3}\text{:RADIX } 10]$

[Fo]r bases greater than 10, the ragits (radix digits) are [se]parated by blanks.

```
[radix(i, 5/24) for i in 11..15]
```

(54)

$[0\ .\ \overline{2\ 3}\text{:RADIX } 11,\ 0\ .\ 2\ 6\text{:RADIX } 12,\ 0\ .\ \overline{2\ 9}\text{:RADIX } 13,$

$0\ .\ 2\ 12\ 11\ \overline{9\ 4}\text{:RADIX } 14,\ 0\ .\ 3\ \overline{1\ 13}\text{:RADIX } 15]$

[Th]ese numbers are bona fide algebraic objects.

```
p := decimal(1/4)*x**2 + decimal(2/3)*x + decimal(4/9)
```

(55) $0.25x^2 + 0.\overline{6}x + 0.\overline{4}$

```
q := pderiv(p, x)
```

(56) $0.5x + 0.\overline{6}$

```
g := gcd(p, q)
```

(57) $x + 1.\overline{3}$

The function `fracRagits` gives the stream of ragits in the fractional part of its argument.

```
rr: RADIX(8) := 3/49
```

(58) $0.\overline{0372615}$

```
fracRagits rr
```

(59) $[0,\overline{3,7,2,6,1,5,0}]$

```
%.30
```

(60) 7

5. Power Series

5.1. Construction via Defining Relation

The functions in the Stream domain and stream packages are particularly suitable for the implementation of algorithms on power series. The domain PowerSeries is provided as a field, and the domain UnivariatePowerSeries and an elementary function package adds to it the functions exp, log, sin, cos, tan, the hypergeometric function, composition, lagrange inversion, reversion together with the solution of linear differential equations in power series.

A general method of producing programs which solve recursion or differential equations in power series by the method of undetermined coefficients has been developed in which the program can be written down almost immediately from the defining relation. In the method of undetermined coefficients a trial series together with an initial value or two is substituted into the recursion or differential equation, and then coefficients of equal powers are equated.

In these programs the trial series is made up of the initial values followed by the as yet unevaluated stream. The tail of the stream is then defined in terms of the whole stream and when elements are required the trial series becomes the resulting stream. The program, because it uses functions that operate on whole streams, rather than stream elements has the same structure as the defining relation.

For example e raised to the power series power $A(x)$, has defining relation

$$(e^{A(x)})' \equiv A'(x)e^{A(x)}$$

The corresponding program for generating the power series exp A, in Scratchpad II, where A is a power series is

```
exp A == integrate(1,pderiv A*exp A))
```

in which `integrate` and `deriv`, respectively, integrate and differentiate power series.

5.2. Examples

The command

```
)set streams calculate n
```

will cause the series to be displayed up to n^{th} order. If s is a variable assigned a series as its value, then one way to view it to higher order is to re-issue the ")set" command with a higher value of n and then re-display the value of s. The following declares x to be a UPS(x,RN), in other words a UnivariatePowerSeries with variable x and with rational number coefficients.

```
x := ps x

  (12)  x


exp x

  (13)
             1  2    1  3    1   4    1   5
    1 + x + (-)x  + (-)x  + (--)x  + (---)x
             2       6       24       120
  +
      1   6     1    7     1     8      1      9
    (---)x  + (----)x  + (-----)x  + (------)x
     720       5040       40320      362880
  +
        1     10      11
    (-------)x   + 0(x  )
     3628800


cos x ** cos x

  (14)
         1  2    7   4     19  6     1597  8
    1 - (-)x  + (--)x  - (---)x  + (-----)x
         2       24       180       40320
  +
      373   10      11
    - (-----)x   + 0(x  )
      32400


x/(exp x-1)

  (15)
         1     1   2     1    4      1     6
    1 - (-)x + (--)x  - (---)x  + (-----)x
         2     12       720       30240
  +
         1      8       1       10      11
    - (-------)x  + (--------)x   + 0(x  )
       1209600       47900160
```

The hypergeometric function:

```
hyp(1/2,1,3/2,-x**2)

  (18)
         1  2    1  4    1  6    1  8    1  10
    1 - (-)x  + (-)x  - (-)x  + (-)x  - (--)x
         3       5       7       9       11
  +
        11
    0(x   )
```

Power series provide a method of solving differential equations when all else fails. The function `lde` solves the nth order linear differential equation, its argument is a list of power series coefficients. The two solutions of

$$y'' + (\cos x)y' + (\sin x)y = 0$$

are

```
lde([sin x,cos x])
  (19)
  [
          1  2    1  4     31  6     379  8
     1 - (-)x  + (-)x  - (---)x  + (-----)x
          2       6       720       40320
   +
        1639   10      11
     - (------)x   + 0(x  )
        907200
   ,

          1  3    1   5     59   7     31   9      11
     x - (-)x  + (--)x  - (----)x  + (----)x  + 0(x  )
          3       10       2520       6480
  ]
```

Power series are also used as enumerating generating functions and the power series may be expanded from its generating function. For example the generating function for the Legendre polynomials is

$$\frac{1}{(1-2xt+t^2)^{1/2}}$$

ith suitable declarations for x and t, it may be ex-
nded directly as follows:

```
(1-2*x*t+t**2)**(-1/2)

   (24)
                        2        2
      1 + x*t + ((3/2)x  - 1/2)t
    +
              3            3
      ((5/2)x  - (3/2)x)t
    +
               4         2        4
      ((35/8)x  - (15/4)x  + 3/8)t
    +
               5         3                5
      ((63/8)x  - (35/4)x  + (15/8)x)t
    +
                 6           4             2        6
      ((231/16)x  - (315/16)x  + (105/16)x  - 5/16)t
    +
                      7            5             3
          (429/16)x  - (693/16)x  + (315/16)x
        +
          - (35/16)x
     *
          7
         t
    +
           8
       O(t )
```

is also possible to expand certain infinite products power series. The function lambert will transform e series into another in which the coefficient A_n of is the sum of the coefficients of the original a_n for i that divide n, including 1 and n. In other words, $f(x)$ is a power series, then $lambert(f(x))$ is the wer series

$$f(x) + f(x^2) + f(x^3) + f(x^4) + \ldots$$

e series for the number of divisors of n is

```
lambert(x/(1-x))

   (20)
             2      3      4      5      6      7      8
      x + 2x  + 2x  + 3x  + 2x  + 4x  + 2x  + 4x
    +
         9      10       11
      3x  + 4x   + O(x  )
```

ing this function it is possible to expand certain inite products as power series. For example the umerating generating function for partitions is

$$\prod_{n=1}^{\infty} \frac{1}{(1-q^n)}$$

```
partitions := exp(lambert(log(1/(1-x))))

   (21)
                  2     3     4     5      6      7
      1 + x + 2x  + 3x  + 5x  + 7x  + 11x  + 15x
    +
         8      9      10       11
      22x  + 30x  + 42x   + O(x  )
```

Euler's theorem is then:

```
1/partitions
                      2    5    7        11
   (22)  1 - x - x  + x  + x  + O(x  )
```

The function h, defined below expands the infinite product

$$\prod_{n=1}^{\infty} g(x^n)$$

where g is a power series with constant term 1.

```
h g == exp lambert log g
```

The coefficient of $y^i z^n$ below is the number of partitions of n into i parts

```
h(1/(1-y*z))
   (7)
                     2        2     3    2        3
      1 + y*z + (y  + y)z  + (y  + y  + y)z
    +
        4    3     2      4
      (y  + y  + 2y  + y)z
    +
        5    4     3     2      5
      (y  + y  + 2y  + 2y  + y)z
    +
        6    5     4     3     2      6
      (y  + y  + 2y  + 3y  + 3y  + y)z
    +
        7    6     5     4     3     2      7
      (y  + y  + 2y  + 3y  + 4y  + 3y  + y)z
    +
        8    7     6     5     4     3     2      8
      (y  + y  + 2y  + 3y  + 5y  + 5y  + 4y  + y)z
    +
         9
      O(z  )
```

Jacobi's celebrated result:

```
h(1-x)**3

                      3     6      10       11
   (11)  1 - 3x + 5x  - 7x  + 9x   + O(x  )
```

The Hermite polynomials:

```
h(1/(1-a*f))*h(1/(1-f/a))

(15)
          2           4    3    2
        a  + 1       a  + a  + a  + a + 1  2
  1 + (------)g + (--------------------)g
          a                  2
                             a
 +
      6    5     4     3     2
     a  + a  + 2a  + 2a  + 2a  + a + 1  3
    (----------------------------------)g
                    3
                   a
 +
      8    7     6     5     4     3     2
     a  + a  + 3a  + 3a  + 4a  + 3a  + 3a  + a + 1  4
    (----------------------------------------------)g
                          4
                         a
 +
      5
    O(g )
```

The examples above illustrate the present capability of writing expressions that denote power series.

6. Continued Fractions

We use the following notations for continued fractions:

$$\mathop{\Phi}_{i=1}^{n} \frac{a_i}{b_i} = \cfrac{a_1}{b_1 + \cfrac{a_2}{b_2 + \dots \cfrac{a_n}{b_n}}}$$

$$= \frac{a_1|}{|b_1} + \frac{a_2|}{|b_2} + \dots + \frac{a_n|}{|b_n}$$

The notation $\mathop{\Phi}_{i=1}^{\infty} a_i/b_i$ may be used to represent the limit of an infinite sequence of convergents.

The function `continuedFraction` provides one method of forming a continued fraction. It takes as arguments the whole part, the partial numerators and the partial denominators.

The continued fraction $\mathop{\Phi}_{i=1}^{\infty} \frac{i}{i}$ which has the value $1/(e-1)$ is entered as

```
s := continuedFraction(0, [1..], [1..])

(46)

 1 |      2 |      3 |      4 |      5 |      6 |
,---' + ,---' + ,---' + ,---' + ,---' + ,---' + ...
 | 1      | 2      | 3      | 4      | 5      | 6
```

If all the numerators are one, the reducedContinuedFraction may be used. Euler discovered the relation $\frac{e-1}{e+1} = \mathop{\Phi}_{i=1}^{\infty} \frac{1}{4i-2}$

```
t := reducedContinuedFraction(0, [4*i-2 for i in 1..])

(47)

 1 |     1 |     1 |      1 |      1 |      1 |
,---' + ,---' + ,----' + ,----' + ,----' + ,----' + ...
 | 2     | 6     | 10     | 14     | 18     | 22
```

Arithmetic on infinite continued fractions is supported. The results are given in reduced form. We illustrate by using the values $s = 1/(e-1)$ and $t = (e-1)/(e+1)$ to recover the expansion for e.

```
e := 1/(s*t) - 1

(48)

      1 |     1 |     1 |     1 |     1 |     1 |
2 + ,---' + ,---' + ,---' + ,---' + ,---' + ,---' + ...
     | 1     | 2     | 1     | 1     | 4     | 1
```

The following command evaluates the 15th convergent to a floating point number.

```
convergents(e).15::F

(49) 2.71828182847
```

Many univariate power series may be transformed to continued fractions. Here we show the conversion of series to continued fractions in normal form, that is a continued fraction in which the partial denominators are one and the partial numerators after the first are monomials of degree 1. The quotient difference algorithm takes the coefficients of the series and produces the partial numerators of the continued fraction.

The first example is the series for $\exp x$:

```
expx := exp ps x

(50)

               1  2     1  3     1   4      1   5       6
  1 + x + (-)x  + (-)x  + (--)x  + (---)x  + O(x )
           2        6       24      120
```

The domain for these continued fractions is abbreviated CFPS.

```
expx::CFPS(x,RN)

                                   1
 (51)  ------------------------------------------
                                   1x
        1 - --------------------------------------
                                     1
                                    (-)x
                                     2
            1 + ----------------------------------
                                       1
                                      (-)x
                                       6
                1 - ------------------------------
                                         1
                                        (-)x
                                         6
                    1 + --------------------------
                                           1
                                          (--)x
                                          10
                        1 - ----------------------
                                             1
                                            (--)x
                                            10
                            1 + ------------------
                                               1
                                              (--)x
                                              14
                                1 - --------------
                                                 1
                                                (--)x
                                                14
                                    1 + ----------
                                                   1
                                                  (--)x
                                                  18
                                        1 - -------
                                            1 - etc
```

nother example is:

```
qq := q::QF UP(q, RN);

[qq**(i**2) for i in 0..]

                4  9  16  25  36  49  64  81
   (53)  [1,q,q ,q ,q  ,q  ,q  ,q  ,q  ,q  ,...]

%::UPS(x,QF UP(q,RN))
   (28)
                 4 2    9 3    16 4    25 5    36 6
     1 + q*x + q x  + q x  + q  x  + q  x  + q  x
   +
      49 7    64 8    81 9    100 10      11
     q  x  + q  x  + q  x  + q   x   + O(x  )
```

```
%::CFPS(x,QF UP(q,RN))

                                    1
 (55)  ------------------------------------------
                                   q*x
        1 - --------------------------------------
                                     3
                                   (q  - q)x
            1 - ----------------------------------
                                       5
                                      q x
                1 - ------------------------------
                                        7    3
                                      (q  - q )x
                    1 - --------------------------
                                           9
                                          q x
                        1 - ----------------------
                                          11    5
                                        (q   - q )x
                            1 - ------------------
                                             13
                                            q  x
                                1 - --------------
                                             15    7
                                           (q   - q )x
                                    1 - ----------
                                                 17
                                                q  x
                                        1 - -------
                                            1 - etc
```

7. Concluding Remarks

We have viewed lazy evaluation and self reference of data as particular techniques for infinite structures and we have shown how these techniques are particularly powerful when used together.

Scratchpad II provides the basic requirements for manipulating infinite structures: the ability to include programs as parts of composite data objects and the ability to create and modify self-referential data objects. These basic facilities have been used to build a variety of abstract data types which provide logically infinite structures.

The additions so far have been to build a number of domains so that infinite sequences (streams), power series, decimal expansions and continued fractions may be treated.

It should be possible in the future to enter differential or recursion equations that define new power series in terms of existing ones as suggested in the example for exp in section 5.1.

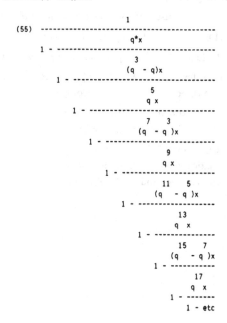

Bibliography

1. H. Rutishauser [1954], *Der Quotienten-Differenzen-Algorithmus*, Z. Angew. Math. Physik 5 233-251.
2. H.B. Curry and R. Feys [1958], *Combinatory Logic*, North Holland, Amsterdam.
3. P. Henrici [1977], *Applied and Computational Complex Analysis, Volume 2*, John Wiley & Sons.
4. R.D. Jenks and B.M. Trager [1981], *A Language for Computational Algebra*, Proc. 1981 ACM Symposium on Symbolic and Albebraic Computation.
5. H. Abelson and G. Sussman (with J. Sussman) [1985], *Structure and Interpretation of Computer Programs*, The MIT Press, Cambridge Mass.
6. R.D. Jenks, R.S. Sutor and S.M. Watt [1986], "Scratchpad II: An Abstract Datatype System for Mathematical Computation" in *Mathematical Aspects of Scientific Software*, J. R. Rice ed., IMA Volumes in Mathematics and Its Applications, Volume 14, Springer-Verlag, New York, 1988.

Application of a structured LISP system to Computer Algebra.

J. Smit S.H. Gerez
R. Mulder
University of Twente, EF9274
PoBox 217, 7500AE Enschede, the Netherlands.

Abstract

A new LISP token-reader, interfaced with an hierarchical name-space organisation has been implemented and used as a structuring tool for large LISP applications which demand high speed execution. The potential of this approach for the improved implementation of REDUCE 3 will be discussed.

1 Introduction

The REDUCE algebra system [1] is, due to its portability, one of the most widely used systems in its sort. It can however be seen by inspection of its code that the techniques used to achieve this portability date back to the early 1970's. Especially the parser contains numerous goto statements, which presumably were introduced for reasons of efficiency. This aspect, in combination with the flat implementation of the algebraic evaluator, in the end, makes REDUCE relatively heavy to maintain. Easy maintenance of symbolic mode code, is however of crucial importance for the REDUCE community as few real applications have been written in algebraic mode.

The recent development in our VLSI design group at the University of Twente of an hierarchically structured LISP system, known nowadays as TULISP from Enschede, has shown that such a structured system can be implemented with little memory overhead and with even improved execution speed, as compared to more traditional implementations.

To achieve portability, we coded the TULISP kernel, on top of which we built a VLSI design language, MoDL [2,3] in C. The accompanying LISP compiler generates highly efficient (portable) C-code. The implementation does not rely on the availability of some free bits in the high part of the encoded LISP-pointers to represent a tagged pointer, or a (30 bit) fixnum, nor does it use an extra LISP cell to represent the tag of a built-in datatype.

2 Background

It was decided to support our Modelling and Design Language, MoDL, which had to cope with numerous internal wires (in the order of 2700 in one specific case) in a given VLSI design, nested in numerous hardware model instances (3300 in that same case), with a more powerful object list structure and reader interface than available in any other LISP system available. This affected several aspects of our VLSI design system:

- LISP variables representing specific signal values can be grouped together according to some criterium.
- Global variables used in the MoDL software system are easily separated from user-variables.
- The highly structured software, which is slightly faster than the code usually developed in LISP, is easier to develop and maintain in the proposed environment.

The approach taken was certainly different from the module concept, used in many LISP systems available, among which COMMON LISP [4] ,which implements the idea of a name-space with import and export privileges, similar to the one supported nowadays in MODULA [5], which assumes that the partitioning of the name-spaces corresponds to the partitioning of the files from which the system is built, whereas the nesting of the name-spaces in TULISP is assumed to correspond to the taxanomy of the problem at hand.

On the other end of the spectrum there is the object-oriented approach used in smalltalk [6] and related systems, which build an object hierarchy which is apparently not reflected in the hierarchy of the name-space used. I.e. the smalltalk classes are not interfaced with the smalltalk syntax, reader and printer.

2.1 A space and time efficient name-space organisation

When implementing the structured name space, we were faced with the problem that little could be said in advance about the statistics of the usage of the nested spaces, this was especially true for VLSI design problems, where we might have 3300 directories in core with 2700 names spread over them, resulting in fewer than one entry on the average per actual oblist, or on the contrary be confronted with one flat object list in which all 2700 names should be allocated. The organisation which proved to be sufficiently flexible and space- and time-efficient was the storage of the directories and the object-lists in the form of balanced- or ALV-trees [7], which need $O(n)$ storage locations and $O(\log n)$ time for the search, insert and delete operations. Only one extra pointer was needed as compared to the more traditional linked list representation with hashed access used to achieve speed. The organisation of all directories and object-lists through balanced trees avoids problems of memory overflow and/or hash table reorganisation.

2.2 Fast runtine organisation

The organisation of the name-space does only influence the speed at which expressions can be read, as the (TU)LISP reader returns atom pointers to names in much the same way as a traditional LISP reader. The search strategy used by the reader resolves inheritance of names and functions at read-time. Functions used for search of names, c.q. application of functions, at run-time exhibit a logarithmic time complexity, due to the nature of the balanced trees used, whilst the more traditional search strategies implemented on property lists are frequently of linear time complexity and on the order $O(n/2)$ in the average.

3 Object oriented approaches in algebra systems

An object oriented approach to algebra system design can be found in the Views system [8], which implements a powerful computer algebra system in the Smalltalk environment. Their paper shows the methods and methodology which can be used for structuring of algebraic domains, but it does not give any details about the performance of the system. We are very eager to get some favorable performance figures from such a computer algebra system, but realize that this may be hard to obtain. In any case if they lag behind the performance of well developed LISP implementations, we would be interested in timing results which factor out the fact that smalltalk normally runs its compiled code from an interpretative, byte-code machine, from the overhead of the object oriented programming paradigm.

Several others use similar methods. For instance Hearn et. al. describe a method to enlarge the domain of computation in REDUCE 3 [9], through the introduction of algebraic domains, represented by a tag on the REDUCE datastructures, specific for a set of domain functions placed on the property list of the tag identifier. It should be noted however that their method of implementation is, at least in the way in which it is described and not so much in the way in which it works, fundamentally different.

We will review here the syntax offered by REDUCE 3 to specify such algebraic domains, and show an alternative based on an hierarchically nested name-space. To this end we will follow the example given in [9] of the introduction of fixed slash arithmetic, as originally specified by Matula [10].

4 Establishing an algebraic domain in REDUCE 3

The fixed slash number system, defines operations on rational numbers with a fixed precision. The procedure fixslash (m, n), defined in [9] returns the best rational approximation (as a dotted pair of integers) to n/m, such that min (n, m) < fixsize.

To include this algorithm in REDUCE 3.2 one has to define a set of primitive functions which define the operations +, *, -, / as well as some predicates and coercion functions, to let the system find and manipulate a canonical representation. The mere inclusion of such an interrelated set of functions is sufficient to include fixedslash numbers in REDUCE. The names and structures used to include fixslash arithmetic in REDUCE are taken from [9].

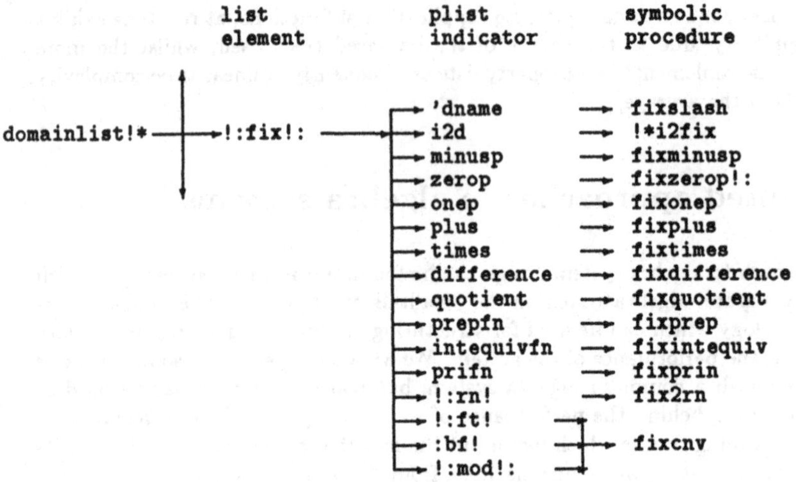

Figure (1)

The statements used to build this structure are of the form:

```
domainlist!* := union ('(!:fix!:), domainlist!*);

put ('!:fix!:, 'i2d, '!*i2fix);
```

...16 more statements of this type follow ...

Finally 16 symbolic procedures of the following kind are introduced:

```
symbolic procedure !*i2fix u;
% integer -> fixed slash. 'mkfix1' constructs a
% domain element from its argument.
mkfix1 (u . 1);
```

...15 more symbolic procedures of this kind follow ...

The basic structure used in the given implementation is:

list-element → plist-indicator → symbolic procedure

We will now study the mapping of the same structure onto a nested name space.

5 Implementing an algebraic domain with the hierachical name-space

The hierarchy on which basically the same functions will be mapped is:

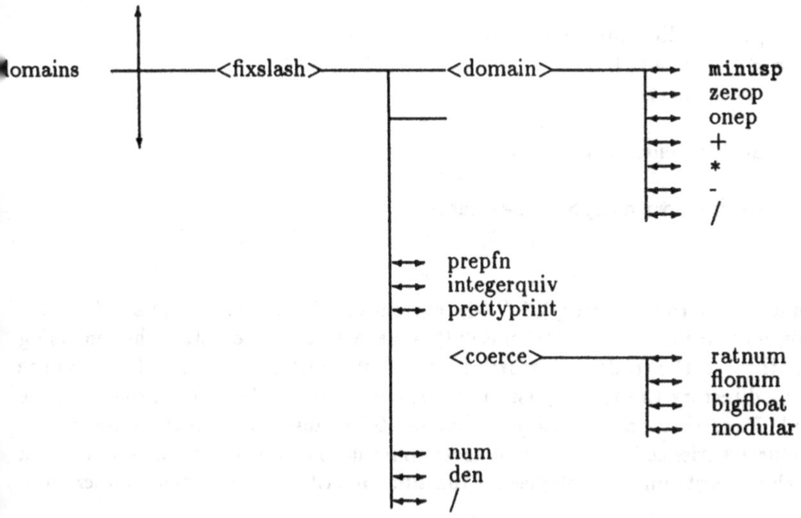

Figure(2)

Note that we do not use fancy names like !:fix!: to denote the fixslash domain, but instead simply the name fixslash. This can be done because of the fact that a reference to this entry in the hierarchy has to be specified as: domains/fixslash. The tree shown contains directory entries, indicated in between a pair of arrows: <- “dir-entry” -> and oblist entries which follow a double arrow: <-> “oblist-entry”.

We have grouped all predicates and domain operations in the subdirectory:

domains/fixslash/domain/

Three local functions “num”, “den” and “/” have been added to increase the readability of the code. It should be noted that these functions do not interfere with any definition of “num”, “den” and “/” in any other part of the system. Before we can introduce the definition of the fixslash domain with the new structuring facilities, we have to be more precise about the syntax of directories and identifiers and the related treatment of these objects by the (TU)LISP reader.

5.1 Naming conventions

The syntax of a directory-entry is given in BNF form as follows:

```
<directory-entry> ::= [ / ] { <dir_name>/ } | ""/
<dir_name>        ::= <basename> | . | ..
<basename>        ::= <identifier>
```

whereas the syntax of an oblist-entry is given by:

```
<oblist-entry>    ::= [ <directory-entry> ] <basename>
```

Names starting with a ‘/’ (the root directory), ‘./’ (the current or working_directory) and ‘../’ (the parent of the working_directory) are called ‘absolute’; they have an absolute path. The remaining names are called ‘relative’, as the reader will search for the first directory-basename found on the path from the working directory to the root, which corresponds with the first directory-basename in the name specified. From then on the path is considered to be absolute from the subdirectory found on the path. Our experience has shown that these conventions are natural and easy to work with. Functions which support similar strategies for run-time inheritance are discussed later on in this section.

5.2 A new primitive datatype, the directory

The fields accessible from within (TU)LISP of an <oblist-entry> and a <directory-entry> are repectively:

<oblist-entry>	<directory-entry>
value	..
print-name	print-name
parent	parent
function	function
plist	directory
oblist	

Three additional fields are used, the left, right field to support the balanced-tree organisation of the object-list and the directories and the discipline to support our object-oriented evaluator. The function entry is not needed in a value cell only environment. There is no value cell for the ¡directory-entry¿, however we did reserve an empty slot at the position of the value cell, as we did not want to include the risk that a compiled, unsafe assignment could overwrite the oblist- or directory- structure.

The function 'parent' can take both an oblist-entry and a directory-entry as an argument and returns the parent-directory. This is why we have included the double arrows in figure 2, indicating that one can traverse the hierarchy from the root to its leaves (oblist) with a selection function and from the leaves to the root with the 'parent' function.

A complete set of functions has been defined to work with the hierachical oblist structure, however only a few can be discussed here.

5.3 Selective application of procedures

The procedures :

apply_dir (dir, name, args) and call_dir (dir, name, arg1,.., argn)

search (in $O(\log n)$ time) for the (expr, lexpr, subr, lsubr or macro) procedure with basename 'name' in the directory 'dir'. The procedure definition given on the directory-entry is taken if no basename is found in the given directory which matches the basename of the name given. This procedure can be written to reinvoke the apply-dir operation in the parent-directory to inherit all function definitions given in the parent directory. This process can be continued to obtain multiple inheritance, but other directories than the parent directory might be used in the next selection step as well or error messages may be issued as needed.

5.4 The fixslash domain, an example

In this example we show the power of the proposed hierarchically organized name-space. In some cases we use new language constructs too, such as the backquote (‘) operator and the related (postfix) ˜ and ˜˜ operators. These will be described in more detail in section 5. Similar remarks hold for the redefinition of the / operator and the explicit reference of the division operation in the /symbolic/ directory with /symbolic/!/. It is assumed in the following example that the proposed new REDUCE parser works like the current MoDL parser which relates the syntax of operators to the basename of the operator, such that for instance domain/* is handled in the same way as the * operator in the root.

```
cd ('/symbolic/domains/fixslash/ );
fixsize := 2 ^ 30;
```

```
% Define two local procedures to access the numerator and denomi-
% nator of a fixslash number.
cp ('second, 'num);
cp ('rrest,  'den);

% A constructor function for fixslash numbers:

procedure / (u, v);
  if minusp m then (-u / -v)
   else '(domains/fixslash (fixslash (u, v)~~);

% Recognizer functions for the fixslash domain:

procedure domain/minusp u;
  minusp num u;

procedure domain/zerop u;
  zerop num u;

procedure domain/onep u;
  num (u) == 1 and den (u) == 1;

% Specific fixslash domain-operations:

procedure domain/+ (u, v);
  (num (u)*den (v) + den (u)*num (v)) / den (u)*den (v);

procedure domain/- (u, v);
  (num (u)*den (v) - den (u)*num (v)) / den (u)*den (v);

procedure domain/* (x, y);
  (if domain/zerop (u)
       and not (domain/zerop (x) or domain/zerop (y))
     then rederr "underflow in fixed slash multiply"
     else u)
  where u = num (x)*num (y) / den (x)*den (y);

procedure domain/!/ (x, y);
  (if domain/zerop (u) and not domain/zerop (x)
     then rederr "underflow in fixed slash divide"
     else u)
  where u = num (x)*den (y) / den (x)*num (y);

% Conversion of fixslash to prefix form

procedure coerce/prefix u;
  if onep den (u) then num (u)
   else '(num (u)~ /symbolic/!/ den (u)~);

% Conversion functions between domains:

procedure coerce/ratnum u;
  '(domains/ratnum rest (u)~~);
```

```
% This shows a proposed syntax extension in which multiple names
% share one function definition.

procedure (coerce/flonum coerce/bigfloat  coerce/modular) u;
  rederr list ('"Conversion between fixed-slash and",
               first (u), "not defined");
procedure prettyprint u;
  <<prin2 num (u);
    if not onep den (u)
     then <<prin2 '/!/ ; prin2 den (u)>>>>;
```

6 Specific new language constructs used in the example

A backquote operator was used in conjunction with the (postfix) ~ and ~~ operators to construct symbolic data with embedded programming constructs. The backquote operator is used to introduce a symbolic constant in a programming construct, without having to resort to prefix notation. This has the advantage that almost all knowledge about the internal representation of datastructures can be kept outside the code, as it is sufficient that this is concentrated in the parser. This aspect of information hiding has great advantages, especially in cases where the surface syntax differs much from the internal representation, such as in 'where constructs' which usually map to lambda constructs. The postfix ~ operator is used to reference a programming construct under a backquote. The postfix ~~ operator merges a list of operands into a backquoted expresssion.

```
Hence :               `domains/fixslash (fixslash (u, v)~~)


is equivalent to :  domains/fixslash . fixslash (u, v)


while :               `(num (u)~ /symbolic/!/ den (u)~)


is equivalent to :  list ('/symbolic/!/, num (u), den (u))
```

This last expression may be slightly confusing for those who see it for the first time, because of the explicit reference to the atom '/ in the directory /symbolic/. The confusion may come from the fact that the division operator / has the same syntax as the directory separator character, which makes it necessary to escape it. We use in TULISP the backslash character for this purpose, or the double quote ". In this text we wanted to adhere to the REDUCE style, which uses the exclamation mark.

Note that we do suggest to abandon the current practice to represent + internally as PLUS, * as TIMES etc. as any implementation using our standard package of functions for the maintenance of the hierarchical name-space might use a sequence of statements of the form mv ('PLUS, '+); mv ('TIMES, '*); etc. to normalize the content of the name-space.

7 Overall impact on computer algebra system design

The techniques shown may be used to implement all algebraic domains in the way shown, but other large parts of REDUCE may benefit from the suggested approach as well. Probably the most striking step might be to create a directory /symbolic/ and a directory /algebraic/ under the root, each holding apropriate operators and variables for that domain. This would make it simple to distinguish between assignment in symbolic mode (SETQ) and assignment in algebraic mode (SETK), but it also offers new chances to reference algebraic values in symbolic mode and vice versa. The so called form-functions, which rewrite (compile) an algebraic mode procedure body in an equivalent form for the normal LISP evaluator, will have less transformations to do when certain algebraic mode features can directly be inherited from the directory in which we are working in.

8 Speed and memory requirements

Assuming that all atoms used as indicator in the existing REDUCE implementation are already allocated and do not contribute to the amount of memory required, and taking the extra words in the new implementation into account we have the following approximate memory requirements:

LISP cells used:

	list	plist	dir	oblist	total
Current REDUCE implementation:	2 +	16*4 +	0 +	14*7 =	164
New structured name-space:	0	0	3*9 +	17*8 =	163

This shows that the amount of memory needed for the new implementation is slightly less than that in the old situation. This is due to the fact that the plist links are no longer needed. The comparison takes even the local functions "num", "den" and "/" which were introduced in the new situation for code readability and clarity into account. The memory taken into account for their oblist entries can be discarded (in TULISP) if these functions are declared to be local. All references to functions within the domain are resolved at read-time in both approaches, so we need only compare the time needed to apply one of the functions put on the plist of the symbol '!:fix!:, versus the time to access the corresponding function in the new situation. This gives average access times, based on an average search-time on a plist of length n proportional to $O(n/2)$ and an average search in a balanced tree with n elements of $O(logn)$ for both implementations as given:

Current REDUCE implementation: $O(n/2)$ with n = 16 $\rightarrow$ 8

New structured name-space: $O(logn1) + O(logn2) \rightarrow$ 5 a 6
with n1 = 8 and n2 = 7
or n2 = 4

These times may become even more in favor if the calling algorithm does not always have to find the operator or the conversion function starting from the domain tag: domains/fixslash/ but instead computes the pointer to one of the directories: domains/fixslash/domain/ or domains/fixslash/coerce/ only once.

The comparison shows that an efficient implementation can go hand in hand with a nicely structured description, fully supported by adequate syntax, without too much worry about overhead.

9 Portability considerations

The features mentioned so far have all been implemented in TULISP, a highly portable LISP implementation, which supports a compiler with C as target language. So far we have only implemented a parser for MoDL, a rather simple surface language for a very powerfull system. The MoDL parser is built using the object-oriented techniques shown in this text. We would most welcome a better structured parser for REDUCE too as this would make it easier to adapt the TULISP environment.

Another solution would be to adapt existing implementations. This would require several major changes, as it is considered inadequate to simply read a name and break it in pieces using existing LISP readers and programming techniques. Another point of concern is the translation process, in which one does not want the code to be immediately bound to the current environment, i.e. we keep all references of the kind: ./, a/../../b in their original form, be it a compilation into LISP format or a compilation to TULISP compatible C-code.

It is our feeling that REDUCE cannot be maintained very much longer in its current form. The flat object-list is too hard to maintain and the syntax inherited from Algol does not support a full duality of programming constructs and symbolic data. This has made it almost unavoidable that a lot of system knowledge, be it only in the way in which (conversion) routines have to be written, has gone into the encoding of the algorithms. This aspect becomes especially hard to ignore if such things as algebriac domains should be specified by users. Important things to consider for existing system implementations are the organisation of the object list, the extension of the datatypes with another atom, the directory entry and the reworking of existing code which did not differentiate between directory-entries and oblist- entries.

The inclusion of the backquote operator may have deep consequences because it assumes that all potential n-ary functions should be implemented as n-ary functions in the interpreter and the compiler.

Anyhow we suggest that a 'transloader' be developed to automatically convert REDUCE code from the new, proposed notation, to the older one or vice versa.

References

[1] Hearn, A.C. (ed) REDUCE Users manual, Version 3.2, The Rand Corporation, Santa Monica, California, April 1983.

[2] Smit, J. O. Herrmann, S.H. Gerez, R. Luchtmeijer, R.J. Mulder and L. Spaanenburg, "Syntactic and Semantic Definition of MoDL", In: "The Integrated Circuit Design Book", P. de Wilde, (Ed.), Delft University Press, The Netherlands, ISBN 90-6275-246-2.

[3] J. Smit, J., B.J.F. van Beijnum, S.H. Gerez and R.J. Mulder, "Proving Correctness of Digital Designs in the Multidimen- sional Design Space", Proceedings of the IFIP Working Conference: "From HDL Descriptions to Guaranteed Correct Circuit Designs", North Holland, September 1986.

[4] Steele, G.L. Common LISP Manual, Digital Press, Cambridge, Massatuchets.

[5] Gutknecht, J., "Separate Compilation in Modula-2: An Approach to Efficient Symbol Files", IEEE Software, Nov. 1986, pp 29-38.

[6] Goldberg, A. and D. Robson, "Smalltalk-80 The Language and its implementation". Addison-Wesley, Reading Mass., ISBN 0- 201-11371-6, 1983.

[7] Andelson-Velskii, G.M. and E.M. Landis, "Soviet Math. #3", pp 1259-1263.

[8] Abdali S.K., G.W. Cherry, N. Soiffer, "An object oriented approach to algebra system design". Proc. of the 1986 Sympo- sium on Symbolic and Algebraic Computation, Symsac '86, July 21-23, Waterloo, Ontario, pp 100-106.

[9] Bradford, R.J., A.C. Hearn, J.A. Padget and E. Schrufer, "Enlarging the REDUCE Domain of Computation", Proc. of the 1986 Symposium on Symbolic and Algebraic Computation, Symsac '86, July 21-23, Waterloo, Ontario, pp 100-106.

[10] Matula, D.W. and Kornerup, P. "Finite Precision Rational Arithmetic: Slash Number Systems". IEEE Transactions on Computers, C-34, Vol. 1, No 1, January 1985, pp 3-18.

NUMBER-THEORETIC TRANSFORMS OF PRESCRIBED LENGTH

Reiner Creutzburg
Akademie der Wissenschaften der DDR
Zentralinstitut für Kybernetik und Informationsprozesse
Kurstr. 33, PF 1298
DDR - 1086 Berlin

Manfred Tasche
Wilhelm-Pieck-Universität Rostock
Sektion Mathematik
Universitätsplatz 1
DDR - 2500 Rostock 1

The number-theoretic transform (NTT) was introduced as a generalization of the discrete Fourier transform over residue class rings of integers in order to perform fast cyclic convolutions without round-off errors.

A large number of transform methods were developed to remove some of the length limitations of conventional Fermat number and Mersenne number transforms [2], respectively. These NTTs, which under certain conditions can be computed via fast transform algorithms, allow the implementation of digital signal processing operations with better efficiency and accuracy than the fast Fourier transform.

However, it is always difficult to find moduli m that are large enough to avoid overflow, and to find primitive N-th roots of unity modulo m with minimal binary weight for transform lengths N that are highly factorizable and large enough for practical applications.

In [1], a useful way is shown to solve this problem by prime factorization of cyclotomic polynomials with integer-valued arguments.

Let z be the ring of integers and $m > 1$ an odd integer. Further let $N > 2$ be an integer. Then $a \in Z$ $(|a| \geq 2)$ is called primitive

N-th root of unity modulo m if

$$a^N \equiv 1 \bmod m \quad \text{and} \quad GCD(a^n-1,\ m) = 1, \quad (n = 1, \ldots, N-1).$$

We denote the N-th cyclotomic polynomial by ϕ_N.

An important question for practical application is the following: Is it possible to find convenient moduli m for NTTs, if a highly factorizable length N and an integer a with small binary weight are prescribed, such that a is a primitive N-th root of unity modulo m?

THEOREM [1]. Let $N > 2$ and $a \in Z$ ($|a| \geq 2$) be given, where $(N,a) \neq (3,-2), (6,2)$. Under these assumptions, a is a primitive N-th root of unity modulo m if and only if $m > 1$ is a divisor of

$$M = \phi_N(a) \;/\; p^{\tau_p},$$

where p denotes the greatest prime factor of N with $p^t | N$ and $p^{t+1} \nmid N$ ($t \geq 1$), and

$$\tau_p = \begin{cases} 1 & \text{if a belongs to the exponent } N/p^t, \\ 0 & \text{otherwise.} \end{cases}$$

Many known results can be obtained as simple corollaries of this new theorem and are helpful in determining parameters of number-theoretic transforms [1].

The method can be applied to the case of residue class rings of Gaussian integers [3] and cyclotomic integers [4], respectively.

REFERENCES

[1] R. Creutzburg & M. Tasche: Number-theoretic transforms of prescribed length. Math. Comp., 47, 1986, pp. 693-701

[2] H. J. Nussbaumer: Fast Fourier Transform and Convolution Algorithms. Springer: Berlin 1981

[3] R. Creutzburg & M. Tasche: Parameter determination for complex number-theoretic transforms using cyclotomic polynomials. Math. Comp. (January 1989)

[4] R. Creutzburg & G. Steidl: Number-theoretic transforms in rings of cyclotomic integers. J. Inf. Process. Cybern. EIK 24 (1988), pp. 573-584

A Hybrid Algebraic-Numeric System ANS and its Preliminary Implementation †

Masayuki Suzuki, Tateaki Sasaki

The Institute of Physical and Chemical Research
2-1, Hirosawa, Wako-shi, Saitama 351-01, Japan

Mitsuhisa Sato

Department of Information Science
University of Tokyo
7-3-1, Hongo, Bunkyo-ku, Tokyo 113, Japan

Yoshinari Fukui

Total Information & System Division
Toshiba Corporation
72, Horikawa-cho, Saiwai-ku, Kawasaki-shi 210, Japan

ABSTRACT

Recently the present authors proposed a simple and promising scheme of linking different computer languages and discussed linking Lisp and FORTRAN in detail from the viewpoint of hybrid algebraic-numeric computation [1]. The algebraic-numeric system based on our linking scheme has been named **ANS**, and we implemented it preliminarily on UNIX †† operating system. This paper explains our linking scheme briefly and describes the **ANS** and its preliminary implementation. Furthermore, we discuss desirable properties of programming languages and operating systems from the viewpoint of hybrid computation.

1. Introduction

As for hybrid algebraic-numeric computation, there is a strong demand from many users and several schemes have been proposed and implemented so far [2, 3, 4, 5]. Some of them seem to be quite effective and useful, but they still contain important defects as we will explain in **2**. Recently we proposed a simple scheme of hybrid system which seems to be quite promising from the viewpoints of application as well as implementation [1].

Our scheme is based on an idea of linking different computer languages by a simple but powerful mechanism. Hence, the scheme allows us to construct not only a hybrid algebraic-numeric system but also general hybrid systems, by linking various application systems written in different computer languages.

In **2**, we discuss a desirable scheme for hybrid algebraic-numeric computation, and survey hybrid algebraic-numeric schemes proposed so far. In **3**, we explain our scheme of linking different languages briefly, then we describe our hybrid system **ANS** and its features. The **ANS** has been implemented preliminarily on UNIX, and **4** explains the implementation. In **5**, we discuss some problems inherent

---

† This work was supported in part by Grant in aid for scientific research of the Ministry of Education under Grant-61780054.

†† UNIX operating system is developed and licensed by AT&T.

to our hybrid scheme and make several proposals to designers of languages and operating systems.

2. Hybrid systems for algebraic-numeric computation

First of all, we clarify the meaning of *hybrid algebraic-numeric computation*. Some users want to call efficient FORTRAN subroutines when using algebraic systems, some users want to activate algebraic computation phase in FORTRAN programs, some users want to generate and execute FORTRAN subprograms in algebraic systems, and so on. Thus, there are quite various demands from users and we define the meaning of the hybrid computation most generally as follows (below, we use FORTRAN as a language representing numeric computation).

(1) We can fully utilize a FORTRAN system and its library as well as an algebraic system and its library in such a way that we can utilize each system separately;

(2) FORTRAN and algebraic programs can be linked together, easily and with minimum changes to the programs, so that we can call the algebraic system from FORTRAN programs and the FORTRAN system from algebraic programs;

(3) Batch processing and interactive processing which are dominant in FORTRAN and algebraic systems, respectively, are allowed for hybrid systems also.

We think a hybrid scheme satisfying the above conditions is sufficiently general for algebraic and numeric computation. However, we assume more conditions as follows.

(4) The desirable hybrid scheme should be applicable not only for FORTRAN and an algebraic language but also for other languages;

(5) If the realization of the hybrid system requires modifications and/or extensions of the languages and systems concerned, the modifications and/or extensions should be minimum.

In order that the hybrid scheme may be accepted widely, it must be accepted by the main-frame makers also who are very conservative for the change of languages and systems. The last condition is imposed from this reason.

Various hybrid computation systems have been constructed so far, most of which have been realized by adding numeric computation facilities to algebraic systems. The hybrid schemes employed in these systems may be summarized as follows.

(a) Providing a facility of generating FORTRAN programs (mostly, function subprograms) in an algebraic system. The FORTRAN programs generated are transferred to FORTRAN systems via files. When using REDUCE [2] for hybrid computation, the user must do in this way. According to our definition of hybrid computation, we cannot view this scheme as hybrid.

(b) Implementing an algebraic system as FORTRAN subroutines. This scheme was adopted by FORMAC [3]. Since algebraic computation is much more general than numeric computation, it is unreasonable to include algebraic computation facilities into FORTRAN.

(c) Extending Lisp (or language in which an algebraic system is implemented) to execute FORTRAN programs. This scheme was adopted by MACSYMA [4], and it allows various hybrid computations actually. However, implementing such a system is quite laborious.

(d) Implementing an algebraic system in a language which allows both numeric and symbolic computations. This strategy was adopted by SMP [5] which was written in C. Although this strategy seems to be reasonable, it should be noted that constructing a general-purpose algebraic system is very time-consuming.

It should be commented that, in any of the above schemes, hybrid computation is considered mostly from the viewpoint of computer algebrist and the viewpoint of FORTRAN user is almost lacking. Currently, numeric computation is much more popular than algebraic computation and most numeric users want to perform algebraic computation in FORTRAN systems. However, using algebraic systems in FORTRAN have been almost ignored so far. We further note that the schemes mentioned above will face a serious problem when extending the algebraic-numeric system to a more general hybrid system. We think that constructing general hybrid systems is an inevitable trend of technology development. Therefore, we should consider the future extension of the system seriously when constructing a hybrid system.

3. Hybrid computation system ANS

We first explain our scheme of linking computer programs written in different languages briefly, then describe the hybrid algebraic-numeric computation system **ANS**.

3.1. A scheme of linking different systems

The largest problem in linking computer programs written in different languages is how to adjust the internal data representations which are not the same in different language systems. For example, even the fixed-precision integers are differently represented internally in FORTRAN and Lisp systems. The conventional hybrid schemes listed in **2** avoid this problem by constructing the whole hybrid system on a single language system.

In our hybrid system, we will use application programs written in the conventional languages as far as possible. In order to do so, we convert the data representation when a datum is sent from one language system to another. Describing more precisely, our scheme for linking different languages is composed of the following five points (we denote the languages to be linked as A and B).

(1) The hybrid program is a mixture of programs written in A and B, with commands indicating transition from one language to another, and it is executed as sequential tasks, i.e., coroutines;

(2) The data commonly used in A and B are declared in advance to the execution and they are allocated to both A and B systems separately. The common data are, however, defined for the user as if they are allocated to one area which is common to both A and B;

(3) When the program execution is moved from A to B, the common data allocated to A are copied to the common data area for B by transforming the internal data representations;

(4) The programs dynamically generated are written into a file shared commonly and transported between A and B systems (the program is compiled immediately and loaded dynamically);

(5) The internal representations of data which are used locally, such as the arguments of functions, are automatically converted even if they are not declared to be common, so far as the conversion is side-effect free.

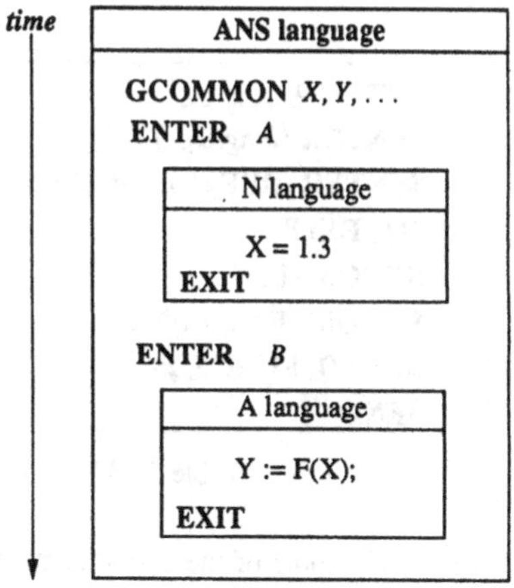

Fig.1. Schematical illustration of hybrid program

The variables declared by **GCOMMON** can be used in both A and B systems. This figure shows the case where no transition is caused in subprogram (subprogram may also be hybrid).

We call our hybrid algebraic-numeric system based on the above scheme **ANS** (Algebraic Numeric System). We use **ANS** as a hybrid language for algebraic-numeric computation also.

We explain our hybrid scheme by figures, where the meanings of some commands are as follows. The transition from a language system *A* to *B* (or from *B* to *A*) is caused by the command **ENTER** *B* (or **ENTER** *A*) and that a datum *X* which is common to different language systems is declared as **GCOMMON** *X*. We also use the command **EXIT** to show the end of program block which is executed in the same level of **ENTER** command. Figures 1 and 2 illustrate schematically a program with no subroutines and the treatment of common data, respectively, in our hybrid system.

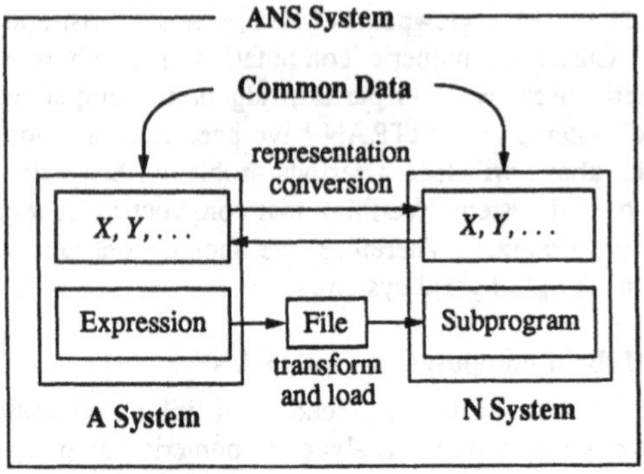

Fig. 2. Illustration of treatment of common data

In executing the **ENTER** command, the common data are copied after conversion of the internal data representation. The programs generated dynamically in A system are transported via common files.

3.2. Hybrid language ANS

As we have mentioned above, the **ANS** system utilizes the conventional language systems as far as possible by linking them by a simple mechanism. Therefore, main part of the **ANS** program is written in conventional languages. The **ANS** language is characterized by a small number of **ANS**'s own commands which occupy only a very minor part of the program.

In the current preliminary version of **ANS**, the following commands are prepared as **ANS**'s own commands; the prefix character **$** is to distinguish them from other commands easily.

Command	Meaning
$ENTER (*language*) [{]	Enter into the *language* mode
$ENTER_DEF (*language*) [{]	define subroutines in *language* and initialization
$[}] EXIT	Exit the *language* mode
$GCOMMON	declare global common variables
$GCOMMON_FUNC	declare global mathematical functions
$LOAD_FUNC (*func*)	transform and load a subprogram *func* generated
$END	terminate the execution of ANS

Table 1. ANS's own commands (preliminary version)

We explain some of the above commands. Commands **$ENTER** and **$ENTER_DEF** declare the language mode, but they works differently. Command **$ENTER**, being paired with **$EXIT**, causes the transition of language mode during the execution. Since **$ENTER** may be issued recursively, the **ANS** system saves the return environment before causing the mode transition. On the other hand, **$ENTER_DEF** is used to define programs before the execution or to initialize system constants. Hence, with **$ENTER_DEF**, the **ANS** saves no environment.

Command **$EXIT** is necessary to process the nested call of language systems correctly. For example, Figure 3 illustrates two different ways of language mode transition; first, the control is in *N* system, then it goes to *A* system by the command "**$ENTER** *A*", and comes back to the *N* system by the command "**$EXIT**" or enters to the *N* system recursively by the command "**$ENTER** *N*". Without **$EXIT** command, we cannot distinguish the recursive call of a system from the return to the original system. We explain **$GCOMMON_FUNC** and **$LOAD_FUNC** in the next subsection.

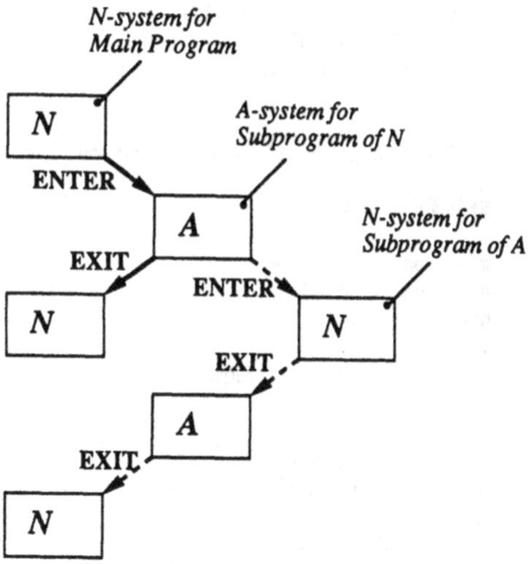

Fig. 3. Two ways of mode transition (—►and - -►)

3.3. Treatment of mathematical functions in ANS

One very desirable property of hybrid algebraic-numeric system is to generate numeric subprograms dynamically by the algebraic system and use them in the numeric system. The **ANS** system provides the user with this facility in such an elegant way that some mathematical functions calculated by the algebraic system are automatically transformed to numeric subprograms.

In the current version of **ANS**, the user must declare the function names which are common to both algebraic and numeric systems by **$GCOMMON_FUNC** before the execution. The **$GCOMMON** function must be generated by the algebraic system before the use in numeric system. The transformation of the **$GCOMMON** function to a numeric subprogram and loading it to the numeric system are done by command **$LOAD_FUNC**. Note that no facility of dynamic linking is necessary in the numeric system because **$GCOMMON_FUNC** registers the names of **$GCOMMON** functions in the numeric system before the execution; only the facility of dynamic loading is required. The transformed subprogram may be an optimized code (this code optimization facility is not implemented yet).

4. Preliminary implementation of ANS

In order to prove effectiveness of the **ANS** scheme and easiness of its realization, we have implemented **ANS** preliminarily on 4.2BSD UNIX/VAX11-750. The complete implementation requires the OS to satisfy several minor conditions as we have mentioned in [1], hence the implementation is rather difficult on the current OS supplied by main-frame makers. By this reason, we implemented **ANS** in the C language on UNIX. Thus, current preliminary version links C and REDUCE and not FORTRAN and Lisp. However, linking C and FORTRAN is quite easy on UNIX, so the current implementation has realized most of the facilities of **ANS** mentioned above.

The **ANS** system is composed of a preprocessor, a data converter, and a server performing inter-process communication.

```
;;
;; Common variables declaration
;;
$GCOMMON ((int CO 10))                   ;; Vector for coeffs of polynomial.
$GCOMMON_FUNC                            ;; Functional variable holding
    ((double RECREL ((double X))))       ;; Newton's recurrence relation.

;;
;; Enter language C mode (Numerical language)
;;
$ENTER (C)

# define EPS 1.0e-6                      ;; Accuracy of coeffs.
# define Max 20                          ;; Maximum count of iteration
main()
{
    double x0, x1, fabs();
    int i;

    for (i=0; i<10; i++)                 ;; Read coeffs. and set them to
        scanf("%i", &CO[i]);             ;; common variables.
;;
;; Enter REDUCE
;;
$ENTER (REDUCE) {`
    F := 0;
    FOR I := 0:9 DO                      ;; Make an equation with
        F := (F*X + CO(9-i));            ;;   coeffs. given by common variables.
    RECREL := X - F /DF(F, X);           ;; Calculate recurrence relation.
$} EXIT
;;
;; Generate new function to calculate the recurrence relation.
;;
$LOAD_FUNC (RECREL)                      ;; compile and load
;;
;; Newton's iteration in C
;;
    x0 = 0.6;
    for ( i=0; i<Max; i++ ) {
        x1 = (*RECREL) (&x0);            ;; Calculate the next approximation.
        if ( fabs(x1-x0)<EPS )           ;; Test of convergence.
            break;
        x0 = x1;
    }
    printf("Result = %lf\n", x1);
    exit_ans();
}
$EXIT
```

Fig.4. Example of **ANS** program
(in the current preliminary version linking C and REDUCE)

4.1. ANS preprocessor

The **ANS** preprocessor scans a hybrid program which has the form illustrated in Fig.1 and separates it into pure algebraic and pure numeric programs. Since the change of the language mode is specified by the command **ENTER**, it is quite easy to perform this separation. The programs separated are stored into their respective program files and complied if necessary.

In addition to the program separation, the preprocessor calls an **ANS**-command processor when it meets **ANS**'s own commands (symbols beginning with the character **$**). The command processor eliminates **ANS**'s own commands and inserts corresponding codes written in the respective languages.

4.2. Data transfer (interprocess communication)

In the **ANS** system, at least two processes are co-existing as coroutines, so the communication between them is crucially important. In the current implementation, the communication is performed by using the interprocess communication (IPC) facility of UNIX [6]. Using the IPC facility, we can easily realize the conversion of data representations and the control of coroutines.

Let us first explain how to convert data representations, which is quite simple as follows. When the control is transferred from *A* system to *B*, the data in **GCOMMON** area of *A* are output into the IPC port by the *write* command, then the data in the port are read into the **GCOMMON** area of *B* system by the *read* command. Then, the internal representations of the data are automatically converted in the stage of writing by *A* system and reading by *B* system.

Next, we note that the actual reading by the *read* command is suspended until the IPC port is supplied with the read data. Therefore, by utilizing the IPC facility, coroutine processing is realized by only issuing *read* and *write* commands in both algebraic and numeric parts of the hybrid program.

It should be mentioned that linking different programs by interprocess communication is not a new idea but already described in [7].

4.3. Simple example

Let us explain the use of **ANS** by a simple example. Figure 4 shows an **ANS** program (in a current preliminary version) for calculating roots of a polynomial equation in a single variable by Newton's method. In the program, the **GCOMMON** variable *CO* is an integer array of size 10 and it is used for inputting coefficients of polynomial. The variable *RECREL* is also declared to be common, and it is used as a function name (the function is a polynomial in this example).

The program is executed as follows: first, the coefficients of polynomial are read in by the C system, then the REDUCE system calculates the function *RECREL* algebraically. Finally, the C system executes Newton's numeric iteration procedure. Note that the numerical evaluation of *RECREL* is executed not by REDUCE but by a C subroutine.

5. Discussions

Our preliminary implementation shows easiness of realizing our hybrid scheme, at least on UNIX. Hence, we think that our **ANS** scheme is quite promising for hybrid computation. However, in order to implement the **ANS** on various operating systems and use it widely, we must solve several problems. In the following, we discuss these problems briefly.

5.1. Problem on precision

The data which are shared by *A* and *B* systems are of only such types that are existing in both *A* and *B*. Since these data are allocated to each system separately, as illustrated by Fig.2, there is no problem of memory allocation. However, in some cases, there are serious problems about conversion of data-representations. In the case of numeric data, difference in the manner of treating precision causes a problem. For example, the precision of integers is fixed in FORTRAN, while it is variable in

Lisp. Therefore, there happens a case that a fixed-precision integer introduced in FORTRAN system is transferred to Lisp system but it cannot come back to FORTRAN system because its precision has increased largely in Lisp system. As this example shows, linking different languages poses us to reinvestigate the data types and their representations.

As for the above-mentioned problem of integer precision, we are proposing to introduce mixed integer-real numbers into FORTRAN. Here, by mixed integer-real number, we mean the numbers which are initially fixed-precision integers and they are converted automatically to fixed-precision real numbers when the integer precision overflows. Introducing such mixed-type numbers into FORTRAN does not change the philosophy of FORTRAN much or decrease the efficiency of FORTRAN much.

5.2. Proposals to OS-designers

In order to use the **ANS** system for actual applications, we have to implement the **ANS** on mainframe computers because most application problems require very high computer power. In order to do so, however, the operating system must be equipped with the following facilities:

(1) Facility of executing multi-tasks as coroutines;

(2) Facilities of run-time conversion and copying of common data allocated to different systems;

(3) Facility of sharing files by different systems;

(4) Facility of loading FORTRAN object codes dynamically (dynamic linking of FORTRAN program is better, but the dynamic loading is almost sufficient).

We think the above requirements are not hard to accept for operating system designers. We propose to OS-designers strongly to equip these facilities in their operating systems.

5.3. On general hybrid system

As we have mentioned in **2**, hybrid systems will surely be extended successively towards general systems which perform not only numeric and algebraic computation but also graphic processing, data base access, etc. Currently, many softwares for these processings are being written in various languages. Therefore, we should design hybrid systems so that they link various language systems easily. We think that the **ANS** scheme is quite suited for constructing such a general hybrid system.

We have mentioned that two problems, one is on operating system and the other is on conversion of data representation, are very important to our **ANS** scheme. When constructing a general hybrid system using **ANS** scheme, another important problem happens, i.e., conversion of data type.

We have already met with an example of data type conversion, i.e., transformation of a function defined by the algebraic system to a numeric subprogram in **3.3**. This is rather a sophisticated kind of type conversion, and many kinds of type conversions will be necessary in general hybrid systems. This is because, in a general hybrid system, most part of the user program will be occupied by commands which call various application operations ranging over many language systems. For example, consider that the command

$$\textbf{Graph}\,(F(x), x{=}0{:}1)$$

is issued to draw a graph of function $F(x)$ in the range of $0 \le x \le 1$. The most efficient way of drawing a graph is to call a FORTRAN subroutine, which requires $F(x)$ to be converted to a FORTRAN function subprogram. In the current **ANS** system, the user must declare F to be **GCOMMON** function name.

In a general hybrid system, it is very desirable that the user need not specify the type conversion explicitly but the system preforms the required type conversion automatically because the type conversion is so frequent. This automatic type conversion is, however, not easy if many different kinds of languages are linked together, and we have a very challenging theme here.

References

1. Sasaki, T., Fukui, Y., Suzuki, M., and Sato, M., "Proposal of a Scheme for Linking Different Computer Languages," *preprint of IPCR*, May, 1986.
2. Hearn, A.C., *REDUCE User's Manual, version 3.2,* The Rand Corporation (U.S.A.), 1985.
3. Sammet, J.E. and Bond, E., "Introduction to FORMAC," *IEEE Trans. Electron. Computers*, vol. EC-13, pp. 386-394, 1965.
4. The MATHLAB group, *MACSYMA Reference Manual, version 9,* Lab. Comput. Sci., MIT, 1979.
5. Cole, C.A., Wolfram, S., et al., *SMP Handbook, version 1,* CALTEC, 1981.
6. Lettler, S.J., Fabry, R.S., and Joy, W.N., *A 4.2BSD Interprocess Communication Primer*, 1983.
7. Purtilo, J.M., "Application of a Software Interconnection System in Mathematical Problem Solving Environment," in *Proc. SYMSAC'86* , pp. 16-23.

THE CALCULATION OF QCD TRIANGULAR FEYNMAN GRAPHS IN THE EXTERNAL GLUONIC FIELD USING REDUCE-2 SYSTEM

L.S.Dulyan

Yerevan Physics Institute, Markarian St.,2,

375036 Yerevan, Armenia, USSR

The ways and methods used in calculation of one class of the QCD Feynman graphs with the help of the REDUCE-2 system [1] are described. The physical essence of the problem as well as results obtained are presented in [2].

To each of the mentioned graphs corresponds the integral

$$I_{\mu\nu\ldots}(s_1,s_2) = \int d^4l \frac{Sp_{\mu\nu\ldots}(p,k,l)}{P^a K^b L^c} \tag{1}$$

where p,k,l are vectors from the 4-dimensional pseudoeuclidean momentum space, and $k^2=0$, $p^2=s_2$, $(p+k)^2=s_1$; $\mu,\nu=0,1,2,3$ are indices numerating the 4-vector components; P,K,L are some functions quadratic over p,k and l; a,b,c are integers, and $Sp_{\mu\nu\ldots}(p,k,l)$ is a trace of product of the Dirac γ-matrices.

In the mathematical sence the problem reduces to finding the derivatives of (1) with respect to s_1 and s_2 at the point $s_1=s_2=0$ of the order of N and K respectively as functions of N and K.

For calculation of 4-dimensional integrals we have written a special program which by successive use of $P^a \to F(a)$ type simple substitutions allowed in REDUCE performs the complex substitutions of $P^a K^b L^c \to F(a,b,c)$ type direct use of which is not allowed in REDUCE.

At the step of derivatives calculation we have introduced a new differentiation operator which calculates the derivatives of order N where N is not a number but a symbolic variable.

In the result of calculations there came out rather compact expressions containing factorials and polynomials of power not higher than fourth. As a test we have reproduced the results of Refs.[3,4], where similar calculations have been made without a computer.

Besides that we have discussed the execution of $X^j \to F(j)$ type substitutions in the case when j is a sum of symbolic variable and a number. The difficulties arise here because the REDUCE separates the symbolic and numerical exponents. For example the expression $X^{N+2}= X^{N}*X^{2}$ transforms into *F(N)*F(2)* whereas we expect *F(N+2)*.

Also we have discussed the features of REDUCE system at execution of loop containing scanning instructions like `FOR I:=M:N DO A(I):=A(I);` (suppose `A(I)` declared operator) when some substitution rule is introduced by means of `LET` operator. In this case the substitution will not be performed, whereas by the explicit instruction `A(1):=A(1); A(2):=A(2);` etc. the substitution is performed. The recipies avoiding this and some oter troubles are given.

References

1. Hearn A.C. REDUCE user's manual, 2nd ed., University of Utah, 1973.
2. Dulyan L.S., Oganessian A.G., Khodjamirian A.Yu. Yad.Fiz., 1986, 44, 746.
3. Khodjamirian A.Yu. Yad.Fiz., 1984, 39, 970.
4. Beilin V.A., Radyushkin A.V. Yad.Fiz. 1984, 39, 1270.

COMPUTER ALGEBRA APPLICATION FOR DETERMINING LOCAL SYMMETRIES OF DIFFERENTIAL EQUATIONS

R.N. Fedorova
Laboratory of Computing Techniques and Automation
Joint Institute for Nuclear Research
Head Post Office, P.O. Box 79, Moscow, USSR

V.V. Kornyak
Institute of Mathematics AN UkrSSR
Repina 3, Kiev 4, USSR

Extended abstract

The symmetries of differential equations depending on local values of dependent variables and their derivatives are called local symmetries. These symmetries are point and contact Lie symmetries and Lie-Backlund symmetries. The local symmetries play an important role in applied mathematics and mathematical physics [1,2]. To obtain the symmetry group of the differential equations it is necessary to carry out the tremendous amount of algebraic computations. It is natural to perform these calculations with computer algebra system.

We proposed three programs for obtaining the determining equations of local symmetries. The first of them is written in REDUCE-2 language and is intended for obtaining point and contact symmetries [3]. The programs for determining Lie-Backlund symmetries are based on the REDUCE-2 [4] and PL /1- FORMAC [5]. The main peculiarity of algorithms of these programs consist of the representation of the variables and derivatives in calculations in the form of linear ordered rows. These representations permit to unify the calculation algorithms and to bring about significant computer storage economy. The element in the row is determined completely by its ordinal number only. To reconstruct all attributes of the element special arithmetical subprograms are used. When computing the symmetries of differential equations the effectiveness of the algorithms and computer algebra system used causes a problem, since the number of derivatives increases rapidly with the order of derivatives. By this reason the PL/1-FORMAC system is more suitable than REDUCE-2.

Some results of the investigation of the symmetry properties of equations of mathematical and theoretical physics with the help of the computer algebra are presented in [5,6].

The systems of determining equations obtained by the programs mentioned above are usually very large in spite of the simplifications performed in the programs. Recently the algorithm for integrating of the system of determining equations is developed and implemented. This algorithm is in agreement with the data representation used in the programs [3-5].

REFERENCES

1. L.V. Ovsiannikov (1978). Group Analysis of Differential Equations. Nauka, Moscow, p. 400 (in Russian).
2. N.H. Ibragimov (1983).Trasformation Groups in Mathematical Physics. Nauka, Moscow, p. 286 (in Russian).
3. V.P. Eliseev,R.N. Fedorova,V.V.,Kornyak (1985).A REDUCE Program for Determining Point and Contact Lie Symmetries of Differential Equations. *Comp. Phys. Comm.* v. 36, No. 4,p. 383-389.
4. R.N. Fedorova, V.V. Kornyak (1987).A REDUCE Program for Calculation of Determining Equation of Lie-Backlund Symmetries of Differential Equation. JINR, P11-87-19, Dubna (in Russian).
5. R.N.Fedorova,V.V. Koryak (1986). Determination of Lie-Backlund Symmetries of Differential Equations Using FORMAC.*Comp. Phys. Comm.* v. 39, N1, p. 93-103.
6. R.N. Fedorova, V.V. Koryak (1985). Application of Algebraic Computation to Determination of Lie-Backlund Symmetries of Differential Equations. In: International Conference on Computer Algebra and its Applications in Theoretical Physics. Dubna, p. 248-261 (in Russian).

TRACE CALCULATIONS FOR GAUGE THEORIES ON A PERSONAL COMPUTER

Johannes Ranft, Holger Perlt
Sektion Physik, Karl-Marx-Universität Leipzig
Karl-Marx-Platz, 7010 Leipzig, GDR

Abstract

We report about a program on a personal computer to evaluate symbolically traces of γ- matrices needed for perturbative calculations of gauge theories.

The kernel of perturbative calculations in gauge theories are the generation and evaluation of Feynman diagrams as determined by the so called Feynman rules. This invitably leads to trace expressions involving γ- matrices which obey the anticommutation rule $\gamma^\mu\gamma^\nu + \gamma^\nu\gamma^\mu = 2g^{\mu\nu}$. The evaluation of such traces is based on the recursion formula

$$\mathrm{tr}(\not a_1 \not a_2) = 4\, a_1.a_2$$
$$\mathrm{tr}(\not a_1 \ldots \not a_{2n}) = \sum_{i=2}^{2n} a_1.a_i\,(-1)^i\, \mathrm{tr}(\not a_2 \ldots \not a_{i-1}\not a_{i+1} \,..\, \not a_{2n}) \quad (1)$$

($\not a = \gamma^\mu a_\mu$) where a_μ is the 4-momentum.

Calculation by hand becomes more and more impossible in higher order perturbation theory, i.e. with increasing number of γ-matrices. On the other hand, recursion formula (1) suggests the use of symbolic program languages. Indeed, this problem belongs to the standard problems solved since a long time by languages like REDUCE /1/, SCHOONSHIP /2/ and others.

We have written a code in PASCAL in order to handle this sort of problems. PASCAL is rather convenient to generate the structures needed for our problem : strings, records, binary trees etc. . For personal computer especially TURBO PASCAL has an efficient compiler and implementation. Presently, our code consists of the following parts:

(1) Calculations of traces

Traces are calculated according to recursion formula (1). The elements of the traces are represented as linked lists whereas the resulting trace is given as a binary tree. Our code needs running time of about 1 second for a 6 element trace, about 10 seconds for a 8 element trace and about 80 - 100 seconds for a 10 element trace using the 8088/8087 microprocessors.

(2) Simplifications

Simpel algebraic expressions can be analysed also. They have - like the traces - binary tree representation . Along the tree we can do

- simplification of brackets
- collecting equal terms
- substitutions of dot-products and symbols

(3) Diagram generation

For simpel scattering processes (like $e^+ e^- \longrightarrow e^+ e^-$, i.e. positron (e^+) - electron (e^-) scattering) our code generates from the input (fermion lines, boson lines and interaction vertices characterising a given diagram) the trace expressions as determined by the Feynman rules and evaluates them.

Kaneko et al. /3/ have developed an efficient algorithm to generate all diagrams up to a certain order for various QED scattering processes. We would like to improve our code by including their method extended by the implementation of electroweak interaction. Further, we plan to transmit the results of our code to REDUCE in order to continue the evaluation (simplifications, symbolic integration,...).

Acknowledgement

We would like to thank Ines Leike for collaborating during some time in this project.

References

/1/ A. C. Hearn, 'REDUCE user's manual', version 3.2, The Rand Corporation, 1985

/2/ M. Veltman, CERN preprint, 1967
H. Strubbe, Comput. Phys. Commun. 8 (1974) 1

/3/ T. Kaneko, S. Kawabata and Y. Shimizu, Comput. Phys. Commun. 43 (1987) 279

EVALUATION OF PLASMA FLUID EQUATIONS COLLISION INTEGRALS USING REDUCE

R. Liska, L. Drska
Faculty of Nuclear Science and Physical Engineering
Technical University of Prague
Brehova 7, 115 19 Praha 1, Czechoslovakia

Abstract

Plasma physics theory ranks among prospering application areas of computer algebra. One of the tasks solved in this branch is computation of collision terms appearing as the six-dimensional integrals in the plasma fluid equations derived from the Fokker-Planck equation. In this paper the collision terms are derived in full 13-moment approximation. The results obtained represent a substantial generalization of the terms published up to now. They are based solely on the general 13-moment approximation conditions, without any other additional limitations used.

The tedious calculations necessary to evaluate the integrals required a carefully designed and efficient symbolic mode REDUCE program, especially as far as its speed is concerned. Several general recommendations for speeding up REDUCE programs are presented.

The possibility of automatic arrangement of computed collision terms according to special physical assumption is shown. The REDUCE statement performing this simplification uses the database of general collision terms created by our calculations.

The complete article will be published in the Journal of Symbolic Computation.

COMPUTERIZED SYSTEM OF ANALYTIC TRANSFORMATIONS FOR ANALYSING OF DIFFERENTIAL EQUATIONS

VLADISLAV L. KATKOV, MICHAIL D. POPOV
Institute of Mathematics Akad. of Sci. of BSSR ul. Surganova 11,
SU-220604 Minsk USSR

In recent years, analytic methods have received widespread attraction of investigators in solving the problems of mathematical phyzics. Among the examples are the group analysis of differential equations [1], derivation of conservation laws [2], reduction of the differential equations systems to passive form, and so on.

Due to the amount of analytic calculations required by these methods the class of equations amenable to manual treatment is greatly restricted. For example, the group analysis of gas dynamics equations (5 dependent and 4 independent variables) creates 680 constitutive equations. As a consequence, the problem of computerizing these investigations becomes urgent. The use of the system of analytic transformations (SAT) releaves the investigator from routine works and improves the reliability of obtained results, because the probability of error is high with the large amount of manual calculations. Besides, the SAT allows investigation of more complex systems of differential equations unamenable to manual treatment.

The objective is:

1) create the library of basic analytic transformations founded on the analysis of algorithms for solving of differential equations;

2) by using the library of basic analytic transformations, implement the series of application programs that ensure automatic solution of the following problems:

- formulate the constitutive equations of Lie groups admitted by a systems of differential equations;
- same for Lie-Backlund group;
- formulate the constitutive equations for conservation laws;
- reduce the systems of differential equations to passive form.

As an example, we describe the algorithm for deriving the Lie group which a systems of differential equations admits of. The input system of differential equations is taken in the form solved for main derivatives:

$$R=\Phi(X,U,P), \quad (1)$$

where X are independent variables, U are dependent variables, R are main derivatives, P are parametric derivatives, Φ are rational functions.

The infinitesimal group operator is of the following form

$$L=a^i(X,U)*d/dx^i+b^k(X,U)*d/du^k, \quad (2)$$

where a^i, b^k are unknown functions (coordinates) of independent and dependent variables, d/dx^i and d/du^k are partial differentiation operators.

Formulation of the constitutive equations involves the following four steps:

1) extend the infinitesimal operator (2) to a derivatives of dependent variables:

$$M=L+A_{Ri}*d/dR^i+B_{Pk}*d/dP^k, \quad (3)$$

where the coordinates A and B of extended operator are calculated by the known formulas which we don't show here;

2) employ the extended operator (3) for the input equations (1):

$$A_{Ri}-a^j*d\Phi^i/dx^j-b^k*d\Phi^i/du^k-B_{Pl}*d\Phi^i/dP^l=0; \quad (4)$$

3) substitute a derivatives R^i for $\Phi^i(X,U,P)$; and

4) split the resulting equations by parametric derivatives p^l to obtain as a result a differential equations for functions a^i and b^k that are referred to as constitutive group equations.

As a rule, a systems of a constitutive equations for Lie groups, Lie-Backlund groups and a conservation laws prove to be strongly overdetermined; their reduction to passive form greatly simplifies their further treatment. The algorithm consists of cyclic iterations of two large transformations:

1) reduction of a system to the orthonorm form with each equation being solved for a main derivative;

2) add a consistency conditions to a system of equations.

As a result, the system of differential equations is either reduced to the passive form or its inconsistency is revealed.

From a logical point of view, our system of analytic transformations can be considered as a hierarchy of virtual machines, each is oriented to a special problem and has an appropriate storage structure, a required set of data types and instructions. The general scheme of virtual machines is represented as an oriented graph with the virtual machines being its nodes (see Figure): M1 is a BESM-6 computer; M2 is a machine with special virtual storage; M3 is a basic analytic transformations machine; M4.1 - M4.4 are problemoriented machines; and M5 is a control machine. The edges outcoming from a virtual machine point to a machines that are used to interpret its operation. Machines M2 - M5 are implemented by programming. The necessity of virtual machine M2 follows from the large volume of intermediate analytic expressions and limited size of BESM-6 main storage.

The core of the system is a basic analytic transformation machine M3 which allows a treatment of a rational expressions and a system of relations, i.e. the sequence of expressions separated by a relation sign (=, /=, <=, >=, <, >). In an expressions, integers, independent and dependent variables as well as special form functions that depend upon all an independent and dependent variables and all derivatives up to the specified order may be used as operands.

In a machine M3, the following transformations can be performed on analytic expressions: parentheses removal, combining like terms, factor out, simplification of expression by its reduction to one of three canonic forms, calculation of derivatives, transition to variety specified by a system of equations in an expression, splitting an equation by the specified parameters and so on.

A fourth-level machines may be considered as a special-purpose virtual machines that are intended for execution of specific algorithm for updating a systems of differential equations. A number of machines on the level is not limited and may be steadily increased, thereby improving the intellectual level of the analytic transformation system as a whole. At present, our system contains four such machines: M4.1 - for deriving the constitutive equations for transformations of Lie groups admitted by a systems of differential equations; M4.2 - the same for the Lie-Backlund group; M4.3 - for deriving a constitutive equations for a conservation laws; M4.4 - for reducing of a differential equations systems to passive form.

```
                                   ------
          ----------------------¦ M5 ¦---------------------
          ¦               ¦        ------       ¦               ¦
          V               V          ¦          V               V
      --------        --------       ¦      --------        --------
      ¦ M4.1 ¦        ¦ M4.2 ¦       ¦      ¦ M4.3 ¦        ¦ M4.4 ¦
      --------        --------       ¦      --------        --------
          ¦               ¦          V          ¦               ¦
          ¦               ¦        ------       ¦               ¦
          --------------------->¦ M3 ¦<---------------------
                                   ------
                                     ¦
                                     V
                                   ------
                                   ¦ M2 ¦
                                   ------
                                     ¦
                                     V
                                   ------
                                   ¦ M1 ¦
                                   ------
```

Scheme of SOPHUS system

To control an applications machines and a basic analytic transformation machine, M5 machine is employes which uses a simple language to describe input data and specify a sequence of their processing by application programs. Besides, it allows to write a simple algorithms for processing analytic expressions and the systems of relations by basic transformations.

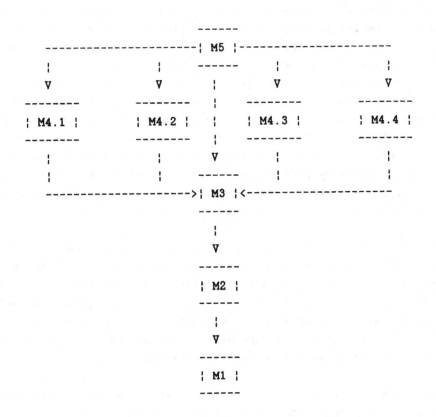

The statements of the language are executed in the interpretive mode and may be entered both in conversational and batch mode. A batch mode may overlap the conversational mode at executing the same job. The usual mode is automatic since transformations of a symbolic data by a specified algorithm may be time consuming. However, interaction with the program at execution period can greatly advance a problem solving than only automatic mode. With a special batch/conversational mode, cumbersome computations will be performed in batch mode. When user intervention is required, problem solving will be interrupted and all the required data saved. Later on the user will be able to resume problem solving in the conversational mode, properly instruct the system and then continue in the batch mode, etc.

The checkpoint mechanism available in the system allows to save the SOPHUS date on disk or tape at different times of job processing and to restore the last saved point later, if necessary, to continue execution. Of the computer fails, checkpoints allow to resume job processing without having to restart the system from the beginning, which is very important for time-consuming calculations. The use of checkpoints and batch/conversational mode of execution allows to automate the group analysis of differential equations, i.e. to find the group of transformations admitted for different values of parametric constants and functions with user intervention.

Irrespective of the fact that the majority of intermediate results are not saved many expressions that are no longer required will accumulate in computer memory in time. To clear the memory from "excess" expressions, special "garbage" collection mechanisms are provided in machines M2 and M3.

To simplify debugging of programs for basic analytic transformations and applications, the SOPHUS system provides for the possibility of accessing control machine M5 in the debugging mode and at the same time to look through and modify any data in BESM-6 main storage or M2 virtual storage. Also, debugging in greatly accelerated by the use of checkpoints.

Statistics gathering feature allows data gathering on memory allocation and any program section operation time. By analyzing the data it may locate critical sections and take more care of their optimization.

A system of analytic transformations is written in a machine-oriented high level language YARMO [4]. A total system size in object form is about 200 Kbytes with up to 50 Kbytes being allotted to a virtual storage machine, 62 Kbytes - to the basic analytic transformations machine, 40 Kbytes to application algorithms machines, 48 Kbytes to control machine.

In closing we provide the table that illustrates some characteristics of our system at solving certain problems (BESM-6 computer has about 1 MIPS, main storage size is 192 Kbytes).

Equations	Order of derivatives	Independent variables	Dependent variables	Equations	Lie group (M4.1)		Passive (M4.4)	
					Equations	Calcul. time	Equations	Calcul. time
Gas dynamics	1	2	3	3	27	49"	38	6'59"
	1	3	4	4	125	4'52"	78	59'48"
	1	4	5	5	332	17'11"	138	2^h39'42"
Lame	2	4	3	3	1206	41'28"	46	2^h 2'35"

References

1. Ovsiannikov L.V. The group analysis of differential equations.-M.: Nauka, 1978, pp. 400.
2. Kaptzov O.V. The constitutive equations of conservation laws for evolution equations.-Dynamics of continuum. Novosibirsk, 1980, N46, pp. 46-57.
3. Ibragimov N.H. On the theory of Lie-Backlund transformations.- Mathematichesky sbornik, 1979, N109, pp.229-253.
4. YARMO Language Specification /Gololobov V.I., Tcheblakov B.G., Tchinin G.D. - Novosibirsk, 1980, 43 pp. (Preprint /VTZ SO AN SSSR).

ABSTRACT

This paper presents the system of analytic transformations SOPHUS that has been intended for solving such problems us calculation of Lie [1] and Lie-Backlund [2] groups admitted by the systems of differential equations, for finding out the conservation laws of the evolution differential equations [3], for reducing the differential equations to passive form, etc.

A distinguishing feature of the SOPHUS system is its implementation as a hierarchy of virtual machines: at the lower levels of hierarchy the BESM-6 computer and the virtual machine for basic analytic transformations are placed, at the higher levels, the virtual machines for various applications (calculation of Lie and Lie-Backlund groups, etc.) are placed. The SOPHUS system was the first to implement the algorithms for calculating the Lie-Backlund groups and the conservation laws.

Operations with the system can be realized in the batch/conversational mode and is oriented to a non-programmer. The convenient input language and sophisticated service facilities such as statistics gathering, checkpoint mechanism, debugging in terms of the entities of analytic transformations, etc., are available.

INTEGRAL EQUATION WITH HIDDEN EIGENPARAMETER SOLVER: REDUCE + FORTRAN IN TANDEM

Eugene Shablygin
Institute for Nuclear Physics,
Moscow State University, Moscow 119899, USSR

1. Introduction

The problem of finding out the analytical solutions of the homogeneous linear Fredholm equations is still quite actual. Although the existence and uniqueness of the solution of such equation can be proved without any troubles, it sometimes appears rather hard to write down the explicit solution. There were a few numerical methods developed to solve the problem [1, 2]. However we should say that most of them can not be used effectively in certain circumstances. One problem arises when one try to solve an equation ot the third kind. Let us recall that the Fredholm equation of the third kind is defined as an equation of the form

$$\varphi(x)\ A(x) = \int_{\Omega} K(x,y)\ \varphi(y)\ dy\ , \tag{1.1}$$

where

$$A(x) = 0 \mid x \in \omega \subset \Omega\ . \tag{1.2}$$

Singularity of the effective kernel of the equation (1.1) leads to the well known problem of correct regularization. This problem is essential but not the only. Here we should define the term «*hidden eigenparameter*» and explain why a numerical method can not satisfy us. Let us consider the equation

$$\varphi(x) = \lambda^{-1} \int_{\Omega} K(x,y,\xi)\ \varphi(y)\ dy\ , \tag{1.3}$$

where ξ is an unknown parameter, and the usual eigenvalue λ should be of certain value. The equation (1.3) may be of the second or the third kind; in the latter case we consider that the kernel K contains all singularities. If the parameter ξ can be factored out of the kernel, then we deal with the usual eigenvalue problem, and it is possible to solve it numerically. There are a few cases, however, when the kernel of the equation depends on the variable ξ some nontrivial way. In such cases we will call the value ξ «hidden eigenparameter» (hep). If the equation (1.3) can be solved exactly, there are no problems with the hep: as far as it were possible to obtain the explicit eigenvalue dependence on hep, namely $\lambda_n = \Lambda_n(\xi)$, we can say that the «hidden spectrum» $\{\xi_n\}$ of the equation (1.3) is determined by the functional equation

$$\Lambda_n(\xi_n) = \lambda_o \ , \tag{1.4}$$

where λ_o is the pre-defined value mentioned above. As for numerical methods, to obtain the implicit function $\Lambda_n(\xi_n)-\lambda_o$ one needs to solve the «inverse problem»: for each possible ξ find out the usual spectrum $\{\lambda_n\}$, and then reconstruct the required spectrum $\{\xi_n\}$. It is clear that this way is not the best one; it is especially inconvenient when the functions $\Lambda_n(\xi)$ appeares to be nonmonotonous. The latter fact leads to multiple-valued dependence $\xi(\lambda)$ and thus produces the special troubles in the numerically calculated data interpretation.

For this reason it were nice to create a direct method for construction analytical solutions of the Fredholm integral equations with hidden eigenparameter, ever approximate. Recent paper is devoted to the method, oriented to computer algebraic system utilization.

2. General Principles

The general method for construction a solution of an eigenvalue problem is the variational one [3]. There are a lot of realizations of this approach, e.g. Ritz, Källog, momenta and some other methods. Each of them based on finding an approximate solution in a specific class of functions. Here we will try to find out the solution of the equation of the form slightly differ from (1.3), namely

$$\varphi(x) = B(x,\xi) \int_{\Omega} K(x,y)\ \varphi(y)\ dy \ , \tag{2.1}$$

where $B(x,\xi)$ is a function contains all the dependence on the hep ξ. Let us consider both the kernel $K(x,y)$ and the function $B(x,\xi)$ have sufficiently good analytical behavior for all transformations done below to be valid, for us not to discuss the problem of the formal algebraic manipulation (e.g. change of the order of summation and integration, etc.) correctness.

If a solution of the equation (2.1) exists, it can be expanded in a series in a complete set of functions in Ω. An apt choice of the set of functions leads to effective calculations of eigenfunctions; if the series tends to the solution quite quicklly, it is possible to replace the infinite series by the finite sum and thus to obtain the approximate solution. On the contrary, incorrect choice of such set can bring to the impossibility to obtain more or less correct solution.

In other words, we will try to find out eigenfunctions of the equation (2.1) in the form

$$\varphi(x) = \sum_{n=1}^{N} a_n \psi_n(x) \ , \tag{2.2}$$

where a_n are the unknown coefficients to be find out.

Let us suppose that the equation

$$\varphi(x) = \lambda \int_{\Omega} K(x,y)\ \varphi(y)\ dy \ , \tag{2.3}$$

which we call «*shortened*» regarding to the equation (2.1), has a set of exact solutions (eigenfunctions) $\{\psi_n(x)\}$ and characteristic values $\{\lambda_n\}$. We will try to construct analytical solutions of equation (2.1) in the form (2.2), using eigenfunctions of equation (2.3) as a basis. By substitution of (2.2) to (2.1), and taking into account the fact that on the set $\{\psi_n\}$ the equation (2.3) turns into identity, we get the equation:

$$\sum_{n=1}^{N} a_n \psi_n(x)\ (1 - B(x,\xi)/\lambda_n) = 0 \ , \tag{2.4}$$

where $\{a_n\}$ is an unknown vector. Now we shall reduce the latter equation to the set of linear equations. There are a lot of different methods to do so. Let us choose the following way: it is possible to differentiate (6) with respect to x in a fixed point as many times as one needs, because (6) should be satisfied by any x. We will differentiate it at x=0, then we obtain the set of equations we look for:

$$\sum_{n=1}^{N} a_n (\psi_n^{(l)} - \lambda_n^{-1} \sum_{k=0}^{l} \psi_n^{(k)} B^{(l-k)}(\xi)\ C_l^k) = 0 \ , \tag{2.5}$$

where $l \in \mathbb{N}$,

$$\psi_n^{(k)} = \frac{\partial^k}{\partial x^k} \psi_n(x) \Big|_{x=0}$$

$$B^{(k)}(\xi) = \frac{\partial^k}{\partial x^k} B(x,\xi) \Big|_{x=0} \qquad (2.6)$$

So as, we got the set of homogeneous linear equations. Its coefficients are depend on hep ξ, and this dependence is quite probable to be non-primitive. For this set of equations to have a non-trivial solution it is necessary and sufficiently to secure that the determinant of its coefficient matrix be equal to zero. It is clear that it is practically impossible to obtain the explicit functional dependence of this determinant on the hep ξ in the general case. The most convenient solution in each particular case, in our mind, is to use a computer based algebraic manipulation system. Such a system allows us to calculate the explicit expression for the determinant of the coefficient matrix of the equation set (2.5).

3. Practical example

When we were faced the problem to solve an equation of the form (2.1), and the method described above may lead to the success, because the exact solution of the shortened equation (2.3) were known, we chose Reduce 3 [4] as the most convenient system. In our particular case the function $B(x,\xi)$ was of the form

$$B(x,\xi) = \frac{\alpha \xi^2}{\sqrt{x^2 + \xi^2}}, \qquad (3.1)$$

where α is a fixed parameter (coupling constant). Derivatives of the function $B(x,\xi)$ are

$$B^{(n)}(\xi) = \begin{cases} \alpha\,(-1)^{n/2}\,\xi^{1-n}, & n = 2k \\ 0, & n = 2k+1,\ k \in \mathbb{N}. \end{cases} \qquad (3.2)$$

By the way, if the shortened equation (2.3) is the well known Schrödinger equation describing the bound state of two non-relativistic particles, the equation (2.1) with the function B of the form (3.1) is the relativistic generalization of the Schrödinger equation, so called «quasipotential» equation [5].

An explicit solution to the problem was obtained by means of Reduce 3 system running on EC-1045 computer (USSR made IBM/370 clone) with 4 Mbytes of RAM. We calculated the characteristic polynomial of the matrix of 9^{th}order - for 10th order there were no enough memory. It appeared that in the case of 9th order determinant calculation, the first three eigenfunctions and heps were calculated with sufficiently good precision. Those who are interested in the physical part of this job, we can refer to our paper [6].

When the Reduce program completes its work, we obtain the analytical answer: a tedious quotient of multivariate polynomials, zeros of which determines the hidden spectrum $\{\chi_n\}$. Unfortunately, current Reduce can not help us to analyze such expressions. The only way for a human being to understand the behavior of zeros of the resulting determinant, is to see its plot. For this reason we were forced to use numerical calculations in the final part of our work. The back side of our program tandem consists of VS Fortran quadruple precision functions genereted by the Reduce program after the determinant calculation, and a few programs perform the plot. The Fortran function generator creates the separate functions for the numerator and denominator of the determinant to ensure the absence of denominator zeros. The final analysis those was performed by the Fortran portion.

Recently the division between analytical calculations by means of Reduce and numerical ones by Fortran seems to be optimal. The future versions of Reduce (or, possibly, some other algebraic system) with integrated graphics and better package for high accuracy numerical calculation (possibly based on the principles described in [7] ?) eliminates the necessity to use Fortran. It is good prospective because of both convenience of utilization of a single system and neglecting the floating point calculation. It should be mentioned here that the usage of VS Fortran compiler (release 1.2.1) with quadruple precision (16 byte) float arithmetics and the optimization level not equal to zero in some cases leads to incorrect programs (!).

4. Conclusion

The approach described above is certainly not the only way to investigate homogeneous Fredholm equations by means of an algebraic manipulation computer system. Utilization of the Reduce 3 system allows to obtain not only approximate but ever exact solutions of a number of equations of the form (1.3), but it is the topic of the other paper.

The author express his gratitude to V.Savrin for his interest in this work and very useful discussions, and to V.Edneral, N.Kabashova and A.Kryukov for their kind assistance during computer operation.

References

[1] K.E.Atkinson. A Survey of Numerical Methods for the Solution of Fredholm Equation of the Second Kind. SIAM, Philadelphia: 1976.

[2] A.F.Verlan, V.S.Sizikov. Integral Equations: Methods, Algorithms, Programs. Naukova Dumka Publishers, Kiev: 1986 (in Russian).

[3] K.Rektorys. Variational Methods in Mathematics, Science and Engineering. Dr. Reidel Publishing Company, Dordrecht: 1980.

[4] A.C.Hearn. Reduce User's Manual, version 3.3. Rand Publ. CP78 Rev. 7/87, Santa Monica, CA: 1987.

[5] A.A.Logunov, A.N.Tavkhelidze. - Nuovo Cim., 1963, 29, #2, pp. 370-400.

[6] V.I.Savrin, E.M.Shablygin. - Theor. Mat. Phys., 1988, 75, #2, pp. 212-217.

[7] A.P.Kryukov, G.L.Litvinov, A.Ya.Rodionov. Using Reduce for High Accuracy Computations. - This issue, pp. ...

COMBINATORIAL ASPECTS OF SIMPLIFICATION OF ALGEBRAIC EXPRESSIONS

A.Ya.Rodionov, A.Yu.Taranov
Institute of Nuclear Physics, Moccow State University,
Moscow 119899, USSR

ABSTRACT. A possible way of developing computer algebra systems is an "education" of the system. By "education" we mean here supplying the system with the capacity for handling new classes of mathematical objects. Tensors or, more generally, functions of several variables are an examples of such objects of practical interest. The main features of these objects which makes nontrivial the problem of incorporating them into a computer algebra system, are the symmetry properties exhibited by tensors (functions) we meet in practice. The simplification of expressions contaning tensor monomials involves two operations: monomial identification and monomial pattern matching. The tensor identification problem when dummy summation indices are included and the tensor pattern matching problem when the pattern has free variables may both be reformulated as combinatorial problems. The first one is reducible to the double coset problem for the permutation group and the second, to a variant of the general combinatorial object isomorphism problem. The backtrack algorithms for solving the above two problems are based on some well-known ideas in computational group theory. A recursive variant of the tensor identification algorithm for a restricted class of symmetry groups and the algorithm implementation are discussed briefly.

1. Introduction

The symplification of algebraic expressions involves two basic operations: identification of some primitive subexpressions (such as monomials) connected by an algebraic transformation and pattern matching of subexpressions. Algebraic expression pattern matching is in a sence an identification of the expresion and a pattern.

An algorithmic implementation of expression identification and pattern matching strongly depends on the nature of mathematical objects involved and their representation. Here we discuss some problems arising when such objects are tensors or, more generally, functions of several variables. When we speak of tensors we mean the so-called "indexed tensors" /1,2/ rather than many-component objecs. The capacity for treating complicated mathematical objects makes some classes of problems more tractable by the system. For example, teaching the system to work with tensors we obtain the possibility to of posing some problems of the quantum field theory /3-5/ in an adequate language. Another example of the mathematical object of practical interest are special functions. In what follows we will not distinguish between tensors of the form $T_{i,\dots,j}$ and functions $T(i, \dots, j)$. The main difficulty in the simplification of tensorial expressions are the symmetry properties of a tensor. By this we mean that the tensor is invariant under some permutations of its indices. Such permutations generate a group - the symmetry group of the tensor. Typical examples of such groups are defined by the identities:

$$S(i,j,k,l) = S(j,i,k,l) = S(k,l,i,j) \tag{1}$$

$$F_3(a,a',b,b',c,x,y) = F_3(b,b',a,a',c,x,y) = \\ = F_3(a',a,b',b,c,y,x) \tag{2}$$

Here S is the symmetric Riemann tensor /6/ and F_3 is the Appel function /7/. The identities (1),(2) define the symmetry groups of 8-th order. Hence identical expressions may be connected by a symmetry transformation and in fact we have to work with equivalence classes rather then with individual objects. Moreover, tensor identification and tensor pattern matching both display symmetries of different nature leading to mixing of different equivalence classes of tensors.

In the present paper we discuss the combinatorial nature of tensor identification and matching problems and partially reformulate them in an algebraic language. Sect. 2 is devoted to matching, and Sect. 3 to tensor identification.

2. Pattern matching

To introduce the problem we give an example from Riemann geometry. This example will be used throughout the paper. Let S(i,j,k,l) be the symmetric Riemann tensor having symmetry properties defined by eq. (1).

Consider tensors T, T_1 and T_2 constructed from S:

$$T(i,j,k,l,p,q,r,s) = S(i,j,k,l)*S(p,q,r,s) \quad (3)$$

$$T_1(i,j,k,l) = S(i,j,K,L)*S(k,l,K,L) \quad (4)$$

$$T_2(i,j,k,l) = S(i,K,j,L)*S(k,l,K,L) \quad (5)$$

We mean here that K and L are dummy summation indices and the precise meaning of S(i,j,K,L)*S(k,l,K,L) is $S(i,j,K,L)*S(k,l,P,Q)*g^{KP}*g^{LQ}$ where g^{ij} is the metric tensor.

In fact T_1 and T_2 are linearly dependent due to the Bianchi identity: $T_2 = -(1/2)*T_1$. Hence the definitions (4),(5) may be written in the form

$$T(i,j,K,L,p,q,K,L) = T_1(i,j,p,q) \quad (6a)$$

$$T(i,K,j,L,p,q,K,L) = -(1/2)*T_1(i,j,p,q) \quad (6b)$$

Suppose now that we wish to simplify an expression containing monomials of the form (3) - (5) using eqs. (6) as the substitution rules. Then the left hand sides of eqs. (6) may be considered as a pattern for matching monomials in the expression under consideration.

For our purpouses it is convenient to represent the objects of the form $F(x_1,...,x_v)$ by a function f defined on the set V = {1,...,v} such that $f(i) = x_i$. We devide the range of f into two nonintersecting subsets R and D where D stands for 'dummy' or 'free' variables. The set R may include numbers, symbols etc. according to the nature of a problem. The set D consists of symbols only. Moreover, each element of D must be destinguishable from those of R. Corresponding to the function f is a group H of permutations acting on the set V = { 1 ,..., v } - the symmetry group of f. The symmetry properties of f manifest themselves as the equvivalence

$$f \sim f \circ \sigma \qquad \sigma \in H \quad (7)$$

After these preliminares we give the following definitions:

Definition 1. A combinatorial object is a pair (f, H) where f : {1,...,v} -> R and H is the symmetry group of f.

We also may regard the combinatorial object as an equvivalence class of f H.

Definition 2. A pattern for the combinatorial object (f,H) is the triple (p,H,C) where p: {1,...,v} -> R U D and $\{c_1 ... c_N\}$ - the set of constraints on values of p.

Definition 3. The matching of (f,H) and (p,H,C) consists in finding $\sigma \in$ H and φ : p(V) $\cap$ D -> R such that

$$\varphi \circ p \circ \sigma = f \tag{8}$$

$$C \circ \Phi \equiv \text{"true"} \tag{9}$$

where Φ: p(V) -> R is defined as follows: $\Phi|_{p(V)} = \varphi$, $\Phi|_R = id|_R$.

The substitution rules (6) lead to the following pattern:

p(i) = J , all J $\in$ D;

The group H is generated by the permutations

σ_1 = (12) σ_2 = (13)(24)

σ_3 = (15)(26)(37)(48)

(the permutations are given in the cyclic representation /8/). The constraints imposed by eq. 6a are

p(3) = p(7), p(4) = p(8)

and on eq. 6b

p(1) = p(6), p(4) = p(8)

So we see that the matching occurs whenever the term considered satisfies one of the whole set of the substitution rules obtained by the symmetry transformation from the initial substitution rule. A natural way to implement eqs. (8),(9) is the backtrack search over the group H. Similar algorithms are widely used in the combinatorial object automorphism and isomorphism problems /9-10/. The problem-dependent part of such a backtrack search must be organized so that the prunning of unsuccessful choices may be made at the earliest possible stage. The main idea of our matching algorithm consists in coding elements of the symmetry group H by the Schreier vectors. Let us remind of some concepts in permutation group theory /11/. Let

$$H(i) = \{ \sigma \in H : \sigma(k) = k, k \leq i \} \text{ and } H(0) = H$$

If $U(i) = H(i-1)/H(i)$ - coset space, then every coset u U(i) is uniquely defined by the value $\sigma(i)$ where σ is an arbitrary element in the coset u. In every $u \in U(i)$ we choose an element h and then identify U(i) with the set of h 's. It is easy to show that every H has a unique expansion of the form

$$\sigma = h_1 \circ h_2 \circ \ldots h_v \qquad (10)$$

where $h_k \in U(k)$. An ordered set $\{ h_i \}$ is called a Schreier vector. The expansion (10) has two properties important in our context. We give them as two statements:

Statement 1. $\sigma(k)$ is defined by $h_1, \ldots h_k$.

Statement 2. Let $\sigma = h_1 \circ \ldots h_v$, $\sigma' = h'_1 \ldots h'_v$
Then $\sigma(k) = \sigma'(k)$, $k \leq i$ iff $h_k = h'_k$ for $k \leq i$.

Now we are in a position to describe the backtrack algorithm for pattern matching of the functions possessing a symmetry group. Input of our algorithm is a pair (f,H) and a pattern (p,H,C). Output is the function φ of eq. (7),(8), or the symbol 'nil' if such a function does not exist. For simplicity, we will describe the algorithm only if explicit constraints in C are absent and all constraints are impicit, i.e. are of the form p(i) = p(j) as is really the case when the substitution rule (6) is considered. The modifications needed for the general case are straightforward.

The symmetry group H is usually specified by a set of generators. So the algorithm consists of two steps.

Step 1 is the construction of the sets of coset representatives u(i). This can be done, for example, using Algorithm 3 from Chapter II of Hoffmann 's book /12/.

Step 2 needs some subalgorithms. We list them here in a functional form.

The function NEXT(u,i) takes a Schreier vector u and a number i and returns a new Schreier vector differing from u in, at least, one k-th component with $k \leq i$. Moreover, NEXT excludes from consideration all Schreier vectors coinciding with u in the first i components. When there are no more Scheier vectors NEXT returns the symbol 'empty'.

The function FREE is a predicate which is true when its argument belongs to D. Function UNMATCH is a predicate wich is true when its argument belongs to D and not have been matched yet.

According to our statement 2 NEXT not only produces the next step in the backtrack search but also performs the prunning.

The function IMAGE (u,i) returns $h_1 \circ \ldots h_i(i)$ where $u = (h_1, \ldots h_v)$ is a Schreier vector. Note, according to the statement 1 we really need $h_1, \ldots h_i$ only.

Now we can represent the step 2 in an ALGOL-like syntax as follows:

```
BEGIN                              % initial value of Schreier vector
   sch := sch0;
   WHILE sch /= 'empty' DO
      BEGIN
         i := 1;
         math :=  'true ';
         WHILE i ≤ v AND match =  'true ' DO
            BEGIN
               v := p(IMAGE(sch,i));      % p is a "pattern"
                                          % cf. definition 2
               IF FREE(v) THEN
                 IF UNMATCH(v) THEN  φ(v) := f(i)
                 ELSE IF  φ(v)  /= f(i) THEN match := i
                      ELSE i := i+1
               ELSE IF v = f(i) THEN i := i+1
                    ELSE match := i;
            END;
         IF match =  'true ' THEN RETURN
         ELSE sch := NEXT(sch,match);
      END;
   RETURN 'nill';
END;
```

The algorithm was implemented in REDUCE. Its run-times for the substitutions (6) are quite satisfactory.

3. Tensor Identification

Now we proceed to the problem of tensor identification. Suppose we treat tensors as many-argument functions; $T_{ij...k} = T\ (i,j...k)$ and the convention of repeated indices summation as described after eq.(5) is adopted. We again represent $T\ (i_1...i_v)$ as a function

$$t: \quad V \to R \cup D \;:\; t(k) = i_k$$

but now D is a set of dummy summation indices. The invariance of a tensor under the renaming of dummy indices may be formulated as the equivalence

$$t \sim \varphi\circ t \tag{11}$$

where $\varphi|_R = id|_R$ and $\varphi|_D$ has the inverse. Together with (7) we get

$$t \sim \varphi\circ t\circ\sigma \;, \qquad \sigma\in H \tag{12}$$

where H is the symmetry group of T (and t). It is clear that the tensor identification can be regarded as canonical projection in equivalence classes defined by eq.(12). Hence we introduce an ordering on R U D:

$$r_i < r_{i+1}\;,\; d_i < d_{i+1}\;,\; d_i < r_j \tag{13}$$

where $r_i \in R$, $d_i \in D$ and we use the symbol < since this does not lead to misunderstanding. The ordering (13) defines the lexicographical order on tthe set of functions $t: V \to R \cup D$. We define the canonical representative t_c of an equivalence class { t } as the minimal element of {t}. Now without loss of generality we may suppose that $t(V)$ q $D = \{d_1,...d_k\}$ if t has k pairs of summation indices and so the function φ of eq.(11) is a premutation on $\{d_1,...d_k\}$. Let us now state the problem in an algebraic language. For a given function t we introduce the function $\tilde{t}$ defined as follows: t and $\tilde{t}$ have the same (multy) set of values and $\tilde{t}(i) > \tilde{t}(j)$ when $i > j$. (Here > has the same meaning as in eq.(13)). It is evident that there exists a (possibly not unique) permutation $\sigma_t \in S_V$ – the symmetric group on V, such that $t = \tilde{t} \circ \sigma_t$. Consider now the group A_t defined by the generators

$$g_l = (2l-1,\ 2l),\quad l = 1,\ldots k$$

$$h_l = (2l-1,\ 2l+1)(2l,\ 2l+2),\quad l = 1,\ldots k-1$$

where k is the number of dummy indices in t . It is easy to show that $\sigma \in A_t$ iff

$$\tilde{t} \circ \sigma = \varphi \circ \tilde{t} \qquad (14)$$

where φ is a function from eq.(11) renaming dummy indices $d_1,\ldots d_k$.
Let now $t_1 \sim t_{11}$. Then $\tilde{t}_1 = \tilde{t}_2 = \tilde{t}$ and $t_1 = \tilde{t} \circ \sigma_{t_1}$, $t_2 = \tilde{t} \circ \sigma_{t_2}$ and $t_2 = \varphi \circ t_1 \circ \sigma_H$ where $\sigma_H \in H$.
From eq. (14) it follows that

$$\tilde{t} \circ \sigma_{t_2} = \tilde{t} \circ \sigma_\varphi \circ \sigma_{t_1} \circ \sigma_H$$

Defining the stability group of $\tilde{t}$ Stab $(\tilde{t})$ by $t \circ \mathrm{Stab}(t) = \tilde{t}$ we get

$$\sigma_t = \sigma' \circ \sigma_\varphi \circ \sigma_{t_1} \circ \sigma_H,\qquad \sigma' \in \mathrm{Stab}(\tilde{t})$$

It is evident that Stab $(\tilde{t})$ contains all permutations acting on the sets $t^{-1}(J)$, $J \in R \cup D$. So the generators g_l lie in Stab$(\tilde{t})$ (t). Denoting by St(t) the group formed by permutations on the sets $t^{-1}(J)$, $J \in R \cup D$ we may suppose in (16) $\sigma' \in St(t)$ Thus we have demonstrated that eq.(16) hold true. The proof of the inverse statement is straightforward. So we get a Theorem: $t_1 \sim t_2$ iff

$$\sigma_{t_1} \in (A_t \otimes St(\tilde{t}))) \circ \sigma_{t_2} \circ H$$

This theorem states the equivalence of the tensor and double coset identification problems.

A detailed discussion of the tensor identification algorithm is outside the scope of the present work. But one possible improvement will, nevertheless, be touched upon. First we note that the pattern matching algoorithm described in Sect.2 is evidenttly time-linear in order of H. This is a consequence of our backtrackscheme. The similarity of the two problems under consideration become clear when we compare their basic equations eq.(8) and eq.(12). So we could modify the algorithm of Sect. 2 for the tensor identification problem. (This is what we have really done). The algorithm so constructed would be time linear in symmetry group order H too. Let us consider now the case of the tensor product of m identical tensors T with symmetry group H

$$(T_{prt})_{i...k...j} = T_{i...}*T_{...k...} * T_{...j}$$

Note that expressions of this form are frequent in applications. Straightforward computations show that the symmetry group of the tensore (17) has the order

$$| H_{prt} | = |H|^m * m!$$

i.e. it drastically grows with m even when H is small. Fortunately the specific structure of the symmetry group H makes possible a recursive treatment of this group. For the lack of space we will not discuss the method. It will be noted, however, that the recursive variant of the tensor identification algorithm is time-linear in

$$m*|H|*m! \quad \text{instead of} \quad |H_{pr}| .$$

4. Conclusions

In this paper we have discussed two closely related problems concerned with the simplification of algebraic expressions possessing symmetry. Both the problems were treated as combinatorial ones and reduced to essentially group-theoretical problems. The implementation of the algorithm in REDUCE /13/ was made and showed satisfactory run-times.

REFERENCES

1. Bogen R.A., Pavelle R. Lett.Math.Phys. 1977, 2, p.55.
2. Bogen R.A., Journ.Symb. Computations, 1985, 1.
3. De-Witt B.S. Dynamical theory of Groups and Fields. Gordon Breach, 1965.
4. Barvinsky A.O., Vilkovysky G.A. Phys.Rep., 1985, v.119, 1-74.
5. Rodionov A.Ya., Taranov A.Yu. Preprint IHEP 86-86, Serputkhov, 1986.
6. Sing J.L. Relativity the General Theory, North-Holland, Amstersdam, 1960.
7. Bateman H., Erdelyi A. Higher transcendental functions. v.1, McGRAW-HILL 1953.
8. Hall M. Theory of groups. Macmillan, New York, 1959.
9. Leon G.S. in Computational group theory. Academic Press, London, 1984, p.321.
10. Butler G., Lam C.W.H. J. of Symbolic Computations. 1985, 1, p.363.
11. Sims I.C. in Computational problems in abstract algebra. Pergamon Press,1970, p.169.
12. Hoffman C.M. Group Theoretic algorithms and Graph Isomorphism Lect. Notes in Computer Science, No.136.
13. Hearn A.C. REDUCE-2 User 's Manual, UCP 19. 1973, Univ. of Utah, USA.

Dynamic Program Improvement

P. D. Pearce
Department of Computing, Plymouth Polytechnic
J. P. Fitch
School of Mathematical Sciences, University of Bath

In this paper we consider the software tools required in a projected environment to support the improvement of user programs for algebra systems, and also the reasons why we think such a system is needed. Previously we have considered tools for applying before execution (Pearce and Hicks, 1981). Here we concentrate on the tools which would be needed to produce improvements to aspects of a programs whose effect cannot be determined prior to execution, in particular space and time usage. Our aim is to create a problem solving environment which is supportive for novices and helpful for the more experienced.

The Current Situation

The user of an algebraic manipulation system is often faced with a program that will not run to completion. All too frequently this causes the user to give up; a more hopeful response is to ask someone with more experience to help. There is a considerable amount to folklore about how to write programs for algebra systems, and previous attempts to codify this have not been particularly successful. The expert programmers for algebra systems have a tendency automatically to write their programs in certain ways and if they fail to execute as required have a list of program modifications to try, but it is hard to get them to explain them. The novice has to learn these techniques for improving the performance of programs by observation, or to rediscover them for themselves the hard way. This has been our experience for a number of years, and at a conference instigated by computer algebra users in the U.K. and organised by the Institute of Mathematics and its Applications held in Bath University during January 1987 several users confirmed the above experience in a disarmingly frank session.

Some program improvements may be performed by pre-processing, for example as reported in Pearce and Hicks(1981). However it is not always possible to predict the performance of programs prior to execution. It is possible to make an informed *guess* about the best form of a program. For example, the best code order might be one that minimises the number of active variables. Following execution an alternative order for the code may be found to be better because it builds smaller data structures. The user trying to improve a program requires an environment supporting software tools such as those described below.

Software Tools

For brevity here we give a list which indicates the range of software tools required for an improvement environment.

1) Execution of a user program.
2) Improvement of a user program by pre-processing optimisations (selected by the user).
3) Inclusion of measurements during execution *e.g.* time to execute a statement, a list of active variables, the size of expressions associated with variables or the state of execution flags.
4) Alteration of code, for example by the inclusion of substitution statements, and restart of programs from the point of execution prior to the altered code without execution of the whole program again.
5) Examination of data structures caused by different orderings of variables.
6) Recording of possibilities tried and measurements made to assist in the decision making.

Conclusions

We are pleading for system builders to pay attention to the experience of so many novice users, and help in the provision of a problem solving environment if it is not possible to provide automatic improvements. The difficulty of using algebra systems is not just because of bad design, but is intrinsic to them. The recent improved interfaces can help, but if the algebra community is to expand and bring help to the mass of potential users then to provision of an environment like the one we have outlined here is essential.

References

Pearce, P. D. and Hicks R. (1981) Proceedings of SYMSAC 81, Snowbird, Utah, pp131-136, published by ACM, New York.

Marti, J. B. (1984) Proceedings of RSYMSAC-2, Riken, Wako-shi, Japan, pp13-34, published by World Scientific, Singapore.

COMPUTER ALGEBRA AND NUMERICAL CONVERGENCE

Karl-Udo Jahn
Pädagogische Hochschule Halle/Köthen, Sektion Mathematik/Physik
Kröllwitzer Strasse 44, DDR-4050 Halle

Computer algebra systems do not yield in each case the exact solution. So by application of MACSYMA to the initial value problem

$$x' + x^2 + (2t + 1)\cdot x + t^2 + t + 1 = 0 , \quad x(1) = 1$$

the finiteness of the used number representation involves that

$$x(t) = -(0.5518192\cdot t\cdot e^t - t - 1)/(0.5518192\cdot e^t - 1)$$

is obtained as solution. However, the exact solution is

$$x^*(t) = -(1.5/e\cdot t\cdot e^t - t - 1)/(1.5/e\cdot e^t - 1) .$$

Furthermore, it is necessary to take care of the domain of the obtained solution. In the above-mentioned example the function x^* is not defined for $t = 1 + \ln(2/3)$, so that x^* is the solution of the initial value problem at most for $t > 1 + \ln(2/3)$.

A similar situation we have w. r. t. the following initial value problem:

$$x' = x^2 \quad , \quad x(0) = 1 .$$

The solution is $x^*(t) = 1/(1 - t)$ which can be obtained by symbolic integration. If it is not possible to get the solution in such a way then numerical methods will be used. In case of this example Picard iteration can be done exactly by symbolic computation with

$$x^{(0)}(t) := 1 \quad , \quad x^{(i+1)}(t) := 1 + \int_0^t (x^{(i)}(s))^2 \, ds$$

and yields polynomials depending on t. Therefore by a careless use one could believe that the solution exists for all real t.

The above-mentioned examples show that it may happen that symbolic computation allone does not suffice, even if it is applicable. Other examples from other mathematical disciplines can be found. Thus additional informations are necessary about the existence of a solution, its uniqueness, whether the algorithm converges and about error estimations.

One possibility in order to attain this consists in the use of

E-methods (which guarantee in case of applicability existence, uniqueness and enclosure of the solution) in connection with dynamic-precision arithmetic and methods of interval mathematics. In the first above-mentioned example one would get the following enclosure $X(6)$ for x^* using a 6-decimal-digit floating point mantissa interval arithmetic:

$$(X(6))(t) = -(\langle 0.551819\ ,\ 0.551820\rangle \cdot t \cdot e^t - t - 1)\ /\ (\langle 0.551819\ ,\ 0.551820\rangle \cdot e^t - 1)\ .$$

In the lecture such E-methods are represented for various problems in linear and nonlinear analysis in order to get enclosures $X(L)$ of x^* depending on the mantissa length L and having the property

$$\lim_{L \to \infty} X(L) = x^* \quad \text{(numerical convergence)}\ .$$

These methods combine numerical and analytical methods, where the termination criteria used in iteration processes are of such a manner, that the computation will be stopped when the first time during the computation characteristic properties of the algorithm will be violated as a consequence of the use of approximate values.

References

[1] Jahn, K.-U.: The importance of 3-valued notions for interval mathematics. In: Interval Mathematics 1980 (ed. K. Nickel), 75 - 98, Academic Press 1980

[2] Jahn, K.-U.: Abbrechkriterien und numerische Konvergenz. WBZ MKR der TU Dresden 58/82, 50 - 60 (1982)

[3] Kaucher, E. W. and W. L. Miranker: Self-Validating Numerics for Function Space Problems. Academic Press 1984

[4] Krückeberg, F.: Arbitrary Accuracy with Variable Precision Arithmetic. Lect. Notes in Comp. Sc. 212, 95 - 101 (1985)

[5] Kulisch, U. and W. L. Miranker: Computer Arithmetic in Theory and Practice. Academic Press 1981

[6] Nickel, K.: Über die Stabilität und Konvergenz numerischer Algorithmen I, II . Computing 15, 291 - 328 (1975)

COMPUTER ALGEBRA AND COMPUTATION OF PUISEUX EXPANSIONS OF ALGEBRAIC FUNCTIONS

V.P. Gerdt, N.A.Kostov, Z.T. Kostova
Laboratory of Computing Techniques and Automation
Joint Institute for Nuclear Research
Head Post Office, P.O. Box 79, Moscow, USSR

Extended abstract

In the present work we study the computer algebra construction of the Puiseux expansions of algebraic functions using the Newton polygon method. Let w be algebraic function defined by the equation $F(z,w)=$ $=\sum a_i(z)w^i=0$, where $a_i(z)=c_i z^{p_i}+O(z^{p_i})$. The Puiseux expansion of w near z=0 is the series

$$w = w_{\varepsilon_1}z^{\varepsilon_1}+ w_{\varepsilon_2} z^{\varepsilon_2}+ w_{\varepsilon_3} z^{\varepsilon_3}+ O(z^{\varepsilon_3}), \quad \varepsilon_1< \varepsilon_2< \varepsilon_3\ldots \qquad (1)$$

which satisfyes the equation $F(z,w)=0$. The construction of expansions (1) is a fundamental building block of Coates algorithm of finding meromorphic functions on algebraic curves [1]. These calculations are performed in [1] using the Norman´s approach . New method of computation of Puiseux expansions was proposed in [2]. The basic operations on these series are realized in [3] using the computer algebra system (CAS) REDUCE-3.

The problem of construction of series (1) reduces to computation of posible values of w_ε and ε.For the moment let us abbreviate the expression (1) to $w=w_\varepsilon z^\varepsilon+w_1$. The necessary conditions for $F(z,w)=0$ are the terms of lowest order cancel. Then we have: (i) the possible values of ε are the solutions of the following linear equations

$$p_j + j\varepsilon = p_k + k\varepsilon, \qquad (\neq p_l + l\varepsilon),\ j,k,l=0,\ldots,n, \qquad (2)$$

and (ii) the possible w are the roots of

$$\sum c_h\alpha^h = 0, \qquad (3)$$

where the summation is over all values h for which $c_h+h\varepsilon=c_j+j\varepsilon$.Because the explicit calculations are tedious the REDUCE-3 program NEWTON has been developed for the construction of Puiseux expansions using the Newton polygon method [5].The programm characteristics are the following: i)computer EC 1061 (IBM 370), operating system TKS, ii) programming language REDUCE-3.2, high speed storage required, depends on the problem, minimum 600K, number of lines 1000. Similar computer algebra (CA) - program was developed in [5,6]. The comparison is not possible because the program characteristics in [5,6] are not given.

Our algorithm is based on the followings theorems: i)after a finite number of transformation steps of the form $z=z_0+t^r, w=w_0+t^s w_1$ a polinomial (3) will be reached which has only simple roots i.e. the process terminates after a finite number of steps (the finitness of algorithm) ii) the algorithm Newton will provide a complete set of non-equivalent branches of w. The complexity of the Newton polygon method was given in [5,6]. The CPU time of some tests we give in Table 1.

<u>Examples:</u>

$$2z^7-z^8-z^3w + (4z^2+z^3)w^2+(z^3-z^4)w^3 -4zw + +7z^5w^5 + (1-z^2)w^6+5z^6w^7+z^3w^8 = 0.$$

Puiseux expansions:

$$w = 2z^4-z^5+\dots,$$

$$w = 1/4\ z -3/64\ z^2+\dots,$$

$$w = \sqrt{2}\ z^{1/2} \pm \sqrt[4]{2}/4\ z^{3/4}+\dots,$$

$$w =-\sqrt{2}\ z^{1/2} \pm \sqrt[4]{2}/4\ z^{3/4}+\dots.$$

TABLE 1.

CURVE	CPU TIME	Degree of Puise expansions	TYPE of singularity
$w^2+z^3-z^2$	2157 ms	5	double point, distinct tangents
w^3-z^3	1969 ms	5	cusp
$w^2z^2-w^2z+w^2+z^4-2z^2$	2664 ms	5	ramphoid cusp
$-w^5-w^3z^2+z^6$	5326 ms	5	one triple tangent two single tangents
$2w^4-3w^2z+z^4-2z^3+z^2$	4407 ms	5	tacnode
$w^6+3w^4z^2+3w^2z^4- -4w^2z^2+z^6$	19478 ms	5	ordinary triple point

REFERENSES

[1] Davenport J.H.(1981). On the Integration of Algebraic functions, *Lect. Notes in Comp.Sci.*, 102, Springer-Verlag.

[2] Chudnovsky D., Chudnovsky G.(1986). Power Series and Fast Numeric Evaluation of Algebraic Functions and Linear Differential Equations, *The Scratchpad II Newsletter*,1,1.

[3] Feldmar E., Kolbig K.S.(1986). REDUCE Procedures for Manipulation of Generalized Power Series, *Comp.Phys.Comm.* ,39,267.

[4] Walker R.J.(1978). Algebraic curves,Springer, New York.

[5] Duval D.(1987). Calcul formel avec des nombres algebriques, These de d´Etat,Universite de Grenoble, p.199.

[6] Davenport J., Siret Y., Tourner E.(1987) Calcul formel, Masson.

BOUNDARY VALUE PROBLEMS FOR THE LAPLACIAN IN THE EUCLIDEAN SPACE SOLVED BY SYMBOLIC COMPUTATION

F. Brackx and H. Serras

Seminar of Mathematical Analysis, State University of Ghent
Sint-Pietersnieuwstraat 39, B-9000 Gent (Belgium)

Abstract

By means of the symbolic manipulation language REDUCE, a complete set of harmonic polynomials in Euclidean space is constructed. The method relies on the theory of monogenic functions defined in $\mathbb{R}^{n+1}$ and taking values in a Clifford algebra A_n. An approximate solution for the Dirichlet and Neumann problems in the form of a harmonic polynomial may be obtained. The method is applied to solving a Dirichlet problem in a cube.

1. Introduction

In [1] R. Wilkerson solved Dirichlet's problem in the plane by calculating symbolically an approximate solution in the form of a harmonic polynomial. The method goes back to the observations made by Zaremba[2] and relies on the construction of a basis of orthogonal harmonic polynomials. Let D be a domain with boundary ∂D and let f be a continuous real-valued function defined on ∂D. Then Dirichlet's problem consists of finding a continuous function $u: \overline{D} \to \mathbb{R}$ such that u is harmonic in D and coincides with f on ∂D. Provided the shape of the boundary and the functions appearing in the boundary conditions satisfy certain mild requirements, existence and uniqueness theorems may be proved and are of course essential to the solution of this problem. Nevertheless the explicit construction of the solution is often not feasible.
Now assume for a moment the existence of a complete set of harmonic polynomials R. First they have to be orthogonalized, using a Gram-Schmidt process, with respect to the inner product defined by

$$<u,v> = \int_D \overline{\nabla}u.\overline{\nabla}v \; d\omega \, .$$

So the obtained orthogonal basis of harmonic polynomials Q satisfies

$<Q_i,Q_j> = 0$ whenever $i \neq j$, while $<Q_i,Q_i> = N_i \neq 0$ for all i.

Now the solution u to Dirichlet's problem, up to a constant, takes the form [3]

$u = \sum_{i=1}^{\infty} c_i Q_i$ where the Fourier coefficients may be computed in terms of the values of u on the boundary ∂D. Indeed, applying Green's Identity yields for all j = 1,2,...

$$N_j c_j = \langle u,Q_j\rangle = \int_D \bar{\nabla}u.\bar{\nabla}Q_j \, d\omega = \int_{\partial D} u \frac{\partial Q_j}{\partial n} ds = \int_{\partial D} f \frac{\partial Q_j}{\partial n} ds$$

where $\frac{\partial}{\partial n}$ stands for the normal derivative.
The above method is now used to solve boundary value problems for the Laplacian in Euclidean three-space, handling the symbolic manipulation language REDUCE. In fact the three main problems to be attacked by REDUCE are:

(i) generation of a basis of harmonic polynomials in three real variables;

(ii) the Gram-Schmidt orthogonalization process;

(iii) calculation of the Fourier coefficients.

Whereas problem (i) causes no restrictions to the data of the problem, steps (ii) and (iii) are responsible for the restrictions on the shape of the domain D considered and the boundary values f imposed; indeed the volume and surface integrals involved are to be within the symbolic integration capabilities of REDUCE.
As to the precision of the obtained approximating polynomial, it may be noticed that harmonic functions attain their extremal values on the boundary of a connected region. So, as the exact solution as well as the approximate polynomial solution both are harmonic, $\sup_D |u_{exact} - u_{appr}| = \sup_{\partial D} |u_{exact} - u_{appr}|$, and hence it is sufficient for calculating the error in the sense of the sup-norm, to consider the values of the approximate polynomial solution on the boundary ∂D. It should be noted that this method also applies to the Neumann-problem.

2. Construction of a generating set of harmonic polynomials in space

The construction of a generating set of harmonic polynomials in the plane is a direct consequence of the theory of holomorphic functions in $\mathbb{C}$. Indeed each holomorphic function in an open Ω containing the origin, may be developed in powers of $z = x + iy$ in a circular domain with radius $d(0,co\Omega)$. As real and imaginary parts of a holomorphic function are harmonic, the real and imaginary parts of z^n, $n = 1,2,...$ provide a set of harmonic polynomials which is complete (see also [4]).
In Euclidean three-space the situation is clearly different. Nevertheless we may rely on a theory of functions defined in arbitrary Euclidean space $\mathbb{R}^{n+1}$ and taking values in a finite dimensional Clifford algebra A_n, which is indeed a direct generalization to higher dimension of the theory of holomorphic functions in $\mathbb{C}$ [5].

If $(e_1, e_2, \ldots, e_n)$ is an orthonormal basis of $\mathbb{R}^n$, then a basis of the Clifford algebra A_n constructed over $\mathbb{R}^n$ and containing $\mathbb{R}^n$ as a subspace, is given by

$$(e_A = e_{h_1 h_2 \ldots h_r} = e_{h_1} \cdot e_{h_2} \cdot \ldots \cdot e_{h_r} \quad : 1 \leq h_1 < \ldots < h_r \leq n)$$

where $e_\emptyset = e_0$ is the identity element and the multiplication is governed by the rules:

$$e_i^2 = -1 \ , \ i = 1, \ldots, n$$

$$e_i e_j + e_j e_i = 0 \quad , \ i \neq j \ .$$

Conjugation in A_n is defined by

$$\bar{\lambda} = [\sum_A e_A \lambda_A]^{-} = \sum_A \bar{e}_A \lambda_A \quad , \ \lambda_A \in \mathbb{R}$$

where $\bar{e}_A = (-1)^{n(A)(n(A)+1)/2} e_A$, $n(A)$ being the cardinality of A.

This associative but non-commutative Clifford-algebra is not a division algebra.

A function f defined in an open set $\Omega \subset \mathbb{R}^{n+1}$ and with values in A_n takes the form $f(x) = \sum_A e_A f_A(x)$ where $f_A : \mathbb{R}^{n+1} \to \mathbb{R}$.

Such a function is called left monogenic if it is in $C_1(\Omega)$ and satisfies the generalized Cauchy-Riemann equations : $Df = 0$ in Ω, where D is the differential operator $D = e_0 \frac{\partial}{\partial x_0} + e_1 \frac{\partial}{\partial x_1} + \ldots + e_n \frac{\partial}{\partial x_n}$.

Now, as $\bar{D}D = D\bar{D} = \Delta$, Δ being the Laplacian in $\mathbb{R}^{n+1}$, it follows that a monogenic function in Ω and all of its components f_A are automatically harmonic functions -and hence C_∞ functions- in Ω.

So a method for constructiong a complete set of harmonic polynomials in $\mathbb{R}^{n+1}$ $(n > 1)$ is at once clear: take the analogues of the "powers of z" in the complex plane; these are homogeneous monogenic polynomials V, the so called spherical monogenics; the components of each such polynomial V are the desired harmonic polynomials in $\mathbb{R}^{n+1}$.

The homogeneous polynomials $V_{l_1 \ldots l_k}$ are given by

$$V_{l_1 \ldots l_k}(x) = \frac{1}{k!} \sum_{\pi(l_1, \ldots, l_k)} z_{l_1} \ldots z_{l_k}$$

where the sum runs over all distinguishable permutations of all of $(l_1, \ldots, l_k)$ and the hypercomplex variables z are defined as follows:

$$z_l = x_l e_0 - x_0 e_l \ , \ l = 1, \ldots, n.$$

Now by means of REDUCE we are able to carry out all algebraic computations in a Clifford algebra of arbitrary dimension [6].

For the given Dirichlet problem (n=2) we constructed the polynomials V in two hypercomplex variables $ye_0 - xe_1$, $ze_0 - xe_2$, the components R(x,y,z) of which are the desired harmonic polynomials. For any $k \geq 1$ we obtain 3(k+1) homogeneous harmonic polynomials of degree k. From these $\sum_{i=1}^{k} 3(i+1) = \frac{3k(k+3)}{2}$ polynomials up to degree k we retain by elimination a complete set of $(k+1)^2$ homogeneous harmonic polynomials. It is important to notice that the above method generalizes at once to higher dimensions, thus providing a basis for the space $H_{k,n}$ of harmonic homogeneous polynomials of degree k in n real variables, for all k. The REDUCE programs leading up to the construction of a complete set of homogeneous harmonic polynomials R(x,y,z) up to the fifth degree run as follows:

```
COMMENT Stucture of the polynomials in the dummy
        operators U(1) and U(2) ;
OPERATOR P, Q, R, H, G, U, E,RR;
NONCOM U , E ;
N:=5 ; % N: highest degree
FOR K:=1:N DO WRITE P(K):=(U(1)+U(2))**K;
FOR I:=1:N DO << FOR J:=1:2**I DO WRITE
        Q(I,J):=PART(P(I),J)>>;
RR(0,0):=0;
FOR K:=1:N DO <<
        L:=2**K ;
        FOR I:=0:K DO <<
                M:=K-I; RR(I,M):=0;
                        FOR J:=1:L DO <<
                        IF DF(Q(K,J),U(1),I,U(2),M) NEQ 0 THEN
                        RR(I,M):=RR(I,M)+Q(K,J)>>;
                WRITE "RR(",I,",",M,"):= ",RR(I,M) >>;
                        >>;

COMMENT  The monogenic polynomials ;
FOR ALL I LET E(I)**2 = -1;
LET E(1)*E(2)+E(2)*E(1)=0;
U(1) := Y - X*E(1);
U(2) := Z - X*E(2);
factor E;
FOR I:=1:N DO <<
        FOR J:=0:I DO  WRITE "RR(",I-J,",",J,"):= ",RR(I-J,J)>>;
```

```
COMMENT   Harmonic components ;
FOR M:=0:N DO << FOR K:=0:N-M DO <<
        WRITE H(M,K,0):=RR(M,K)-LTERM(RR(M,K),E(1))
                -LTERM(RR(M,K),E(2));
WRITE H(M,K,1):=LCOF(RR(M,K),E(1));
WRITE H(M,K,2):=LCOF(RR(M,K),E(2))>>;>>;
FOR I:=1:N DO << L:=0 ;
        FOR J:=0:I DO <<
                FOR K:=0:2 DO <<
                L:=L+1;
                 G(I,L):=H(I-J,J,K)>>;>>;>>;


COMMENT From the 3*N*(N+3)/2 harmonic polynomials G(I,K),
        I=1,...,N      K=1,...,3*(I+1) we select a basis
         R(L), L=0,...,(N+1)**2-1     R(0)=1        ;
L:=0;
FOR I:=1:N DO << M:=0;
        J:=3*(I+1);
        FOR K:=1:J-2 DO <<
                M:=M+1;
                IF M<3 THEN
                        <<L:=L+1;
                         R(L):=G(I,K);
                        >> ELSE M:=0>>;>>;
```

3. Test of the described method to Dirichlet's problem in a ball

We applied the REDUCE-method as explained in the introduction to several boundary value problems in Euclidean space. A decisive test was furnished by solving the following Dirichlet problem:
find u continuous on $\overline{B}(0,1)$ such that u is harmonic in $\mathring{B}(0,1)$ and u = f on the sphere $\partial B(0,1)$ where, in spherical coordinates,

$$f(\phi,\theta) = 2 + P_3^2(\cos\theta).\cos 2\phi$$
$$= 2 + \frac{15}{4}(\cos\theta - \cos 3\theta)\cos 2\phi,$$

P_3^2 being the associated Legendre function.
By solving this problem by the REDUCE method the following observations were made:
(i) substituting spherical coordinates in the basis of harmonic polynomials turned them into linear combinations of the well known spherical

harmonics, being already orthogonal on the ball;
(ii) the calculated Fourier coefficients turned out to be zero except for c_{10} and c_{14}, which generated the exact solution:

$$u(\rho,\phi,\theta) = 2 + \frac{15}{4}\rho^3(\cos\theta - \cos 3\theta)\cos 2\phi .$$

4. Illustrative example

The Dirichlet problem

$$\Delta u = 0 \text{ in the cube }]0,1[\times]0,1[\times]0,1[,$$
$$u = \sin\pi x \sin\pi y \text{ for } (x,y) \in]0,1[\times]0,1[, z = 1,$$
$$u = 0 \text{ elsewhere on the boundary,}$$

has the exact solution:

$$u(x,y,z) = \frac{\sin\pi x \sin\pi y \, \mathrm{sh}(\sqrt{2}\pi z)}{\mathrm{sh}\sqrt{2}\pi} .$$

By the explained method we obtained an appoximate solution in the form of a harmonic polynomial of the fifth degree. In spite of this low degree the absolute maximal error is at most 0.15 as is shown in the following table where u_{exact} is the value of the exact solution and u_{appr} is the value of the calculated approximating harmonic polynomial.

x	y	z	u_{appr}	u_{exact}	$\lvert u_{exact}-u_{appr}\rvert$
0.000	0.000	0.000	0.00000	0.00000	0.00000
0.000	0.333	0.000	0.09524	0.00000	0.09524
0.000	0.667	0.000	0.09524	0.00000	0.09524
0.000	1.000	0.000	0.00000	0.00000	0.00000
0.333	0.000	0.000	0.09524	0.00000	0.09524
0.333	0.333	0.000	0.04575	0.00000	0.04575
0.333	0.667	0.000	0.04575	0.00000	0.04575
0.333	1.000	0.000	0.09524	0.00000	0.09524
0.667	0.000	0.000	0.09524	0.00000	0.09524
0.667	0.333	0.000	0.04575	0.00000	0.04575
0.667	0.667	0.000	0.04575	0.00000	0.04575
0.667	1.000	0.000	0.09524	0.00000	0.09524
1.000	0.000	0.000	0.00000	0.00000	0.00000
1.000	0.333	0.000	0.09524	0.00000	0.09524
1.000	0.667	0.000	0.09524	0.00000	0.09524
1.000	1.000	0.000	0.00000	0.00000	0.00000
0.000	0.000	0.333	0.07910	0.00000	0.07910
0.000	0.333	0.333	0.05949	0.00000	0.05949
0.000	0.667	0.333	0.05949	0.00000	0.05949
0.000	1.000	0.333	0.07910	0.00000	0.07910
0.333	0.000	0.333	0.05949	0.00000	0.05949
0.333	0.333	0.333	0.09628	0.03679	0.05949
0.333	0.667	0.333	0.09628	0.03679	0.05949
0.333	1.000	0.333	0.05949	0.00000	0.05949
0.667	0.000	0.333	0.05949	0.00000	0.05949
0.667	0.333	0.333	0.09628	0.03679	0.05949
0.667	0.667	0.333	0.09628	0.03679	0.05949
0.667	1.000	0.333	0.05949	0.00000	0.05949

1.000	0.000	0.333	0.07910	0.00000	0.07910
1.000	0.333	0.333	0.05949	0.00000	0.05949
1.000	0.667	0.333	0.05949	0.00000	0.05949
1.000	1.000	0.333	0.07910	0.00000	0.07910
0.000	0.000	0.667	0.14128	0.00000	0.14128
0.000	0.333	0.667	0.05548	0.00000	0.05548
0.000	0.667	0.667	0.05548	0.00000	0.05548
0.000	1.000	0.667	0.14128	0.00000	0.14128
0.333	0.000	0.667	0.05549	0.00000	0.05549
0.333	0.333	0.667	0.22720	0.17013	0.05707
0.333	0.667	0.667	0.22720	0.17013	0.05707
0.333	1.000	0.667	0.05548	0.00000	0.05548
0.667	0.000	0.667	0.05549	0.00000	0.05549
0.667	0.333	0.667	0.22720	0.17013	0.05707
0.667	0.667	0.667	0.22720	0.17013	0.05707
0.667	1.000	0.667	0.05549	0.00000	0.05549
1.000	0.000	0.667	0.14129	0.00000	0.14129
1.000	0.333	0.667	0.05550	0.00000	0.05550
1.000	0.667	0.667	0.05549	0.00000	0.05549
1.000	1.000	0.667	0.14129	0.00000	0.14129
0.000	0.000	1.000	-0.08890	0.00000	0.08890
0.000	0.333	1.000	0.11483	0.00000	0.11483
0.000	0.667	1.000	0.11484	0.00000	0.11484
0.000	1.000	1.000	-0.08890	0.00000	0.08890
0.333	0.000	1.000	0.11484	0.00000	0.11484
0.333	0.333	1.000	0.77721	0.75000	0.02721
0.333	0.667	1.000	0.77721	0.75000	0.02721
0.333	1.000	1.000	0.11484	0.00000	0.11484
0.667	0.000	1.000	0.11485	0.00000	0.11485
0.667	0.333	1.000	0.77721	0.75000	0.02721
0.667	0.667	1.000	0.77722	0.75000	0.02722
0.667	1.000	1.000	0.11486	0.00000	0.11486
1.000	0.000	1.000	-0.08888	0.00000	0.08888
1.000	0.333	1.000	0.11485	0.00000	0.11485
1.000	0.667	1.000	0.11485	0.00000	0.11485
1.000	1.000	1.000	-0.08887	0.00000	0.08887

5. REDUCE programs

The examples mentioned above were worked out with the REDUCE 3.2 installed on a VAX 750. The interested reader can obtain the full REDUCE programs for a nominal fee, by simple request to the first author.

6. Conclusion

The importance of the described method lies in the fact that once a complete set of harmonic polynomials is available it may be applied to any domain in three-space as long as the volume integrals and the surface integrals involving this domain, its boundary and the boundary-values may be calculated with the symbolic manipulation language available.

Generating the necessary harmonic polynomials is the main problem: this is linked to a higher dimensional function theory and needs algebraic computation in a non-commutative Clifford-algebra.

REFERENCES

[1] R.W. WILKERSON, Symbolic Computation and the Dirichlet problem, Eurosam 84, Lecture Notes in Computer Science 174, 59-63.

[2] S. ZAREMBA, L'équation biharmonique et une classe remarquable de fonctions fondamentales harmoniques, Bull. Inter. de l'Acad. Scie. de Cracovie (1907), 147-196.

[3] S. BERGMANN, Über die Entwicklung der harmonischen Funktionen der Ebene und des Raumes nach Orthogonalfunktionen, Math. Annalen 86 (1922), 238-271.

[4] J. WALSH, The approximation of harmonic functions by harmonic polynomials and by harmonic rational functions, Amer. Math. Soc. Bull. 35 (1929), 499-544.

[5] F. BRACKX, R. DELANGHE and F. SOMMEN, Clifford Analysis, Pitman London, 1982

[6] F. BRACKX, D. CONSTALES, R. DELANGHE and H. SERRAS, Clifford Algebra with REDUCE, Rend. Circ. Mat. Palermo, Ser. II, 16, (1987), 11-19.

THE METHODS FOR SYMBOLIC EVALUATION OF DETERMINANTS AND THEIR REALIZATION IN THE PLANNER-ANALYTIC SYSTEM

Vitali A. Eltekov and Vladimir B. Shikalov
Department of Physics, Moscow State University
Moscow 117234, USSR

The necessity for a determinant to be evaluated symbolically, i.e. to be presented as an algebraic expression with all the fessible simplifications made, arises in the numerous applied analysis problems whose solutions or solvability conditions can be expressed through determinants. If the order of the determinant is high enough, the search for the most effective technique of its symbolic evaluation gets especially urgent because the computer time involved increases rapidly, depending on the order of the determinant. When finding the effectiveness criteria, it should be borne in mind that in most of the practically interesting cases a considerable fraction of the determinant elements can be zeroes, while the non-zero elements can be expressions of various complexities. Besides, many of the minors of the examined determinants may prove to be zeroes. Theese factors make it very difficult to derive an universal effectiveness criterion.

1. Review of the various methods for symbolic evaluation of determinants

1.1. The Gaussian elimination method

The main drawback of the Gaussian method in the analytical transformations is that the algorithm contains the division operation which is much more tedious compared with multiplication. In case of division, the greatest common divisor for two expressions has to be found, whereas in case of multiplication it is as a rule sufficient to collect like terms.

A modification of the Gaussian method exists, however, where the greatest common divisor has not to be found. The modification was proposed [1] for calculating integer-valued determinants, but can readily be used in analytical transformations [2,3].

In the latter case the matrix elements of the k-th elimination are found from the formula

$$a_{ij}^{(k)} = \frac{a_{kk}^{(k-1)} a_{ij}^{(k-1)} - a_{kj}^{(k-1)} a_{ik}^{(k-1)}}{a_{kk}^{(k-2)}} \qquad (1)$$

where $i > k$, $j > k$, $a_{oo}^{(-1)} = 1$.

It can be demonstrated that $a_{ij}^{(k)}$ is a minor of the form

$$a_{ij}^{(k)} = \begin{vmatrix} a_{11}^{(o)} & \dots & a_{1k}^{(o)} & a_{1j}^{(o)} \\ \dots & \dots & \dots & \dots \\ \dots & \dots & \dots & \dots \\ \dots & \dots & \dots & \dots \\ a_{k1}^{(o)} & \dots & a_{kk}^{(o)} & a_{kj}^{(o)} \\ a_{i1}^{(o)} & \dots & a_{ik}^{(o)} & a_{ij}^{(o)} \end{vmatrix} \qquad (2)$$

so the division in (1) is exact.

1.2. The direct summation method

The method is based on the presentation of a determinant as

$$\det [a_{ij}] = \sum_{d} (-1)^{N(d)} a_{d_1 1} a_{d_2 2} \dots a_{d_n n} \qquad (3)$$

where the summation is made with respect to all $n!$ permutations taken n elements at a time; $N(d)$ is the number of disorders, i.e. such pairs $(i.j)$ that we get $i < j$ and $d_i < d_j$ simultaneously.

To make the calculations less tedious, we used the following considerations: if the first k elements of permutation $(d_1, d_2, \dots d_k)$ are fixed, the product $a_{d_1 1} a_{d_2 2} \dots a_{d_k k}$ will enter $(n - k)!$ terms which can all be obtained by adding the rest $n - k$ elements to the set of $(d_1, d_2, \dots d_k)$ permutations. If the element $a_{d_k k}$ is zero, the rest $(n - k)!$ permutations have not to be generated because the associated products will be zeroes. Obviously, the saving of computer time will be most substantial if a zero appears in the first column of the determinant. Therefore, to evaluate a high-order determinant by this particular method, it is necessary beforehand to order the matrix columns with respect to lack of increase in the number of the non-zero elements.

It should be noted that the above-described method is not effective when the simplicity of the expression resulting from the symbolic evaluation of a determinant arises from the fact that some of its minors are zeroes rather than from the presence of a great number of zero-valued elements in its matrix. This can be exemplified by the determinant whose matrix consists of identical expressions. Although the minors of such a determinant are higher than the first order, the required computer time does not decrease. The particular examples illustrating the realization of this method can be found in [3,4].

1.3. The method of expansion in the line elements

The method is based on the following expansion of a determinant:

$$\det [a_{ij}] = \sum_{i=1}^{n} (-1)^{i+1} a_{i1} M_{i1} \qquad (4)$$

where M_{i1} are the determinant minors obtainable by deleting the first column and i-th line. Since the problem of evaluating the minor M_{i1} symbolically is again the problem of evaluating the determinant whose order is the initial one less unity, the relation (4) is recurrent, so the method is readily realizable using a recursive algorithm.

If the operation of the algorithm in the symbolic evaluation of a sufficiently high-order determinant is analysed, the symbolic evaluation of the minors M_{i1} made independently of each other will appear to lead to multiple repetition of the same operations because M_{i1} contains identical minors of lower orders. This is why the given algorithm is but insufficiently effective.

1.4. The reverse method of expansion in the line elements

To avoid repeating the identical operation arising in the preceding method, the author of [5] suggested that the above-described algorithm should be reversed, the symbolic evaluation must be started from the second-order minors located in the first column pair of the matrix, and the k-th order minors must be evaluated symbolically using the elements of the k-th column and the ready expressions for the (k-1)-th order minors from the first k-1 columns. In this case, as usual, the increase in the transformation speed involves an increased volume of the memory required. The direct algorithm requires that not more than n minors should be stored, whereas the reversed algorithm involves not more than $C_n^{n/2}$ minors.

1.5. The Laplace theorem based methods

According to the Laplace theorem (see, for example, [6]) a determinant can be presented as

$$\det [a_{ij}] = \sum_{i_1 \dots i_k} (-1)^{i_1+i_2+\dots+j_k} M_{j_1\dots j_k}^{(i_1\dots i_k)} \overline{M}_{j_1\dots j_k}^{(i_1\dots i_k)} \quad (5)$$

where the summation is taken with respect to all C_n^k subsets from k lines of the determinants. The main advantage of the algorithms based on the relation (5) is that, if the minor $M_{j_1\dots j_k}^{(i_1\dots i_k)}$ is zero, its supplementary minor $\overline{M}_{j_1\dots j_k}^{(i_1\dots i_k)}$ may not be evaluated symbolically. Obviously, the subsets of the columns must be discriminated using the successive binary expression, thereby reducing the recursive depth substantially.

2. Estimates of the effectiveness

The effectiveness of the various algorithms for symbolic evaluation of a determinant is fairly difficult to estimate because of the features of analytical transformations when they are made with computer. Contrary to the numbers, the computer transformed analytical expressions have a variable lehgth which is defined by the particular operations made on them during the preceding stages of the transformations.

With the purposes of the comparative analysis of the algorithms for symbolic evaluation of a determinant, it is neccessary to select the type of elements which reflects to a sufficient extent the particular features of analytical transformations and, simultaeously, is sufficiently simple for the number of the operations required ahd for the memory involved to be estimated theoretically. These requirements are made best satisfied by the polynomials which, besides, constitute one of the most extensively used types of expressions in mathematical analysis.

Such an analysis was made in [3] for the direct summation algorithms and for the modified Gaussian method. The total number of operations was estimated to be

$$N_1 = \frac{12m^{2q}n^{2q+3}}{(2q+1)(2q+2)(2q+3)} \quad (6)$$

for the modified Gaussian method and

$$N_2 = \frac{2m^{2q}n^{q+1}}{q+1} n! \quad (7)$$

for the direct summation method. Here, q is number of variables, m is power of polynomials, n is order of the determinant. Obviously, the adopted assumption that but few of the polynomial coefficients are non-zero is rather far from being realistic.

By making the above estimates more rough, we obtain that the ratio of the numbers of operations is

$$N_1/N_2 \approx n^{q+2}/(q^2 n!) \tag{8}$$

whence it follows that each of the methods has its own applicability domain, namely, the Gaussian method is most effective at high values of n, while the direct summation method effectiveness is peaking at high values of q.

The comparison among the rest methods mentioned in Sec. 1 does not require that such detailed estimates should be made because they involve essentially the same operations, but in different orders. When estimating the productivity gain in going over from direct to reverse method of expansion in the column, the number of multiplications proves to be $n!$ in the direct method and $\sum_{k=1}^{n} kC_n^k \approx n2^{n-1}$ in the reverse method. So, the reverse method, as compared with the direct method, is always more effective and gives a substantial gain already at comparatively low values of n.

The comparison of the reverse expansion method with the direct summation method shows that they differ only in the order of addition and multiplication. At the same time, the reverse method unvolves a much smaller number of additions which are, besides, simpler. In the method, however, the volume of operations is reduced but slightly when the determinant contains zero-value elements.

The Laplace theoram based methods are extremely complicated and prove to be to advantage only under highly specific conditions. Besides they require a substantial volume of memory.

Thus, we can single out the following three methods each of which proves to be most effective under certain conditions imposed on the determinant matrix:

(i) the modified Gaussian method proves to be most effective in case of large matrices with relatively simple, but non-zero elements;

(ii) the direct summation method is most effective in case of a rarefied matrix of the determinant. The method should also be applied when the memory volume is constrained;

(iii) the reverse method of expansion in the line elements proves to be most effective in the rest cases.

3. Realization of the algorithm for symbolic evaluation of a determinant

We have chosen to realize the algorithm for reverse expansion in the line elements because it is most effective in case of symbolic evaluation of the determinants whose matrix does not exhibit a definite specificity. The algorithm has been realized in the Planner-analytic system [7-9] whose input and instrumental language is Planner [10].

The realization of the algorithm gives rise to the following two interrelated problems: generation of all the subsets of the set of columns consisting of k elements and storage of the minors of order k to be used in the next stage of the algorithm. Obviously, the minors can best be stored using a many-level list because the Planner language offers ample opportunities of operating with lists. In particular, the necessary minor may be selected through a single reference to the function INDEX.

The elemental matrix of the determinant to be evaluated symbolicly is set to be of the form of two-level list:

((A11 A12 ... A1K) (A21 A22 ... A2K) ... (AK1 AK2 ... AKK).

Presented below in the Planner language is the basic part of the algorithm for symbolic evaluation of the matrix A determinant.

```
[DEFINE  DET  (LAMBDA (A)
```

% The function is determined with identifier DET and argument A ;

```
[PROG  (N A1 (K 2) B) [SET  N  [LENGTH  .A]]
```

% The local variables A1, N, K, B are introduced, with K being given the initial value 2 and N being the value of the order of the determinant;

```
[SET  B [APL  '$' [1  .A] ] ]
```

% B is the list of second-order minors;

```
A [SET  A1  [.K  .A]] [RPLAC  :MINORS  .K  .B]
```

% Origin of the cycle, the k-th line of the elements is sent to A1 , in the glogal list MINORS the k-th element is replaced by B ;

```
[SET  B [ML  .K  ()  [-  .N  .K]]]
```

% Forming the set of minors in the k-th step;

```
[COND  ( [LT  .K  .N] [ ADD1  K] [GO  A])
```

% If K is smaller than N , K is raised by unity, return to the cycle origin;

```
(T [RETURN [ =&= [EVAL [FORM [INDEX .B <ARRAY .N 1>]]]]] )]
```

% otherwise return to the result and end of determining the function DET ;

```
])]
```

In the above presented algorithm text written in the Planner language, the chains of symbols between % and ; are comments. The functions '$' and =&= are determined in the Planner-analytic system and correspond to relocation of the expression from external to internal representation. The functions APL and ML are else determined (they are not described here). Before calling the function DET , it is necessary to compile the reference list MINORS consisting of N empty lists, where N is the order of the determinant:

```
[CSET MINORS [ARRAY N ()]]
```

The above-described algorithm was presented as a program package and showed a sufficiently high effectiveness.

REFERENCES

1. Wouch,E.V., Dwyer,P.S.: Compact computation of the inverse of a matrix, Ann. Math. Statistics, 1945, v.18, No.3, pp.259-271.

2. Chubarov,M.A.: Computation of polynomial determinants by the modified Gaussian method with selecting the basic element in the line, in: Computational Mathematics, Programming, and Experimental Data Processing, Naukova Dumka, Kiev, 1979, pp.47-52.

3. Poddashkin,A.V., Chubarov,M.A.: Computation of polynomial determinants in the Sirius hierarchical system, ibid, pp.72-80.

4. Zarkhin,Yu.G., Kovalenko,V.N.: Search the solution of the system of two algebraic equations, Scientific Centre for Biological Research, Pushchino, 1978.

5. Smit,J.: New recursive minor expansion algorithms, a presentation in a comparative context, in: Symbolic and Algebraic Computation, Eurosam '79, LNCS 72, Springer, 1979, pp.74-87.

6. Voevodin,V.V., Kuznetsov,Yu.A.: Matrices and Computation, Nauka, Moscow, 1984.

7. Eltekov,V.A.: Analytical computation in the Planner language, in: Proc. of Conf. on Systems and Techniques of Analytical Computing and Their Applications in Theoretical Physics, JINR, Dubna, 1983, pp.34-39.

8. Sveshnikov,A.G., Eltekov,V.A.: Teaching the symbolic computation om the basis of the Planner-analytic system, in: Proc. of Int. Conf. onComputer Algebra and its Applications in Theoretical Physics, JINR, Dubna, 1985, pp.416-420.

9. Shaposhnikov,N.N., Eltekov,V.A.: Application of the matching method to realizing the analytical computation, Programmirovanie, 1986, No.5, pp.84-87.

10. Pilshchikov,V.N.: The Planner Language, Nauka, Moscow, 1983.

TRANSFORMATION OF COMPUTATION FORMULAE IN SYSTEMS OF RECURRENCE RELATIONS

Zima E.V.
Facul. Computat. Math. and Cybernetics, Moscow State University
SU - II9899, Moscow, USSR

Implementation of various numerical methods often needs organization of computation using complex iterative formulae, i.e. cycles. The time for performing a program on a computer and the necessary storage capacity are largely dependent on form of these formulae. One should take into account, that these formulae may be obtained as the solution of some problems using systems of symbolic computations, and as consequence may be unwieldy and inconvenient for efficient computation. Besides, computations with frequent precision, used in systems of symbolic computations, makes arithmetic operations very expansive. Hence, the problem of economizing the number of arithmetic operations in this case is still very essential.

It is desirable to construct computations so that each iterative step might use the results obtained at the previous steps as completely as possible. This implies the association of the given formulae with the recurrence relations, that bring about the same result but economize the arithmetic operations.A special algebraic method has been created that provides for the automatic construction of such recurrence relations.

In [I] a certain class of systems of recurrence relations is described and special operations on systems from this class are defined. A general algorithm of constructing systems of recurrence relations on the basis of an expression that defines some iterative function is created. The construction is done in one sending of the binary tree which defines this expression in endorder. The possibility of this construction is based on the fact that any leaf node of this tree can be related to the elementary system of recurrence relations, and on the usage of operations which the internal nodes of the tree contain to the systems we already got. The result of the algorithm is an expression, whose operands are systems of recurrence relations. The methods of cycle construction using such expression are proposed. Note that such often used optimizing program transformations, as code motion and the transformation of inductive variables, can be obtained as a result of constructing elementary systems of recurrence relations.

Some transformations of expressions associated with systems of recurrence relations have been proposed in [2]. The problem of constructing the transformation for a given expression that reduces the number of operations performed in a cycle has been considered. If one should examine all the variants in order to solve this problem then the number of variants would be proportional to the factorial of the expression length. The properties of expressions and systems of recurrence relations, that provides for reduction of numbers of variants, has been formulated; some of them make it possible to choose a single variant.

Besides, a class of arithmetic expressions for which the problem of the expression equivalence can be solved using the systems of recurrence relations construction has been defined.

The methods proposed can be used in program optimization systems, real-time systems and in systems of symbolic computation.

REFERENCES

I. Zima E.V. Automatic construction of systems of recurrence relations. Computat. math. and math. phys., I984, Vol. 24,N 6, p. I93-I97.

2. Zima E.V. Transformations of expressions associated with systems of recursion relations. Moscow University computat. math. and cybernetics, I985, N I, p. 60-66.

"DIMREG"
THE PACKAGE FOR CALCULATIONS IN THE DIMENSIONAL REGULARIZATION WITH 4-DIMENSIONAL γ^5-MATRIX IN QUANTUM FIELD THEORY

V.A.Ilyin and A.P.Kryukov

Institute for Nuclear Physics, Moscow State University
Moscow 119899, USSR

ABSTRACT.

It is therefore highly desirable for modern quantum field theory to have well developed methods of derivation of expressions containing the γ^5-matrix. The dimensional regularization remains the most simple and effective procedure for this purposes. There is principle obstacle for γ^5 inclusion in dimensional regularization. It is convenient to use the 4-dimensional γ^5-matrix in this procedure.

In the paper we describe the realization of dimensional regularization with the γ^5-matrix in REDUCE.

1. Introduction.

The supersymmetric and string models are most important in the modern quantum field theory. Many of these models are chiraly asymmetric. It is therefore highly desirable to have well developed methods of derivation of expressions containing the γ^5-matrix. The dimensional regularization remains so far the most simple and effective procedure. However, from the very beginning this scheme involves difficulties of including the γ^5-matrix. Here it is customary to speak about the ambiguity of continuation of 4-dimensional absolutely antisymmetric tensor into higher dimensions. According to ref [1] it is convenient to use the 4-dimensional γ^5-matrix commuting with γ-matrix corresponding to additional dimmensions. In this case the symmetry reduces from O(d) to O(4)×O(d-4) where the parameter d is the space-time dimension. Here and in the folowing we mean the quantum field theory in Euclidean metrics. The reduction in symmetry complicates considerably the calculations and, therefor, this scheme has not found wide use. the paper [2] suggested more simple variant with a d-dimensional γ^5-matrix. But this scheme is inconsistent [3]. Later there have appeared [4] consistent variants of inclusion of the γ^5-matrix into the dimensional regularization. A general comment is to be made here. The object chosen as the γ^5-matrix is

$$\gamma^5(d) = F_{\mu_1 \dots \mu_d}(d) \quad \gamma^{\mu_1} \dots \gamma^{\mu_d} \tag{1}$$

where $\mu = 1, \dots d$; are d-dimensional γ-matrices; F is the d-dimensional tensor, which is antisymmetric,at least, in the components corresponding to the physical 4-dimensional subspace. In other words, it is the d d matrix with a 4-dimensional absolutely antisymmetric tensor in its upper left angle. Note that the variants [1] and [2] are particular cases of (1). the only O(d) invariant tensor of this type is the d-dimensional absolutely antisymmetric tensor, i.e. the unacceptable variant [2]. Therefore the consistent choice of a tensor F in (1) will break the symmetry O(d). It is clear from the aforsaid that the variant [1] with the 4-dimensional γ^5-matrix is typical of the consistent dimensional regularizations, at least, in the sense of the symmetry reduction. And this variant has the most simple formulation.

In the present paper we realize the scheme of dimensional regularization with the 4-dimensional γ^5-matrix in REDUCE.

The question which arises first is how to realize the vector algebra with the symmetry O(4)×O(d-4). It is convenient to introduce two scalar products which we denote by # and ?, and define as

$$k\#p = k_1 p_1 + \dots + k_d p_d \, , \quad k?p = k_1 p_1 + \dots + k_4 p_4 - k_5 p_5 - \dots - k_d p_d \tag{2}$$

Here k and p are vectors in d-dimensional space-time. The metric tensor corresponding to these scalar products will be designated as $\mu \# \nu$, $\mu ? \nu$ (here μ and ν are vector indices). Then we may record the relations

$$\begin{aligned} &\mu\#\mu = d \, , \quad \mu?\mu = 8 - d \, , \\ &(\mu\#\nu) \ (\nu\#\lambda) = \mu\#\lambda \, , \quad (\mu\#\nu) \ (\nu?\lambda) = \mu?\lambda \, , \\ &(\mu?\nu) \ (\nu?\lambda) = \mu\#\lambda \, . \end{aligned} \tag{3}$$

These formulae represent the whole set of operations on dummy indeces.

To calculate the loop integrals it is necessary to introduce the gradient-type operator

$$\frac{\partial}{\partial p_\mu}(p\#k) = k\#\mu , \qquad \frac{\partial}{\partial p_\mu}(p?k) = k?\mu . \tag{4}$$

The second problem is to calculate traces of product of γ-matrices.

In Eucledean metric the γ-matrix algebra is given by the formula [5]

$$\gamma^\mu \cdot \gamma^\nu + \gamma^\nu \cdot \gamma^\mu = - 2 \cdot \mu\#\nu .$$

The relations used to calculate traces are

$$(\gamma^\mu)^2 = -1 , \qquad (\gamma^5)^2 = (\gamma^1 \cdot \gamma^2 \cdot \gamma^3 \cdot \gamma^4) = 1 ,$$

$$\mathrm{Tr}(A\gamma^\mu \cdot \gamma^\nu B) = - \mathrm{Tr}(A\gamma^\nu \gamma^\mu B) - 2\, \mu\#\nu\, \mathrm{Tr}(AB) . \tag{5}$$

If an expression contain γ^5-matrices we use the relation

$$\gamma^5 \cdot \gamma^\mu = - \gamma^\nu \cdot \gamma^5 \cdot (\nu?\mu) \tag{6}$$

to reduce them to 0 or 1. In the latter case γ^5 is replaced by

$$\gamma^\mu \gamma^\nu \gamma^\sigma \gamma^\delta \, e^1_\mu \, e^2_\nu \, e^3_\sigma \, e^4_\delta , \tag{7}$$

where $e^1,\dots,e^4$ are orthonormal basis of physical 4-dimensional sub-space.

The loop integrals are calculated using the main formula of the dimensional regularization [1]

$$\int \frac{d^d p}{(p^2 + 2(pk) + M^2)^a} = \pi^{\frac{d}{2}} \frac{\Gamma(a-d/2) \cdot (M^2 - k^2)^{\frac{d}{2} - a}}{\Gamma(a)} \tag{8}$$

where Γ is Euler's function. To use this formula in the Feynman amplitude calculations, it is necessary to transform the denominator in the integrand to the standard form $(p^2 + 2(pk)+M^2)$. This is done by using recurrently the formula

$$\frac{1}{a^{\alpha}\cdot b^{\beta}} = \frac{\Gamma(\alpha+\beta)}{\Gamma(\alpha)\cdot\Gamma(\beta)}\int\frac{dx \quad x^{\alpha-1}\cdot(1-x)^{\beta-1}}{[ax+b(1-x)]^{\alpha+\beta}}, \tag{9}$$

which introduces integrals over the Feynman variables x.

Using the gradient operator (4), the loop integrals with a polynomial numerator are recurrently reduced to (8).

Upon taking the loop integrals, we expand the obtained functions into a Laurant series at d=4 and then identify a finite truncated series. The analytic functions here are polynomials, power functions, logarithms and Euler's Γ-functions. Before expanding the latter into a Laurant series, we transform them to the canonical form

$$\Gamma(1 + A) , \tag{10}$$

where A is either a number from the [1,0) interval or an algebraic expression.

2. Realization in REDUCE.

Note, that the REDUCE high-energy package cannot be used in view of the necessity of introducing the second scalar product (2). Besides to be effective the vector algebra requires free indices (see the formulae (4) and (6), for example). In REDUCE this is not possible.

To solve the above problems we have written five subprograms:

VALG, LNRX, SMPL, GAMM, RINT.

The package is called DIMREG. It contains 735 cards in REDUCE.

The VALG subprogram contains 180 cards in RLISP language. VALG is a service program in which, in particular, the declaration operators of vector objects - vectors and indices, are realized. The vectors and indices are represented in an algebraic mode by the sentences

```
LVECTOR    p,k,...;
LINDEX     mu,nu,l,...;                                    (11)
```

The possibility of generating vectors, indeces and scalar variables in an algebraic mode is provided for. For example,

```
x := MK!_LVECTOR();
y := MK!_LINDEX();                                         (12)
```

```
z := MK!_VAR();
```

The values of x,y,z will be

```
lvectorN
lindexN
variableN
```

where N is the ordinal number of call to the functions (12)

The predicates for vectors, indeces and vector objects (vectors or indeces) are

```
LVECTORP U;          LINDEXP U;
LOBJP U;             LOBJECTP U;
```

whose values are if U is the corresponding object, other wise, false.

The difference between the latter predicates is that LOBJP gives TRUE if only U is vector or index declared in an explicit form through (11). The predicate LOBJECTP gives TRUE even if U is of the form $a_1p_1+...+a_np_n$, where a_i are scalars and p_i – vectors.

The symmetric operators of the scalar products (2) are given in the infix form (# and ?).

The gradient operators (4) is realized using the standard differentiation function DF.

The program LNRX consists of 80 cards in RLISP language. It determines the linear properties of arbitrary operator of any number of arguments. On and off for the simplificator of expressions which uses the linear properties of # and ? are provided for. To do this the flag LIN is used.

The standard linear operator in REDUCE specifies the property

$$A(a\cdot U,x) = a\cdot A(U,x),$$

where a is x-independent. In LNRX realized the property

$$A(\ldots,a\cdot U,\ldots) = a\cdot A(\ldots,U,\ldots) ,$$

if a does not satisfy the special predicate. For these purposes LNRX uses the predicate LOBJP. The predicate can be changed acording to the nature of the problem in question. A specific illustrative example of using LNRX is the equality ((p#p)·p)#q = p#p·p#q.

The program SMPL consists of 135 cards in RLISP language. It is intended to simplify vector expressions. This program is, in fact, a convolution function of pairs of dummy indeces taking into account two scalar products (2). It follows the arguments of the operators # and ? and searches for the pairs of the same indeces. Futher simplification goes in accord with (3). On and off of the simplifier is done using the flag VSIMP. When the simplifier is on, the convolution of dummy indices is performed after reassigning the expression.

The program GAMM consists of 90 cards in an algebraic mode. To record the reccurent rule (5) and also (6) and (7) in a compact form we employ the program, suggested in ref.[6], LETT enables the user to use the LETT-substitutions for the operators with arbitrary number of arguments (&-variables).

The program RINT consists of 250 cards in an algebraic mode. The loop integral is calculated after the call to RINT(U,p). Here p is the loop momentum (integration variable), U - the integrand. U should be a rational function of p. The denominator may be a product, any factor of which should be a polynomial of power < 3. All the scalar products in the denominator must be of type #. If U does not satisfy these requirements, the output is

UNUSUAL CASE FOR RINT(U,p)

If the Feynman parametrization (9) is not used, the RINT output is the ε-analytic function V, otherwise

FINT(V,X).

Here $d=4-2\cdot\varepsilon$, X is the formal product of the integration variables x (Feynman parameters), V is the x -dependent ε-analytic function, FINT designates definite integrals over x between 0 and 1. The integrals can be computed either by loading the integrator or by using a table. Here the integrands are polynomials and logarithms of x and (1-x).

At the last step RINT performs the Laurant series expansion in ε at a point $\varepsilon=0$. The number of terms is regulated by the parameter LAURORDER. For example, at (-1) the output will be all pole terms, and at 2 all pole terms with terms in ε up to order 2 included. At LAURORDER=INFINITY the starting analytic function is produced. Acces to the Laurant series expansion program may be direct.

In LAURANT the pole power has an upper bound (<11). This, specifically, prevents infinite looping in case of a essential singular point of function U. In some cases the essential singular is diagnosed as division by zero, for example, for $\sin(1/\varepsilon)$.

LAURANT is intended to operate with polynomials, power functions, logarithms etc, i.e. with all functions for which the standard differentiation function DF is defined and also with Euler's Γ-functions. The latter, befor the expansion, are reduced to the canonical form (10).

3. Effectiveness of the package and tests

The suggested package has a number of advantages over the REDUCE high-energy package. Besides free indices, it is worthwhile to mention the possibility of on and off for the linear properties of scalar products and the vector simplifier, which can substantially reduce the consumption of time.

To calculate the loop integrals we have obtained the results for RINT(U,p) with LAURORDER=INFINITY

U = 1/(p#p+2p#k+m)	10000 ms
p#k/(p#p+2p#k+m)	14000 ms
p#p/(p#p+2p#k+m)	35000 ms
1/(p#p(p#p+m))	40000 ms
1/(p#p(p#p+m)(p#p+m))	80000 ms

We present time required for calculation using ES1045 computer running VM/SP operation sytstem (USSR made IBM 370 compatible clone with 0.5 Mflop productivity).

4. General comments

After some modifications the vector algebra suggested can be applied to the groups $O(d_1, d_2)$, specifically, for the Minkowskian space. Besides, the vector algebra is natural for the problems with the spherical symmetry O(d) in which a vector is fixed or some vectors are knownto belong to the subspace.

The program LAURANT was applied with success in ref.[7] where the pole terms of 200 complex expressions (rational functions and Euler,s -functions) had to be calculated.

The program were prepared using the dialogue subsystem ROS [8] which accelerated the development and debugging.

The authors express their gratitude to A.Rodionov for helpful discussions.

5. References.

1. G.'tHooft, M.Veltman. Nucl.Phys., B44(1972) 189.
   P.Breitenlohner, D.Maison. Comm.Math.Phys., 52(1977) 11.
2. W.Siegel. Phys.Lett., 84B(1979) 193.
   M.T.Grisaru, W.Siegel, M.Rocek. Nucl.Phys., B159(1979) 429.
3. W.Siegel. Phys.Lett., 94B(1980) 37.
   L.Avdeev, A.Vladimirov. Nucl.Phys., B219(1983) 262.
4. G.Thompson, Hoi-Lai Yu. Phys.Lett., 151B(1985) 119.
5. C.Itzykson, J.B.Zuber. Quantum Field Theory, McGraw-Hill, New York, 1983.
6. A.Kryukov,A.Rodionov. In Proc. SIMSAC'86, pp. 91-93.
   A.Kryukov,A.Rodionov. Preprint IHEP 86-78, Serpuchov, 1986.
7. E.Boos,A.Davyidyichev. Preprint IHEP 86-92, Serpuchov, 1986.
8. A.Kryukov,A.Rodionov. SIGSAM Bull., 19(1985) 43.
   A.Kryukov. In Proc. SIMSAC'86, pp. 107-109.

CTS - ALGEBRAIC DEBUGGING SYSTEM FOR REDUCE PROGRAMS

A.P.Kryukov & A.Ya.Rodionov
Institute for Nuclear Physics, Moscow State Univesity,
Moscow, 119899, USSR

This paper describes Conversational Trace System (CTS) intended to debug programs written in REDUCE. The system permits the user to trace algebraic operators, procedures, global and local variables in both the dialogue and batch modes. In dialogue mode human intervation of calculations is permissible.

Introduction.

REDUCE [1-4] is one of the most popular systems of computer algebra. In fact, it is realized in almost all types of modern computers, supercomputers and some personal computers [5,6].

This system has a wide use in many fields of physics, mathematics and mechanics. A convinient source language, dialogue means and a wide variety of mathematical tools make it rather attractive. An important contribution to REDUCE was the fast loading of compiled programs. The result was the organization of program libraries [7,9].

It will be noted that the debugging remains a weak point in the system. Besides the functions TRACE and UNTRACE of the Standard LISP [10], REDUCE has no any other debugging means. This shortage is especially acute if the user is not familiar with the LISP language. The user do not enable to trace changes of local and global variables. The aforesaid points to the necessity of supplying REDUCE with the adequate debugging means. The dynamic debugger [11] of the interactive REDUCE version [2] is one of such attempts. This rather powerfull debugging system makes it easy to follow step by step the execution of procedures. The output information in this case is in the usual algebraic form. With all its advantages the system, however, has a number of faults. We mention two of them.

1. Impossibility to trace the evaluation of algebraic operators.

2. Much time consumption because of the length enlargement of procedure codes.

The origin of the above faults is the modification of procedures during translation from RLISP to LISP language. This modification makes the procedures self-tracing by inserting special tracing routines. Thise ideology led, in our opinion, to the most serious drawback in the dynamic debugger - the strong attachment to the second version of REDUCE. Adaptation of the dynamic debugger to more recent versions is a rather difficult task.

Taking into consideration these drawbacks we have developed a new debugging system for the programm in REDUCE which we will describe below.

1. CTS - Conversational Trace System.

When realizing the CTS we had the following goals in mind.

1. The system should provide the user with the information on all types of objects used in REDUCE: operators, procedures, global and local variables.

2. The output information should be given to the user in the usual algebraic forms.

3. The system must work under any REDUCE version. Expenses must be minimal.

4. The system must work both in batch and dialogue modes. In dialogue mode the user's intervation of computations is permissible.

Alebraic expressions are evaluated in REDUCE by means of algebraic processor whose main function is SIMP. Therefor the CTS operation is based on the interception of control at the moment of call to this function (Fig.1).

```
      !                                            !
      !          +------------------------+        !
      !          !                    CTS !        !
      !          !     +---------------+  !        !
      !          !  +->!read command   !<-!--------+
      !          !  !  +---------------+  !
      !          !  !          !          !
      V          !  !          V          !
+-----------+    !  !  +---------------+  !  +-----------+
!   SIMP    !    !  +--!process command!<-!->!   SIMP    !
+-----------+    !     +---------------+  !  +-----------+
      !          !             !          !
      !          !             V          !
      !          !     +---------------+  !
      !          !     !print value    !--!--------+
      !          !     +---------------+  !        !
      !          !                        !        !
      !          +------------------------+        !
      !                                            !
      V                                            V

      a)                       b)
```

Figure 1. Evaluation of algebraic expresions by REDUCE (a) and CTS (b).

This enables the user to trace algebraic operators, procedures and global variables. An exception is local variables. To trace local variables we redefined the function SETQ.

The implementation of this scheme permitted us to reduce expenses. Specifically, the bodies of procedures are not modified.

The dialogue with the user is realized through the CTS monitor, which gives the user information on the evaluated expressions and also enables the user to intervene computations.

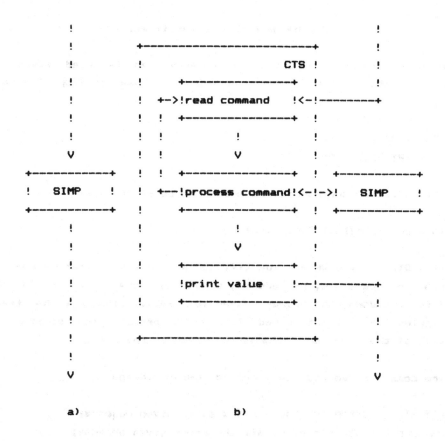

2. The main CTS capabilities.

When the CTS is loaded, two new commands can be used i.e. the command to start algebraic tracing and the command to stop algebraic tracing

```
ATR P1,P2,...,PN;
UNATR P1,P2,...,PN;
```

where P1,P2,...,PN are parameters which can be of the form

```
ID   or   ID(ID1,ID2,...,IDK)
```

where ID,ID1,ID2,...,IDK are identificators. The identificators can be the names of operators, procedures, global and local variables. If the second form is used, the objects named ID1, ID2,...,IDK will be trace only inside the object named ID, which permits the substantial reduction of the redundant information given to the user.

The computer can also obey the following commands.

```
ATR NIL   - stop tracing all the above-given objects;
UNATR NIL - start tracing all the above-given objects;
ATR T     - output of the list of objects to be traced;
UNATR T   - clear the list of traced objects.
```

It is sometimes usefull to use the command

```
ATR ID(ALL);
```

which enable the user to trace all algebraic objects encountered in the calculations.

As an example, Appendix 1 presents the computation of Hermitean polynomials in the tracing mode.

3. The CTS commands.

When the command to start tracing is involved and the inputed expression is printed , the control is passed to the user. There are 10

CTS commands: information commands, commands to control computations, commands to interact with REDUCE.

All CTS commands have 2 symbols: the name and parameter, both on a single line. The total list of CTS commands is given in Appendix 2.

Information command.

Two information commands are available: the command "*"(print) and "?"(help). By the command "*" the computer can re-display the input expression or its argument or the computed results. Besides, this command is used to print the tracing stack. The command "?" enables the user to obtain brief information on any CTS command and its syntax and also on all commands at once.

Control computation commands.

This set of commands consist of 5 commands enabling the user to deeppen the tracing, to a given number of levels or to go back to a higher level of tracing. It is also possible to interrupt computations to get at a given trace level. In particular, the computation can be ceased.

Commands to interact with REDUCE.

This group contains of 3 commands. The first is the command "!!"(to suspend) which suspends computation to enable the user to perform some computations in REDUCE. Specifically, the user can display the values of variables. In which case the command OFF CTS is automatically executed. By the command END, the computer goes back to the tracing mode.

There exist 3 global variables used to exchange information between CTS and REDUCE. These are !:EXPR, !:VAL,!:ARG. The command ":"(to put) permits the input expression, its value or one of its arguments to be assigned to one of these variables. Analogously, the command "%"(to get) permits the value of these variables to be use as input expression, its value or one of its arguments. Thus the user can correct the course of computations or to retain some intermediate results for further work.

Batch and demonstration mode.

Besides the dialogue mode the CTS can operate in batch and demonstration modes. In this case the system gives information but does not request a command. This enables the user to trace the course of computations and to print it for further analysis.

This operating mode is controlled by the flags INT (to set on for interactive mode), DEMO (to set on for demonstration mode) and variable MAXTRLEVEL (the initial value is 10). This variable restricts the depth of tracing to a given number of levels in order to reduce the amount of output information.

Collection of statistics.

The second version of CTS offers an additional function collection of statistics. To collect the statistics of calls to the objects named ID1, ID2, ..., IDK, the user must print the command

```
SCOLLECT ID1,ID2,...,IDK;
```

After that the flag STAT will be on. The collection of statistics is ceased when the flag STAT is turned off or the user input the command

```
SCOLLECT OFF;
```

To output the result of collection of the statistics the user employs the command

```
SPRINT TIME;     or     SPRINT CALLS;
```

At first case will be output the time statistics, and the other case the call statistics.

The collection of statistics is illustrated in Appendix 3.

Conclusion.

The CTS proved to be rather efficient. For example, when developing the package for the quantum chromodynamics calculations one of the authors (A.K.) could not for a long time find an error in the calculation algorithm. He nested many times the statement WRITE without any success and the function TRACE was not easy to use because of cumbersome expressions. The CTS, when applied, enabled to find at once the cause of the error. Fortunately, the error was easy to pass over.

The CTS is currently in use at the Institute for Nuclear Physycs, Moscow State University, Joint Institute of Nuclear Research. It contains near 1000 cards in RLISP [1,4] and requires 10K bytes binary program space. At present time the CTS system available under REDUCE-3.3.

Acknowledgements.

The authors wish to thank V.A. Il'in for fruitfull discussions and V.A.Rostovtsev for valuable remarks and help with the implementation of the CTS in REDUCE-3.2 at JINR.

References.

1. Hearn A.C. REDUCE-2 User' Manual, UCP 19. 1973, Univ. of Utah, USA.
2. Edneral V.F.,Kryukov A.P., Rodionov A.Ya. Symbolic calculation language REDUCE (in russian), Moscow, MSU, part 1 - 1983, part 2 - 1986.
3. Bogolubskaya A.A., Zhidkova I.E. and Rostovtsev V.A. REDUCE-2 programming system. JINR 1-11-83-5/2, 1983, Dubna.
4. Hearn A.C. REDUCE User's Manual. Version 3.3, The Rand publication CP78 (Rev. 7/87), 1987, USA.
5. Marti J.B., Hearn A.C. REDUCE as a LISP Benchmark. SIGSAM Bull., 1985, v. 19, no.3, pp.8-16.
6. Petrov P.M. Realization of REDUCE-2 on small computers. Proc. of the Workshop on systems and methods of computer-aided analitic calculations and its applications in theoretical physics. D11-83-511, 1983, pp.64-69, Dubna.
7. Butenko V.A., Kryukov A.P., Rodionov A.Ya. Authomatic loading of procedures in the REDUCE system. Programming, 1985, no.4, pp.89-92.
8. Hearn A.C. User's library formed. REDUCE News Letter, 1985, no.10, pp.3-4.

9. Bogolubskaya A.A., Gerdt V.P. and Tarasov O.V. Package loading of the SCHOONSCHIP and REDUCE systems. Proc. of the Workshop on systems and methods of computer-aided analitic calculations and its applications in theoretical physics. D11-85-791, 1985, pp.82-89, Dubna.

10. Marti J.B., et al. Standard LISP report. SIGPLAN Notices, 1979, v.14, no.10, pp.48-68.

11. Kryukov A.P., Rodionov A.Ya. Dynamic debugging system for REDUCE programs. SIGSAM Bull., 1985, v.19, no.2, pp.34-37.

12. Kryukov A.P., Rodionov A.Ya. Interactive REDUCE. SIGSAM Bull., 1985, v.19, no.3, pp.43-45.

Appendix 1.

Calculation of Hermitean polynomials under CTS by the recurrent formula:

$$H_n(x) = 2(xH_{n-1}(x) - (n-1)H_{n-2}(x)), \qquad n>1;$$
$$H_0(x) = 1;$$
$$H_1(x) = 2x.$$

The arrow (->) marks the input lines.

```
-> OPERATOR H;
-> H(0,X) := 1¤
-> H(1,X) := 2*X¤
-> FOR ALL N SUCH THAT N>1
->    LET H(N,X) = 2*(X*H(N-1,X)-(N-1)*H(N-2,X));
-> ATR H;
-> ON DEMO;
-> H(3,X);

   0.1>>H(3,X)
      0.2>>H(2,X)
         0.3>>H(1,X)
         0.3<<2*X
         0.3>>H(0,X)
         0.3<<1
```

```
                 2
   0.2<<2*(2*X -1)
   0.2>>H(1,X)
   0.2<<2*X
                2
0.1<<4*X*(2*X -3)


           2
4*X*(2*X -3)
```

In this example, the first digit before "<<" or ">>" corresponds to the level of the command "!!"(to suspend) and the second one, to the trace level. The flag DEMO service to turn on the demonstration operating mode of the CTS.

Appendix 2.

The CTS commands.

Type	Assinment	Name	Para-meter	Comments
I	Help	?	?	To output the list of all
n				commands.
f			C	To output the format of
o				command C.
r	Print	*	>	the input expression
m			<	the output expression
.			N	the N-th argument
			*	the trace stack
C	Upwards	<	<	on one level
o			N	on N level
n	Down	>	>	on one level
t			N	on N level
r	To cease	#	#	all levels and return to
o	compu-			REDUCE
l	tation		N	all levels starting with N
		X	X	one level
			N	N levels

```
!       ! To compute!  =   !   =   ! all levels and return to      !
!       !           !      !       ! REDUCE                        !
!       !           !      !   N   ! all levels starting with N    !
+-------+-----------+------+-------+-------------------------------+
!   I   ! To suspend!  !!  !       ! calculations and go to        !
!   n   !           !      !       ! REDUCE. Return by END         !
!   t   !           !      !       ! command.                      !
!   e   ! To put    !  :   !   >   ! to assign !:EXPR the value of!
!   r   !           !      !       ! input expression.             !
!   a   !           !      !   <   ! to assign !:VAL the value of  !
!   c   !           !      !       ! output expression.            !
!   t   !           !      !   N   ! to assign !:ARG the value of  !
!   i   !           !      !       ! N-th argument of input        !
!   o   !           !      !       ! expression.                   !
!   n   ! To get    !  %   !   >   ! set input expression to       !
!       !           !      !       ! !:EXPR.                       !
!       !           !      !   <   ! set the value to !:VAL.       !
!       !           !      !   N   ! set the N-th argument to      !
!       !           !      !       ! !:ARG.                        !
+-------+-----------+------+-------+-------------------------------+
```

Notes: 1) the parameter C designates the name of any command;
2) the parameter N designate the digit (0 through 9);
3) for the commands to interact with REDUCE N equal to zero means name of operator or procedure.

Appendix 3.

The result of statistics collection for REDUCE test. For collection was selected next operarots: FAC, F1, G1, DF, SIN, COS.

```
4:
SCOLLECT FAC,F1,G1,DF,SIN,COS;
5:
IN TEST¤
6:
SPRINT TIME;
***************** TIME STATISTICS ***************
NAME        MS     %  0....+....1....+....2....+....3....+....4
FAC         10     0  !
F1          61     1  !
```

```
G1          159    3  !**
DF          2537   54 !****************************************
SIN         928    20 !***************
COS         928    20 !***************
-----------------
TOTAL:      4633      ONE STAR(*) CORRESPONEDS TO (1 . 35)%
******************* END STATISTIC ***************
7:
SPRINT CALLS;
**************** CALLS STATISTICS ***************
NAME        N      %  0....+....1....+....2....+....3....+....4
FAC         2      0  !
F1          50     13 !**********
G1          50     13 !**********
DF          210    54 !****************************************
SIN         37     9  !*******
COS         35     9  !******
-----------------
TOTAL:      384       ONE STAR(*) CORRESPONEDS TO (1 . 35)%
******************* END STATISTIC ***************
```

Applications of Computer Algebra in Solid Modelling

Adrian Bowyer, James Davenport, Philip Milne, Julian Padget, and Andrew Wallis

Schools of Mechanical Engineering and Mathematical Sciences
University of Bath, Bath, AVON BA2 7AY, UK

1. Introduction

Existing computer algebra systems are not particularly well suited to the solution of geometrical problems. This is not because the underlying theory suffers from any inherent limitation, but because the need to provide algorithms and systems for the solution of such problems has not been addressed by the computer algebra research community. In this abstract we outline some of the problems we have identified in solid modelling where computer algebra may be applied, then discuss some specific solutions to some of those problems. We describe briefly the strategy we are employing (based on the REDUCE algebra system) to help us in our work and how this will lead to a widely applicable package of geometric algebraic algorithms. More details of the solid modelling part of the system may be found in [Bowyer *et al.*, 1987].

A large number of engineering processes require an unambiguous description of the shapes of the components with which they have to deal. The two most common techniques for shape representation are *boundary description* and *set-theoretic solid modelling* (also called CSG). We are concerned with the latter in which solids are combined using set-theoretic operations. The approach summarised in this paper allows arbitrary polynomial inequalities to be used to define the primitive solid shapes. Hence all semi-algebraic varieties can be represented and manipulated, whereas all current commercial systems are limited to a much narrower range of primitives.

2. The Problems

Classification of Polynomials against Planes and other surfaces

The generation of a divided structure (a tree of sub-volumes of the original volume surrounding the solid model) requires the classification of model primitives against sub-volumes. A primitive may either contribute entirely *air* or *solid* to a sub-volume or may lie completely or partly inside the sub-volume. This classification is easy in the case of planar surfaces but significantly harder when the surface is curved. Surprisingly, an exact classification is not necessary, but a *conservative* strategy, which may conclude that a primitive and a sub-volume intersect, even if they do not (but never *vice versa*), works well.

Ray-casting: Root finding

Ray casting is a fundamental operation in interrogating models (in displaying images or in automating manufacture). It requires descending the tree generated by the division process to determine if the ray intersects any of the sub-volumes. If the primitive is represented by a polynomial inequality the parametric ray equation may be substituted into the polynomial half-space expression to generate a polynomial equation in the ray parameter; the roots of this equation give the intersections of the ray and the half-space. Unfortunately, the polynomials defining the surfaces generated by blending techniques [Zhang & Bowyer, 1986] (for the construction of fillets) are frequently of very high degree — we have observed 24. Numerical techniques are not well suited to isolating the roots of such polynomials. Some symbolic methods are described by Davenport [1985].

Line and Section drawing

For faster feedback in the design loop, it is also useful to be able to produce skeleton drawings (wireframes). In addition the sectional representation has a long tradition in engineering. In both cases, it is necessary to produce the edges of the set-theoretic model. We describe the use of discriminant and resultant calculations to help with this in the next section.

3. Algebraic Solutions

Testing for the intersection of a line with a surface

This is the crux of the implementation of ray-casting. Given a (parametric) line in three-space, substitution into the equation of a surface renders a polynomial of which the roots identify the points where it meets the surface. Although this is a familiar operation in computer algebra systems, it is usually done by hand and then coded in solid modelling systems. Sturm sequences (examining the difference in the number of sign changes between two evaluation points plugged into polynomial remainder sequence derived from the polynomial) provide a powerful method for finding the roots analytically.

Eliminating a variable from two equations: Resultants

A *resultant* is the single polynomial which results from the elimination of a variable from two polynomials which contain it. So starting from $P(x,y,z)$ and $Q(x,y,z)$ and eliminating z we get a polynomial in x and y which is zero iff P and Q have a common zero for these values of x and y. Geometrically, it is the projection of the intersection. The algorithm for finding a resultant is described in [Davenport, 1985].

Testing for intersection between two surfaces

This is the next level of complexity up from ray-casting and plays a crucial role in the model division process. It can be done by taking the resultant of the defining equations of the two surfaces: if there are no zeroes of the resulting bivariate which are real in both variables, the surfaces do not intersect.

The Generic Ray

Ray-casting is an expensive process, since the cost is proportional to the number of pixels on display area. Conventional methods compute the colour and intensity of *each* pixel. An algebraic approach can provide the *generic ray*, whereby the 3-D model is mapped to a set of 2-D regions bounded by polynomials. The colour of each pixel within one of those regions is defined by a single function and can be evaluated very efficiently.

4. Implications of Using Algebra

Conventional solid modelling systems are being overwhelmed by the complexity of current problems, so although the numerical techniques used to date have been effective, they do not scale sufficiently. Algebra has several advantages, not the least of which is generality, this also gives rise to accuracy, since in using an analytical solution the only errors that arise are in its evaluation, not in its derivation. Algebra is not without its problems however, there are major issues in system building (mostly relating to storage management) to be handled and in the representation of the algebraic expressions. The implementation effort involved should not be underestimated.

5. Conclusions

Work is currenty progressing, using REDUCE for prototyping. This paper has outlined the scope of a new research project, whose necessity arises both from the computational imperative of applying algebra to CAD and the difficulties of applying existing systems to this problem. The authors are currently working on the detailed specification of the memory management system, and the model domain. During the life of the project the authors expect to investigate the topics discussed in the paper and to derive and implement Computer Algebra based techniques for solving them.

6. References

[Bowyer, 1987] Bowyer A., Davenport, J.H., Milne, P.S., Padget, J.A. & Wallis, A.F., *The Design of a Geometric Algebra System*, to appear in the Proceedings of the UK conference on Geometric Reasoning, Oxford University Press, 1988.

[Davenport, 1985] Davenport, J.H. *Algebra for Cylindrical Algebraic Decomposition* TRITA-NA-8511, NADA, KTH, Stockholm, Sept. 1985. Bath Computer Science Technical report 88-10.

[Zhang & Bowyer, 1986] Zhang, D. & Bowyer, A. *CSG Set-theoretic Solid Modelling and NC machining of Blend Surfaces* Proc. 2nd ACM Symposium on Computational Geometry. New York June (1986)

Implementation of a Geometry Theorem Proving Package in SCRATCHPAD II

K. Kusche, B. Kutzler, H. Mayr

Research Institute for Symbolic Computation (RISC-LINZ)
Johannes Kepler University
A-4040 Linz, Austria

The problem of automatically proving geometric theorems has gained a lot of attention in the last two years. Following the general approach of translating a given geometric theorem into an algebraic one, various powerful provers based on characteristic sets and Gröbner bases have been implemented by groups at Academia Sinica Bejing (China), U. Texas at Austin (USA), General Electric Schenectady (USA), and Research Institute for Symbolic Computation Linz (Austria). So far, fair comparisons of the various provers were not possible, because the underlying hardware and the underlying algebra systems differed greatly. This paper reports on the first uniform implementation of all of these provers in the computer algebra system and language SCRATCHPAD II. We summarize the recent achievements in the area of automated geometry theorem proving, shortly review the SCRATCHPAD II system, describe the implementation of the geometry theorem proving package, and finally give a computing time statistics of 24 examples.

1 AUTOMATED GEOMETRY THEOREM PROVING

So far, two different strategies how to do mechanical proofs of geometry theorems have been explored: By an appropriate *axiomatization*, the geometric problem can be translated into a proof problem in a first order theory, which can be attacked by theorem provers. This approach goes back to (Gelernter 1959), who implemented a "geometry theorem proving machine". Another milestone following this paradigm is (Coelho and Pereira 1979), realizing a geometry theorem prover in Prolog. So far, only rather simple theorems could be confirmed using these provers. By an appropriate *algebraization*, the geometric problem can be translated into a corresponding algebraic problem, which can be attacked by computer algebra methods. This approach turned out to be quite powerful in practice and is further explained in the sequel.

An algebraization requires the introduction of a system of coordinates and hence the assignment of coordinate variables to all points involved in the geometric problem. Then, by simple analytic geometry, the hypotheses and conjectures, respectively, can easily be expressed by algebraic relations in these variables. As is necessary for the methods discussed in this paper, we restrict on those geometric problems that can be expressed by polynomial equations only, i.e. geometric problems that can algebraically be expressed in the following form:

$$(\forall\alpha)(H(\alpha) = 0 \Longrightarrow c(\alpha) = 0)$$

(Here $H(\alpha) = 0$ stands for $h_1(\alpha) = 0 \wedge \ldots \wedge h_n(\alpha) = 0$, where the $h_i(\alpha) = 0$ are the polynomial equations corresponding to the hypotheses of the geometric problem, $c(\alpha) = 0$ is the polynomial equation corresponding to it's conjecture. α is a vector of numbers. Clearly, one can restrict attention to a single conjecture.)

The following theorem is of that type: "Given a circle. Let AB and CD be two secants of equal length. Then the secants' midpoints have equal distance from the circle's center."

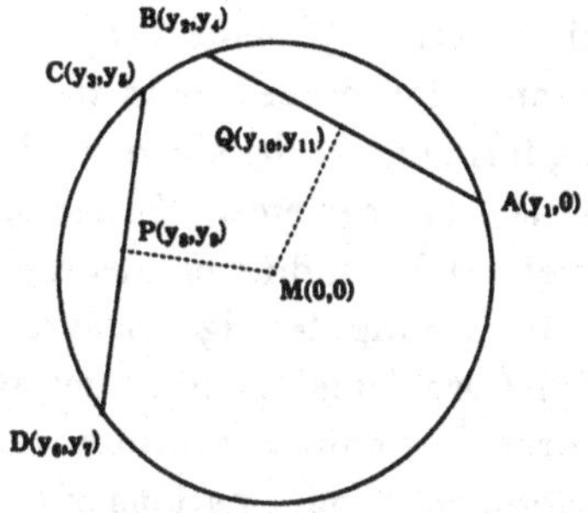

An algebraization is obtained by introducing variables for coordinates as shown in the above figure and expressing the hypotheses and the conjecture by the following polynomial equations:
$h_1 := y_2^2 + y_4^2 - y_1^2 = 0$ (length of AM = length of BM),
$h_2 := y_3^2 + y_5^2 - y_1^2 = 0$ (length of AM = length of CM),
$h_3 := y_6^2 + y_7^2 - y_1^2 = 0$ (length of AM = length of DM),
$h_4 := (y_3 - y_6)^2 + (y_5 - y_7)^2 - (y_2 - y_1)^2 - y_4^2 = 0$ (length of CD = length of AB),
$h_5 := (y_8 - y_6)(y_5 - y_7) - (y_9 - y_7)(y_3 - y_6) = 0$ (P lies on CD),
$h_6 := (y_8 - y_6)^2 + (y_9 - y_7)^2 - (y_3 - y_8)^2 - (y_5 - y_9)^2 = 0$ (length of DP = length of PC),
$h_7 := (y_{10} - y_1)y_4 - y_{11}(y_2 - y_1) = 0$ (Q lies on AB),
$h_8 := (y_{10} - y_1)^2 + y_{11}^2 - (y_2 - y_{10})^2 - (y_4 - y_{11})^2 = 0$ (length of AQ = length of QB),
$c := y_8^2 + y_9^2 - y_{10}^2 - y_{11}^2 = 0$ (length of PM = length of QM).

In the above formulation, the conjecture polynomial does not vanish on *all* zeros of the hypotheses. The reason is, that in case A and B happen to be identical, the above algebraization allows any point in the plane to be Q. Hence the theorem does not hold in this degenerate situation. One often has to exclude certain degenerate cases that can be described by one or more polynomial inequalities $d_1 \neq 0, \ldots, d_s \neq 0$, yielding

$$(\forall\alpha)(H(\alpha) = 0 \wedge d_1(\alpha) \neq 0 \wedge \ldots \wedge d_s(\alpha) \neq 0 \Longrightarrow c(\alpha) = 0).$$

The traditional question in theorem proving is to *decide* the above formula for given $H, d_1, \ldots, d_s, c$. In contrast to that, (Wu 1984) asked for *finding* "consistent" $d_1, \ldots, d_s$ for given H, c, such that the above formula holds. (A nondegeneracy polynomial d is consistent, if it does not vanish on all common zeros of H.) Since the non-degeneracy conditions heavily depend on the algebraization, this approach is more appropriate than the mere decision problem. However, there is an ongoing discussion whether Wu's formulation of the problem is adequate, cf. (Chou and Yang 1987).

The first paper describing a successful application of a computer algebra system to proving a geometry theorem is (Cerutti and Davis 1969), who used FORMAC. They did not develop a general method but used the system's capabilities of handling large polynomials and performing simplification steps. During the last decade, various methods for solving different variants of the above problems have been developed, all based on three general purpose methods in computer algebra: Ritt's characteristic set method (Ritt 1950), Buchberger's Gröbner bases method (Buchberger 1965, 1985), and Collins' cylindrical algebraic decomposition method (Collins 1975). Collins' method is the core technique for a new quantifier elimination procedure in the theory of real closed fields, hence the application for deciding formulae of the above kind is straightforward and not only restricted to this particular class of problems. (Kutzler and Stifter 1986a) describe an algorithm for using Collins' method for finding nondegeneracy conditions and report on some experiments using the SAC-2 computer algebra system. But the method is still too complex for attacking nontrivial problems, the limit lies at 5 to 6 variables. Characteristic sets and Gröbner bases provide complete solutions for the above problems over algebraically closed fields. Therefore, only confirmations of true geometric theorems can be achieved by these methods, since, normally, geometry is allied to real numbers. Nevertheless, the implementations turn out to be very powerful practical tools for confirming true geometric theorems. A lot of rather difficult theorems, some involving 10 or more points, have successfully been proved. One of the hardest is Morley's theorem: "For a triangle ABC the neighboring trisectors of the three angles of the triangle will intersect to form 27 triangles in all, of which 18 are equilateral", as is reported in (Wu 1984).

2 EXISTING IMPLEMENTATIONS

We have confined ourselves to the provers based on characteristic sets or Gröbner bases. The interest is also restricted to the problem of finding consistent nondegeneracy conditions that have to be added to the hypotheses in order to be able to confirm the conjecture (called the "Finding Problem" in the sequel), which is widely regarded as the more interesting one. In this section the various methods are sketched and a survey on existing implementations is given. In the sequel let the set of polynomials corresponding to the hypotheses, H, and the polynomial corresponding to the conjecture, c, be given as input to the provers. The output will either be a set D of polynomials corresponding to the nondegeneracy conditions or `'not confirmed'` in case the prover fails to confirm the theorem. y is the tuple of coordinate variables.

2.1 The Provers of Wu and Chou

Based on Ritt's characteristic set method for decomposing an algebraic variety into nondecomposable components and for deciding, whether a polynomial vanishes on "most" zeros of a nondecomposable set, the Chinese mathematician Wu Wen-Tsün developed an algorithm for solving the Finding Problem (Wu 1978, 1984). The prover of Wu, roughly, has the following structure:

$\{H_1, \ldots H_t\}$:= *DECOMP(H)*
for all $1 \leq i \leq t$ *do*
    (A_i, D_i) := *VANISH?*(H_i, c)
if $\exists i A_i$ = `'true'` *then return* $D := \bigcup_{i=1,\ldots,t} D_i$
else return `'not confirmed'`

Here, *DECOMP* returns, for a given set of polynomials H, an uncontractible decomposition $H_1, \ldots, H_t$ of H such that none of the H_i can be decomposed any further (or the empty set in case it is contradictory). *VANISH?* returns, for a given nondecomposable set of polynomials H_i and a polynomial c, the answer `'true'` together with a set of consistent nondegeneracy polynomials $d_1, \ldots, d_t$ in case there exists a solution for the $H_i - c$–instance of the Finding Problem, and the answer `'false'` together with a set of polynomials distinguishing H_i from the original set H otherwise. (H is considered as a global variable here). Details can be found in (Wu 1984) or (Kutzler 1988).

This complete procedure for solving the finding problem is not practical because Ritt's decomposition method (i.e. the procedure *DECOMP*) requires factorization over successive extension fields, which is extremely complex. All reported implementations are incomplete in one way or the other. (First, all use a weaker notion of characteristic sets or just a triangularized set, and second, none involves factorization for the general case.) But, since even a complete version could only be used to confirm true geometric theorems, this does not significantly effect the power of the provers. An incomplete implementation, actually, requires a distinction of y into independent and dependent variables as an additional input. (Independent variables describe points that can be chosen arbitrarily, dependent variables describe points that are constructed subject to conditions. This notion of independence coincides with the notion of algebraic independence in algebraic geometry.)

(Wu 1984) on a HP 9835A, (Ko and Hussain 1985) using MACSYMA on a Symbolics LISP machine and (Kutzler and Stifter 1986a) using SAC-2 on an IBM 4341 do not involve decomposition at all, (Chou 1985) using MACSYMA on a Symbolics LISP machine developed a fast factorization algorithm that is only applicable for the quadratic case. (Chou 1986) is the most extensive collection of geometric theorems that have been proved using Wu's method and documents mechanical proofs of 360 theorems.

2.2 The Prover of Kapur

Based on Buchberger's Gröbner bases method for deciding the radical membership problem (c.f. for example Buchberger 1987), D. Kapur developed an algorithm for solving the Finding Problem (Kapur 1986). The prover of Kapur, roughly, has the following structure:

$\{g_1, \ldots, g_t\}$:= *RADICAL?(H,c)*
if $g_1 = 1$ *then return* *D:={}*
else for all $1 \leq i \leq t$ *do*
$\{h_1, \ldots, h_s\}$:= *RADICAL?(H,g_i)*
if $h_1 \neq 1$ *then return* $D := \{g_i\}$
return `'not confirmed'`

Here, *RADICAL?* returns, for a given set of polynomials H and a polynomial c (H, c elements in $\mathbb{Q}[y]$), a (reduced) Gröbner basis G for $Ideal(H \cup \{z \cdot c - 1\}) \cap \mathbb{Q}[y]$ (z a variable not occuring in y). The standard interpretation is $G = \{1\}$ iff c lies in the radical of H. Kapur found that if there exists a solution of the Finding Problem for H and c, i.e. a consistent nondegeneracy polynomial, then there must exist a consistent nondegeneracy polynomial in (the finite set) G. (The consideration can always be restricted to a single nondegeneracy polynomial.) Details can be found in (Kapur 1986).

Although this algorithm does not involve factorization at all, it provides a complete solution of the Finding Problem over an algebraically closed field (and hence still allows only confirmations of valid geometric theorems). As a matter of experience, the nondegeneracy polynomials found by Kapur's prover are in general less restrictive than those found by Wu's prover. Also, by determining all consistent nondegeneracy polynomials lying in $\{g_1, \ldots, g_t\}$ and forming their disjunction, one obtains a "weak" (though not necessarily the weakest possible) nondegeneracy condition. Originally, Kapur proved the completeness of his algorithm according to the pure lexical ordering, implementations were reported in (Kapur 1986) on a Symbolics LISP machine and in (Kutzler and Stifter 1986a) using SAC-2 on an IBM 4341. Later, Kapur generalized his method for certain mixed orderings.

2.3 The Prover of Kutzler and Stifter

Based on Buchberger's Gröbner bases method for deciding the ideal membership problem, B. Kutzler and S. Stifter developed an algorithm for solving a slightly different finding problem (Kutzler and Stifter 1986b). Their algorithm also requires a distinction into independent and dependent variables. Like the implementations of Wu's prover that do not involve factorization, the prover of Kutzler and Stifter demands the list of independent variables u and the list of dependent variables x as an additional input. Their algorithm decides, whether there exists a polynomial d in the independent variables u, such that $d \cdot c$ lies in the ideal generated by H. (Such a d, by construction, is a consistent nondegeneracy polynomial.) The prover, roughly, has the following structure.

$\{\mathcal{A}, F\}$:= *IDEAL?(H,c)*
if $\mathcal{A}$ =`'true'` *then return* *D := all denominators occuring in F*
else return `'not confirmed'`

Here, *IDEAL?* returns, for a given set of polynomials H and a polynomial c (H, c are elements in $\mathbb{Q}[y]$, but are regarded as elements in $\mathbb{Q}(u)[x]$), the answer `'true'` together with a set F containing the factors necessary for representing c in terms of the elements of H ($F = \{f_1, \ldots, f_n\} \subset \mathbb{Q}(u)[x]$, where $c = \sum_{i=1,\ldots,n} h_i \cdot f_i$) in case c lies in the ideal generated by H, and the answer **'false'** otherwise. Details can be found in (Kutzler and Stifter 1986b).

This algorithm is structurally very simple and, in fact, resembles Wu's algorithm without decomposition. Together with a decomposition of H, it would as well give a complete solution of the Finding Problem. An implementation using SAC-2 on an IBM 4341 is reported in (Kutzler and Stifter 1986b), where also an alternative algorithm is described, in which the computations in $\mathbb{Q}(u)[x]$ are all replaced by computations in $\mathbb{Q}[u, x]$ by using a newly defined pseudo-reduction. (Kutzler and Stifter 1986c) is a collection of protocols of proofs of 75 theorems. Their method over $\mathbb{Q}(u)[x]$ has later also been proposed in (Chou and Schelter 1986).

2.4 Our Goals

Most of these implementations differ greatly in hardware, the underlying algebra system and, of course, programming style. Hence a fair comparison of the different approaches was not possible. Our goal was to provide a uniform implementation of all of the above provers (including variants) and to compare the various methods on the basis of examples from the literature. Additionally, we wanted to gain experiences with SCRATCHPAD II, which is propagated to be one of the most sophisticated computer algebra systems today.

3 THE SCRATCHPAD II COMPUTER ALGEBRA SYSTEM

SCRATCHPAD II is a computer algebra language and system currently under development at IBM Thomas J. Watson Research Center in Yorktown Heights. A recent reference is (Jenks et al 1987). It runs on IBM mainframes under the VM/CMS operating system. However, it is not yet a commercial product. The current version is based on LISP/VM, but this fact is perfectly hidden from the user who will not need any knowledge about LISP. Roughly, SCRATCHPAD II consists of three parts: the compiler, the interpreter, and the interactive environment.

The SCRATCHPAD II computer algebra system provides its own programming language. Basically, this language was designed to express mathematical algorithms in a compact and easy-to-read fashion. However, it is a general purpose programming language. All the algebra code of SCRATCHPAD II is expressed using this language; currently, even an implementation of the whole system (including the compiler) in its own language is under development. Syntactically, SCRATCHPAD II programs share the usual Algol-Pascal-Ada style, conventional programmers will feel right at home. Moreover, many of the advanced features of LISP and APL were added to make the language especially suited for mathematical applications. In contrast to LISP, the language is strongly typed, it enforces strict typechecking at compile- and runtime.. It also supports abstract data types in their most general way: They are constructed at runtime and, therefore, may take parameters. Together with abstract data types, generic operators are provided. For example, there are more than one hundred "−" currently defined (unary and binary, for integers, complex numbers, matrices, polynomials, etc.), but in a program always the same "−"-sign is written (i.e. no calls to "idiff", "qdiff", "polydiff", etc. is necessary as in many other systems). The compiler will automatically select the correct "−" and find the code library where it was defined. Similarly, there is only one sorting algorithm, which works on any domain covering a total ordering "<". In addition, SCRATCHPAD II provides convenient ways to structure code into manageable units and it offers safe ways of separate compilations: It keeps track of the code dependencies between modules and enforces recompilation.

In SCRATCHPAD II, there are three different types of code with quite different semantics: categories, domains and packages. A *category* states certain properties which all domains belonging to that category must have. It defines all operations offered (i.e. the functions by their names and the number and type of their arguments and their result), together with the attributes which are true for objects and operations (e.g. "*" is commutative, "<" is total, etc.). Examples for categories are `Set`, `Group`, `Ring`, `Euclidean Domain`, or `Vector Space`. A *domain* creates new realizations of a category by defining the representation of the objects and providing the actual code of the functions. However, both the actual representation and the implementation of the functions remain hidden in the domain, objects and functions are provided as black boxes to the outside world. Examples of domains are `List`, `Integer`, `Sparse Matrix`, `Distributed Multivariate Polynomial`. Categories and domains are objects, too. They may be assigned, passed as arguments or returned by functions. Thus it is easy to build "domain towers" of arbitrary size (e.g. lists of polynomials over the integers: `List Polynomial Integer`). *Packages* are very similar to domains: They are a collection of (hidden) functions. However, they do not introduce a new data type but operate only on existing domains. Examples are the Gröbner bases package, factorization, multivariate gcd's, etc.

Basically, the interpreter uses the same language as the compiler. However, it was optimized to allow mathematical calculations to be done as with paper and pencil. It is easy to use even for novice users with no programming experience. There is no need to give explicit declarations or to care about details of typing, because sophisticated type analysis facilities will usually find out what type was intended. Additionally, there are quite relaxed ways of handling (even infinite) sequences of numbers or expressions, and of defining rules, functions, and macros. Furthermore, it is possible to call compiled code from the interpreter; it even performs automatic loading of libraries as soon as need arises.

The SCRATCHPAD II interactive environment provides the odds and ends to make the work with the system comfortable. It allows calling the compiler, LISP/VM or CMS, debugging functions interactively and collect detailed execution statistics, saving and restoring sessions on disk, executing prepared input files or produce output files, querying the "knowledge base", defining and look up abbreviations for almost everything, and accessing the online documentation.

4 OUR PROVER PACKAGE IN SCRATCHPAD II

Our geometry theorem proving package, essentially, consists of two parts: a preprocessor for facilitating the translation of a geometric problem into an algebraic one, and a collection of (various variants of) the provers based on characteristic sets and Gröbner bases.

4.1 THE GEOMETRIC PREPROCESSOR

After having formulated the hypotheses and the conjecture(s) of a geometric theorem in terms of "geometric" predicates like *is-midpoint-of*, *lies-on*, *are-cocircular*, etc., the translation of a geometric problem into an algebraic one is done in the following steps: (1) Assign coordinate variables to all points involved, (2) express the hypotheses and the conjecture(s) in terms of polynomial equations in the coordinate variables, (3) distinguish between independent and dependent variables (not necessary for Kapur's prover).

Step (1) is one of the few "creative" parts that remain to be done by the user when applying our geometry theorem proving package. It could, of course, be also automated by just assigning each point systematically generated coordinate variable names. But, in fact, the choice heavily effects the computing time required for the proof and hence the efficiency of the whole method. (Besides choosing favorable positions for the origin and the axes of the coordinate system, one can also minimize the number of variables by utilizing special geometric configurations.) In his collection of examples (Chou 1986) has used some programmed (but not explicitely described) heuristics for finding a "good" choice, but our investigations showed that most of his algebraizations could be "improved" easily.

Step (2) is straightforward using well known techniques from elementary analytic geometry, but it is timeconsuming and errorprone. The actual power of our preprocessor lies in automating this task. A similar translator has already been used in (Chou 1984), but was not very comfortable to use. (To express certain facts in Chou's prover, one can not use points but only indices of single variables as arguments. For example 'B 3 4 5 6 1 2 7 8 11 12 9 10' means that the angles BCA and DFE are congruent, where $A(x_1, x_2)$, $B(x_3, x_4)$, $C(x_5, x_6)$, $D(x_7, x_8)$, $E(x_9, x_{10})$, $F(x_{11}, x_{12})$.) As is demonstrated later, our realization allows a much more convenient way of entering geometric properties and also contains many more such functions. Still, the concrete choice of how to express a geometric situation as hypotheses and conjectures using geometric predicates allows many alternatives and also effects the computing time required for the proof.

Step (3) could also be automated in principle using computer algebra methods (although not very efficiently), but as has been demonstrated in (Chou and Yang 1987), not every choice of algebraically independent variables is a "good" choice in the context of the underlying geometric theorem.

The whole problem of finding a suitable and "efficient" algebraization of a geometric situation is of crucial importance for applying algebraic methods to geometric reasoning problems. A systematic investigation of this problem area is one of the most urgent goals for the future.

Our preprocessor allows, for instance, two- and three-dimensional points (specified by their coordinates) as primitive objects, which are realized as SCRATCHPAD II domains. Step (1) is done by defining the points involved in a geometric problem as concrete objects in these domains, choosing names for the coordinate variables. Following the choice of coordinates in our sample problem from section 1, the input would have the following form.

```
m := pp(0,0)
a := pp(y1,0)
b := pp(y2,y4)
c := pp(y3,y5)
d := pp(y6,y7)
p := pp(y8,y9)
q := pp(y10,y11)
```

Here, pp is a binary function returning two-dimensional points with the function's arguments as coordinates. For space reasons the system's responses are not displayed.

For expressing geometric properties (hypotheses, conjectures), the preprocessor provides functions which take points as arguments and return lists of polynomials corresponding to the respective geometric properties. For our sample problem, the input is:

```
h1 := eql(m,a,m,b)
h2 := eql(m,a,m,c)
h3 := eql(m,a,m,d)
h4 := eql(a,b,c,d)
h5 := on(p,c,d)
h6 := eql(d,p,p,c)
h7 := on(q,a,b)
h8 := eql(a,q,q,b)
conj := eql(p,m,q,m)
```

Here `eql` returns, for given four points, a polynomial in the four points' variables corresponding the the fact that the (square of the) distance of the first two points equals the (square of the) distance of the last two points. on returns, for given three points, a polynomial expressing the fact that the first point lies on the line defined by the last two points. Again, the system's responses (i.e. the polynomials) are not displayed.

The preprocessor contains about 60 such functions, which cover most of the geometric properties used to formulate the examples in the two collections (Chou 1986) and (Kutzler and Stifter 1986c). Additionally, the domain "circle" is offered and many more useful features were implemented, all details can be found in (Mayr 1988).

4.2 The Provers

Based on an existing implementation of Buchberger's Gröbner bases algorithm, we implemented the following provers:

name	method	remarks
WU	Wu	triangularized set, no factorization
WUR	Wu	characteristic set, no factorization
WUF	Wu/Chou	triangularized set, factorization in the quadratic case
WUFR	Wu/Chou	characteristic set, factorization in the quadratic case
LKAP	Kapur	pure lexical ordering, "all" nondegeneracy conditions
LKAP1	Kapur	pure lexical ordering, one nondegeneracy condition
KKAP	Kapur	mixed ordering, "all" nondegeneracy conditions
KKAP1	Kapur	mixed ordering, one nondegeneracy conditions
LRED	Kutzler/Stifter	pure lexical ordering over $Q(u)$
TRED	Kutzler/Stifter	total degree ordering over $Q(u)$
LPRED	Kutzler/Stifter	pure lexical ordering, pseudoreduction over Q
TPRED	Kutzler/Stifter	total degree ordering, pseudoreduction over Q

For all provers there exists a verbose and a nonverbose version. Technical details can be found in (Mayr 1988).

4.3 A Computing Time Statistics

The following table compares the computing times of 24 examples, most of which appeared in the literature on algebraic approaches to mechanical proofs of geometric theorems. The entries are in seconds, obtained on an IBM 3081. ∞ indicates that the proof was not found after one hour CPU-time, □ indicates that the computation exceeded the available 4MB memory. A timing preceeded by † indicates an answer `'not confirmed'`.

example	reference	WUFR	WUF	KKAP1	LRED	TPRED
Altitudes concurrency	-	.26	.16	12.67	1.12	3.90
Euler's line	-	.15	.12	6.13	.77	.80
Ceva's thm.	-	8.55	.93	568.66	23.80	19.30
Simson's thm. 1	Chou 84, p.264	.88	.19	1234.30	2.37	34.43
Simson's thm. 2	Chou 84, p.265	.63	.87	878.79	2.17	65.57
Simson's thm. 3	Chou 84, p.261	3.30	.84	∞	7.63	6073.48
Nine point circle 1	Chou/Schelter 86, p.12	97.83	1.06	347.41	2.93	32.36
Nine point circle 2	-	.99	.22	3.33	2.80	3.07
Pappus' thm.	Chou 84, p.266	.76	.27	87.00	2.73	20.86
Butterfly thm. 1	Chou 84, p.269	□	78.94	∞	352.04	∞
Butterfly thm. 2	Chou 84, p.269	5.32	2.86	∞	†284.59	∞
Pappus' dual thm.	-	112.44	12.07	3124.92	9.95	948.14
Pascal's thm.	Chou/Schelter 86, p.13	11.23	.69	∞	397.45	∞
Circle's secants	-	79.27	9.21	1484.83	678.08	∞
Isosceles midpoint	Chou/Schelter 86, p.18	.66	.55	479.06	1.82	20.40
Equidistant secants 1	section 1	13.87	71.71	64.19	11.68	105.60
Equidistant secants 2	-	1.04	.49	1.37	2.78	1.43
Pappus' point thm.	Chou/Schelter 86, p.16	571.52	6.76	∞	18.11	□
Gauss' thm.	Chou/Schelter 86, p.18	.17	.13	.42	.87	.41
Wang's thm. Gauss lines	Chou/Schelter 86, p.19	774.85	4.12	3499.79	6.40	621.78
Parallelogram 1	Chou 84, p.261	.19	.10	2.34	.66	.43
Parallelogram 2	-	.14	.08	.99	.59	.24
Square	Chou 84, p.267	.13	.10	3.14	1.05	.53
Triangle inscribed circle	-	.75	.62	59.54	26.00	20.60

A detailed statistics on these and more examples with timings of the single steps for each method as well as an analysis of the results is contained in (Kutzler 1988).

4.4 Experiences Using SCRATCHPAD II

We found SCRATCHPAD II a marvellous tool for implementing new algebra code. The abstract data type philosophy and the powerful language constructs were helpful means for implementing the geometry theorem proving package. On the other hand, SCRATCHPAD II performs satisfactory only on very large machines. (The IBM 4341 of the U. Linz computing center, for example, does not suffice for serious work, if used by other people simultaneously.) Also, due to the fact that the system ist still under development and therefore still contains severe errors, a project like ours, currently, can only be finished successfully with a close cooperation with the SCRATCHPAD group in Yorktown Heights.

5 CONCLUDING REMARKS

So far, emphasis in the field of automated geometry theorem proving was put on finding efficient inference methods, assuming a "suitable" algebraization of the considered problem. In some sense, our project also aimed at summarizing the various achievements made by different groups in China, USA and Austria in this particular respect.

Playing around with the preprocessor we found that computing times for proofs of different algebraizations of the same problem vary drastically (factors of 100 and more occur). Therefore, we think that a future concentration on the investigation of the translation process would be worthwhile, in particular the development of criterions for measuring the "quality" of an algebraization and the development of methods for finding "good" algebraizations are of highest interest.

6 ACKNOWLEDGEMENT

The authors are indebted to Dr. R.D. Jenks for enabling a visit at the IBM T.J. Watson Research Center in Yorktown Heights, which was crucial for the success of our project. The authors are also greatful to Dr. S.C. Chou for helpful hints on the implementation of Wu's prover and his factorization algorithm. This work was partially supported by VOEST ALPINE AG Linz and the European Community (Project COST 13).

7 REFERENCES

Buchberger B. 1965. *An algorithm for finding a basis for the residue class ring of a zero-dimensional ideal.* PhD thesis. Univ. Innsbruck, Austria. (In German)

Buchberger B. 1985. Gröbner bases: An algorithmic method in polynomial ideal theory. In *Multidimensional Systems Theory*, ed. N. K. Bose, Dordrecht: D. Reidel Publ. Comp., 1:184–232

Buchberger B. 1987. Applications of Gröbner bases in non-linear computational geometry. *Lect. Notes Comp. Sci.* 296:52–80

Cerutti E., Davis P.J. 1969. FORMAC meets Pappus. *Am. Math. Monthly*, vol. 76, pp. 895–905

Chou S. C. 1984. Proving elementary geometry theorems using Wu's Algorithm. *Contemp. Math.*, vol. 29, pp. 243–286

Chou S. C. 1985. *Proving and discovering geometry theorems using Wu's method.* PhD thesis. Univ. Texas, Austin. 142pp.

Chou S. C. 1986. A collection of geometry theorems proved mechanically. *Inst. f. Comput. Sci. Tech. Rep. 1986-50.* Univ. Texas, Austin

Chou S. C., Schelter W. F. 1986. Proving geometry theorems with rewrite rules. *J. Autom. Reason.* 2:253–73

Chou S. C., Yang J. G. 1987. On the algebraic formulation of geometry theorems. *Inst. f. Comput. Sci. Tech. Rep.* Univ. Texas, Austin

Coelho H., Pereira L.M. 1979. GEOM: A PROLOG geometry theorem prover. *Laboratorio Nacional de Engenharia Civil mem. no. 525*

Collins G. E. 1975. Quantifier elimination for the elementary theory of real closed fields by cylindrical algebraic decomposition. In *Lect. Notes Comput. Sci.* 33:134–83

Gelernter H. 1959. Realization of a geometry theorem-proving machine. *Proc. Int. Conf. Info. Process., Paris, 1959*

Jenks R.D., Sutor R.S., Watt S.M. 1987. SCRATCHPAD II: An abstract datatype system for mathematical computation. *Lect. Notes Comput. Sci.* 296:12–37

Kapur D. 1986. Geometry theorem proving using Hilbert's Nullstellensatz. *Proc. Symp. Symb. Algebraic Comput., Waterloo, Canada, July 21–23, 1986*, ed. B.W. Char, pp. 202-8

Ko H. P., Hussain M. A. 1985. ALGE-Prover: An algebraic geometry theorem proving software. *Gen. Elect. Tech. Rep.* 85CRD139. 28pp.

Kutzler B. 1988. *Algebraic methods for geometric theorem proving.* PhD thesis. Univ. Linz, Austria

Kutzler B., Stifter S. 1986a. New approaches to computerized proofs of geometry theorems. *Proc. Computers & Mathematics, Stanford, USA, July 1986*, eds. R. Gebauer, J. Davenport

Kutzler B., Stifter S. 1986b. Automated geometry theorem proving using Buchberger's algorithm. *Proc. Symp. Symb. Algebraic Comput., Waterloo, Canada, July 21–23, 1986*, ed. B.W. Char, pp. 209–14

Kutzler B., Stifter S., 1986c. Collection of computerized proofs of geometry theorems. *Res. Inst. Symbol. Comput. Tech. Rep. 86-12.0*, Univ. Linz, Austria

Mayr H. 1988. Geometry theorem proving package in SCRATCHPAD II: A primer. *Res. Inst. Symbol. Comput. Tech. Rep. 88-1.0*, Univ. Linz, Austria

Ritt J. F. 1950. *Differential Algebra*, Colloq. Publ. vol. 33. New York: Am. Math. Soc. 184pp.

Wu W. T. 1978. On the decision problem and the mechanization of theorem proving in elementary geometry. *Sci. Sinica* 21:159–72 also *Contemp. Math.* 29:213–34

Wu W. T. 1984. Basic principles of mechanical theorem proving in elementary geometries. *J. Syst.Sci. & Math. Sci.* 4:207–35

Collision of Convex Objects

Bernhard Roider, Sabine Stifter
Research Institute for Symbolic Computation
Johannes Kepler University
A-4040 Linz, Austria

Abstract.

We treat one of the many geometrical problems that face him who simulates tasks for industrial robot manipulators: given two convex objects (in 2 or 3 dimensions), determine whether they intersect. Our iterative algorithm will either find a wedge which contains all of one object but nothing of the other or find a point which belongs to both objects or say it cannot tell. The last alternative occurs only if the objects are not more than (a given) ε apart. The wedge or the common point, respectively, can be found iteratively by constructing touching-lines (touching-planes, in space) from a point on one object to the other object. This geometrical idea at the base of the algorithm is new.

Statement of the Problem.

The problem of testing two moving objects for collision is often solved by considering snapshots, (see e.g. Boyse (1979), Freund, Hoyer (1985)). Furthermore, in practice a motion will not be performed if the objects approach each other too closely. Hence, we isolate the following mathematical problem:

Given: $\mathcal{A}$, $\mathcal{B}$, compact, convex sets, having inner points, $\varepsilon > 0$.

Answer: "disjoint" or "intersect" or "cannot tell" subject to the conditions:

a) if objects intersect, report "intersect" or "cannot tell";
b) if objects are disjoint, report "disjoint" or "cannot tell";
c) if objects are more than ε apart, report "disjoint".

Solution in the Plane.

Suppose for $\mathcal{C}\epsilon\{\mathcal{A},\mathcal{B}\}$ we are given algorithms (i) to decide whether a point is in the interior of $\mathcal{C}$, (ii) to construct touching-lines to $\mathcal{C}$ that meet a given point and touching-points on them, (iii) to intersect a straight line with the boundary of $\mathcal{C}$, and (iv) to find a "starting point" on the boundary of $\mathcal{C}$.

We present the method by giving an example, (for a detailed description see Roider, Stifter (1987), Stifter, Roider (1987)):

Consider the objects of Fig. 1. Let P be a "starting point" on the boundary of $\mathcal{A}$. Find the touching-lines to $\mathcal{B}$. Determine T_1, T_2, S_1, S_2 as in Fig. 1. Decide whether $\mathcal{A}$ has a point in common with wedge $T_1S_1T_2$. (This can be done by considering the placement of the 5 points P, S_1, S_2, T_1, T_2.) Since it has, take S_2, P, T_2 as new P, S_1, T_1, and determine T_2, S_2 as shown in Fig. 2. Now, $\mathcal{A}$ is disjoint with wedge T_1PT_2, which, hence, is a "witness to disjointness". In case $\mathcal{A}$ and $\mathcal{B}$ intersect the same procedure will find a point common to $\mathcal{A}$ and $\mathcal{B}$, a "witness to intersection".

However, this is only the rough form of the algorithm. In this form the method may not terminate if objects are disjoint and $\mathcal{A}$ has a cusp "near" $\mathcal{B}$, or objects intersect and the boundary of $\mathcal{B}$ has a straight line segment "near" $\mathcal{A}$, or objects touch. While the first situation can be handled by changing the roles of the objects (subject to an easy algorithmic criterion), the latter two are handled by stoping the algorithm with the answer "cannot tell" as soon as distance between

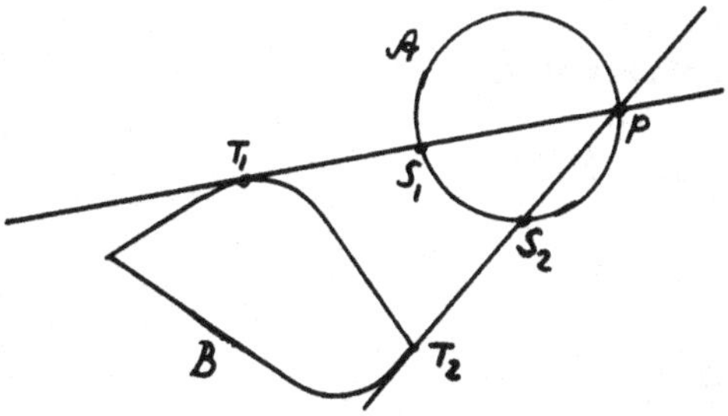

Fig. 1. An uninformative start.

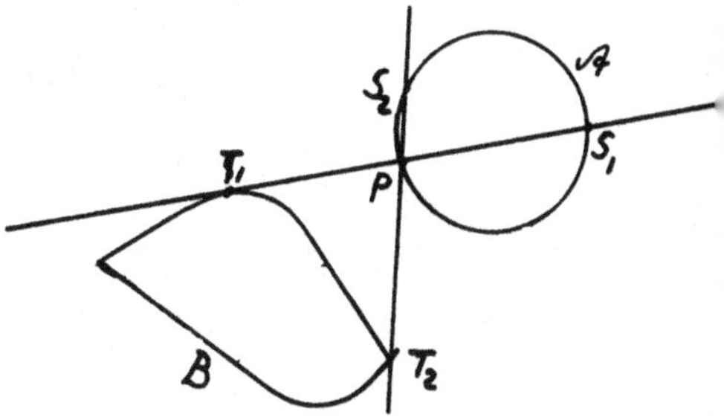

Fig. 2. "Witness to disjointness".

We have implemented the algorithm for $\mathcal{A}$ and $\mathcal{B}$ ellipses. Computing times (on a Micro VAX II) are in the range of a few seconds for quite non-trivial examples (about 3000 iterations). Of course, it would be unfair to compare these computing times with those reported by investigators of Kahan's problem, see e.g. Kahan (1975), Arnon, Mignotte (1988). For, these authors set themselves a rather more ambitious aim. In industrial practice, however, our answers are just as useful.

Solution in the Space.

The idea of the algorithm can be carried over from plane to space. The two touching-lines are replaced by three touching-planes, i.e. $\mathcal{B}$ is enclosed into a three-faced pyramid. There are 13 points whose placements suffice to tell whether $\mathcal{A}$ has a point inside the three-faced pyramid containing $\mathcal{B}$. These points are the touching- and intersection-points on the touching-lines through the intersection-point of the touching-planes and corresponding points on the intersection-lines of these planes. Because there are more than three touching-planes to an object that meet a given point, some problems arise that do not arise in the plane. To handle the choice of the three touching-planes in each iteration step is not worked out in detail, today.

Acknowledgements.

We wish to thank Professor Bruno Buchberger who enabled us to work together. We have been supported by VOEST-Alpine AG, division FAF.

References.

Arnon, D.S., Mignotte, S.F., (1988). On mechanical quantifier elimination for elementary algebra and geometry; Journ. of Symbolic Computation, vol. 5, no. 2, pp. 237 – 260.

Boyse, J.W., (1979). Interference detection among solids and surfaces; Comm. ACM, vol. 22/1, pp. 3–9.

Freund, E., Hoyer, H., (1985). A method for automatic collision avoidance for robots (German); Robotersysteme, vol. 1, pp. 67–73.

Kahan, W., (1975). An ellipse problem; ACM SIGSAM Bulletin, vol. 9/3, p. 11.

Roider, B., Stifter, S., (1987). Collision of convex objects; Techn. Rep. RISC-87-2.0, Univ. Linz, Dept. of Mathematics.

Stifter, S., Roider, B., (1987). Collision of convex objects in the plane; Proc. 10th Tagung Berichte aus Informatikinstituten, Vienna, Austria, Oldenburg Verlag, pp. 359–368.

Solving Algebraic Equations via Buchberger's Algorithm

S.R. Czapor

Department of Applied Mathematics
University of Waterloo
Waterloo, Ontario, Canada N2L 3G1

ABSTRACT

It is demonstrated that, when using Buchberger's Algorithm with the purely lexicographic ordering of terms, it is not generally feasible to inter-reduce basis polynomials during the progress of the algorithm. A heuristic is obtained (for polynomials over the rationals) which improves the efficiency of the reduction sub-algorithm, when the basis is not inter-reduced. Some improvements are made to a recent scheme for combining Buchberger's Algorithm with multivariate factorization. We present a hybrid variant of this scheme, in which extraneous sub-problems are detected outside of the lexicographic/elimination algorithm. Through this approach, the reduced solution bases for dense systems (previously impossible in the lexicographic ordering) may be found.

1. Introduction

For a system of algebraic equations having finitely many solutions, the Gröbner basis with respect to the graduated (or *total degree*) ordering of terms can be used to construct a set of univariate polynomials which give a finite inclusion of these solutions. This is done (when the structure of the basis indicates that there do exist only finitely many solutions) *without* an explicit elimination process. (See [3] for the details.) Hence, this procedure is often preferable to the triangularization provided by the lexicographical ordering of terms, which is sensitive to permutations of the variable ordering and can produce much more complex calculations.

Nonetheless, it is clear that this latter method cannot be ignored, since not all systems of interest have finitely many solutions. Moreover, there exist systems (*e.g.* those of Trinks Hairer, Butcher found in [1]) for which the Gröbner basis is actually *easier* to compute with the lexicographic ordering (and a suitably chosen ordering of variables) because of special structure. Finally, there are systems (*e.g.* that of Rose, in [1]) for which computing the univariate polynomials [from the total degree basis] may be difficult because of the structure of the solution manifold - even though the basis itself is relatively easy to compute.

In this paper we discuss ways to improve the viability of using Buchberger's Algorithm with the lexicographic ordering of terms to solve algebraic equations. Consider a set F of polynomials in $K[x_1, ..., x_n]$, where K is a field and $x_1 > ... > x_n$, and let p, p_1, ... be an

This research was supported by the Natural Sciences and Engineering Research Council of Canada under Grant A8967.

ther such polynomials . We denote the leading monomial of p with respect to the lexico-raphic ordering by $M(p)$. We then denote the corresponding coefficient and term by coeff(p) and *hterm*(p), respectively, and define

$$M(p_1, \ldots, p_m) \equiv lcm(M(p_1), \ldots, M(p_m)) , \tag{1.1}$$

$$Spoly(p_1, p_2) \equiv M(p_1, p_2) \left[\frac{p_1}{M(p_1)} - \frac{p_2}{M(p_2)}\right] . \tag{1.2}$$

inally, we say that p is in *normal form modulo F* if no headterm of any polynomial in F ivides any monomial in p. Then the simplest variant of Buchberger's Algorithm (including he improvements of [2]) to compute the reduced Gröbner basis of F is as follows.

Algorithm 1 ([2]):

> $G \leftarrow F$; $k \leftarrow length(F)$; $B \leftarrow \{[i,j] \mid 1 \leq i < j \leq k\}$;
> **while** $B \neq \emptyset$ **do**
> $[i,j] \leftarrow Select\_Pair(B, G)$; $B \leftarrow B - \{[i,j]\}$
>   **if** $Criteria([i,j], B, G)$ **then**
> $h \leftarrow Normal\_Form(\, Spoly(G_i, G_j), G)$
>   **if** $h \neq 0$ **then**
>   (cutpoint 1)
> $k \leftarrow k+1$; $G \leftarrow G \cup \{h\}$; $B \leftarrow B \cup \{[i,k] \mid 1 \leq i < k\}$
> $G \leftarrow Fully\_Reduce(G)$

he pair-selection procedure and criteria for avoiding zero-reductions are described in [3]. lso described is the procedure *Fully_Reduce*, which transforms the basis so that each polyno-ial is fully reduced modulo the others. This procedure could be invoked after *each* new olynomial is created; this results in another variant, which we refer to as Algorithm 2 in the equel. (See [3], p. 6.13 .) In the next section it is seen that there is a considerable advantage the approach of Algorithm 1, both with respect to the complexity of subsequent -polynomials, and the extra freedom allowed during their reduction. Following this, we dis-uss some improvements to the scheme given in [7], which combines Algorithm 1 with mul-variate factorization at cutpoint 1 (by making recursive calls to the scheme when "h" factors). n example is given of an adaptation of this scheme to a simple yet difficult class of problems namely, dense systems of low degree).

. Tuning the Reduction Algorithm

Clearly the cost of Algorithm 1 is dominated by the steps which perform polynomial rithmetic, namely the forming and reduction of S-polynomials. Hence it is easily seen why riteria that detect zero-reductions *a priori* are so important. Consider an ideal basis $G = \{g_1,$ $\ldots, g_k\}$ of polynomials in an ordered set of variables, and let p be another polynomial in these determinates. We say that *p reduces modulo G* if the set

$$R_p \equiv \{f \in G \text{ such that } hterm(f) \,|\, hterm(p)\} \subset G$$

is non-empty; if p reduces (to p'), we then write

$$p \gg p' = p - [M(p)/M(g)]\, g \;, \text{ for } g \in R_p$$

(although it is better in practice to avoid explicit fractions if working over a fraction field; see [6]). If we have $R_p = R_{p_1} = \ldots = \emptyset$, where $p_1 \equiv p - M(p)$ and $p_{i+1} \equiv p_i - M(p_i)$, then p is in normal form modulo G. Without loss of generality, let us consider only R_p. The degree of freedom offered when R_p has *many* elements poses a problem: how should we select a particular $g \in R_p$? There are two obvious possibilities: we can either choose g such that $hterm(g)$ is maximal among the headterms in R_p, or choose g such that it is minimal. If we were using the total degree ordering, the sensible choice would appear to be the latter strategy. This follows from the fact that the total degree of g gives a bound on the number of terms in g, which in turn bounds the number of term operations in the reduction. However, using the lexicographic ordering, there are several reasons why this strategy may perform poorly in practice. Consider, for example, the reduction of $p = x^2y^2z^2 + x^2y^2z + \cdots + 1$ with respect to either

$$g_1 = c_{11}x^2yz + \cdots + c_{12}y^2 + \cdots + c_{13}z^4 + \cdots + c_{14} \;,$$

or

$$g_2 = c_{21}x + \cdots + c_{22}y^5 + \cdots + c_{23}z^{11} + \cdots + c_{24} \,.$$

If we use g_2, the result may be more complex since the correction term must be larger, and since the degree of g_2 in the subordinate variables y, z is greater. Therefore we would continue the reduction process with a more complicated object. In practice, the basis polynomials in G used for reduction are themselves the products of a reduction/elimination process; so comparisons such as this are quite realistic. Moreover, there is no reason why g_2 should have fewer terms than g_1 (and in fact, it will often have more). Finally, since g_2 must have appeared at a later point in the algorithm, and since the coefficient growth during reduction is linear, the coefficents in g_2 may be much larger than those of g_1.

It is also clear that even though $Spoly(p, g_1)$ and $Spoly(p, g_2)$ are equivalent under the "normal selection strategy" given in [3], we should choose the former. Hence, variants of Buchberger's Algorithm which discard "redundant" polynomials (such as g_1) are at a disadvantage. When several S-polynomial pairs (equivalent under the normal selection strategy) exist, a general strategy to choose from among them should somehow take into account the degrees of the polynomials in the pair. But since the situation described above is common in practice, the S-polynomial formed from the *earliest* pair created is almost always the optimal choice. In what follows, we use this simple approach implicitly. Note, however, that this degree of freedom is lost in Algorithm 2.

Let us now examine more closely the reduction process in the case of polynomials over the rationals (*i.e.*, all coefficient arithmetic is done over $\mathbf{Z}$). Suppose we perform a (re-scaled) reduction

$$p \gg hcoeff(g_i)\, p \;-\; hcoeff(p)\, \frac{hterm(p)}{hterm(g_i)}\, g_i \,, \quad g_i \in R_p \,, \tag{2.1}$$

'here

$length(hcoeff(g_i)) = N'$, $length(\|g_i\|_\infty) = N$, g_i contains ν terms,

$length(hcoeff(p)) = P'$, $length(\|p\|_\infty) = P$, p contains π terms.

'urther suppose that the times required to multiply and add integers of lengths M, N are c_1MN) and $c_2(M+N)$, respectively, and that the time required to multiply two terms in n ariables is bounded by (c_3n). (Note that, practically speaking, the degrees of all monomials hould be reasonably small.) Then the total cost of the reduction (2.1) is bounded by

$$C = c_1\pi N'P + c_1\nu P'N + c_2\, min(\nu, \pi)(N'P + P'N) + c_3\nu n \quad . \tag{2.2}$$

ome simple experiments reveal that for the Maple system ([5]) we have $c_1 \simeq .1\, c_2$ and $c_3 \ll$ $_2$. Now, if $N' \simeq N$, then (2.2) is approximately

$$N[c_1\pi P + c_1\nu P' + c_2\, min(\nu, \pi)(P + P')\,] \, . \tag{2.3}$$

o, as ν becomes large, only the contribution of the second term in (2.3) increases. pparently, the cost is more strongly dependent on N than ν. We might therefore use a uantity such as

$$N\,\nu^{1/2} \simeq (N\,N'\,\nu)^{1/2}$$

s a heuristic measure of the cost of using g_i for the reduction. If $N' \ll N$, then (2.2) is pproximately

$$c_1P'\nu N + c_2\, P'\, min(\nu, \pi) N \quad . \tag{2.4}$$

Ioting that NN' is not much larger than N, and that $P' \leq P$ (*i.e.* (2.4) is smaller than (2.3)), he quantity $NN'\nu$ again seems appropriate. A heuristic reduction strategy which takes this nto account is the following: associate with each basis polynomial a "complexity" value, say

$$complexity(f) \;\equiv\; length(\|f\|_\infty)\; length(hcoeff(f))\; (\#\text{ of terms in } f) \;, \tag{2.5}$$

nd sort G in order of ascending complexity. Then for p we may choose $g_i \in R_p$ such that its omplexity is minimal. If $\nu \gg \pi$ or $\nu \ll \pi$, the quantities νN and N', respectively, are better neasures of the reduction cost. However, there is no way to take this into account without onsidering also the object p. Moreover, if p is an S-polynomial formed from the basis conaining g_i, we expect that $\pi \sim O(\nu)$.

Let us now compare reduction strategies (in the Maple language) as follows. Consider irst Algorithm 1 (which keeps all polynomials in original form), and have it reduce with espect to a basis which is sorted in descending lexicographic order, in ascending lexicographic rder, or in ascending order by complexity (2.5). Then, to demonstrate the importance of sing the "redundant" polynomials *at least* in forming S-polynomials, consider also Algorithm . This has been implemented so that *Fully_Reduce* will reduce first the smallest (lexicographically) of the polynomials which can be reduced (*i.e.*, like the normal selection strategy for S-polynomials), and to reduce using our heuristic strategy. The following timings (and those in the sequel) were computed using Maple Version 4.0 on a VAX/785 processor. The test problems are described in the Appendix.

Problem		Algorithm 1			Algorithm 2
		ascending basis	descending basis	ascending by complexity	
1	time (sec) space (Kb)	570 1164	350 1172	280 1057	467 1270
2 (Butcher)	time space	646 1221	632 1222	636 1122	646 1336
3(a) (Fee (a))	time space	3164 1417	2858 1417	1979 1319	49426 6218
4(a)	time space	7675 2458	12000 3202	5826 3080	>186000 >17000
5 (Trinks)	time space	10911 2687	13753 2507	6061 2376	>168000 >17000
6 (Katsura)	time space	19490 2556	43448 2744	17195 2742	>200000 ≥14000
3(b) (Fee (b))	time space	85271 4072	128724 4956	47175 3396	>200000 ≥14401
7 (Rose)	time space	119770 7600	207602 9636	90238 8296	72011 8700
8	time space	>72000 >17000	>90000 >17000	>83000 >17000	>92000 >17000

One reason for the superiority of the heuristic approach has not yet been mentioned: if we use for reduction a g_i with small coefficients, the bound on the coefficient growth for the step (2.1) is also small. Hence, the [significant] cost of removing the integer content *after* the reduction should also be minimized. Typically, the reduction strategies based on headterms perform well when their correlation with the heuristic is high.

3. Using Multivariate Factorization

Despite the improvement possible with a heuristic reduction strategy, Algorithm 1 still requires long running times for (apparently) modest problems. This is particularly true when the initial extent of triangularization is small. For example, Problem 2 is relatively easy because, in a suitably chosen ordering of variables, the input polynomials are close to full separation of variables. By comparison, Problem 8 (with fewer variables) is difficult because it is more homogeneous in structure.

Fortunately, the intermediate results produced by the algorithm permit multivariate fac-
orization surprisingly often. In [7], we presented an efficent recursive variant of Algorithm 1 o exploit this fact. This scheme (which we henceforth refer to as Algorithm 3) displayed a ast improvement in efficency compared to Algorithm 1, when both used a lexicographically escending basis in the reduction sub-algorithm. We now make some further improvements which, implicitly, include use of the new reduction scheme), and give an example of how to xploit the extra flexibility offered by this approach. (The structure of Algorithm 3 will ecome clear when we present this modification.)

The results of the previous section show why Algorithm 1 was chosen as the framework f the new scheme. However, Algorithm 2 might also perform well *if* combined with a heuris-ic to determine when (and to what extent) the entire basis should be reduced. One obvious pecial case is the appearance of a univariate polynomial. When factoring is introduced, it is ommon for one or more univariate polynomials (usually in the base variable "x_n") to appear - t *any* point in the algorithm - when a factorization succeeds. This is so even for systems with nfinitely many solutions, since these often give rise to sub-problems with finitely many solu-ions. We note the following (obvious) fact:

emma: Let F be a set of polynomials in $K[x_1, \ldots, x_n]$ which contains $p \in F \cap K[x_j]$. If p oes not factor and $Ideal(F) \neq (1)$, then p is the polynomial in $Ideal(F) \cap K[x_j]$ of lowest egree.

o, if another univariate in the same variable appears, the sub-problem is extraneous and can e abandoned without further reductions (or factorizations). It is also clear that a univariate olynomial should be used to reduce previous polynomials, since it is an optimal element of e reduction basis. The resulting "improved" version of Algorithm 3, although not much dif-erent, can be much faster (as we shall see).

So far, we have not exploited the fact that, when a factorization produces several sub-roblems, we may apply entirely different methods to them. For example, the total degree ethod mentioned in Section 1 may become applicable. Or, a sudden drop in the degree of a articular variable may suggest the use of pseudo-remaindering and resultants in the manner f [9]. Since no single scheme will be viable for all problems, we will discuss one possible daptation of Algorithm 3. This was motivated by systems such as Problem 4(b), which is nly of total degree 2 in four variables, but dense. The Gröbner basis using the total degree rdering of terms is relatively easy to obtain, irrespective of the density. But let us suppose at (for some reason) we wish to obtain the solutions by the lexicographic-elimination ethod. It happens (*because* of the density) that, during the elimination and factoring, very rge extraneous factors - and hence sub-problems - are produced. These inevitably cause me/space limits to be exceeded before the more modest computation of actual solutions can roceed.

However, if the total degree Gröbner basis is already known, then each factor can be sted for consistency by using it to extend the basis in the total degree ordering. In this way, nly "consistent" polynomials need be passed to the elimination scheme. Let $TG(F)$ and $NF(p, F)$ denote the Gröbner basis of F and the normal form of p modulo F, respectively, in e total degree ordering. Then a hybrid scheme may be constructed as follows.

Algorithm 4:

```
sol_bases ← ∅ ; append_bases(F) ,  where we define
append_bases ← procedure(F, B, newpoly, nonzero, H)
  if #arguments = 1 then
    k ← length(F) ; B ← {[i,j] | 1≤i<j≤k} ; G ← F
    nonzero ← ∅ ; H ← TG(F)
  else
    k ← length(F)+1 ; G ← F ∪ {newpoly} ; B ← B ∪ {[i,k] | 1≤i<k}
  while B ≠ ∅ do
    [i,j] ← Select_Pair(B, G) ; B ← B − {[i,j]}
    if Criteria([i,j], B, G) then
      p ← Normal_Form( Spoly(G_i, G_j), G)
      if p ≠ 0 then
        P ← {distinct factors of p}
        if P ≠ {p} then
          for m from 1 to length(P) do
            if P_m ∈ nonzero then next
            q ← TNF(P_m, H)
            if q ≠ 0 then
              H̃ ← TG( H∪{q} )
              if H̃ = {1} then next
            else
              H̃ ← H
            append_bases(G, B, P_m, nonzero ∪ {P_1,...,P_{m-1}}, H̃ )
          return
        else
          k ← k+1 ; G ← G ∪ {p} ; B ← B ∪ {[i,k] | 1≤i<k}
  sol_bases ← sol_bases ∪ { Fully_Reduce(G) }
  return
```

(Note that Algorithm 3 is obtained by removing the steps involving the total degree ordering.) Here, each factor not marked as nonzero is reduced modulo H, the total degree Gröbner basis corresponding to the current (irreducible, lexicographic) sub-problem. If the factor is part of a *larger* ideal, we use its reduced form to extend H in the total degree ordering. (This will produce {1} if the factor is extraneous.) If the factor is consistent, we continue in the lexicographic ordering. Again, although it is not stated explicitly above, we should use any univariate polynomials which appear to reduce the rest of the basis. In fact, since we no longer have to consider extraneous sub-problems in the lexicographic part of the scheme, we can go much further. Define variable x_k to be *separated* (*in F*) if:

(a) $\exists$ irreducible $p \in F \cap K[x_k]$, or

(b) $\exists\, p \in F$ such that $hterm(p) = x_k$, and all variables in $p - M(p)$ are separated.

Then, by the Lemma, any polynomial containing only separated variables must reduce to zero. The regular algorithm ensures separation of variables by actually carrying out an exhaustive set of S-polynomial reductions (only some of which will be avoided by the standard "criteria"), often *after* full separation has occurred. Since we are working with irreducibl olution components, even the simple criterion above often allows us to avoid very large zero-eductions, when the sub-problem contains only finitely many solutions.

Let us now compare the improved and "basic" (*i.e.* without special handling of univariates r optimized reduction) versions of Algorithm 3 with an implementation of Algorithm 4 including the above feature); in the latter case, the times include the separate computation of he total degree Gröbner basis.

Problem		Algorithm 3		Algorithm 4
		basic	improved	(hybrid)
1	time (sec)	297	265	234
	space (Kb)	1286	1310	1188
2	time	362	251	3778
(Butcher)	space	1393	1114	1498
3(a)	time	885	629	732
(Fee (a))	space	1786	1572	1548
4(a)	time	3460	1904	2178
	space	3031	2384	2392
5	time	2354	1590	1667
(Trinks)	space	2220	2032	1744
6	time	4386	2830	2470
(Katsura)	space	2712	2350	2146
3(b)	time	22773	10352	>200000
(Fee (b))	space	3048	2621	≥4400
7	time	14340	7189	11705
(Rose)	space	3876	3891	4174
8	time	191132	78045	52913
	space	11592	11076	6889
4(b)	time	>168000	>141000	60924
	space	>17000	>17000	9887

The expected asymptotic improvement of Algorithm 3 over Algorithm 1 is clear. We also ote that Algorithm 4 is the only scheme which is able to cope with the dense problem 4(b). his is of interest because most problems tend to become denser as the first few variables are iminated. It is surprising that Algorithm 4 is competitive on many of the simpler problems, nce use of the total degree ordering to "filter" extraneous factors is less than an ideal general lution. Although modular methods for Gröbner bases are elusive (see [8]), the development such an algorithm for this specific task seems a worthwhile problem for future study. It ould also be possible to weaken somewhat our definition of "full separation of variables".

Appendix: List of Test Problems

Problem 1 :

$$F_1 = 343 + 804z - 919z^2 - 996y + 705yz - 312y^2 - 252x + 526xz + 202xy - 761x^2,$$
$$F_2 = -869 + 443z - 175z^2 - 335y - 58yz + 893y^2 - 566x + 544xz - 174xy + 915x^2,$$
$$F_3 = -494 + 642z + 98z^2 - 213y + 527yz - 880y^2 - 743x - 380xz - 768xy - 465x^2,$$

using $x > y > z$.

Problem 2 : Butcher's system (see [1]) of 8 equations in 8 unknowns, using the ordering $b_1 > a_{32} > b_2 > b_3 > a > c_3 > c_2 > b$.

Problem 3 : Fee's system (see [6]) of 4 equations in 5 unknowns.

(a) Substitute $b = 2$, and use the poor ordering $q > c > p > d$.

(b) Use the ordering $c > d > q > b > p$.

Problem 4:

$$F_1 = 4 + 8w - 10w^2 - 10z + 7zw - 3z^2 - 3y + 6yw + 2yz + c_1y^2$$
$$- 9x + 5xw - 2xz - 4xy + d_1x^2,$$
$$F_2 = 9 - 6w + 6w^2 - 2z + 10zw - 5z^2 + 7y + yw - 2yz + c_2y^2$$
$$- 9x - 8xw - 4xz - 8xy + d_2x^2,$$
$$F_3 = -9 - 2w + 10w^2 - 9z + 8zw + 10z^2 - 3y - 2yw - 2yz + c_3y^2$$
$$- 2x + 3xw - 9xz + 7xy + d_3x^2,$$
$$F_4 = -7 + 9w + 2w^2 + 3z + 10zw - 2z^2 + 8y + 4yw - 3yz + c_4y^2$$
$$+ 3x - 6xw + 9xy + d_4x^2.$$

(a) Substitute $c_1 = c_2 = c_3 = c_4 = d_1 = d_2 = d_3 = d_4 = 0$, and use the ordering $x > y > z > w$.

(b) Substitute $c_1 = -8$, $c_2 = 6$, $c_3 = -3$, $c_4 = -6$, $d_1 = -1$, $d_2 = -5$, $d_3 = 6$, $d_4 = -3$, and use $z > x > w > y$.

Problem 5 : Trinks system [10], using the poor variable ordering $b > t > s > w > p > z$.

Problem 6 : Katsura's system (see [1]) of 5 equations in 5 unknowns, using the ordering $u_4 > u_0 > u_3 > u_2 > u_1$.

Problem 7 : Rose's system (see [1]) of 3 equations in 3 unknowns, using the ordering $u_4 > u_3 > a_{46}$.

Problem 8:

$$F_1 = 4 + 8z - 10z^2 - 10z^3 + 7y - 3yz - 3yz^2 + 6y^2 + 2y^2z - 8y^3 - 9x + 5xz - 2xz^2 - 4xy - xyz + 9xy^2 - 6x^2 + 6x^2z - 2x^2y ,$$

$$F_2 = -5 + 7z + z^2 - 2z^3 + 6y - 9yz - 8yz^2 - 4y^2 - 8y^2z - 5y^3 - 9x - 2xz + 10xz^2 - 9xy + 8xyz + 10xy^2 - 3x^2 - 2x^2z - 2x^2y ,$$

$$F_3 = -2 + 3z - 9z^2 + 7z^3 + 6y - 7yz + 9yz^2 + 2y^2 + 3y^2z + 10y^3 - 2x + 8xz + 4xz^2 - 3xy - 6xyz + 3xy^2 - 6x^2 + 9x^2y ,$$

using $x > y > z$.

References

[1] W. Böge, R. Gebauer, H. Kredel: "Some Examples for Solving Systems of Algebraic Equations by Calculating Gröbner Bases", J. Symbolic Computation, Vol. 2, No. 1, 1986.

[2] B. Buchberger: "A criterion for detecting unnecessary reductions in the construction of Gröbner bases", Proc. EUROSAM '79, Marseille, June 1979, (W. Ng, ed.), *Lecture Notes in Computer Science*, Vol. 72, 1979, pp. 3-21.

[3] B. Buchberger: "Gröbner Bases: An Algorithmic Method in Polynomial Ideal Theory", in *Progress, directions and open problems in multidimensional systems theory*, (N.K. Bose, ed.), D. Reidel Publishing Co., 1985, pp. 184-232.

[4] J.C. Butcher: "An application of the Runge Kutta space", BIT Computer Science Numer. Math., Vol. 24, 1984, pp. 425-440.

[5] B.W. Char, G.J. Fee, K.O. Geddes, G.H. Gonnet, M.B. Monagan: "A Tutorial Introduction To Maple", J. Symbolic Computation, Vol 2, No. 2, 1986.

[6] S.R. Czapor, K.O. Geddes: "On Implementing Buchberger's Algorithm For Gröbner Bases", Proc. SYMSAC '86, (B.W. Char, ed.), Waterloo, July 1986, pp. 233-238.

[7] S.R. Czapor: "Solving algebraic equations: combining Buchberger's algorithm with multivariate factorization", to appear in J. Symbolic Computation.

[8] G.L. Ebert: "Some Comments On The Modular Approach To Gröbner Bases", ACM SIGSAM Bull. 17, No. 2, May, 1983.

[9] M.E. Pohst, D.Y.Y. Yun: "On Solving Systems of Algebraic Equations via Ideal Bases and Elimination Theory", Proc. SYMSAC '81 , (P.S. Wang, ed.), Utah, Aug. 1981, pp. 206-211.

[10] W. Trinks: "Über B. Buchbergers Verfahren, Systeme algebraischer Gleichungen zu lösen", J. Number Theory, Vol. 10, No. 4, Nov. 1978, pp. 475-488.

Primary Ideal Decomposition

Heinz Kredel
Gesellschaft für Schwerionenforschung,
Planckstraße 1, D–6100 Darmstadt, F.R.G. *

1 Introduction

In a polynomial ring (a Noetherian Ring) it is known, that every ideal $\mathcal{A}$ can be decomposed into finitely many primary components $\mathcal{Q}_i$:

$$\mathcal{A} = \mathcal{Q}_1 \cap \ldots \cap \mathcal{Q}_l$$

This fact was discovered by Lasker, Macaulay [Lasker 1905, Macaulay 1916], and in the special case of zero dimensional ideals already by M. Noether. By E. Noether this fact is true for every ring satisfying the so called 'Teilerkettensatz' which is equivalent to Hilberts Basis Theorem [E. Noether 1921].

In 1926 G. Hermann made an attempt to discribe a constructive way to find the primary decomposition of a given polynomial ideal over fields which allow constructive factorization. However she was not aware of the fact, that in general not for all ground fields, the polynomials can be factorized constructively. Seidenberg gave an investigation of these topics [G. Hermann 1926, Seidenberg 1974].

The early constructive approaches were based on the method of solving linear equations in a module over the polynomial ring. Today we are equipped with the powerful method of Gröbner Bases, as investigated by B. Buchberger [Buchberger 1965, 1985].

In 1976 R. Schrader proposed a way for zero dimensional polynomial ideals by constructing univariate polynomials, factorization and computing the primary ideals from the prime ideals by $\mathcal{Q} = \mathcal{A} + \mathcal{P}^\sigma$, where σ is the exponent of $\mathcal{P}$ [Schrader 1976]. However his primality test and the generation of prime ideals is hard to realize.

D. Lazard gave an algorithm to compute prime components and multiplicites for pure equidimensional ideals. In 1985 he published a careful study of the primary decomposition in the case on a polynomial ring in two variables [Lazard 1982, Lazard 1985]. Kandri-Rodi studied constructive ways to compute the radical of an ideal via the computation of associate isolated prime ideals [Kandri-Rodi 1984].

P. Gianni, B. Trager & G. Zacharias developed a set of algorithms for primary decomposition in 1984. Theire method uses ideals in 'generic position' and factorization, instead of field extensions by adjoining new variables. However they use a very complicated primality test, involving mappings to polynomial rings over field extensions of the ground field. [Gianni, Trager & Zacharias 1984]

In this paper we present a set of algorithms which are already implemented in the SAC-2 Computer Algebra System, and a complete example of a primary decomposition using these programs. We compute the prime ideals by first decomposing the ideal according to Schrader.

*Since 1988: Universität Passau, Innstraße 27, D–8390 Passau, F.R.G

An algorithm has been published by Böge, Gebauer and Kredel [Schrader 1976, Böge *et al.* 1986]. Next we transform the generated ideals by some algebraic extension of the ground field until we reach a prime basis for the ideals. This gives us an easy primality test, and the primary ideals can now be determined by $\mathcal{Q} = \mathcal{A} + \mathcal{P}^{\sigma}$.

Our method works for zero dimensional ideals, but we conjecture that it can be extended to cover also the general non zero dimensional case. At the moment the algorithms are implemented for the rational numbers as ground field, but the method applies also to the rational functions as ground field. Since field extensions are involved, the method is restricted to fields of characteristic zero.

The method uses:

- Gröbner base calculations
- construction of univariate polynomials and univariate polynomial factorisation
- algebraic field extensions of the ground field
- an easy to handle ideal primality test

The method produces for a given zero dimensional ideal:

- the associated prime ideals
- the primary ideals
- the exponents of the associated prime ideals

The plan of the paper is as follows: Section 3 presents the ideal decomposition achievable by univariate polynomial factorization. Section 4 introduces the ideal decomposition using algebraic field extensions of the ground field. Section 6 gives the algorithms for the determination of the primary ideals. Section 7 presents a short example for the computational power of the method.

We are indebted to V. Weispfenning, W. Böge and A.Kandri-Rodi for valuable discussions on the subject. We would also like to thank the **GSI** for computing time support.

2 Preliminarys

Let $\mathbf{P} = \mathbf{K}[\mathcal{X}]$ be the polynomial ring over some field $\mathbf{K}$ (of characteristic zero) in r variables: $\mathcal{X} = \{X_1, \ldots, X_r\}$. Small letters will denote elements of $\mathbf{P}$, capital letters denote finite subsets of $\mathbf{P}$, p.e. $A = \{a_1, \ldots, a_l\}$ with $a_i \in \mathbf{P}$ for $i = 1, \ldots, l$ so $A \subset \mathbf{P}$. Caligraphic letters will denote infinite subsets of $\mathbf{P}$, usually ideals in $\mathbf{P}$, p.e. $\mathcal{A} = (a_1, \ldots, a_l)$ with $a_i \in \mathbf{P}$ for $i = 1, \ldots, l$. We write also *ideal*(A) for $\mathcal{A}$ when $A = \{a_1, \ldots, a_l\}$.

In the sequel we adopt the notation of [v.d. Waerden 1971, 1967] and [Buchberger 1985], if not otherwise stated.

For the primary ideal decomposition we need a constructive method to solve for example the problem, whether a polynomial belongs to a given ideal. In 1965 Buchberger introduced the concept of Gröbner bases to solve this problem (among many others). We will denote the Gröbner base of an ideal $\mathcal{A}$ with $GBASE(\mathcal{A})$ and $GBASE(A)$ if $\mathcal{A} = ideal(A)$. We assume the Gröbner bases beeing reduced, but we will indicate this whereever ist is especially important. For an overview about properties and applications of Gröbner bases see [Buchberger 1985].

For the computation of Gröbner bases Buchberger gave an algorithm in 1965. In [Buchberger 1985] several improvements of the original algorithm are discussed. We are computing Gröbner bases using Buchberger's algorithm as it was implemented in SAC-2 by [Gebauer, Kredel 1984].

It is known that, the polynomial ring over a field $\mathbf{K}$ is a unique factorisation domain. [Samuel, Zariski, I] Especially if $\mathbf{K} = \mathbf{Q}$ or $\mathbf{K} = \mathbf{Q}(Y_1, \ldots, Y_n)$ the factorization can be carried out constructively. See [Seidenberg 1976] for a discussion of the properties $\mathbf{K}$ must satisfy in order to allow a constuctive factorisation in the polynomial ring over such a field.

In our case we use the algorithm for univariate polynomial factorization over $\mathbf{Q}$ from the SAC-2 factorization package [Collins, Loos 1980]. Descriptions of factorization algorithms can be found in [Kaltofen 1982].

3 Ideal decomposition using univariate Polynomials

In this section we will first define the dimension of a polynomial ideal and give some characterizations of ideals having dimension 0. Then we discuss an algorithm for constructing univariate polynomials of minimal degree which are contained in a given ideal. Finally the ideal decomposition using univariate polynomials is described.

3.1 Dimension of an Ideal

The dimension of an ideal is classically defined as the maximum over all transcendence degrees of the residue class rings $\mathbf{P}/\mathcal{P}$ for all associated prime ideals $\mathcal{P}$ of $\mathcal{A}$. Gröbner gave a definition of dimension which does not use the primary decomposition of the ideal [Gröbner 1970]. Renschuch showed the equivalence of the two definitions [Renschuch 1976].

Zero dimensional ideals can be characterized as containing univariate polynomials for each variable. See [Kredel 1985] for a further characterization of zero dimensional ideals by the set of common zeros.

The cases of dimension -1, 0 and > 0 can be determined using Buchberger's criteria [Buchberger 1965, 1970], see Problem 6.11 in [Buchberger 1985]. The dimension $d > 0$ can be determined by a method of [Kredel, Weispfenning 1986].

3.2 Construction of univariate polynomials

In zero dimensional ideals there exist univariate polynomials in every variable X_i ($i = 1, \ldots, r$). Such polynomials with minimal degree can be computed using Method 6.11 of [Buchberger 1985].

The next algorithm UPM realizes this construction. UPM can use a Gröbner base G in any admissible termordering $\leq_T$.

However the computation of the univariate polynomial in $X_i \in \mathcal{X}$ involves only elements of the vectorspace $\mathbf{P}/\mathcal{A} = \mathcal{Y}$ which correspond to powerproducts which are $\leq_T X_i$. The number of these base elements is bounded by $\prod_{j=1}^{i} n_j$ where n_j is the degree of the polynomial in G with univariate headterm in X_j. So the systems of linear equations have only few equations when i is small. We experienced that using a Gröbner base with respect to the graded lexicographical ordering $\leq_G$ resulted in much bigger systems of linear equations to be solved than in the pure lexicographical case. So the advantage of fast Gröbner base computation in the $\leq_G$-case will disappear when construction of univariate polynomials has to be done.

Algorithm $f \leftarrow UPM(G, X)$.

Univariate polynomial in X of minimal degree.
Given: G = Gröbner base for an ideal $\mathcal{A} \subset \mathbf{K}[X_1, \ldots, X_r]$, with dim $\mathcal{A} \leq 0$.
$X \in \{X_1, \ldots, X_r\}$ a variable for which the univariate polynomial is to be constructed.
Find: $f = f(X)$ a univariate polynomial of minimal degree contained in *ideal*(G).

3.3 Ideal decomposition using univariate Polynomials

Lemma 3.3.1 *Let $\mathcal{A} \subset \mathbf{K}[X_1, \ldots, X_r]$ be an ideal and let $f = q \cdot p^e \in \mathcal{A}$ where e is a natural number, then*

$$1 \in \mathcal{A} + (p) \Longrightarrow q \in \mathcal{A}.$$

Proposition 3.3.2 *Let f be a univariate polynomial in the ideal as constructed in section 3.2. For all non trivial factors p of f we have*

$$(\mathcal{A} + (p)) \neq (1).$$

Proof From $(\mathcal{A} + (p)) = (1)$ it would follow from Lemma 3.3.1 that the cofactor of this factor is also contained in the ideal and is of lower degree than f. This would contradict the degree of the univariate polynomial being minimal. ⊗ See also [Kredel 1985] Section I.3.

Proposition 3.3.3 *For any two distinct irreducible factors g, h of such a univariate polynomial with minimal degree $f = g \cdot h \cdot f', g \neq h$ we have*

$$\mathcal{A} \subset \left([\mathcal{A} + (g)] \cap [\mathcal{A} + (h)]\right) \tag{1}$$
$$[\mathcal{A} + (g)] + [\mathcal{A} + (h)] = (1) \tag{2}$$

Proof From $\mathcal{A} \subset [\mathcal{A} + (g)]$ and $\mathcal{A} \subset [\mathcal{A} + (h)]$ follows (1). Since the polynomial ring in one variable over a field is an euclidean domain and g and h are irreducible there exisist $f_1, f_2 \in \mathbf{K}[X] \subset \mathbf{P}$ such that $1 = f_1 g + f_2 h$ so (2) is true. ⊗

Adjoining all distinct irreducible factors of univariate polynomials to $\mathcal{A}$ we obtain a decompositon of the radical ideal *radical*$(\mathcal{A})$ [Schrader 1976] [Böge *et al* 1986].

Algorithm $L \leftarrow ZDID(G, M)$
Zero-dimensional ideal decomposition.
Given: G = Gröbner base for an ideal $\mathcal{A} \subset \mathbf{K}[X_1, \ldots, X_r]$.
$M = \{X_k, \ldots, X_r\}$ a set of variables for which the univariate polynomials are to be constructed $(1 \leq k \leq r)$.
Find: $L = \{L_1, \ldots, L_s\}$ a maximal list of Gröbner bases for radical ideals $Q_i = ideal(L_i) \subset \mathbf{K}[X_1, \ldots, X_r]$ such that:
ideal$(G) \subset$ *ideal*$(L_1) \cap \ldots \cap$ *ideal*(L_s).
ideal(L_i) contains a univariate irreducible polynomial in the variable X_j $(1 \leq i \leq s; k \leq j \leq r)$.
ideal(L_i) + *ideal*$(L_j) = (1)$ $(1 \leq i < j \leq s)$.
ideal$(L_i) \neq (1)$ $(1 \leq i \leq s)$.

4 Ideal decomposition by algebraic field extensions

Using univariate polynomial factorization for ideal decomposition gives only a set of radical ideals associated with $\mathcal{A}$. To obtain the prime ideals associated with $\mathcal{A}$ additionally algebraic field extensions of the ground field $\mathbf{K}$ have to be considered.

4.1 Computing in algebraic extension fields

By the construction of an algebraic extension field over $\mathbf{K}$ we know $\mathbf{K}(\alpha) \cong \mathbf{K}[E]_{/(p(E))}$ with $p(\alpha) = 0$ and $deg(p) = n \geq 1$ when $p(E)$ is irreducible over $\mathbf{K}$. This gives us the isomorphism

$$\mathbf{K}(\alpha)[X_1, \ldots, X_r] \cong \mathbf{K}[E, X_1, \ldots, X_r]_{/(p(E))}.$$

And a one to one correspondence of ideals $\mathcal{A}' \longleftrightarrow [\mathcal{A} + (p)]$, where $\mathcal{A}' \subset \mathbf{K}(\alpha)[X_1, \ldots, X_r]$ and $[\mathcal{A} + (p)] \subset \mathbf{K}[E, X_1, \ldots, X_r]$. I.e. computing in algebraic extension fields can be done simply by computing in a polynomial ring with one more variable and adjoining the irreducible polynomial for the primitive element to the ideals. More precisely we have for the sum of two ideals

$$\mathcal{A}' + \mathcal{B}' \longleftrightarrow [\mathcal{A} + (p)] + [\mathcal{B} + (p)] = [\mathcal{A} + \mathcal{B} + (p)].$$

For the product of two ideals we have $\mathcal{A}' \cdot \mathcal{B}' \longleftrightarrow [\mathcal{A} + (p)] \cdot [\mathcal{B} + (p)] + (p) = [\mathcal{A} \cdot \mathcal{B} + (p)]$.

4.2 Primitive elements for algebraic extension fields

Definition 4.2.4 *For a reduced Gröbner base G of a zero dimensional ideal let $deg(G, X) = deg(p, X)$ where $p \in G$ and $HT(p) \in \mathbf{K}[X]$.*

p is uniquely determined for a reduced Gröbner base of a zero dimensional ideal.

By the theorem of the primitive element we can find $t \in \mathbf{K}$ such that for the 'extension ideal'

$$\mathcal{A}' = \mathcal{A} + (E - X_i - tX_j) \subset \mathbf{K}[E, X_1, \ldots, X_r] \text{ with } \mathcal{A} = \mathcal{A}' \cap \mathbf{K}[X_1, \ldots, X_r]$$

we have $deg(G', X_i) = deg(G', X_j) = 1$, if G' denotes the Gröbner base for $\mathcal{A}'$ and $1 \leq i, j \leq r$ [Böge *et al* 1985]. For the computation of t see [Loos 1982] especially section 5 'Constructing Primitive Elements'.

Algorithm $F \leftarrow FE(G, M)$
Field extension for ideal.
Given: $G =$ Gröbner base for an ideal $\mathcal{A} \subset \mathbf{K}[X_1, \ldots, X_r]$.
$M = \{X_i, X_j\} \subset \{X_1, \ldots, X_r\}$
where $deg(G, X_i) > 1$ and $deg(G, X_j) > 1$.
Find: $F =$ Gröbner base for the extension ideal $\mathcal{A}' \subset \mathbf{K}[E, X_1, \ldots, X_r]$
where $deg(F, X_i) = deg(F, X_j) = 1$.
comment Introduction of a new variable E.
$H \leftarrow [G \in \mathbf{K}[E, X_1, \ldots, X_r]]$.
comment Field extension.
$t \leftarrow 0$.
repeat $t \leftarrow t - 1$.
$f \leftarrow E - X_i - tX_j$.
$F \leftarrow GBASE(H \cup \{f\})$
until $deg(F, X_i) = deg(F, X_j) = 1$.
return.

4.3 Prime basis of Ideals over algebraic extension fields

Over suitable algebraic extension fields zero dimensional prime ideals have the form

$$(X_1 - \beta_1, \ldots, X_r - \beta_r) \subset \mathbf{K}(\alpha)[X_1, \ldots, X_r].$$

Where α denotes the primitive element of the algebraic extension over **K**. These so called monoidal prime bases are discused for example by [Gröbner 1970] Volume II, Chapter II, Section 3. The prime basis over $\mathbf{K}(\alpha)$ corresponds to a prime basis over $\mathbf{K}[E]_{/(g)}$:

$$G = \{g(E), X_1 + g_1(E), X_2 + g_2(E), \ldots, X_r + g_r(E)\}.$$

We will call such Gröbner bases P-bases:

Definition 4.3.5 *A P-base is characterized by:*

1. *it is a reduced Gröbner base of a zero dimensional ideal with respect to the lexicographical termordering,*
2. *it exists exactly one irreducible polynomial in one variable,*
3. *all other polynomials depend on at most two variables, and have univariate linear headterms.*

This fact is used for our ideal primality test.

Proposition 4.3.6 *Let G be the Gröbner base for A. If $deg(G, X_i) = 1 \quad (i = 1, \ldots, r)$ and the univariate polynomial g in E is irreducible over* **K** *then A is prime.*

The specification of the used primality test is as follows:

Algorithm $M \leftarrow PITEST(G)$
Ideal primality test.
Given: $G =$ Gröbner base for an ideal $A \subset \mathbf{K}[X_1, \ldots, X_r]$.
Find: $M = \{X_i, X_j\} \subset \{X_1, \ldots, X_r\}$
where $deg(G, X_i) > 1$ and $deg(G, X_j) > 1$.
if no such M exist, return $M = \emptyset$, i.e. $ideal(G)$ is prime.
comment Use 'some' strategie to determine such M.

4.4 Computing the associated prime basis

The algorithm FE from section 4.2 is iteratively applied to the extension ideals. Together with the factorization of the univariate polynomials over **K** and building the decomposition ideals we reach several prime bases P_i. The P_i are Gröbner bases in $\mathbf{K}[E_{i1}, \ldots, E_{ij_i}, X_1, \ldots, X_r]$, where $E_{i1} = E_i$ by definition.

$$P_i = \{h_{1i}(E_i), E_{2i} + h_{2i}(E_i), \ldots, E_{li} + h_{li}(E_i), X_1 + g_{1i}(E_i), \ldots, X_r + g_{ri}(E_i)\}.$$

The variables E_{ki}, $(k = 2, \ldots, l)$ and the corresponding polynomials $E_{2i} + h_{2i}(E_i), \ldots, E_{li} + h_{li}(E_i)$ can be removed from the P_i. So we get

$$P_i = \{h_{1i}(E_i), X_1 + g_{1i}(E_i), \ldots, X_r + g_{ri}(E_i)\}.$$

Now $deg(P_i, X_i) = 1$ $(i = 1, \ldots, r)$ and h_{1i} is irreducible over **K** so $ideal(P_i)$ is prime.

By this approach we may have to add at most $r - 1$ new variables but for every prime ideal only the smallest algebraic extension field of **K** is required.

The next algorithm actually computes the associated prime ideals.

Algorithm $L \leftarrow PIA(G)$
Prime ideal associated with ideal.
Given: $G =$ Gröbner base for an ideal $\mathcal{A} \subset \mathbf{K}[X_1, \ldots, X_r]$.
Find: $L = \{P_1, \ldots, P_l\}$. Where each
P_i is a Gröbner base for a prime ideal $\mathcal{P}_i \subset \mathbf{K}[E_i, X_1, \ldots, X_r]$
associated with $\mathcal{A}$ $(i = 1, \ldots, l)$.
comment Ideal decomposition without field extensions.
$S \leftarrow ZDID(G)$. $L \leftarrow \emptyset$.
comment Compute ideal decomposition.
repeat select P from S. $S \leftarrow S \setminus \{P\}$.
$M \leftarrow PITEST(P)$.
if $M = \emptyset$
then $L \leftarrow L \cup \{P\}$
else begin
$P' \leftarrow FE(P, M)$.
$T \leftarrow ZDID(P')$.
$S \leftarrow S \cup T$ **end**
until $S = \emptyset$.
comment Remove unnecessary variables and polynomials.
return.

4.5 Computation of prime ideals over K

From the prime ideals $\mathcal{P} \subset \mathbf{K}[E, X_1, \ldots, X_r]$ the prime ideals $\mathcal{P}' \subset \mathbf{K}[X_1, \ldots, X_r]$ can be found by the ideal intersection:

$$\mathcal{P}' = (\mathcal{P} \cap \mathbf{K}[X_1, \ldots, X_r]).$$

Proposition 4.5.7 *$\mathcal{P}$ is prime $\Longrightarrow$ $\mathcal{P}$' is prime.*

Proof $\mathcal{P}$' not prime would imply $\mathcal{P}$ not prime by the construction of $\mathcal{P}$'. ⊗

Proposition 4.5.8 *$\mathcal{P}$is associated with ideal $\mathcal{A}$.*

Proof Since $\mathcal{A}$ and $\mathcal{P}$ are zero dimensional, and from $\mathcal{A} \subset \mathcal{P}$ in $\mathbf{K}[E, X_1, \ldots, X_r]$ and $\mathcal{A} \subset \mathbf{K}[X_1, \ldots, X_r]$ it follows $\mathcal{A} \subset (\mathcal{P} \cap \mathbf{K}[X_1, \ldots, X_r]) = \mathcal{P}'$. ⊗

5 Example

To illustrate the method of field extension we consider the following ideal in $\mathbf{Q}[X, Y]$:

$$\mathcal{A} = (X^2 - 2, Y^2 - 2)$$

One field extension with $t = -2$ i.e. adjoining $E - X + 2Y$ to the ideal leads to a Gröbner base in $\mathbf{Q}[E, X, Y]$:

$$\{E^4 - 20E^2 + 36, X - 1/24E^3 + 13/12E, Y - 1/12E^3 + 7/6E\}$$

The univariate polynomial in E can be factored into: $E^4 - 20E^2 + 36 = (E^2 - 2) \cdot (E^2 - 18)$
This gives the two Gröbner bases:

$$\{E^2 - 2, X + E, Y + E\} \text{ and } \{E^2 - 18, X + 1/3E, Y - 1/3E\}$$

The Gröbner bases in $\mathbf{Q}[X, Y, E]$ are

$$\{X^2 - 2, Y - X, E + X\}, \quad \{X^2 - 2, Y + X, E + 3X\}$$

From this we can read off the desired ideal intersection over $\mathbf{Q}$:

$$\mathcal{A} = (X^2 - 2, Y - X) \cap (X^2 - 2, Y + X).$$

6 Primary ideal decomposition

6.1 Computation of corresponding primary ideals

In the case of a zero-dimensional ideal the primary components are isolated and their exponents e can be computed using the corresponding prime ideals:

$$\mathcal{Q} = \mathcal{A} + \mathcal{P}^e \text{ with } e \text{ minimal such that } \mathcal{P}^e \subset [\mathcal{A} + \mathcal{P}^{e+1}]$$

See [v.d.Waerden 1967] Volume II.

Algorithm $PYAPI(P, G.Q, e)$
Primary ideal for prime ideal associated with ideal.
Given: $G =$ Gröbner base for an ideal $\mathcal{A} \subset \mathbf{K}[X_1, \ldots, X_r]$.
$P =$ Gröbner base for a prime ideal $\mathcal{P} \subset \mathbf{K}[E, X_1, \ldots, X_r]$
associated with $\mathcal{A}$.
Find: $Q =$ Gröbner base for a primary ideal $\mathcal{Q} \subset \mathbf{K}[E, X_1, \ldots, X_r]$.
$e =$ exponent of $\mathcal{Q}$.
comment Initialize.
$q' \leftarrow P$. $e \leftarrow 0$.
comment Determine maximal e with $ideal(G) \subset ideal(P)^e$.
repeat $q \leftarrow q'$.
$e \leftarrow e + 1$. $q' \leftarrow q \cdot P$.
until $ideal(G) \not\subset ideal(q')$.
comment Determine minimal e with
$ideal(p^e) \subset [ideal(G) + ideal(p^{e+1})]$.
$Q \leftarrow GBASE(G, q)$.
repeat $Q' \leftarrow GBASE(G, q')$.
$t \leftarrow [ideal(q) \subset ideal(Q')]$.
if *not* t **then begin**
$q \leftarrow q'$. $Q \leftarrow Q'$.
$e \leftarrow e + 1$. $q' \leftarrow q \cdot P$ **end**
until t.
return.

6.2 Primary ideal decomposition over $\mathbf{K}(\alpha)$

By computing all associated prime ideals over some algebraic field extension $\mathbf{K}(\alpha)$ and by determining the corresponding primary ideals and theire exponents we finally reach the primary decomposition of $\mathcal{A}$.

Algorithm $L \leftarrow PIDA(G)$
Decompositon of an ideal over an algebraic extension field over $\mathbf{Q}$.

Given: $G =$ Gröbner base for an ideal $\mathcal{A} \subset \mathbf{K}[X_1, \ldots, X_r]$.
Find: $L = \{(P_1, Q_1, e_1), \ldots, (P_l, Q_l, e_l)\}$.
Where each P_i is a Gröbner base for a prime ideal $\mathcal{P}$ associated with $\mathcal{A}$ and $P_i \subset \mathbf{K}[E_i, X_1, \ldots, X_r]$ for some new variable E_i $(i = 1, \ldots, l)$.
Each Q_i is a Gröbner base for the primary ideal for $ideal(P_i)$ $(i = 1, \ldots, l)$ such that $\mathcal{A} = [ideal(Q_1) \cap \ldots \cap ideal(Q_l)]$.
The e_i are the exponents of the respective primary ideals $(i = 1, \ldots, l)$.
comment Get prime ideals.
$F \leftarrow PIA(G)$. $L \leftarrow \emptyset$.
comment Compute primary ideals associated with $ideal(G)$.
repeat select P from F. $F \leftarrow F \setminus \{P\}$.
$PYAPI(P, G.Q, e)$.
$L \leftarrow L \cup \{(P, Q, e)\}$.
until $F = \emptyset$.
return.

6.3 Primary ideal decomposition over K

The prime ideals over **K** can be computed using the ideal intersection with $\mathbf{K}[X_1, \ldots, X_r]$.

Algorithm $L \leftarrow PID(G)$
Decompositon of an ideal over **Q**.
Given: $G =$ Gröbner base for an ideal $\mathcal{A} \subset \mathbf{K}[X_1, \ldots, X_r]$.
Find: $L = \{(P_1, Q_1, e_1), \ldots, (P_l, Q_l, e_l)\}$.
Where each P_i is a Gröbner base for a prime ideal $\mathcal{P}$ associated with $\mathcal{A}$ and $P_i \subset \mathbf{K}[X_1, \ldots, X_r]$ $(i = 1, \ldots, l)$.
Each Q_i is a Gröbner base for the primary ideal for $ideal(P_i)$ $(i = 1, \ldots, l)$ such that $\mathcal{A} = [ideal(Q_1) \cap \ldots \cap ideal(Q_l)]$.
The e_i are the exponents of the respective primary ideals $(i = 1, \ldots, l)$.
comment Get prime ideals over some field extension over **Q**.
$F \leftarrow PIA(G)$. $L \leftarrow \emptyset$.
comment Compute primary ideals associated with $ideal(G)$.
repeat select P from F. $F \leftarrow F \setminus \{P\}$.
$P' \leftarrow (P \cap \mathbf{K}[X_1, \ldots, X_r])$.
$PYAPI(P', G.Q, e)$.
$L \leftarrow L \cup \{(P', Q, e)\}$.
until $F = \emptyset$.
return.

7 Example

To illustrate the method of primary decomposition we consider the ideal of two Lissajous curves and their partial derivatives in $\mathbf{Q}[X, Y, Z]$ [Schrader 1976, Winkler *et al* 1985]. With our programms we are now able to determine the second prime ideal explicitly. The computing times are on an IBM 3090-200 VF running MVS/XA.

The Gröbner Base (computed in 880 milliseconds with 41896 storage cells used) is:

```
( X**12 -2 X**10 +2 X**8 )
( X**11 Y -2 X**9 Y +2 X**7 Y )
```

```
( X Y**2 -1/2 X**9 + X**7 - X**3 )
( Y**3 - X**10 Y +3/2 X**8 Y - X**6 Y - X**2 Y )
```

The first step of the decomposition is the univariate polynomial factorization, which leads to two ideals:

```
RADICAL IDEAL NO 1                  RADICAL IDEAL NO 2

( X**4 -2 X**2 +2 )                 X
( Y**2 + X**2 -2 )                  Y
```

The first ideal has no real roots, the second ideal has exactly one real root. The field extension step yields again 2 ideals. One field extensions is necessary, one to get linear headterms in 'X' and 'Y' (denoted by 'EXY').

```
EXTENSION IDEAL NO 1

( EXY**8 - 20 EXY**6 + 104 EXY**4 - 40 EXY**2 + 1156 )
( X - 7/33456 EXY**7 + 155/16728 EXY**5 - 1639/16728
      EXY**3 + 428/697 EXY )
( Y - 7/16728 EXY**7 + 155/8364 EXY**5 - 1639/8364 EXY**3
      + 159/697 EXY )

EXTENSION IDEAL NO 2 = RADICAL IDEAL 2
```

From the prime ideals over the field extension over **Q** we compute the prime ideals over **Q**. They are identical with the ideals above, thus proving that they are prime ideals.

In the last step we compute the primary ideals and their exponents:

```
PRIMARY IDEAL NO 1             PRIMARY IDEAL NO 2

( X**4 -2 X**2 +2 )            X**8
( Y**2 + X**2 -2 )             X**7 Y
                               ( X Y**2 + X**7 - X**3 )
                               ( Y**3 - X**6 Y - X**2 Y )

THE EXPONENT IS 1              THE EXPONENT IS 8
```

Statistics : Computing time: 3790 ms, 78661 storage cells used.

Experience shows that most of the computing time is spend during the computation of the ideal intersection $\mathcal{P} \cap \mathbf{K}[\mathcal{X}]$. and during the computation of the primary ideals.

Univariate polynomial factorization seems to be very fast. Moderate time is used to do the algebraic field extensions, the introduction of new variables and consecutive computation take minor influence on the computing time.

8 Directions for Future Research

To extend our method from zero dimensional ideals to the general non zero dimensional case we suggest to attempt the following approach:

Factorize the polynomials in the Gröbner base using multivariate polynomial factorization algorithms. Thereby avoid losing imbedded prime ideals. If all polynomials in a Gröbner base are irreducible (wrt. UFD) then the generated ideal should be equidimensional.

Now using the method of [Kredel, Weispfenning 1986] one can determine the exact dimension > 0 of an ideal. Further more we can compute maximal independent sets M of variables modulo $\mathcal{A}$. Use transcendental field extensions by M to get zero dimensional ideals and apply our method to $\mathbf{K}(M)[\mathcal{X} \setminus M]$.

The computation of the primary ideals can then be done over $\mathbf{K}(M)$. Then 'lift' the prime and primary ideals back to $\mathbf{K}[\mathcal{X}]$. The case where some primary ideals are containing the intersection of some others (over $\mathbf{K}$) has still to be considered.

9 References

B. Buchberger 1965, "Ein Algorithmus zum Auffinden der Basiselemente des Restklassenringes nach einem nulldimensionalen Polynomideal",
Dissertation, University of Insbruck 1965.

B. Buchberger 1970, "Ein algorithmisches Kriterium für die Lösbarkeit eines algebraischen Gleichungssystems", Aequ. Math. 4, pp. 374-383.

B. Buchberger 1985, "Gröbner bases: An algorithmic method in polynomial ideal theory", In:(N.K.Bose ed.) Progress, directions and open problems in multidimensional systems theory. pp. 184-232. Dordrecht: Reidel Publ. Comp. (1985)

W. Böge, R. Gebauer, H. Kredel 1985, "Gröbner bases using SAC-2" Proc. EUROCAL '85 European Conference on Computer Algebra. Linz 1985, Springer Lect. Notes Comp. Sci. 204, pp. 272-274, 1986.

W. Böge, R. Gebauer, H. Kredel 1986, "Some Examples for Solving Systems of Algebraic Equations by Calculating Gröbner Bases." J. Symbolic Computation (1986) 1, pp. 83-98.

G. E. Collins, R. G. Loos 1980, "SAC-2 - Symbolic and Algebraic Computation version 2, a computer algebra system, ALDES - ALgorithm DEScription language", ACM SIGSAM Bulletin 14, No. 2, 1980.

R. Gebauer, H. Kredel 1983, "Buchberger algorithm system" Technical Report, Institut für Angewandte Mathematik, Universität Heidelberg, 1983 (see also ACM SIGSAM Bulletin 18, No. 1, 19).

R. Gebauer, H. Kredel 1984, "Real solution System for algebraic equations" Technical Report, Institut für Angewandte Mathematik, Universität Heidelberg, 1984.

P. Gianni, B. Trager, G. Zacharias 1986, "Gröbner Bases and Primary Decomposition of Polynomial Ideals" J. Symbolic Computation, to appear 1988.

W. Gröbner 1968, 1970, "Algebraische Geometrie I, II",
Bibliographisches Institut, Mannheim 1968, 1970.

G. Hermann 1926, "Die Frage der endlich vielen Schritte in der Theorie der Polynomideale" Math. Ann. Bd. 95 (1926) P. 736-788.

E. Kaltofen 1982, "Factorization of polynomials" in B. Buchberger, G. E. Collins, R. G. Loos, "Computer algebra - symbolic and algebraic computation" Vienna: Springer, 1982.

A. Kandri-Rody 1984, "Effective Methods in the Theory of Polynomialideals", Ph. D. Thesis, RPI, Troy, NY, May 1984.

H. Kredel 1985, "Über die Bestimmung der Dimension von Polynomidealen",
Diplomarbeit, Fakultät für Mathematik Universität Heidelberg, 1985.

H. Kredel, V. Weispfenning 1986, "Computing Dimension and Independent Sets for Polynomial Ideals" J. Symbolic Computation, to appear 1988.

E. Lasker 1905, "Zur Theorie der Moduln und Ideale" Math. Ann. Bd. 60 (1905) P. 20-116.

D. Lazard 1982, "Commutative algebra and computer algebra", Springer Lect. Notes Comp. Sci. 144, P. 40-48.

D. Lazard 1985, "Ideal Bases and Primary Decomposition: Case of Two Variables" J. Symbolic Computation (1985) 1, pp. 261-270.

R. Loos 1982, "Computing in Algebraic Extensions" in B. Buchberger, G. E. Collins, R. G. Loos, "Computer algebra - symbolic and algebraic computation" Vienna: Springer, 1982.

F. S. Macaulay 1916, "Algebraic Theory of modular systems" Cambridge Tracts in Mathematics, Vol. 19, Cambridge 1916.

E. Noether 1921, "Ideal Theorie in Ringbereichen" Math. Ann. Bd. 83 (1921) P. 24-66.

B. Renschuch 1976, "Elementare und praktische Idealtheorie" VEB Deutscher Verlag der Wissenschaften, Berlin, 1976.

R. Schrader 1976, "Zur konstruktiven Idealtheorie" Diplomarbeit, Mathematisches Institut II, Universität Karlsruhe, 1976.

A. Seidenberg 1974, "Constructions in Algebra", Trans. Amer. Math. Soc. 197, 1974, pp. 273-313.

W. L. Trinks 1978, "Über Buchbergers Verfahren Systeme algebraischer Gleichungen zu lösen". J. Number Theor. 10, pp. 475-488, 1978.

F. Winkler, B. Buchberger 1985, F. Lichtenberg, H. Rolletschek 1985, "An algorithm for constructing canonical bases (Gröbner bases) of polynomial ideals".
ACM/TOMS 11, pp. 66-78, 1985.

v.d.Waerden 1971, 1967, "Algebra I, II" Springer, Heidelberg 1971, 1967.

O. Zariski, P. Samuel 1958, "Commutative algebra", vol I, van Nostrand, 1958.

Solving Systems of Algebraic Equations by Using Gröbner Bases

Michael Kalkbrener
Research Institute for Symbolic Computation (RISC)
Johannes Kepler Universität Linz, Austria

September 26, 1988

Abstract

In this paper we give an explicit description of an algorithm for finding all solutions of a system of algebraic equations which is solvable and has finitely many solutions. This algorithm is an improved version of a method which was deviced by B. Buchberger. By a theorem proven in this paper, gcd-computations occuring in Buchberger's method can be avoided in our algorithm.

1 Introduction

The method of Gröbner bases was introduced by B. Buchberger in his 1965 Ph.D.thesis. This work is accessible in [1]. His method, as its central objective, solves the simplification problem for polynomial ideals and, on this basis, gives easy solutions to a large number of other algorithmic problems.

In the present paper we use Gröbner bases for the solution of systems of algebraic equations. In particular, we deal with the following problem:

Given: *F, a finite set of polynomials in the indeterminates $x_1, \ldots, x_n$ over a field K such that F is solvable and has finitely many solutions. (A solution of F is an element b of $\bar{K}^n$ such that $f(b) = 0$ for all $f \in F$, where $\bar{K}$ is the algebraic closure of K.)*

Find: *all solutions of the system F.*

A first algorithm for reducing this multivariate problem to a univariate one by using Gröbner bases appears in [1].

A second algorithm (see [2], Method 6.10) makes use of the fact that if the purely lexicographical ordering is used every Gröbner basis G of a zero-dimensional ideal I consists of finitely many polynomials

$$
\begin{array}{rcl}
G_{1,1} & \in & K[x_1], \\
G_{2,1} & \in & K[x_1, x_2], \\
\cdots & & \cdots \\
G_{2,car_2} & \in & K[x_1, x_2], \\
\cdots & & \cdots \\
G_{n,1} & \in & K[x_1, \ldots, x_n], \\
\cdots & & \cdots \\
G_{n,car_n} & \in & K[x_1, \ldots, x_n],
\end{array}
$$

where $car_2 \geq 1, \ldots, car_n \geq 1$, and the i-th elimination ideal of I is the ideal generated by $\{G_{1,1}, G_{2,1}, \ldots, G_{i,car_i}\}$ (see [4]). Method 6.10 finds a solution $(b_1, \ldots, b_i, c)$ of the $i+1$-th elimination ideal by adjoining a zero c of the polynomial

$$gcd(G_{i+1,1}(b_1, \ldots, b_i, x_{i+1}), \ldots, G_{i+1,car_{i+1}}(b_1, \ldots, b_i, x_{i+1}))$$

to the solution $(b_1, \ldots, b_i)$ of the i-th elimination ideal.

In this paper we prove a theorem which states that there exists a $d \in \bar{K}$ and an $r \in \{1, \ldots, car_{i+1}\}$ such that

$$d \cdot G_{i+1,r}(b_1, \ldots, b_i, x_{i+1}) = gcd(G_{i+1,1}(b_1, \ldots, b_i, x_{i+1}), \ldots, G_{i+1,car_{i+1}}(b_1, \ldots, b_i, x_{i+1}))$$

and that the polynomial $G_{i+1,r}$ can be easily found by a test for zero in an extension field of K. Therefore, this theorem leads to an improved version of Method 6.10, in which the gcd-computation is avoided.

In section 2 we introduce a few definitions. In section 3 a specification of the problem and the explicit descriptions of the algorithm in [2] and of our improved version are given. Furthermore, the theorem on which our method is based is presented. In section 4 we prove this theorem.

2 Definitions

Throughout the paper K denotes an arbitrary field and $\bar{K}$ the algebraic closure of K.

Let n be a natural number. By $K[x_1, \ldots, x_n]$ we denote the ring of all polynomials over K in n indeterminates.

Let f be an element of $K[x_1, \ldots, x_n]$ and r an element of $\{1, \ldots, n\}$.

We denote the *degree of* f in the variable x_r by $deg(f, r)$. For a non-constant f, there is a first s such that $f \in K[x_1, \ldots, x_s]$. Considering f as a polynomial in x_s, we denote the *leading coefficient* by $lc(f)$.

Let H be a finite subset of $K[x_1, \ldots, x_n]$ and I an ideal in $K[x_1, \ldots, x_n]$.

By $Ideal(H)$ we denote the *ideal generated by* H. The set

$$V(I) = \{\, a \in \bar{K}^n \mid f(a) = 0 \text{ for all } f \in I \,\}$$

is called the *variety of I*. We denote the *radical of I* by $\sqrt{I}$ and the set $I \cap K[x_1, \ldots, x_r]$ by $I_{/x_r}$.

Let n be greater than 1 and b an element of $\bar{K}^{n-1}$.

By $I(b, x_n)$ we denote the set

$$\{ h(b, x_n) \in K(b)[x_n] \mid h \in I \}.$$

By Hilbert's basis theorem, we can choose a finite subset $F = \{f_1, \ldots, f_m\}$ of $K[x_1, \ldots, x_n]$ such that $Ideal(F) = I$. Clearly,

$$I(b, x_n) = Ideal(\{f_1(b, x_n), \ldots, f_m(b, x_n)\}),$$

where the ideal on the right-hand side is formed in $K(b)[x_n]$.

For h, an element of $K(b)[x_n]$, the following conditions are equivalent:

1. $h = gcd(\{f_1(b, x_n), \ldots, f_m(b, x_n)\})$,
2. $I(b, x_n) = Ideal(\{h\})$ and h is normed.

We denote the uniquely determined $h \in K(b)[x_n]$ which satiesfies these conditions by $gcd(I, b)$.

Throughout the paper we fix the "purely lexicographical ordering" of the power products of $x_1, \ldots, x_n$. We denote it by $\ll$. Furthermore, we assume

$$x_1 \ll x_2 \ll \ldots \ll x_n.$$

We refer to [2] for the definitions of *LeadingPowerProduct, SPolynomial, Gröbner basis*, and *reduced Gröbner basis*.

Let G be a reduced Gröbner basis.

Let $G_{r,1}, \ldots, G_{r,car_r}$ be the polynomials in G that belong to $K[x_1, \ldots, x_r]$ but not to $K[x_1, \ldots, x_{r-1}]$. We suppose the order chosen in such a way that

$$LeadingPowerProduct(G_{r,j}) \ll LeadingPowerProduct(G_{r,k}) \text{ for } j < k.$$

3 Solving Systems of Algebraic Equations by Using Gröbner Bases

Throughout the following sections n is a natural number greater than 1.

In this paper we want to solve the following problem:

Given: F, a finite subset of $K[x_1, \ldots, x_n]$ such that I is zero-dimensional, where $I := Ideal(F)$.

Find: $V(I)$.

In [2] B. Buchberger presents the following algorithm for this problem:

Algorithm 1

input: F, a finite subset of $K[x_1, \ldots, x_n]$ such that I is a zero-dimensional ideal in $K[x_1, \ldots, x_n]$, where $I := Ideal(F)$.

output: X_n, a finite subset of $\bar{K}^n$ such that $X_n = V(I)$.

$G := GB(F)$, where $GB(F)$ is the uniquely determined reduced Gröbner basis such that $Ideal(GB(F)) = Ideal(F)$.

Comment: It is proven in [4] that $Ideal(G) \cap K[x_1, \ldots, x_r] = Ideal(G \cap K[x_1, \ldots, x_r])$ (for $r = 1, \ldots, n$), where the ideal on the right-hand side is formed in $K[x_1, \ldots, x_r]$. Therefore, the polynomials in G have their variables "separated". G contains exactly one polynomial in $K[x_1]$ (actually, it is the polynomial in $Ideal(G) \cap K[x_1]$ with smallest degree). According to the definition in section 2 we denote it by $G_{1,1}$.

The successive elimination can, then, be carried out by the following process:

$X_1 := \{ c \mid c \in \bar{K} \text{ and } G_{1,1}(c) = 0 \}$
for $r := 1$ *to* $n - 1$ *do*
  $X_{r+1} := \emptyset$
  for all $b \in X_r$ *do*
    $H := \{ G_{r+1,s}(b, x_{r+1}) \mid s \in \{1, \ldots, car_{r+1}\} \}$
    $q :=$ greatest common divisor of the polynomials in H
    $X_{r+1} := X_{r+1} \cup \{ (b, c) \mid c \in \bar{K} \text{ and } q(c) = 0 \}$

Upon termination, X_n will contain all the solutions.

The improved version of this algorithm is based on the following new theorem which we prove in the next section.

Theorem 1 *Let l be an element of $\{2, \ldots, n\}$, I a zero-dimensional ideal in $K[x_1, \ldots, x_n]$, G the reduced Gröbner basis in $K[x_1, \ldots, x_n]$ such that $Ideal(G) = I$, and $b \in V(I_{/x_{l-1}})$. Then there exists a $d \in \bar{K}$ such that*

$$d \cdot G_{l,min_b}(b, x_l) = gcd(\{G_{l,1}(b, x_l), \ldots, G_{l,car_l}(b, x_l)\}),$$

where min_b denotes the minimum of the set

$$\{ r \mid r \in \{1, \ldots, car_l\} \text{ and } lc(G_{l,r})(b) \neq 0 \}.$$

Therefore, we can replace the instructions

$H := \{ G_{r+1,s}(b, x_{r+1}) \mid s \in \{1, \ldots, car_{r+1}\} \}$
$q :=$ greatest common divisor of the polynomials in H

in Algorithm 1 by the instruction

$q := G_{r+1,min_b}(b, x_{r+1})$

and obtain the following algorithm:

Algorithm 2

input: F, a finite subset of $K[x_1,\ldots,x_n]$ such that I is a zero-dimensional ideal in $K[x_1,\ldots,x_n]$, where $I := Ideal(F)$.

output: X_n, a finite subset of $\bar{K}^n$ such that $X_n = V(I)$.

$G := GB(F)$
$X_1 := \{ c \mid c \in \bar{K} \text{ and } G_{1,1}(c) = 0 \}$
for $r := 1$ *to* $n-1$ *do*
  $X_{r+1} := \emptyset$
  for all $b \in X_r$ *do*
    $q := G_{r+1,min_b}(b, x_{r+1})$
    $X_{r+1} := X_{r+1} \cup \{ (b,c) \mid c \in \bar{K} \text{ and } q(c) = 0 \}$

Note that for computing min_b, where $b \in V(I_{/x_{r-1}})$, one has to check only whether

$$\begin{aligned}(lc(G_{r,1}))(b) &= 0,\\ (lc(G_{r,2}))(b) &= 0,\\ \ldots & \;.\end{aligned}$$

till the first s is found such that

$$(lc(G_{r,s}))(b) \neq 0.$$

4 Proof of Theorem 1

For proving Theorem 1 we first show a stronger result for reduced Gröbner bases of zero-dimensional primary ideals:

Theorem 2 *Let l be an element of $\{2,\ldots,n\}$, Q a zero-dimensional primary ideal in $K[x_1,\ldots,x_n]$, and G the reduced Gröbner basis in $K[x_1,\ldots,x_n]$ such that $Ideal(G) = Q$. Then*

$$G_{l,1}(b,x_l) = \ldots = G_{l,car_l-1}(b,x_l) = 0 \tag{1}$$

for all $b \in V(Q_{/x_{l-1}})$.

Proof:

We first show that (1) holds for some $b \in V(Q_{/x_{l-1}})$:

We assume, to the contrary, that

$$\begin{gathered}\text{for every } b \in V(Q_{/x_{l-1}})\\ \text{there exists an } r \in \{1,\ldots,car_l - 1\} \text{ with } G_{l,r}(b,x_l) \neq 0.\end{gathered} \tag{2}$$

In this proof we denote $(G \cap K[x_1,\ldots,x_l]) \setminus \{G_{l,car_l}\}$ by F.
Let $f_1, f_2 \in F$.

By Method 6.9 in [2], there exists a natural number s such that

$$LeadingPowerProduct(G_{l,car_l}) = x_l^s.$$

Therefore,

$$deg(f_1, l) < deg(G_{l,car_l}, l) \text{ and } deg(f_2, l) < deg(G_{l,car_l}, l).$$

From this and the definition of the S-polynomial we obtain

$$deg(SPolynomial(f_1, f_2), l) \leq \max\{deg(f_1, l), deg(f_2, l)\} < deg(G_{l,car_l}, l).$$

Thus, $SPolynomial(f_1, f_2)$ reduces to zero modulo F. By Theorem 6.2 in [2],

$$F \text{ is a Gröbner basis.} \tag{3}$$

Obviously,

$$F \text{ is reduced.} \tag{4}$$

$G \cap K[x_1, \ldots, x_{l-1}]$ is a reduced Gröbner basis because $SPolynomial(g_1, g_2)$ reduces to zero modulo $G \cap K[x_1, \ldots, x_{l-1}]$ for all $g_1, g_2 \in G \cap K[x_1, \ldots, x_{l-1}]$. By Lemma 6.8 in [2],

$$Ideal(G \cap K[x_1, \ldots, x_{l-1}]) = Q_{/x_{l-1}}.$$

Thus, by Method 6.9 in [2],

$$V(Q_{/x_{l-1}}) \text{ is finite.} \tag{5}$$

Let $(c_1, \ldots, c_l) \in V(Ideal(F))$. Then

$$f(c_1, \ldots, c_{l-1}) = 0 \text{ for every } f \in G \cap K[x_1, \ldots, x_{l-1}].$$

So, by Lemma 6.8 in [2],

$$(c_1, \ldots, c_{l-1}) \in V(Q_{/x_{l-1}}).$$

From assumption (2) we know that there exists an $r \in \{1, \ldots, car_l - 1\}$ with

$$G_{l,r}(c_1, \ldots, c_{l-1}, x_l) \neq 0.$$

Thus,

$$\{\, a \mid a \in \bar{K} \text{ and } (c_1, \ldots, c_{l-1}, a) \in V(Ideal(F)) \,\} \text{ is finite.}$$

By this fact and (5),

$$V(Ideal(F)) \text{ is finite.} \tag{6}$$

Thus, by (3), (4), and (6), F is a reduced Gröbner basis and $V(Ideal(F))$ is finite.

On the other hand, there exists no polynomial f in F such that

$$LeadingPowerProduct(f) \in K[x_l].$$

This is a contradiction to Method 6.9 in [2].

Thus, in contrast to assumption (2), there exists a $b' \in V(Q_{/x_{l-1}})$ with

$$G_{l,1}(b', x_l) = \ldots = G_{l,car_l - 1}(b', x_l) = 0.$$

From this we now deduce that (1) holds for all $b \in V(Q_{/x_{l-1}})$:

Let $b'' \in V(Q_{/x_{l-1}})$.

It is easy to prove that $Q_{/x_{l-1}}$ is a zero-dimensional primary ideal in $K[x_1, \ldots, x_{l-1}]$. Hence,

$$\sqrt{Q_{/x_{l-1}}} \text{ is a zero-dimensional prime ideal.}$$

Let $k \in \{1, \ldots, car_{l-1}\}$.

We write $G_{l,k}$ in the form

$$p_j(x_1, \ldots, x_{l-1})x_l^j + \ldots + p_0(x_1, \ldots, x_{l-1}),$$

where $j := deg(G_{l,k}, l)$. As

$$V(Q_{/x_{l-1}}) = V(\sqrt{Q_{/x_{l-1}}})$$

(see [5], section 131, p. 167), b' and b'' are elements of $V(\sqrt{Q_{/x_{l-1}}})$. Thus,

$$p_s(b') = 0 \text{ iff } p_s \in \sqrt{Q_{/x_{l-1}}} \text{ iff } p_s(b'') = 0 \text{ for all } s \in \{0, \ldots, j\}$$

(see [5], section 129, p. 162). Hence,

$$G_{l,1}(b'', x_l) = \ldots = G_{l,car_l-1}(b'', x_l) = 0. \bullet$$

Corollary 1 is an easy consequence of the previous theorem.

Corollary 1 *Let l be an element of $\{2, \ldots, n\}$, Q a zero-dimensional primary ideal in $K[x_1, \ldots, x_n]$.*

Then there exists an $f \in Q_{/x_l}$ such that

$$f \in K[x_1, \ldots, x_l] \backslash K[x_1, \ldots, x_{l-1}],\ lc(f) = 1, \text{ and } gcd(Q_{/x_l}, b) = f(b, x_l) \text{ for all } b \in V(Q_{/x_{l-1}}).$$

Proof: Let G be the reduced Gröbner basis in $K[x_1, \ldots, x_n]$ such that

$$Ideal(G) = Q.$$

By definition,

$$G_{l,car_l} \in K[x_1, \ldots, x_l] \setminus K[x_1, \ldots, x_{l-1}].$$

We have proven that

$$lc(G_{l,car_l}) = 1.$$

Furthermore, by Lemma 6.8 in [2] and Theorem 2,

$$gcd(Q_{/x_l}, b) = gcd(\{G_{1,1}(b_1), \ldots, G_{l,car_l}(b, x_l)\}) = G_{l,car_l}(b, x_l) \text{ for all } b \in V(Q_{/x_{l-1}}). \bullet$$

A generalization of Corollary 1 is the next theorem.

Theorem 3 *Let l be an element of $\{2,\ldots,n\}$, I a zero-dimensional ideal in $K[x_1,\ldots,x_n]$, and $b \in V(I_{/x_{l-1}})$.*

Then there exists an $f \in I_{/x_l}$ such that

$$f \in K[x_1,\ldots,x_l] \setminus K[x_1,\ldots,x_{l-1}],\ (lc(f))(b) \neq 0,\ \text{and } gcd(I_{/x_l}, b) = f(b, x_l).$$

Before we give a proof of Theorem 3 we show the following lemma, which is required in this proof.

Lemma 1 *Let m be a natural number, J a zero-dimensional ideal in $K[x_1,\ldots,x_m]$, and*

$$J = Q_1 \cap \ldots \cap Q_r$$

a reduced primary decomposition of J. Then

$$V(Q_s) \cap V(Q_{s'}) = \emptyset \text{ for } s \neq s'.$$

Proof: We assume that there exists a

$$b \in V(Q_s) \cap V(Q_{s'}) \text{ for some } s, s' \in \{1,\ldots,r\}.$$

As $V(Q_s) = V(\sqrt{Q_s})$ and $V(Q_{s'}) = V(\sqrt{Q_{s'}})$,

$$b \in V(\sqrt{Q_s}) \cap V(\sqrt{Q_{s'}}).$$

As $\sqrt{Q_s}$ and $\sqrt{Q_{s'}}$ are zero-dimensional, b is a generic zero of $\sqrt{Q_s}$ and $\sqrt{Q_{s'}}$.

Let $f \in K[x_1,\ldots,x_m]$. From

$$f \in \sqrt{Q_s} \quad \text{iff} \quad f(b) = 0 \quad \text{iff} \quad f \in \sqrt{Q_{s'}}$$

we obtain

$$\sqrt{Q_s} = \sqrt{Q_{s'}}.$$

Hence,

$$s = s',$$

because we assumed the primary decomposition to be reduced. •

Proof of Theorem 3:

Let $Q_1,\ldots,Q_r$ be zero-dimensional primary ideals in $K[x_1,\ldots,x_n]$ such that $I_{/x_l} = Q_{1/x_l} \cap \ldots \cap Q_{r/x_l}$ is a reduced primary decomposition of $I_{/x_l}$. From

$$I_{/x_{l-1}} = Q_{1/x_{l-1}} \cap \ldots \cap Q_{r/x_{l-1}},$$

we have

$$V(I_{/x_{l-1}}) = V(Q_{1/x_{l-1}}) \cup \ldots \cup V(Q_{r/x_{l-1}}).$$

Without loss of generality, we assume that the primary ideals $Q_1,\ldots,Q_r$ are ordered in such a way that there exists an $s \in \{1,\ldots,r\}$ with

$$b \in V(Q_{1/x_{l-1}}),\ldots,b \in V(Q_{s/x_{l-1}}), b \notin V(Q_{s+1/x_{l-1}}),\ldots,b \notin V(Q_{r/x_{l-1}}).$$

We define $h_t \in Q_{1/x_l} \cap \ldots \cap Q_{t/x_l}$ such that $h_t \in K[x_1, \ldots, x_l] \setminus K[x_1, \ldots, x_{l-1}]$, $h_t(b, x_l) = gcd(Q_{1/x_l} \cap \ldots \cap Q_{t/x_l}, b)$, and $lc(h_t) = 1$ for every $t \in \{1, \ldots, s\}$:

By Corollary 1, there exists an $f \in Q_{1/x_l}$ with

$$f \in K[x_1, \ldots, x_l] \setminus K[x_1, \ldots, x_{l-1}],\ lc(f) = 1, \text{ and } f(b, x_l) = gcd(Q_{1/x_l}, b).$$

Set $h_1 := f$.

We assume that $t \in \{1, \ldots, s-1\}$ and that h_t is already defined.

Let $f \in Q_{t+1/x_l}$ such that

$$f \in K[x_1, \ldots, x_l] \setminus K[x_1, \ldots, x_{l-1}],\ lc(f) = 1, \text{ and } f(b, x_l) = gcd(Q_{t+1/x_l}, b).$$

Set $h_{t+1} := h_t \cdot f$.

From

$$gcd(Q_{1/x_l} \cap \ldots \cap Q_{t+1/x_l}, b) \in (Q_{1/x_l} \cap \ldots \cap Q_{t/x_l})(b, x_l) \text{ and}$$
$$gcd(Q_{1/x_l} \cap \ldots \cap Q_{t+1/x_l}, b) \in Q_{t+1/x_l}(b, x_l)$$

we obtain

$$h_t(b, x_l) \text{ divides } gcd(Q_{1/x_l} \cap \ldots \cap Q_{t+1/x_l}, b) \text{ and}$$
$$f(b, x_l) \text{ divides } gcd(Q_{1/x_l} \cap \ldots \cap Q_{t+1/x_l}, b).$$

Assume that there exists a $c \in \bar{K}$ such that

$$h_t(b, c) = f(b, c) = 0.$$

From the fact that $h_t(b, x_l)$ divides every element of $(Q_{1/x_l} \cap \ldots \cap Q_{t/x_l})(b, x_l)$ and that $f(b, x_l)$ divides every element of $Q_{t+1/x_l}(b, x_l)$ we obtain

$$(b_1, \ldots, b_{l-1}, c) \in V(Q_{1/x_l} \cap \ldots \cap Q_{t/x_l}) \cap V(Q_{t+1/x_l}).$$

As

$$V(Q_{1/x_l}) \cup \ldots \cup V(Q_{t/x_l}) = V(Q_{1/x_l} \cap \ldots \cap Q_{t/x_l}),$$

we have a contradiction to Lemma 1.

Therefore, $h_t(b, x_l)$ and $f(b, x_l)$ are relatively prime. Thus,

$$h_{t+1}(b, x_l) \text{ divides } gcd(Q_{1/x_l} \cap \ldots \cap Q_{t+1/x_l}, b).$$

Furthermore,

$$h_{t+1} \in K[x_1, \ldots, x_l] \setminus K[x_1, \ldots, x_{l-1}] \text{ and } lc(h_{t+1}) = 1.$$

From this and $h_{t+1}(b, x_l) \in (Q_{1/x_l} \cap \ldots \cap Q_{t+1/x_l})(b, x_l)$, it follows

$$h_{t+1}(b, x_l) = gcd(Q_{1/x_l} \cap \ldots \cap Q_{t+1/x_l}, b).$$

We define $q \in K[x_1, \ldots, x_l] \setminus K[x_1, \ldots, x_{l-1}]$ such that there exists an $e \in \bar{K}$ with $e \cdot q(b, x_l) = gcd(I_{/x_l}, b)$ and $deg(q, l) = deg(q(b, x_l), l)$:

We choose a $p_t \in Q_{t/x_{l-1}}$ for every $t \in \{s+1, \ldots, r\}$ such that

$$p_t(b) \neq 0.$$

This is always possible, because $b \notin V(Q_{t/x_{l-1}})$ for all $t \in \{s+1, \ldots, r\}$.

Set $q := p_{s+1} \cdot \ldots \cdot p_r \cdot h_s$.

Obviously, $q \in I_{/x_l}$. As

$$q(b, x_l) = p_{s+1}(b) \cdot \ldots \cdot p_r(b) \cdot gcd(Q_{1/x_l} \cap \ldots \cap Q_{s/x_l}, b) \text{ and } p_{s+1}(b) \cdot \ldots \cdot p_r(b) \in \bar{K} \setminus \{0\},$$

we know that

$$q(b, x_l) \text{ divides } gcd(Q_{1/x_l} \cap \ldots \cap Q_{s/x_l}, b).$$

From $h_s \in K[x_1, \ldots, x_l] \setminus K[x_1, \ldots, x_{l-1}]$ and $lc(h_s) = 1$ we obtain

$$deg(q, l) = deg(h_s, l) = deg(h_s(b, x_l), l) = deg(q(b, x_l), l). \tag{7}$$

As $I_{/x_l}$ is a subset of $Q_{1/x_l} \cap \ldots \cap Q_{s/x_l}$,

$$gcd(Q_{1/x_l} \cap \ldots \cap Q_{s/x_l}, b) \text{ divides } gcd(I_{/x_l}, b).$$

Thus,

$$q(b, x_l) \text{ divides } gcd(I_{/x_l}, b) \text{ and } q \in I_{/x_l}.$$

Hence, there exists an $e \in \bar{K}$ such that

$$e \cdot q(b, x_l) = gcd(I_{/x_l}, b). \tag{8}$$

From $q \in K[x_1, \ldots, x_l] \setminus K[x_1, \ldots, x_{l-1}]$, (7), and (8) we obtain that

$$e \cdot q \in K[x_1, \ldots, x_l] \setminus K[x_1, \ldots, x_{l-1}],\ lc(e \cdot q)(b) \neq 0, \text{ and } (e \cdot q)(b, x_l) = gcd(I_{/x_l}, b).$$

By means of this theorem it is relatively easy to prove Theorem 1:

Proof of Theorem 1: Let $q \in I_{/x_l}$ such that

$$q \in K[x_1, \ldots, x_l] \setminus K[x_1, \ldots, x_{l-1}],\ (lc(q))(b) \neq 0, \text{ and } gcd(I_{/x_l}, b) = q(b, x_l).$$

We know that

$$g(b) = 0 \text{ for all } g \in G \cap K[x_1, \ldots, x_{l-1}],$$
$$q(b, x_l) \neq 0, \text{ and}$$
$$q \text{ reduces to zero modulo } G.$$

Thus, there exists an $f \in G \cap K[x_1, \ldots, x_l] \setminus K[x_1, \ldots, x_{l-1}]$ such that

$$f(b, x_l) \neq 0 \text{ and } deg(f, l) \leq deg(q, l).$$

Therefore,

$$deg(f(b, x_l), l) \leq deg(f, l) \leq deg(q, l) = deg(q(b, x_l), l).$$

As $q(b, x_l)$ divides $f(b, x_l)$, there exists an $e \in \bar{K}$ such that

$$e \cdot f(b, x_l) = q(b, x_l) = gcd(I_{/x_l}, b).$$

From

$$deg(f(b, x_l), l) = deg(q(b, x_l), l) = deg(q, l) \geq deg(f, l).$$

we obtain

$$lc(f)(b) \neq 0.$$

Thus,

$$deg(G_{l,min_b}(b, x_l), l) \leq deg(G_{l,min_b}, l) \leq deg(f, l) = deg(f(b, x_l), l).$$

On the other hand, $f(b, x_l)$ divides $G_{l,min_b}(b, x_l)$. Hence, there exists a $d \in \bar{K}$ such that

$$d \cdot G_{l,min_b}(b, x_l) = e \cdot f(b, x_l) = gcd(I_{/x_l}, b).$$

From Lemma 6.8 in [2],

$$d \cdot G_{l,min_b}(b, x_l) = gcd(\{G_{l,1}(b, x_l), \ldots, G_{l,car_l}(b, x_l)\}).$$

In the case of two variables this result can be easily deduced from Lazard's structure theorem (see [3]).

Acknowledgement: I want to thank B. Roider. He suggested Theorem 1 might be true and stimulated me to prove this.

References

[1] B. Buchberger: *Ein algorithmisches Kriterium für die Lösbarkeit eines algebraischen Gleichungssystems*, Aequationes Math. 4/3, 374-383 (1970)

[2] B. Buchberger: *Gröbner Bases: An Algorithmic Method in Polynomial Ideal Theory*, in Recent Trends in Multidimensional Systems Theory, N.K. Bose (ed.), D. Reidel Publ. Comp. 184-232 (1985)

[3] D. Lazard: *Ideal Bases and Primary Decomposition: Case of Two Variables*, J. of Symbolic Computation 1, 261-270 (1985)

[4] W. Trinks: *Über B. Buchbergers Verfahren, Systeme algebraischer Gleichungen zu lösen*, J. of Number Theory 10/4, 475-488 (1978)

[5] B.L. van der Waerden: *Algebra II*, Springer-Verlag (1967)

Properties of Gröbner Bases under Specializations

Patrizia Gianni
University of Pisa

Abstract

In this paper we prove some properties of Gröbner bases under specialization maps. In particular we state sufficient conditions for the image of a Gröbner basis to be a Gröbner basis. We apply these results to the resolution of systems of polynomial equations. In particular we show that, if the system has a finite number of solutions, (in an algebraic closure of the base field K), the problem is totally reduced to a single Gröbner basis computation (w.r.t. purely lexicographical ordering), followed by a search for the roots of univariate polynomials and a "few" evaluations in suitable algebraic extensions of K.

1. Definitions

In this paper we will denote by K a field, of any characteristic, by $\overline{K}$ the algebraic closure of K, by $R = K[x_1, \dots, x_n, y_1, \dots, y_m] = K[\overline{X}, \overline{Y}]$ the ring of polynomials in n+m variables with coefficients in K. We will call a specialization any homomorphism $\varphi : R \mapsto \overline{K}[\overline{X}]$ such that $\varphi(K[\overline{Y}]) \subset \overline{K}$, and $\varphi(x_i) = x_i$, $\forall x_i$.

Example 1. Let $\alpha = (a_1, \dots, a_m) \in \overline{K}^m$ the "*evaluation*" homomorphism $\varphi_\alpha(f(\overline{X}, \overline{Y})) = f(\overline{X}, \alpha)$ is a specialization homomorphism.

We recall now some definitions and properties of Gröbner bases that will be used in what follows. For a more detailed exposition of general properties of Gröbner bases see [1]).

Definition 1. *Given an ordering "$>$" on the monomials of R, if* $f \in K[\overline{X}, \overline{Y}]$, $f \neq 0$, *we can write* $f = c\overline{X}^A\overline{Y}^B + f'$, *with* $A \in \mathbf{N}^n$, $B \in \mathbf{N}^m$, $c \in K$, $c \neq 0$, *and* $\overline{X}^A\overline{Y}^B$ *biggest monomial in f. We define:*

$$\mathrm{lt}(f) = c\overline{X}^A\overline{Y}^B, \textit{the } \text{leading term } \textit{of } f.$$

$$\deg(f) = (A, B) \in \mathbf{N}^{n+m} \textit{ the } \text{degree} \textit{ of } f.$$

For a subset G of R we define :

$Lt(G)$ = the ideal generated by the $\mathrm{lt}(g)$, for $g \in G$, the leading term ideal of G.

Definition 2. *We will say that* $f \in R$ *is* reducible *modulo* $G \subset R$, *if f is non-zero and* lt(f) $\in$ Lt(G). *Otherwise we say that f is* reduced *modulo G.*

Definition 3. *A subset G of an ideal I in R is a* Gröbner basis *for I, if* Lt(G) = Lt(I).

Proposition 1. *If G is a Gröbner basis for* $I \subset R$ *then* $f \in I$ *if and only if f reduces to zero modulo G.*

Definition 4. *Given a specialization φ, we will say that an ordering ">" on the monomials of R is "φ-admissible" if all the x_i's are bigger than any of the y_j. More precisely if we denote by $>_1$ (resp. $>_2$) the restriction of the ordering to the $\overline{X}$ variables (resp. to the $\overline{Y}$ variables), $>$ is φ-admissible if it satisfies the following condition:*

$$\overline{X}^A\overline{Y}^B > \overline{X}^C\overline{Y}^D \Leftrightarrow \overline{X}^A >_1 \overline{X}^C \text{ or } \overline{X}^A = \overline{X}^C \text{ and } \overline{Y}^B >_2 \overline{Y}^D, \ \forall A, C \in \mathbf{N}^n,\ B, D \in \mathbf{N}^m$$

From now on we will fix a φ-admissibile ordering on the monomials in R.
Let us consider a specialization φ, a φ-admissible ordering on the monomials of R and an ideal I in R. If G is a Grōbner basis for I (w.r.t. the φ-admissibile ordering), we want to study the behavior of G under the action of the specialization map φ, more precisely we want to state sufficient conditions for φ(G) to be a Grōbner basis for φ(I). For this reason it is very natural to set the following definitions.

Definition 5. *For any $f \in R$, $f \neq 0$, write $f = c(\overline{Y})\overline{X}^A + f'$, with $c(\overline{Y}) \in K[\overline{Y}]$, $c \neq 0$, and $\overline{X}^A$ the biggest monomial in $\overline{X}$ appearing in f, we define:*

$$\text{lt}_\varphi(f) = c\overline{X}^A, \textit{the } \text{leading term } \textit{of } f, \ (w.r.t.\ \varphi)$$

$$\deg_{\overline{X}}(f) = A, \textit{the } \text{degree } \textit{of } f, \ (w.r.t.\ \varphi).$$

For a subset G of R we define :

$$Lt_\varphi(G) = \text{the ideal generated by the } \text{lt}_\varphi(g), \text{ for } g \in G, \text{ the leading term ideal of G.}$$

Example 2. Let $f \in K[x, y]$, $(n = 1, m = 1)$, $f(x, y) = x^2y^2 + x^2 + xy + 1$. If we consider the lexicographical ordering we have:

$$\text{lt}(f) = x^2y^2 \quad \textit{and} \quad \text{lt}_\varphi(f) = (y^2 + 1)x^2.$$

If G is a Grōbner basis w.r.t. a φ-admissible ordering then the Lt_φ(G) has the following property.

Corollary. *If G is a Grōbner basis for an ideal $I \subset R$ then :*

$$\text{Lt}_\varphi(\text{G}) = \text{Lt-}\varphi(\text{I}).$$

Proof. For every polynomial $f \in R$ we have that $\text{lt}(\text{lt}_\varphi(\text{f})) = \text{lt}(\text{f})$. So we have :

$$\text{Lt}(\text{Lt}_\varphi(\text{G})) = \text{Lt}(\text{G}) = \text{Lt}(\text{I}) = \text{Lt}(\text{Lt}_\varphi(\text{I})).$$

Since $\text{Lt}_\varphi(\text{G}) \subset \text{Lt}_\varphi(\text{I})$, the fact that their leading term ideals are equal implies that the two ideals are equal.

Remark 1. For every specialization φ and for every ideal I we have $\varphi(\text{Lt}_\varphi(I)) \subset \text{Lt}(\varphi(I))$.

Theorem 1. *Let $I \subset R$ be an ideal, φ a specialization and G a Grōbner basis G for I, (w.r.t. a φ-admissible ordering). Let us suppose that the following condition holds:*

$$(*) \ \ \text{Lt}(\varphi(I)) = \varphi(\text{Lt}_\varphi(I)).$$

Then $\varphi(G)$ is a Grōbner basis for $\varphi(I)$.

Proof. By remark 1 and the hypothesis we have:

$$\mathrm{Lt}(\varphi(I)) = \varphi(\mathrm{Lt}_\varphi(I)) = \varphi(\mathrm{Lt}_\varphi(G)) \subseteq \mathrm{Lt}(\varphi(G)) \subseteq \mathrm{Lt}(\varphi(I)).$$

So $\mathrm{Lt}(\varphi(G)) = Lt(\varphi(I))$ and φ(G) is a Gröbner basis for φ(I).

Remark 2. The condition in theorem 1 is only sufficient a condition. In the following counterexample we have a specialization that doesn't preserve the leading term ideal, altough the image of the Gröbner basis is still a Gröbner basis.

Example 3 Let us consider the ideal $I = (x_1{}^2y + x_1x_2)$. If we evaluate for y at 0 we obtain $\varphi(I) = (x_1x_2)$, $\mathrm{Lt}(\varphi(I)) = (x_1x_2)$ while $\varphi(\mathrm{Lt}_\varphi(I)) = (0)$.

2. Applications

We want to apply some of the results stated in the previous section to the resolution of systems of polynomial equations.

Let us consider a zero-dimensional ideal $I = (f_1, \ldots, f_s)$, (i.e. an ideal such that the system determined by the f_i's has a finite number of solutions) and a specialization φ such that it specialises all the variables but one, so φ is of the following form: $\varphi : K[x, y_1, \ldots, y_m] \mapsto \overline{K}[x]$.

With these hypothesis we have:

Theorem 3. *There exists a polynomial $g \in I$ such that:*

$\varphi(g)$ generates $\varphi(I)$.
$\deg_x(g) = \deg_x \varphi(g)$.

Proof. First we prove the theorem for primary zero-dimensional ideals. In this case if G is the reduced Gröbner basis for I (w.r.t lexicographical ordering) we have ([2]) :

$$G = (g_{01}, \ldots, g_{0s_0}, g_{11}, \ldots, g_{1s_1}, \ldots, g_{n1})$$

where (for simplicity of notation we use y_0 for x):

$$g_{i1} \in G \cap K[y_i, \ldots, y_m] \text{ are monic polynomials in } y_i,\ \forall i$$
$$g_{ij} \equiv 0 \ mod \ \sqrt{G \cap K[y_{i+1}, \ldots, y_m]}\ \forall i, \text{ for } j > 0.$$

We distinguish two cases:

1. $\exists g_{ij} \in G \cap K[y_i, \ldots, y_m]$ such that $\varphi(g_{ij}) \neq 0$.

In this case $\varphi(I) = (1) = (\varphi(g_{ij}))$ and the condition on the degree is satisfied because $\deg_x(g_{ij}) = \deg_x(\varphi(g_{ij})) = 0$.

2. $G \cap K[y_i, \ldots, y_m] \subset$ **Ker** φ.

In this case by the previous remark on the structure of G we have:

$$(\varphi(I)) = (\varphi(G)) = (\varphi(g_{01}, \ldots, \varphi(g_{n1})) = (\varphi(g_{01})).$$

where g_{01} is a monic polynomial in x hence, in particular, $\deg_x(g_{01}) = \deg_x(\varphi(g_{01}))$.

In general let us consider $I = \bigcap I_i = \prod I_i$ an irredundant primary decomposition. From the first part of the proof we know that in every I_i there exists a polinomial g_i such that $\varphi(I_i) = (\varphi(g_i))$ and $\deg_x(g_i) = \deg_x(\varphi(g_i))$. Let us call $g = \prod g_i$ and $\gamma = \varphi(g) = \prod \varphi(g_i)$. By construction $\deg_x(g) = \deg_x(\gamma)$ and hence since in our hypothesis $\varphi(\prod I_i) = \prod(\varphi(I_i))$ we obtain that $\varphi(I) = (\gamma)$.

Remark 3. It is very important that under our hypothesis the intersection of the primary ideals is exactly the product, otherwise the specialization wouldn't commute.

Example 4. Let $I = (x_1, x_2)$ and $J = (x_3, x_2 + x_4 y)$ be two ideals and let us consider the evaluation φ obtained by putting y=0. If we consider the intersection $I \cap J = (x_1 x_3, x_2 x_3, x_1(x_2 + x_4 y), x_2(x_2 + x_4 y))$ and its evaluation we obtain $\varphi(I \cap J) = (x_1 x_3, x_2 x_3, x_1 x_2, x_2^2)$ while if we take $\varphi(I) = (x, y)$ and $\varphi(J) = (x_3, x_2)$ we obtain $\varphi(I) \cap \varphi(J) = (x_1 x_3, x_2)$.

Theorem 4. *Let $I \subset R = K[x, y_1, \ldots, y_m]$ be a zero-dimensional ideal, $\varphi : R \mapsto \overline{K}[x]$ be a specialization map, that specialize all the variables but one, then :*

$$\text{G Gröbner basis for I} \Longrightarrow \varphi(\text{G}) \text{ Gröbner basis for } \varphi(\text{I}).$$

Proof. We know from Remark 1 that for every φ, $\varphi(Lt_\varphi(I)) \subset Lt_\varphi(I)$. In order to apply Theorem 1 we have to show the opposite inclusion. Since K[x] is a principal ideal domain, this fact follows from the previous theorem that guarantees that there exists $g \in I$ such that $\varphi(g) = \varphi(I)$ and $\deg_x(g) = \deg_x(\varphi(g))$. We can write $g(x, y_1, \ldots, y_m) = c(y_1, \ldots, y_m)x^r + h$ with h of lower degree in x and $\varphi(c) \neq 0$.
Now $Lt(\varphi(I)) = (x^r)$ and $c(y_1, \ldots, y_m)x^r \in Lt_\varphi(I) \Longrightarrow x^r \in \varphi(Lt_\varphi(I)) \Longrightarrow Lt(\varphi(I)) \subset \varphi(Lt_\varphi(I))$.

Remark 4. The condition that the specialization specializes all the variables but one is essential. The following example shows that if we don't specialize all the variables but one, it isn't always true that the specialization of a Gröbner basis is still a Gröbner basis, (even in the zero dimensional case).

Example 5. Let us consider the following ideal: $I = (x_1^2, x_1 x_2, x_1 x_3^2, x_1 y + x_2, x_2 x_3 - x_3^2, x_2 y, x_3^3, x_3^2 y, y^2)$. This is a Gröbner basis for this ideal, if we specialize for y=0 we obtain :$\varphi(I) = (x_1^2, x_1 x_2, x_1 x_3^2, x_2, x_2^2, x_2 x_3 - x_3^2, x_3^3)$ while the Gröbner basis for $\varphi(I)$ is (x_1^2, x_2, x_3^2).

Finally we can state the result for the solution of a system of polynomial equations. It is known ([1],[5]) that if the system has a finite number of solution, a Gröbner basis w.r.t. the purely lexicographical ordering of the corresponding ideal has a "triangular" form. So the solutions can be computed with a back substitution process and the problem is totally reduced to finding zeros of univariate polynomials over suitable algebraic extensions followed either by computations of gcd's or by evaluations (in algebraic extensions). By applying the previous result we show that no gcd computations are necessary and only a "few" evaluations have to be performed.

Corollary. *Let $I \subset K[x_1, \ldots, x_n]$ be a zero-dimensional ideal, $G = (g_1, \ldots, g_s)$ be its Gröbner basis w.r.t. purely lexicographical ordering with $x_1 > \ldots > x_n$ and le*

$\alpha_k,\ldots,\alpha_n) \in \overline{K}^{n-k+1}$ *be a zero of* $I \cap K[x_k,\ldots,x_n]$. *Let* $I_k = (g_{k_1},\ldots,g_{k_r}) = (G \cap K[x_{k-1},\ldots,x_n]) \setminus K[x_k,\ldots,x_n]$. *Let* $g \in I_k$ *be a polynomial of minimal degree in* x_{k-1} *whose leading coefficient* $c(x_k,\ldots,x_n)$ *(w.r.t.* x_{k-1}*) doesn't vanish when evaluated in* $(\alpha_k,\ldots,\alpha_n)$. *We have*

$$g_\alpha = g(x_{k-1},\alpha_k,\ldots,\alpha_n) = gcd((g_{k_1}(x_{k-1},\alpha_k,\ldots,\alpha_n),\ldots,g_{k_r}(x_{k-1},\alpha_k,\ldots,\alpha_n)).$$

In particular if α_{k-1} *is a zero of* g_α *then* $(\alpha_{k-1},\alpha_k,\ldots,\alpha_n)$ *is a zero of* $I \cap K[x_{k-1},\ldots,x_n]$.

Proof. The proof follows immediately from the previous theorem by considering the specialization $\varphi : I \cap K[x_{k-1},\ldots,x_n] \mapsto \overline{K}[x_{k-1}]$, defined by $\varphi(f(x_{k-1},\ldots,x_n)) = f(x_{k-1},\alpha_k,\ldots,\alpha_n)$.

Final Remarks. The previous Corollary in the two variables case could be proved by using Lazard's structure theorems for Standard Bases,([4]) .
The same result has been independently shown by Kalkbrener, ([3]).

Aknowledgements. I want to thank Teo Mora and Barry Trager for many helpful discussions.

References

[1] B.Buchberger, "Gröbner bases: an algorithmic method in polynomial ideal theory", *in N.K.Bose (ed.): Recent trends in multidimensional systems theory, D.Rheidel Publ. Comp., Chapter 6.*

[2] P.Gianni, B.Trager, G.Zacharias, "Gröbner bases and primary decomposition of polynomial ideals", *to appear in J.of Symb. Comp..*

[3] Kalkbrener, "Solving Systems of Algebraic Equations by using Gröbner Bases", *In this Volume.*

[4] D.Lazard, "Ideal Bases and Primary Decomposition: case of two variables", *J.of Symb. Comp.*, Vol.1, n.3, September '85.

[5] W.Trinks, " Über B.Buchbergers Verfharen, Systeme algebraisher Gleichungen zu losen, *J.of Number Theory* **10**,*pp. 475–488.*

The Computation of Polynomial Greatest Common Divisors Over an Algebraic Number Field

Lars Langemyr*
NADA, Royal Institute of Technology, S–100 44 Stockholm, Sweden.

Scott McCallum†
The Australian National University, Research School of Physical Sciences
G.P.O. Box 4, Canberra, ACT 2601, Australia

We present an algorithm for computing the greatest common divisor of two polynomials over an algebraic number field. We obtain a better computing time bound for this algorithm than for previously published algorithms solving the same problem. We have also performed empirical run time tests which have confirmed this. Our motivation for seeking the algorithm of the present paper stems from our interest in the cylindrical algebraic decomposition (cad) algorithm (Collins [3], Arnon *et al.* [1]). The cad algorithm makes essential use of algebraic polynomial gcd computations, which often dominate the cost of the algorithm.

Our algorithm is a generalization of the modular algorithm developed by Brown [2] and Collins [4] independently which computes the greatest common divisor for polynomials over the field $\mathbf{Q}$ of rational numbers. Our result generalizes a result by Rubald [7] who gave a modular algorithm that can be used if the algebraic number field can be generated by α, a root to an irreducible monic polynomial, $r(y)$ over $\mathbf{Z}$, for which $\mathbf{Z}[\alpha]$ constitutes a unique factorization domain. He also required that the polynomial $r(y)$ be irreducible modulo infinitely many primes. We remove these restrictions by applying some theory expounded by Weinberger and Rothschild [8].

Modular Algorithm

Let $\alpha \in \mathrm{C}$ be an algebraic integer with minimal polynomial $r(y)$ over $\mathbf{Q}$. Let $f(x)$ and $g(x)$ be non-zero polynomials over $\mathbf{Q}(\alpha)$. We present an algorithm that computes the greatest common divisor of $f(x)$ and $g(x)$. We take the monic associates of $f(x)$ and $g(x)$ over $\mathbf{Q}(\alpha)$, clear integer denominators, remove integer contents and replace $f(x)$ and $g(x)$ by the results. We thus have integral leading coefficients in both $f(x)$ and $g(x)$ which we denote a and b respectively. Let Δ be the discriminant of $r(y)$. We compute the associate $h(x) \in \mathbf{Z}[\alpha][x]$ of $\gcd(f, g)$ with leading integer coefficient $c = \gcd(a, b)\Delta$. Using the technique of [8] it can be shown that the gcd of $f(x)$ and $g(x)$ over $\mathbf{Q}(\alpha)$ always has an associate that can be written on this form.

Now, consider the following for a prime p such that p divides neither a, b nor Δ. Let $\tilde{r}(y) = r(y) \bmod p$. Let $\tilde{R} = \mathbf{Z}_p[y]/(\tilde{r}(y))$ and let $\phi : R \to \tilde{R}$ denote the natural "mod p" homomorphism. (Denote the natural extension of ϕ to $R[x]$ by ϕ also.) Let $\tilde{f} = \phi(f)$ and $\tilde{g} = \phi(g)$. (Note that $\deg(\tilde{f}) = \deg(f)$ and $\deg(\tilde{g}) = \deg(g)$, as p divides neither a nor b, and that $\tilde{r}(y)$ is square-free, as $p \not| \operatorname{discr}(r(y))$.) Even though $\tilde{R}$ is not necessarily an integral domain (as $\tilde{r}(y)$ may factor non-trivially over $\mathbf{Z}_p$), one can show that $\tilde{f}(x)$ and $\tilde{g}(x)$ have a gcd $h^*(x)$ over $\tilde{R}$. To do this we use the "Chinese Remainder" isomorphism $\psi : \tilde{R} \to \tilde{R}_1 \times \cdots \times \tilde{R}_t$, where $\tilde{r}_1(y), \ldots, \tilde{r}_t(y)$ are the distinct, monic irreducible factors of $\tilde{r}(y)$ over $\mathbf{Z}_p$, and $\tilde{R}_i$ is the field $\mathbf{Z}_p[y]/(\tilde{r}_i(y))$. Denote the natural extension of ψ to $R[x]$ by ψ also. We compute the unique gcd $h^*(x)$ of $\tilde{f}(x)$ and $\tilde{g}(x)$ satisfying $\psi_j(h^*) = \gcd(\psi_j(\tilde{f}), \psi_j(\tilde{g}))$, for $1 \le j \le t$ (where the ψ_j are the component functions of ψ). We show that the degree of $h^*(x)$ could be greater than the degree k of $h(x)$: in this case, p is called *unlucky*, (and otherwise p is called *lucky*).

We compute a list of polynomials $\mathtt{H} = (h_1^*, \ldots, h_n^*)$, all having the same degree k^*, using different primes $\mathtt{P} = (p_1, \ldots, p_n)$, such that the product π of primes in $\mathtt{P}$ $\pi \ge 2K$, where $K = \max(K_1, |R_1|)$. K_1 bounds the integer coefficients in the representation of $h(x)$. This bound can be obtained using results in Weinberger and Rothschild [8] and Landau [5]. R_1 is defined below.

For $1 \le i \le n$ let $c_i = c \bmod p_i$. Solve the system of congruences $h'(x) \equiv c_i h_i^*(x) \pmod{p_i}$, $1 \le i \le n$, for element $h'(x) \in \mathbf{Z}_\pi[y, x]$ of degree k^* using the Chinese remainder algorithm. (In the above system we regard the $h_i^*(x)$ and $h'(x)$ as polynomials in $\mathbf{Z}[y, x]$. Thus the polynomial congruence "$h'(x) \equiv c_i h_i^*(x) \pmod{p_i}$" is really a system of integer congruences, one for each integer coefficient.) Set $h(x) \leftarrow h'(x)$.

Now $p_1, \ldots, p_n$ are either all lucky or all unlucky. We show that $\phi(h)$ divides h^* in $\tilde{R}[x]$ and that we thus have $\deg(h^*) \ge k$. We also show that $\deg(h^*) > k$ if and only if p divides the non-zero integer $R_1 = \operatorname{res}(r(y), \operatorname{psc}_k(f, g))$,

*Supported by STU and NSERC. *Internet:* `larsl%nada.kth.se@uunet.uu.net`
†Supported by NSERC (Grant 3-640-126-30) and VOEST-ALPINE A.G. (Abt. GCT3). *Internet:* `scott@phys4.anu.oz.au`

so if $p_1, \ldots, p_n$ are all unlucky, then by the above π divides the non-zero integer R_1, (as the p_i are distinct). Hence $\pi \leq 2K$, a contradiction. Therefore $p_1, \ldots, p_n$ are all lucky. We must prove that $h'(x) = h(x)$. As K is a strict bound for the absolute values of the integer coefficients of $h(x)$, $h(x) \in \mathbf{Z}_{2K}[y,x]$. Hence, as $2K \leq \pi$, $h(x) \in \mathbf{Z}_\pi[y,x]$. We show that if p is lucky then $\phi(h) = \phi(c)h^*$. By the above and the Chinese remainder theorem, $h'(x) = h(x)$. We are thus able to establish the correctness of the algorithm.

Refinements to Algorithm

We present two refinements that we have made to the algorithm of the previous section. The first refinement is a simplification of the gcd computation over $\tilde{R}$. In our paper we prove that we can compute h^* directly over $\tilde{R}[x]$ using a Euclidean remainder sequence assuming $\tilde{R}$ is a field and thus $\tilde{R}[x]$ a Euclidean domain. If the assumption fails *i.e.*, we get a non-invertible leading coefficient, we discard p and restart the computation for another prime. We show that we can only fail a finite number of times, and thus that the algorithm terminates. The second refinement is the replacement of the termination test using an a priori bound K by a trial division method together with an iterative version of the Chinese remainder algorithm. Further refinements are to appear in [6].

Time Analysis

We have analyzed the complexity of the algorithm, using fixed size primes, and assuming constant time for performing operations in the finite fields given by these primes. Considering an implementation where only the first refinement from above is used, we obtain a computing time of $O(n^4 m^3 \log^2(dn\hat{d}^m))$, where n is a bound of the degree of $f(x)$ and $g(x)$, and d is an absolute value bound on the integer coefficients in $f(x)$ and $g(x)$. Similarly m is the degree of $r(y)$ and $\hat{d}$ is an absolute value bound on the integer coefficients in $r(y)$. The bound is superior to previously given bounds from [7], and [5]. Further improvements of the computing time are to appear in [6].

Empirical Results

The algorithm has been tested on some previously infeasible runs on the quantifier elimination package in the SAC-2 computer algebra system. Our tests indicate that the algorithm improves the performance of the whole cad algorithm significantly. We have also separately tested the algorithm with dense random polynomials as input, and also in this case we have obtained a significant performance increase compared to the monic p.r.s. algorithm currently present in the SAC-2 computer algebra system.

We are grateful to George Collins and Erich Kaltofen for pointing to the work of Weinberger and Rothschild. Thanks to Joachim von zur Gathen for suggesting the first refinement to our algorithm.

References

[1] D. S. Arnon, G. E. Collins, and S. McCallum. Cylindrical algebraic decomposition. *SIAM Journal on Computing*, 13:865–877, 878–889, 1984.

[2] W. S. Brown. On Euclid's algorithm and the computation of polynomial greatest common divisors. *Journal of the ACM*, 18(4):478–504, October 1971.

[3] G. E. Collins. *Quantifier Elimination for Real Closed Fields by Cylindrical Algebraic Decomposition*, pages 134–183. Volume 33 of *Lecture Notes in Computer Science*, Springer-Verlag, 1975.

[4] G. E. Collins. *The SAC-1 Polynomial GCD and Resultant System*. Technical Report University of Wisconsin, MACC Tech. Report 27, University of Wisconsin, Madison, 1972.

[5] S. Landau. Factoring polynomials over algebraic number fields. *SIAM Journal on Computing*, 14:184–195, 1985.

[6] L. Langemyr. *Computing the GCDs of Polynomials Over Algebraic Number Fields*. PhD thesis, Royal Institute of Technology, 1988. to appear.

[7] C. M. Rubald. *Algorithms for polynomials over a real algebraic number field*. PhD thesis, University of Wisconsin, Madison, January 1974.

[8] P. J. Weinberger and L. P. Rothschild. Factoring polynomials over algebraic number fields. *ACM Transactions on Mathematical Software*, 2(4):335–350, December 1976.

AN EXTENSION OF BUCHBERGER'S ALGORITHM TO COMPUTE ALL REDUCED GRÖBNER BASES OF A POLYNOMIAL IDEAL

Klaus-Peter Schemmel
Department of Mathematics
University of Technology Dresden, GDR

Abstract

In this paper we present for the bivariate case an algorithm to compute all possible reduced Gröbner bases of a polynomial ideal. This algorithm is an extension of Buchberger's one, which is based on the possibility to classify and to handle easy all term orderings in case of two variables. The constructed algorithm is interesting for the study of complexity of constructing Gröbner bases in dependence of the chosen term ordering and may lead to new insights on the question which is the best term ordering for quick termination.

0. Introduction.

The following considerations are based on B. Buchberger's algorithm to compute special bases of polynomial ideals, so called Gröbner bases (GB), introduced in his thesis 1965, see also Buchberger,B. (1976,1979,1983).
The form of a Gröbner basis depends in general on the selected term ordering. Different term orderings may lead to different Gröbner bases and also to different behaviour of complexity for the GB-algorithm.
For the lexicographical ordering for instance Buchberger's algorithm yields GB-elements with few variables, but of high degree, whereas the total degree ordering leads to better upper bounds for the degree of the GB-elements and requires less computing steps in general (see Buchberger,B. (1983), Lazard, D. (1983) and Giusti, M. (1985)).

Now the following questions and tasks arise:

(1) to classify and describe explicitly all term orderings.

(2) to give an algorithm to obtain all reduced Gröbner bases of a given polynomial ideal.

(3) to describe the influence of the choice of the term ordering on the complexity of the algorithm and to find criteria or at least strategies for a "favourable" selection of term ordering.

The first part of this paper offers some explanations to question (1), see Robbiano, L. (1985).
In the second part an algorithm to compute all reduced Gröbner bases of a polynomial ideal in $k[x,y]$ is described; while a discussion to question (3) and proposals for the improvement of the complexity of GB-algorithms, also in the case of F. Mora's algorithm (see Mora, F. (1982)) and for special applications, are given in another place.

1. About term orderings.

In the following we identify terms and n-tuples of natural numbers by means of the isomorphism $X^I \dashrightarrow I$ of the multiplicative half group of the terms in n indeterminates on the additive half group $\mathbb{N}^n$. We use the notation $I=(i_1,...,i_n)$, $J=(j_1,...,j_n)$.

Definition 1. An ordering $(\mathbb{N}^n,<)$ is called a t e r m o r d e r i n g if it holds
(N) $0 < I$ for all $I \in \mathbb{N}^n$, and
(P) $I < J$, then $I+K < J+K$ follows for all $I,J,K \in \mathbb{N}^n$.

The two classical examples are the p u r e l e x i c o g r a p h i c a l o r d e r i n g $<_L$, that means $I<_L J$ iff there exists a k, $1 <= k <= n$, such that $i_k < j_k$ and $i_s = j_s$ for all s, $1 <= s < k$, and the so-called d i a g o n a l o r d e r i n g $<_D$ with $I <_D J$ iff $i_1+...+i_n<j_1+....+j_n$ or $i_1+...+i_n=j_1+....+j_n$ and $I<_L J$.
For example, with n=2 we have
$(0,0) <_L (1,0) <_L (2,0) <_L ...$
$<_L (0,1) <_L (1,1) <_L (2,1) <_L ...$
$<_L (0,2) <_L (1,2) <_L (2,2) <_L ...$, and
$(0,0) <_D (1,0) <_D (0,1) <_D (2,0) <_D (1,1) <_D (0,2)$
$<_D (3,0) <_D (2,1) <_D (1,2) <_D ...$.

Definition 2. $<_R$ is called the t e r m o r d e r i n g b e l o n g i n g t o t h e o r d e r m a t r i x $R \in \mathbb{M}_n(\mathbb{R}^+)$ if the following holds

(i) R is a regular matrix with nonnegative real entries.

(ii) $I <_R J$ iff $R.I^t <_L R.J^t$, whereas $<_L$ means the pure lexicographical ordering on $\mathbb{R}^n$ and $R.I^t$ the product of R with the transponed n-tuple of I.

Theorem 1. For every term ordering there exists a corresponding order matrix.

Proof. This is a consequence of Robbiano,L.(1985), who gives a complete classification of term orderings. In our case, however, the following holds.

Remark 1. The mapping of the set of order matrices onto the set of term orderings is not injective. For instance, if the entries $(r_{11},\ldots,r_{1n})$ of the first row of the order matrix are linearly independent over the rationals, then $(r_{11},\ldots,r_{1n}).I^t \neq (r_{11},\ldots,r_{1n}).J^t$ holds for every pair of terms $I,J \in \mathbb{N}^n$, $I \neq J$. That means, in this case the term ordering depends only on the first row of the order matrix (see also Kollreider,C. (1978) and Galligo,A. (1979)). For practical use we notice the following lemma.

Lemma 1. For every finite set of terms we can realize every possible order of these terms occurring in a term ordering by a linear form $L: \mathbb{N}^n \longrightarrow \mathbb{N}$ with positive integral coefficients, not all of them zero.

Proof. Every real number and, therefore, also every vector of reals can be approximated by rationals with any exactness. We take the rationally approximated order matrix and multiply every row with the common main denominator of its entries and with 'suitable' positive integral weights and add them to obtain the announced linear form.

Examples. The pure lexicographical ordering for terms of degree less than 100 in case n=3 is obtained by the order matrix

$$R = \begin{pmatrix} 0 & 0 & 1 \\ 0 & 1 & 0 \\ 1 & 0 & 0 \end{pmatrix}$$ and by the following linear form

$$L(i,j,k) = 1 * i + 100 * j + 10000 * k.$$

The total degree ordering in the same case is realized by

$$R = \begin{pmatrix} 1 & 1 & 1 \\ 0 & 0 & 1 \\ 0 & 1 & 0 \end{pmatrix}$$ and by

$L(i,j,k) = 10000 * i + 10001 * j + 10100 * k$. For all terms of degree less than 100 it holds that if $I \neq J$ then $L(I) < L(J)$ or

$L(J) < L(I)$, that means for those terms we can regard L as an order homomorphism in the set of reals.

Now we consider the case of two variables, $n=2$.

<u>Lemma 2.</u> In case $n=2$, the term orderings can be mapped uniquely onto the order matrices of the following types

$$\text{(i)} \quad \begin{pmatrix} 1 & s \\ 0 & 1 \end{pmatrix} \qquad s \in \mathbb{R}^+$$

$$\text{(ii)} \quad \begin{pmatrix} r & 1 \\ 1 & 0 \end{pmatrix} \qquad r \in \mathbb{Q}^+$$

We call a matrix of one of these types *an order matrix in normal form* (shortly in NF).

<u>Proof.</u> Firstly we show that every regular matrix can be transformed into one of the above types without changing the corresponding term ordering and, secondly, that for every two distinct order matrices R and R' in NF there exist two terms I and J with $I <_R J$ and $J <_{R'} I$.

Let $R = \begin{pmatrix} a & b \\ c & d \end{pmatrix}$ be any order matrix, then we have two kinds of transformations of the matrix, which do not change the corresponding term ordering

(1) multiplication of any row by a nonzero positive real number
(2) addition or subtraction of the i-th row from the j-th row since $i<j$, and the resulting row has positive elements.

We distinguish two cases:

<u>Case 1.</u> $\det R > 0$. Then, since $a>0$ and $d-c.b/a>0$ we have the following sequence of equivalent order matrices

$$\begin{pmatrix} a & b \\ c & d \end{pmatrix} \xrightarrow{(1)} \begin{pmatrix} 1 & b/a \\ c & d \end{pmatrix} \xrightarrow{(2)} \begin{pmatrix} 1 & b/a \\ 0 & d-cb/a \end{pmatrix} \xrightarrow{(1)} \begin{pmatrix} 1 & b/a \\ 0 & 1 \end{pmatrix}.$$

Now with $s=b/a$ we have type (i).

<u>Case 2.</u> $\det R<0$. It follows that $b\neq 0$, $ad<bc$ and

$$\begin{pmatrix} a & b \\ c & d \end{pmatrix} \xrightarrow{(1)} \begin{pmatrix} a/b & 1 \\ c & d \end{pmatrix} \xrightarrow{(2)} \begin{pmatrix} a/b & 1 \\ c-ad/b & 0 \end{pmatrix} \xrightarrow{(1)}$$

$$\begin{pmatrix} a/b & 1 \\ bc-ac & 0 \end{pmatrix} \xrightarrow{(1)} \begin{pmatrix} a/b & 1 \\ 1 & 0 \end{pmatrix}.$$

Now, with $r=a/b$ we get type (ii), however we must exclude those cases where (i) and (ii) yield the same term ordering. But this is clear due to our last remark. The order matrices of type (i) with $s=1/r$ yield the same term ordering as the corresponding matrices of type (ii) iff 1 and s are linearly independent over the rationals. Therefore, either in (i) or in (ii) irrational values of r and s, respectively must be excluded.

For the second step of proof it is clear now, that the crucial point to show is that an order matrix

$R = \begin{pmatrix} a & b \\ c & d \end{pmatrix}$ determines the term ordering by the value a/b (including the case b=0 , that means $a/b = \infty$) and by the sign of the determinant of R. Let R and R' be two different order matrices in NF of type (i) resp. (ii). In case they have the same values a/b, this value must be rational and nonzero, say m/n, and the signs of their determinants different, then I=(n,0) and J=(0,m) are the desired terms. In the remaining nontrivial case (r=0 and s=0 are obvious) let us assume that $a/b < a'/b'$, then we can find positive integers m,n such that $a/b < m/n < a'/b'$ and again I=(n,0) and J=(0,m) are the desired terms.

<u>Remark 2.</u> It is clear that for any finite set of terms every possible order of these terms occurring in a term ordering can be generated by order matrices of type (ii). In the following we identify such matrices

$R = \begin{pmatrix} r & 1 \\ 1 & 0 \end{pmatrix}$ with the parameter $r \in \mathbb{R}^+$ and write instead $<_R$ also $<_r$ for the corresponding order.

Now we look for the classification of the possible orderings of a finite set of terms.

<u>Theorem 2.</u> Let $T = \{t_1, \ldots, t_m\}$, $t_i = (x_i, y_i) \in \mathbb{N}^2$, be a finite set of terms, then we can find a partition OS(T) of $\mathbb{R}^+$ into a finite number of classes (order intervals) with the following properties. Let K,K', K≠K', be any classes in OS(T), then

(i) there exist $\alpha, \alpha' \in \mathbb{Q}^+ \cup \{\infty\}$, such that $K = [\alpha, \alpha')$,

(ii) let $r, r' \in K$, then $(T, <_r) = (T, <_{r'})$ follows,

(iii) if $r \in K$ and $r' \in K'$, then $(T, <_r) \neq (T, <_{r'})$ follows, i.e. there exist $I, J \in T$ such that $I <_r J$ and $J <_{r'} I$.

__Definition 3.__ Let T be defined like above, then we call a partition of $\mathbb{R}^+$ with the properties of the theorem an *o r d e r s p e c t r u m O S (T) o f T*. Any partition of $\mathbb{R}^+$ into finite many intervals of the form (ii) of the theorem is called *a n o r d e r s p e c t r u m*. The set of all bounds α, α' of the intervals of an order spectrum OS(T) is denoted by OB(T) and is assumed to be ordered in the following way: $0=\alpha_1<\alpha_2<\ldots<\alpha_t=\infty$, that means $OB(T)=\{\alpha_i, i=1,\ldots,t\}$ determines OS(T) uniquely.

__Proof of the theorem.__ (by induction over m). For m=1 we have $OS(T) = \{\mathbb{R}^+\} = \{[0,\infty)\}$ and everything is clear.
In the induction step the order spectrum of the first m terms $T_m = \{t_i \mid i=1\ldots m\}$, $t_i=(x_i,y_i)$, will be refined by adding new interval bounds α_i, which we obtain by comparison of the (m+1)-th term with all the other terms. For every pair (t_{m+1},t_i) three cases are possible

1. t_i/t_{m+1} (t_i divides t_{m+1}), then it holds for every term ordering r that $t_i <_r t_{m+1}$,
2. t_{m+1}/t_i, then always $t_{m+1} <_r t_i$,
3. in the remaining case, $(x_{m+1}-x_i)(y_{m+1}-y_i)<0$ and with $\alpha_i=(x_i-x_{m+1})/(y_{m+1}-y_i)$ we obtain the classes $[0,\alpha_i)$ and $[\alpha_i,\infty)$, which represent the two possible successions of t_{m+1} and t_i.

Now we form $OB(T_{m+1}) := OB(T_m) \cup \{\alpha_1,\ldots,\alpha_m\}$ with all α_i obtained in the third case and renumber the elements of $OB(T_{m+1})$ with respect to real order. For the classes $[\alpha_i,\alpha_{i+1})$, $\alpha_i \in OB(T_{m+1})$, the properties (ii) and (iii) are easy to verify, looking on the proof of lemma 2.

This procedure of refinement of order spectrums is the basis of the announced algorithm.

2. The algorithm.

In the following let

$X = (x_1,\ldots,x_n)$	- n variables (in case n=2, X=(x.y))
k	- a field, in which the operations are effectively practicable, for instance $\mathbb{Q}$ - the rational numbers, $\mathbb{F}_p$ - finite field or a finite algebraic extension of them
$k[X]$	- the ring of polynomials
$(f_1,\ldots,f_m)$	- the ideal generated from $f_1,\ldots,f_m$

$T(f)$ - the set of all terms of $f \in k[X]$

$H_m(f)$ - the headterm of f with respect to $<_m$.

<u>Definition 4.</u> Let $F=\{f_1,\ldots,f_m\} \in k[X]$ and $<_m$ be any fixed term ordering ,then $G=\{g_1,\ldots,g_r\}$ is called a Gröbner basis (GB) of F with respect to $<_m$, if for every $f \in F$ there exists a $g \in G$ such that $H_m(g)$ divides $H_m(f)$.

A Gröbner basis G is called a minimal reduced GB of $(f_1,\ldots,f_m)$ w.r.t. $<_m$ if no term of a GB-element is divisible by the headterm of any other GB-element, and the leading coefficient (that is the coefficient of the headterm) of any GB-element is 1.

The following theorem (Buchberger,B.(1976)) is well known.

<u>Theorem 3.</u> For a given polynomial ideal and any fixed term ordering there exists a minimal reduced Gröbner basis which is unique up to the order of elements.

Now we return to case n=2.

Let $F = \{f_1,\ldots,f_m\} \subset k[x,y]$ be any basis of an ideal and $f \in k[x,y]$, then we introduce the following notations

OHS(f) - the head spectrum of f is the order spectrum with the following property: if $K,K' \in OHS(f)$, $K \neq K'$, and $r,s \in K$, $r' \in K'$, then $H_r(f)=H_s(f)$ and $H_r(f) \neq H_{r'}(f)$,

OHS(F) - the head spectrum of F is defined as the order spectrum with $OHS(F) = OHS(f_1) \cup \ldots \cup OHS(f_m)$, that means all $<_r$ in a class of OHS(F) define the same set of head terms of the polynomials of F, whereas, for $<_r$ and $<_{r'}$ in distinct classes of OHS(F), it holds that $(H_r(f_1),\ldots,H_r(f_m)) \neq (H_{r'}(f_1),\ldots,H_{r'}(f_m))$ as m-tuple, although the sets of the head terms may be equal.

Given an order class K, then we say that a polynomial f respects K if for any $r,r' \in K$ it holds that $H_r(f) \neq H_{r'}(f)$. In this case we can divide f into a headterm w.r.t. K and a rest $f = H_K(f) + R_K(f)$ (Attention! $R_K(f)$ is not uniquely ordered by K.). Analogously we define that F respects K if any $f \in F$ respects K.

K is called m-respected by f (maximally respected) if for any term ordering $<_r$ with $r \notin K$ it holds that $H_r(f) \neq H_K(f)$.

A class K is m-respected by F if for any term ordering $<_r$ with $r \notin K$ there exists an $f \in F$ such that

$H_{K}(f) \neq H_{K'}(f)$.

<u>Lemma 3.</u> K is m-respected by F iff for any $f \in F$ there exists an m-respected order class K_f such that $K = K_{f_1} \cap \ldots \cap K_{f_m}$ holds.

The following is also obvious.

<u>Remark 3.</u> The head spectrum of f (respectively F) is a partition into m-respected order classes by f (respectively F).

The algorithm to obtain all reduced Gröbner bases rests on a stepwise refinement of the head spectrum of the basis during the performance of the classical algorithm such that finally, to every class K of this refined head spectrum, there belongs an up to order unique Gröbner basis G_K with the property that for every K' distinct from K there exists an $f \in G_K$ and for all $f' \in G_{K'}$ it holds that $H_K(f) \neq H_{K'}(f')$.

<u>Lemma 4.</u> Let K be an m-respected order class by F, h a reduced polynomial with respect to F, then $K \cap OHS(h)$ consists of m-respected order classes of $F \cup \{h\}$.

<u>Proof.</u> Let $K=[\alpha,\alpha']$ and $OHS(h)=\{[\alpha_j,\alpha_{j+1})/j=1,\ldots,q\}$, then there exist s and t such that
$K \cap OHS(h) = \{[\alpha,\alpha_s),[\alpha_s,\alpha_{s+1}),\ldots,[\alpha_t,\alpha']\}$.
Obviously every class of $K \cap OHS(h)$ is respected by F and by h, too. Lemma 3 implies that every such class is also m-respected by $F \cup \{h\}$ because every class of $K \cap OHS(h)$ is the intersection of m-respected classes of all elements of $F \cup \{h\}$.

Now we can formulate the algorithm which is an extension of B.Buchberger's one.

```
F := {f1,...,fm}         (* start basis *)
OHG := ∅                 (* GB head spectrum *)
OS := OHS(F)             (* head spectrum of start basis *)
AS := OS
WHILE AS ≠ ∅ DO         (* initialisation *)
  TAKE K∈AS
  AS := AS - {K}
  GK := F
  BK := P(F)             (* P(F) := {{f,f'} / f,f'∈F}  *)
WHILE OS ≠ ∅ DO
  TAKE L∈OS
  OS := OS - {L}
  TAKE {f,f'}∈BL
  BL := BL - {{f,f'}}
  h := SL(f,f')          (* S-polynomial of B. Buchberger *)
  h := SL(h,GL)          (* reduction of the S-polynomial *)
```

```
    IF h ≠ 0 THEN
      Oh := L ∩ OHS(h)   (* refinement of order classes *)
      OS := OS U Oh
      WHILE Oh ≠ φ DO   (* initialisation of new cases *)
        TAKE K∈Oh
        Oh := Oh - {K}
        B_K := B_L U {{f,h} / f∈G_L}
        G_K := G_L U {h}
    ELSE
      IF B_L = φ THEN
        OHG := OHG U {L}
      ELSE
        OS := OS U {L}
  AG := OHG
  WHILE AG ≠ φ DO        (* reduce all GB *)
    TAKE K∈AG
    AG := AG - {K}
    G_K := S_K(G_K,G_K)
```

Remark 4. 1) $S_L(f,f')$ is B.Buchberger's S-polynomial with respect to the head terms $H_L(f)$ and $H_L(f')$.
2) $S_L(h,G_L)$ is the completely reduced form of h with respect to G_L that means, no term of $S_L(h,G_L)$ is divisible by $H_1(g)$ for every element $g \in G_L$.
3) $S_L(G_L,G_L)$ is the reduced form of the GB G_L.
4) The index $_L$ always refers to the head terms w.r.t. L.

Theorem 4. The algorithm terminates and delivers every possible reduced Gröbner basis of the ideal of F.

Proof. For termination we refer to the fact that the number of distinct reduced GB's of an ideal is finite because two different GB's must have different sets of head terms, however, this number must be finite because of the known bounds for the degree of a GB, cf. Buchberger,B.(1983a).
The crucial point in our algorithm for getting only order classes with different sets of head terms is the refinement step

$$Oh := L \cap OHS(h)$$
$$OS := OS \cup Oh$$

of order classes, see lemma 4.
The second assertion of the theorem follows from the fact that OHG - the set of order classes for which has found a GB - is a complete order spectrum at the end of the second WHILE-statement of the algorithm, that is, after reduction we have a reduced GB for every term ordering, hence all reduced GB.

Remark 5. OHG is the order spectrum of all reduced GB's of F, that is, for $K,K' \in OHG$, $K \neq K'$, and the corresponding GB's G_K, $G_{K'}$ it holds that $G_K \neq G_{K'}$ in the sense that there is an $f \in G_K$ such that $H_K(f)$ is not a head term of an element from

$G_{K'}$. In this sense we get, however, three different GB's for the ideal $(x^3+y^3+x^2y^2)$ which differ only by the chosen head terms.

Conclusion. The algorithm as described above should only be the proof of the possibility of effective construction of all reduced GB's of a polynomial ideal. A further note about an effective implementation of the algorithm and some illustrative examples are given elsewhere. A generalization of the algorithm for n>2 must generalize the idea of the order spectrum of a set of terms.

Acknowledgement. I would like to thank Prof.Dr.G.Pfister for his permanent encouragement and discussions. I'am also very thankful to Prof.Dr.N.J.Lehmann for his promotion, which enabled this work.

References.

Buchberger,B. (1965), An algorithm for finding a basis for the residue class ring of a zerodimensional polynomial ideal (German), Ph.D. thesis, Univ. of Insbruck (Austria), Math. Inst.

Buchberger,B. (1976), A theoretical basis for the reduction of polynomials to canonical form, ACM SIGSAM Bull. 10/3, 19-29 and 10/4, 19-24.

Buchberger,B. (1983), Gröbner bases: an algorithmic method in polynomial ideal theory, Techn. Report CAMP-83.29, Univ. of Linz, Math. Inst., to appear as chapter 6 in: Recent Trends in Multidimensional System Theory (ed. by N.K.Bose), Reidel, 1985.

Buchberger,B. (1983a), A note on the complexity of constructing Gröbner bases, Proc. EUROCAL 83, Springer LNCS 162, 137-145.

Galligo,A. (1979), The divisition theorem and stability in local analytic geometry (French), Extrait des Annales de l'Institut Fourier, Univ. of Grenoble, 29/2.

Giusti,M. (1985), A note on the complexity of constructing standard bases, Proc. EUROCAL 85, Springer LNCS 204, 411-412.

Kollreider,C. (1978), Polynomial reduction: the influence of the ordering of terms on a reduction algorithm, Techn. Rep. CAMP-78.4, Univ. of Linz, Math. Inst.

Lazard,D. (1983), Gröbner bases, Gaussian elimiation and resolution of systems of algebraic equations, Proc. EUROCAL 83. Springer LNCS 162, 146-156.

Mora,F. (1982), An algorithm to compute the equations of tangent cones, Proc. EUROCAM 82, Springer LNCS 144, 158-165.

Robbiano,L. (1985), Term orderings on the polynomial ring, Proc. EUROCAL 85, Springer LNCS 204, 513-517.

<u>Remark.</u>

At the EUROCAL'87 conference (June 1987, Leipzig) I was told that R.Robbiano and V.Weispfenning worked about the same problem and that the first author solved the problem of constructing all Gröbner bases for arbitrary n. For details, see

Mora,F. and Robbiano,L. (1987), The Gröbner Fan of an Ideal, preprint.

Weispfenning,V. (1987), Constructing universal Gröbner bases, extended abstract.

Singularities of Moduli Spaces

B. Martin, G. Pfister
Humboldt-Universität Berlin
Sektion Mathematik
Postfach 1297
DDR-1086 Berlin

Let (X,o) be the germ of an irreducible curve singularity in the complex plane C^2 defined by the equation $f=0$, f being holomorphic in a neighbourhood of o. Let $B_d=\{(x,y), |(x,y)|<d\}$ be a small ball and $K_d=\{(x,y), |(x,y)|<d, f(x,y)=0\}$. For small d the homeomorphy class of (B_d,K_d) does not depend on d and is called the topological type of the singularity. Denote by $T_{a,b}$ the topological type of the singularity defined by $f=x^a+y^b$, a and b relatively prime. The moduli space $M_{a,b}$ of all germs of plane curve singularities with the same topological type $T_{a,b}$ is a disjoined union of analytic varieties $M_{a,b,t}$, $t=1,\dots,g$. The generic component $M_{a,b,g}$ of $M_{a,b}$ is an algebraic variety, locally an open subset in a weighted projective space P_w. There are examples that the other components are not algebraic varieties (cf.<1>). $M_{a,b}$ is constructed in the following way: Consider the family $p:X \longrightarrow C^r$, X the subset of $C^r \times C^2$ defined by the equation

$$F = x^a+y^b+\sum_B t_{ij}x^i y^j,$$

$B=\{(i,j), ib+ja>ab, i<a-1, j<b-1\}$, p the projection.

This family has the following properties: If (Y,o) is any germ of an irreducible plane curve singularity with the topological type $T_{a,b}$ then there is an t in C^r such that $(p^{-1}(t),o)=(Y,o)$. Using the relative Kodaira-Spencer map of this family one can compute the analytically trivial subfamilies: The kernel K of the Kodaira-Spencer map is a sub Lie-algebra of the derivation of C^r. Along the integral manifolds of K the family $X \longrightarrow C^r$ is trivial. It turns out that $M_{a,b}=C^r/K$ and $M_{a,b,i}=S_i/K$, S_i the set of all points t in C^r such that the integral manifold through t has dimension i, (cf.<3>). Let $\{a_i\}_{i=1,\dots,m}$ be a free base of $C[\![t,x,y]\!]/(\partial F/\partial x, \partial F/\partial y)$ as a $C[\![t]\!]$-module and let $a_i F = \sum_B h_{i,kl}x^k y^l \bmod (\partial F/\partial x, \partial F/\partial y)$ then $\delta_i = \sum_B h_{i,kl}\partial/\partial t_{kl}$, generate K as $C[\![t]\!]$-module. Moreover S_i is determined by $\operatorname{rank}(h_{i,kl}(t))=i$. There is an algorithm to compute generators of K, (cf.<2>).

The generic component $M_{a,b,\varphi}$ od $M_{a,b}$ may have singularities, according to singularities of the corresponding weighted projective space P_w. The singular points of $M_{a,b,\varphi}$ are of special intrest because they correspond to singular curves with a non connected automorphism gruop G, the quotient of G by its connected component turns out to be the isotropy group of the corresponding point in P_w, (cf.<5>). These singularities are charaterized by the weights and can be computed in a combinatorial way.

The weights of P_w depend only on a and b and can also be computed in a combinatorical way, (cf.<2>,<3>). $M_{a,b,\varphi}$ is defined in P_w by the non vanishing of one or two polynomials with integer coefficients. These polynomials are certain minors of the matrix $(h_{i,ki})$ and can also be described in a combinatorial way in terms of a and b, (cf.<2>,<3>).

The components of the singular locus of P_w are suitable intersections of coordinate hyperplanes such that the weights of the corresponding coordinates have a non trivial common factor. Now using Buchbergers algorithm (cf.<4>) we get the singular locus of $M_{a,b,\varphi}$.

REFERENCES

<1> Martin,B., Pfister,G.: The moduli of irreducible curve singularities with the semigroup $\Gamma = \langle 5,11 \rangle$, Rev. Roumaine Math. pures Appl.. 33(1988), 4, 359-368

<2> Martin,B., Pfister,G.: An algorithm to compute the kernel of the Kodaira-Spencer map for an irreducible plane curve singularity, to appear in J. of Symbol. Computation

<3> Laudal,O.A., Martin,B., Pfister,G.: Moduli of plane curve singularities with C*-action, to appear in the publ. ser. of the Banach Centre, Warsaw, Vol. 20

<4> Buchberger,B.: An algorithm for constructing canonical bases of polynomial ideals, ACM Trans. Math. Software 11 (1985) 1, 66-78

<5> Schönemann,H., Automorphismengruppen von semi-quasihomogenen Kurvensingularitäten, Preprint 175, HUB, 1988

Radical Simplification using Algebraic Extension Fields

Trevor J. Smedley

University of Waterloo

Introduction

We present a method for simplifying radicals which are algebraic numbers — i.e. do not contain variables. /e also discuss how this can be extended to handle algebraic functions. The algorithm has been implemented in ie Maple symbolic computation system[2], and the descriptions are presented in a pseudo-language similar to laple.

The Algorithm

In the following presentation, a particular representation is used for a multiple algebraic extension field of ie rationals. There is a single parameter — called *Extension* — which defines the extension. It is a list of three ems: the first is the number of algebraics in the extension; the second is the names of the variables used; and the iird gives the minimal polynomials. Thus if the following assignments are made:

```
n := Extension[1];
alg := Extension[2];
f := Extension[3];
```

hen $f_i(alg_i)$ for $i=1, \cdots, n$ are the minimal polynomials for the extension field $Q(alg_1, \cdots, alg_n)$.

seudo-code Algorithm

The following is an algorithm which puts an expression involving radicals, nested to any depth, in a normal r canonical form.

```
adical_simplify(expr)
  rads_left := get_list_of_radicals(expr);
  Extension[1] := 0;
  Extension[2] := alg;
  Extension[3] := f;
  i := 0;
  substitutions := {};
  while rads_left <> [] do
     rad := head(rads_left);
     rads_left := tail(rads_left);
     if not_in_extension(rad, Extension, substitutions) then
        i := i + 1;
        Extension[1] = i;
        f_i := minimal_polynomial(rad, Extension, alg_i, substitutions);
        substitutions := substitutions union {alg_i=rad};
        expr := subs(rad=alg_i, expr);
        rads_left := subs(rad=alg_i, rads_left);
     else
        s := express_in_extension(rad, Extension, substitutions);
        expr := subs(rad=s, expr);
        rads_left := subs(rad=s, rads_left);
     endif;
  enddo;
  expr := alg_normal(expr, Extension);
  # use alg_canonical for a canonical form
  expr := full_subs(substitutions, expr);
eturn(expr);
```

escription

The above algorithm works as follows: It gets all of the radicals in the expression in order of their depen- encies (i.e. $5^{1/2}$ would come before $(5^{1/2}+1)^{1/3}$); then it goes through the list, checking if each radical is ready in the extension; if it is not, then it is added to the extension by calculating its minimal polynomial; if it is en its representation in the extension is calculated, and a substitution is made; when this is done, the resulting pression is simplified in the extension, and then a substitution is made to put the expression back in terms of e original radicals.

The subroutine *full_subs* performs full substitution of a set of substitutions in an expression. That is, the substitutions are performed repeatedly until there are no further changes in the expression. *full_subs*({$x{=}y$, $y{=}z$}, $x + y^2$) would yield $z + z^2$.

Subroutines

The subroutines *not_in_extension*, *minimal_polynomial*, and *express_in_extension* are all very similar. They first create a polynomial over the current extension which has *rad* as a zero. This polynomial is factored over the current extension, and then it is determined which of these factors has the radical as a root, under certain assumption as to which branch of the radicals is used. The routine *not_in_extension* returns *true* if and only if this factor is not linear, *minimal_polynomial* returns the calculated polynomial, and *express_in_extension*, which is only called if the polynomial is linear, solves it, and returns the expression of *rad* in the current extension. The problem of determining which of the factors to use is non-trivial. The current implementation works as follows. Each of the factors is put back in terms of the original input radicals, and then the input *rad* is substituted for the variable. Under the assumption that we are to use the branch of each radical which contains the positive real axis each factor is evaluated to a floating point number. It is almost always possible to determine which factor is actually zero from these numbers. An improvement to this which is under investigation is to use interval arithmetic as in Moore[4], and iteratively refine intervals containing each of the substituted factors until only one of the intervals contains 0. Using this method we get a true algorithm, and not just a heuristic method.

Extension to Handle Variables

The previous algorithm was designed to work with algebraic numbers only. When the radicals involve variables we are dealing with algebraic functions. To adapt the algorithm to handle these cases, all that must be changed is the routine which determines the minimal polynomial. The test to determine which substituted factor is zero must be modified, as they now involve variables. It is also not obvious which factor should be chosen. For example, should $\sqrt{x^2}$ be x or $-x$? A discussion of this problem is beyond the scope of this paper, and can be found in Caviness & Fateman[1].

Once these problems have been resolved, the routine can be implemented using concepts similar to those in Gonnet[3], concerning determination of equivalence of expressions, or possibly these concepts combined with interval arithmetic as mentioned previously.

Comments and Improvements

For expressions with only a few radicals the simplifications are carried out quickly. However, if there are many radicals the algorithm has trouble in the algebraic factorisation. It is possible to greatly improve the efficiency by avoiding as many factorisations as possible.

To avoid doing a factorisation you must know either that an expression is independent of the current extension, or that it is in the extension, and you are able to calculate its representation. An example of the first method is to add all the prime roots of prime numbers first, since it is known that none of these can depend on the others. An example of the second method involves noting that expressions resulting from solving a cubic equation have certain algebraic numbers in them which can be expressed in terms of others also appearing in the expression.

Caviness & Fateman[1] give other useful methods for avoiding factorisations, but these apply only to radicals which are not nested.

When attempting to simplify an expression which involves many different radicals, there comes a point when the algebraic factorisations which would have to be done become essentially impossible. At this point it would be wise to issue a warning to the effect that not all algebraic dependencies may be noted, and carry on without further factorisations. If the order of adding the radicals to the extension is chosen intelligently, it may be possible to do this without missing many algebraic dependencies.

References

1. B.F. Caviness and R.J. Fateman, "Simplification of Radical Expressions," *Proceedings SYMSAC 1976*, pp. 329-33[?] Association for Computing Machinery, (1976).
2. B.W. Char, K.O. Geddes, G.H. Gonnet, and S.M. Watt, *Maple User's Guide*, WATCOM (1985).
3. G.H. Gonnet, "New Results for Random Determination of Equivalence of Expressions," *Proceedings SYMSAC '86*, pp. 127-131 Association for Computing Machinery, (1986).
4. R.E. Moore, *Interval Analysis*, Prentice-Hall (1966).

Hermite Normal Forms for Integer Matrices

R.J.Bradford
School of Mathematical Sciences
University of Bath
Claverton Down
Bath
England BA2 7AY

Abstract.
We present a new algorithm for the computation of Hermite Normal Forms of integer matrices that proves advantageous in certain important cases.

Introduction.

The use of unimodular row operations to transform a non-singular integer matrix to triangular form was known by Hermite in 1851. Similarly Smith used both row and column operations to generate a diagonal form as long ago as 1861. Since then these forms have been found to be useful in many branches of linear mathematics, from abelian groups to lattices to diophantine equations. See [Bradley 1971] for pointers to a number of applications.

Our interest in Hermite forms of matrices [Bradford 1988] arose in the context of the Round Two algorithm for the computation of integral bases for algebraic extension fields [Ford 1978] [Trager 1984], an algorithm that requires many reductions of non-square matrices. So we were eager to find a method of reduction that was efficient on the type of matrices that arose in this algorithm.

The traditional algorithm given in mathematics textbooks (subtracting multiples of a row from other rows, and then repeating) is subject to the most appalling intermediate expression swell, even though the final Hermite form may have fairly modestly sized elements.

More Advanced Methods.

In [Kannan & Bachem 1979] (reviewed [Alagar & Roy 1984], improved in [Chou & Collins 1982]), there is an algorithm that proceeds by placing successive leading minors in HNF, rather than the usual row-by-row techniques. This method proves to be very good in practice, as it reduces the swell fairly strongly. [Bradley 1971] suggests a method that involves computing the extended gcd with cofactors for an entire column at a time, and using the cofactors to clear a column in one pass. Unfortunately, the cost of computing the cofactors is quite high, and this method does not reduce the intermediate swell. Kannan & Bachem is geared to square matrices, but can be extended to rectangular matrices quite easily. Bradley is directly applicable to rectangular matrices.

A Method Based on GCDs.

An element on the diagonal of a Hermite reduced form is just the gcd of the elements in its column in the original matrix. Working from this we produced the following algorithm:

1. set U to be a $n \times n$ unit matrix.
2. for $c := 1$ to m do

2.1. find the row out of rows c to n with the smallest non-zero absolute value in column c, and swap it with row c. Swap the same rows in U.

2.2. for each row r from $c+1$ to n do

2.2.1. if $M_{cc} | M_{rc}$ then

2.1.1. replace row r by M_{rc}/M_{cc} times row c. Replace row r of U by the same multiple of row c of U.

2.1.2. else by means of the extended Euclidean algorithm (or otherwise) find $g = \gcd(M_{cc}, M_{rc})$, and integers λ and μ such that $\lambda M_{cc} + \mu M_{rc} = g$.

2.1.3. and replace row c by $\lambda(\text{row } c) + \mu(\text{row } r)$, and row r by $\frac{M_{cc}}{g}(\text{row } r) - \frac{M_{rc}}{g}(\text{row } c)$. Replace the same rows of U in the same manner.

2.3. if $M_{cc} < 0$, negate that entire row, and negate row c in U.

3. for $c := 1$ to m do
3.1. for each row r from 1 to $c-1$ subtract $\lfloor M_{rc}/M_{cc} \rfloor$ times row c from row r. Subtract the same multiple of like rows in U.

Step 2.2.1.3 is valid since

$$\det \begin{pmatrix} \lambda & \mu \\ \frac{-M_{rc}}{g} & \frac{M_{cc}}{g} \end{pmatrix} = \frac{\lambda M_{cc} + \mu M_{rc}}{g} = 1,$$

by the definition of λ and μ, so the transformation is unimodular.

This is superficially similar to the algorithm in [Bradley 1971], but it appears to be more efficient in the our case: Bradley's method requires the computation of n simultaneous cofactors to the gcd of n integers, whereas the above method takes advantage of the fact that in most cases in the 2.2 loop the diagonal entry M_{cc} soon converges to the gcd of the column, and straight division suffices from then on. See [Bradford 1988] for a fuller discussion.

Comparisons.

We ran some tests on the various algorithms on various types of matrix—details are in [Bradford 1988]. On small ($< 8 \times 8$) matrices, The Kannan & Bachem (K & B) algorithm and the above (RJB) algorithm were about equal, with Bradley consistently slower. On larger dense matrices K & B had the winning edge over RJB, but Bradley was far behind—computing cofactors for many large elements is very expensive. On rectangular ($2n \times n$) dense matrices the story was much the same: K & B and RJB similar, with the former gaining advantage with size, and Bradley in poor third place. However, when we considered matrices of the type that appears in the Round Two algorithm, we came to some very different conclusions. These are rectangular, often about 50% dense, and decidedly non-random in the relations between its coefficients. This time RJB was a clear winner, and Bradley was consistently better than K & B—the relative sparseness of the matrices is a great help in computing extended cofactors.

Conclusions.

Kannan and Bachem seems to be the best all-round algorithm for the reduction of modestly-sized matrices. However, in the class of matrix that appears in the Round Two algorithm (rectangular, semi-sparse, non-random elements), the method of the above section can make good use of the special form of the coefficients to produce a superior reduction algorithm.

References.

[Alagar & Roy 1984] *A Comparative Study of Algorithms for Computing the Smith Normal Form of an Integer Matrix*, Alagar V.S., and Roy A.K., Int. J. Systems Sci., vol. 15(7), 1984, pp. 727–744, 1984.

[Bradley 1971] *Algorithms for Hermite and Smith Normal Matrices and Linear Diophantine Equations*, Bradley G.H., Mathematics of Computation, vol. 25(116), pp. 879–907, October 1971.

[Bradford 1988] "On the Computation of Integral Bases and Defects of Integrity," Bradford R.J., Ph.D. Thesis, University of Bath, 1988.

[Chou & Collins 1982] *Algorithms for the Solution of Systems of Linear Diophantine Equations*, Chou T-W.J., and Collins G.E., SIAM J. Computing 11(34), pp. 687–708, November 1982.

[Ford 1978] "On the Computation of the Maximal Order in a Dedekind Domain," Ford D.J., Ph.D. Thesis, Ohio State University, 1978.

[Kannan & Bachem 1979] *Polynomial Algorithms for Computing the Smith and Hermite Normal Forms of an Integer Matrix*, Kannan R., and Bachem A., SIAM J. Computing, vol. 8(4), pp. 499–507, November 1979.

[Trager 1984] "Integration of Algebraic Functions," Trager B.M., Ph.D. Thesis, MIT, 1984.

Mr. Smith Goes to Las Vegas: Randomized Parallel Computation of the Smith Normal Form of Polynomial Matrices*

Erich Kaltofen
M. S. Krishnamoorthy

Rensselaer Polytechnic Institute
Department of Computer Science
Troy, New York 12181

B. David Saunders

University of Delaware
Department of Computer and Information Sciences
Newark, Delaware 19716

Extended Abstract

1. Introduction.

The different normal forms of matrices, Hermite, Smith and Jordan Normal Forms are widely used in many different branches of science and engineering. Sequential algorithms for computing these normal forms have been given previously. With advances in parallel hardware and software, development of parallel algorithms is not only an intellectual exercise but also a practical feasibilty.

This paper is third in a series on canonical forms of matrices [5], [4]. Here we offer a new randomized parallel algorithm that determines the Smith normal form of a matrix in $F[x]^{m \times n}$. The algorithm has two important advantages over our previous one. The multipliers relating the Smith form to the input matrix are computed and the algorithm is of Las Vegas type, that is, the result is guaranteed correct, probability only enters in speed considerations. The Smith form algorithm is also a good sequential algorithm, faster than previous methods in the worst case. Its speed is that of the Hermite form algorithm on which it depends. One can use any of the algorithms by Kannan & Bachem [7], Kannan [6], Chou & Collins [1], or Iliopoulos [3].

A sequential solution to the Smith normal form problem proceeds by iterating Hermite normal form computations on the matrix (see, e.g., [7]). Although in practice usually two Hermite iterations suffice, there are input matrices for which the number of iterations is at least linear in the dimension of the matrix. Here we prove that by multiplying the input matrix with a certain randomly chosen matrix, the new randomized matrix will require with high probability only two Hermite steps before the Smith normal form appears. The proof of this fact uses ideas similar to those for our Monte Carlo Smith normal form algorithm [5], but is more complicated. An "unlucky" premultiplication is discovered immediately if after two Hermite steps we do not obtain a Smith normal form. The point is now that if we do, we must have the unique Smith normal form of the input matrix together with the unimodular pre- and post-multipliers. Since the Hermite normal form algorithms are

* This material is based upon work supported by the National Science Foundation under Grant No. DCR-85-04391 (first author), Grant No. MCS-83-14600 (second and third author), and by an IBM Faculty Development Award (first author).

deterministic [4], the entire algorithm is Las Vegas. We refer to [2] for the definition of the complexity classes NC and RNC of problems (probabilistically) solvable by uniform families of Boolean circuits of poly-logarithmic depth and polynomial size. We note that since the class RNC requires us to perform field operations on Boolean circuits, the previous claim is precise only for concrete fields such as the rationals Q or F_q, the finite field with q elements. Our algorithms are randomized in the Las Vegas sense, that is they can fail but they will never give an incorrect answer.

2. Echelon (Hermite) Forms.

In this section we give a fast parallel echelon form algorithm built from our Hermite form algorithm [5]. This algorithm is needed for the parallel version of the Smith form algorithm of the next section. First we give some definitions and facts needed to prove the echelon and Smith form algorithms.

A matrix in $F[x]^{n \times n}$ is *unimodular* if its determinant is a nonzero element of F. Unimodular matrices are precisely the ones with an inverse in $F[x]^{n \times n}$. Matrices, A and B, in $F[x]^{m \times n}$ are *column equivalent* if there exists unimodular Q such that $AQ=B$. A matrix H is in *column echelon form* if

(1) nonzero columns precede zero columns,

(2) the leading nonzero element in each column is monic,

(3) in each row which contains the leading nonzero element of some column, the entries preceding that entry are of lower degree.

Figure 1: Layout of a column echolon form of a matrix of rank 5. The • entries are monic, the o entries are residues with respect to them, and the * entries are the remaining possibly non-zero entries.

•				
*				
o	•			
o	o	•		
*	*	*		
o	o	o	•	
*	*	*	*	
o	o	o	o	•
*	*	*	*	*
*	*	*	*	*

We denote by C_i^n the set of all length i subsequences of (1, ..., n) and by $A_{I,J}$, $I \in C_i^n$, $J \in C_j^n$, the $i \times i$ determinant of the submatrix in the rows I and columns J.

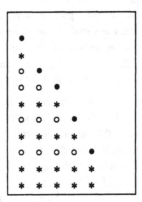

A variant of the echelon form is the Hermite normal form, defined for square matrices and having the zero columns interspersed among the nonzero ones in such a way as to have the leading nonzero elements on the diagonal [8]. Hermite originally developed the form for nonsingular square matrices over the integers. The textbook treatment of echelon forms is for matrices over a field. In that case condition (3) is that all other entries in the row are zero. We do not know of a treatment in the literature giving the canonical form for column or row equivalence of rectangular matrices of arbitrary rank over a PID (over the integers would do). For that reason we state one part of this standard theorem.

2.1 Theorem.

(1) Column equivalent matrices have the same left kernel (row dependencies).

(2) Let I be a fixed sequence of i rows. Column equivalent matrices have the same greatest common divisor of all $i \times i$ minors in the rows I, i.e., $d_I := \text{GCD}_{J \in C_i^n} (\det(A_{I,J}))$ is an invariant.

(3) Each matrix A in $F[x]^{m \times n}$ is column equivalent to a unique matrix H in column echelon form. If the rank of A is m, the unimodular cofactor Q, such that $AQ=H$, is also unique. □

We defer the proofs of most theorems to the full paper.

Most algorithms for the Hermite form have been described for the nonsingular case but extend naturally to echelon form algorithms for the general case. However, our deterministic parallel algorithm (in NC^2, cf. Cook [2] for a description of this parallel computation model) requires a bit more effort. We offer an extended algorithm here.

2.2 Algorithm $(H, Q) \leftarrow CEF(A)$.
[Column Echelon Form. This is a fast parallel algorithm when F is a finite algebraic extension of a prime field.]
Input: $A \in F[x]^{m \times n}$, F a field.
Output: Unimodular $Q \in F[x]^{n \times n}$ and column echelon form $H \in F[x]^{m \times n}$ such that $AQ=H$.

(1) [Find leading independent rows:]
$A' \leftarrow$ the $(m+n) \times n$ matrix $[A^T \; I_n]^T$.
$A'' \leftarrow$ the first n independent rows of A'.
(Compute in parallel the rank [9] of each matrix consisting of the first i rows of A'. Then include the i-th row in A'' if the i-th rank is greater than the $i-1$st rank (the 0-th rank is 0)).
[A'' is $n \times n$. If r is the rank of A, the first r rows of A'' are from A and the remaining $n-r$ from I_n.]

(2) [Hermite form - nonsingular column echelon form:]
$(H'', Q) \leftarrow$ the Hermite form of A'' and the corresponding unimodular cofactor. (Computed by the parallel algorithm of [5]).

(3) [Column echelon form:] $H \leftarrow AQ$. Return Q and H. □

2.3 Theorem. Algorithm CEF to compute the column echelon form and associated unimodular cofactor of a matrix is correct and is in NC^2 when F is a finite algebraic extension of a prime field. □

Row echelon forms are defined by transposing everything in the above. Specifically, the row echelon form and unimodular $m \times m$ cofactor may be computed as follows:

2.4 Algorithm $(Q, H) \leftarrow REF(A)$.
[Row Echelon Form]

(1) $(Q', H') \leftarrow CEF(A^T)$.

(2) $H \leftarrow H'^T$. $Q \leftarrow Q'^T$. $[QA=H]$ Return Q and H. □

3. A Smith Form Algorithm.

A matrix in $F[x]^{m\times n}$ is in *Smith normal form* if it is diagonal, the diagonal entries are monic or zero, and each divides the next. Matrices A and B in $F[x]^{m\times n}$ are ***equivalent*** if there exist unimodular matrices U in $F[x]^{n\times n}$ and V in $F[x]^{m\times m}$ such than $A = UBV$.

3.1 Theorem.

(1) Equivalent matrices have the same determinantal divisors. The i-th *determinantal divisor* of a matrix is the greatest common divisor of all $i\times i$ minors of the matrix. We denote it by s_i^*.

(2) The Smith normal form is a canonical form for equivalence, that is, there is one and only one matrix in Smith form equivalent to a given matrix. The diagonal entries of the Smith form are called the *invariant factors* of the matrix. The i-th invariant factor is $s_i = s_i^*/s_{i-1}^*$ ($s_1 = s_1^*$). □

3.2 Algorithm. $(U, S, V) \leftarrow SNF(A)$.
[Smith Normal Form. Randomizing algorithm.]
Input: A, a matrix in $F[x]^{m\times n}$, where F is a field.
Output: U, S, and V, such that $UAV=S$, S in $F[x]^{m\times n}$ is in Smith form, U in $F[x]^{m\times m}$ is unimodular, and V in $F[x]^{n\times n}$ is unimodular.
Constant: ε, $0<\varepsilon<1$, the probability of failing on one try.

(1) [Randomize:] $d \leftarrow \max_{i,j} \deg(a_{i,j})$.
$R' \leftarrow$ a strictly lower triangular $n\times n$ matrix whose entries are chosen at random from C, a subset of F of size $c = 2d \min(m, n)^3/\varepsilon$, where ε is a constant less than 1. [If F has characteristic 0, C may be the integers 1 to c. The size c guarantees that the probability of having to repeat the algorithm is less than ε. If F is finite of insufficient size, C may be a subset of an algebraic extension of F.]
$R \leftarrow I+R'$ [an invertible matrix].
$A' \leftarrow AR$.

(2) [Row operations:] $(U, H) \leftarrow REF(A')$ [row echelon form: $H = UA'$ $(= UAR)$.]
[The diagonal entries of H are now almost surely the invariant factors sought.]

(3) [Column operations:] $(S, V') \leftarrow CEF(H)$ [column echelon form: $S = HV'$ $(= UARV')$].
[This is expected to be an especially simple echelon form computation. For the most part exact divisions are needed, not GCD's. V' will be very nearly unit upper triangular.]

(4) If S is in Smith form (that with probability $\geq 1-\varepsilon$), $V \leftarrow RV'$. Return U, S, and V.
[One could repeat with S as the input to take advantage of progress made. However, our point is that repetition will not be necessary.] □

Notice that the choice of a lower triangular unitary random multiplier R makes the proof of the following theorem substantially more complicated. However, this choice is preferable, since then one never needs to check R for invertibility and one needs fewer random elements.

3.3 Theorem. Algorithm SNF is correct. It requires repetition only with probability $< \varepsilon$. Hence it is in Las Vegas $\mathbf{RNC}^2$ and runs sequentially in expected time O(CEF time) when F is a finite algebraic extension of a prime field.

Proof. It is clear from the construction that the output conditions are satisfied when the algorithm terminates. The algorithm terminates in k or fewer repetitions with probability $1-\varepsilon^k$, which converges to 1 exponentially fast in k.

It remains to show that the probability that S is not in Smith form is less than ε. We do this with the aid of some lemmas.

From the first, we see that S, computed in step (3), will be in Smith form if H has a certain property. The remaining lemmas enable us to conclude that H, computed in step (2), has that property unless the random entries of R, chosen in step 1, form a root of a certain polynomial π. By a lemma of Schwartz [10], the probability that we pick such an unlucky root is $\deg(\pi)/c$.

A suitable π is the product of the polynomials π_i of lemma 3.7, for $i = 1, ..., r-1$. Each π_i is of degree bounded by $2i^2d+i$. Thus we may bound the degree of π by $2N^3d$ for $N = \min(n, m)$. Since we choose $c = 2N^3d/\varepsilon$ in the algorithm, we obtain the desired probability, ε.

Since the expected number of repetitions is $\varepsilon + 2\varepsilon^2 + 3\varepsilon^3 + \cdots = \varepsilon/(1-\varepsilon)^2$, a constant, and the time of one repetition is dominated by the time for echelon form computation, the parallel and sequential running times are those of CEF. □

3.4 Lemma (A condition under which one more echelon form suffices). Let H be a row echelon form of rank r with $h_{i,i}$ the leading nonzero entry of row i, for $i = 1, ..., r$. Let s_i be the i-th invariant factor of H. If $h_{i,i} = s_i$ for $i = 1, ..., r-1$, then the column echelon form of H is the Smith normal form of H. □

3.5 Substitution Lemma. Let $f_1, ..., f_t$ be polynomials in $F[\bar{\rho}, x]$, $\bar{\rho}$ a list of new variables, with $\deg(f_i) \le e$. Then for some $\bar{e} \le 2e$, there exists an $\bar{e} \times \bar{e}$ determinant Δ in $F[\bar{\rho}]$, whose entries are coefficients of f_i, such that for any evaluation $\bar{\rho} \to \bar{r}$, where $\bar{r}$ a list of corresponding field elements that are not a root of Δ, $\mathrm{GCD}_{i=1,...,t}(f_i(\bar{r})) = (\mathrm{GCD}_{i=1,...,t}(f_i))(\bar{r})$. (Cf. [5], Proof of Lemma 4.1.) □

3.6 Irreducibility Lemma. Let $n \ge 2$, and let

$$R = \begin{bmatrix} 1 & & & \\ \rho_{2,1} & 1 & & \\ \vdots & & \ddots & \\ \rho_{n,1} & \cdots & \rho_{n,n-1} & 1 \end{bmatrix} \in F[\bar{\rho}]^{n \times n},$$

where $\bar{\rho} = (\rho_{j,k})_{j>k}$ is a vector of indeterminants and F is a field. Then for all i, $1 \le i \le n$, $\mathrm{G} \subset C_i^n$, $\mathrm{G} \ne \emptyset$, and for all families of polynomials $f_J(x) \in F[x] \setminus \{0\}$, $J \in \mathrm{G}$,

$$\sum_{J \in \mathrm{G}} f_J R_{J,I} = \mathrm{GCD}_{J \in \mathrm{G}}(f_J)\, p,$$

where $I = \{1, ..., i\} \in C_i^n$ and where $p \in F[x, \bar{\rho}]$ is either an irreducible polynomial in $F[\bar{\rho}, x] \setminus F[x]$ or is 1. □

Note that in this lemma it is crucial that the selected columns are the ones in I. Otherwise, the lemma is not true, and therefore the proof of lemma 3.7 must enforce the additional condition that H be triangular.

3.7 Lemma. Let A be a matrix in $F[x]^{m \times n}$ of rank r and with the degrees of the entries bounded by d, and let $i \in \{1, ..., r-1\}$. Then there is a polynomial π_i in $n(n-1)/2$ variables such that if

(1) R in $F[x]^{n \times n}$ is unit lower triangular,

(2) H is the row echelon form of AR,

(3) s_i^* is the i-th determinantal divisor of A,

then H is upper triangular and $s_i^* = \prod_{j=1}^{i} h_{j,j}$, unless the $n(n-1)/2$ entries below the diagonal in R form a root of π_i. The degree of π_i is no more than $2i^2d+i$. □

Incidentally, we have resolved a question on the coefficient size of multipliers.

3.8 Corollary. For polynomial matrices over the rational numbers, there exist unimodular pre- and post multipliers for the Smith normal form, whose entries have coefficients of binary length polynomial in the dimensions and coefficient lengths of the input matrices. □

4. Conclusion

We have provided a parallel solution for the well-known Smith normal form problem. Our method employs randomization as a tool to remove the iterations along the main diagonal in the classical sequential algorithms, and as such might be useful in similar settings, as well as may speed the sequential methods themselves.

References

1. Chou, T. J. and Collins, G. E., "Algorithms for the solution of systems of diophantine linear equations," *SIAM J. Comp.*, vol. 11, pp. 687-708, 1982.
2. Cook, S. A., "A taxonomy of problems with fast parallel algorithms," *Inf. Control*, vol. 64, pp. 2-22, 1985.
3. Iliopoulos, C. S., "Worst-case complexity bounds on algorithms for computing the canonical structure of finite Abelian groups and the Hermite and Smith normal forms of an integer matrix," Manuscript, Purdue Univ., 1986.
4. Kaltofen, E., Krishnamoorthy, M. S., and Saunders, B. D., "Fast parallel algorithms for similarity of matrices," *Proc. 1986 ACM Symp. Symbolic Algebraic Comp.*, pp. 65-70, 1986.
5. Kaltofen, E., Krishnamoorthy, M. S., and Saunders, B. D., "Fast parallel computation of Hermite and Smith forms of polynomial matrices," *SIAM J. Alg. Discrete Meth.*, vol. 8, pp. 683-690, 1987.
6. Kannan, R., "Polynomial-time algorithms for solving systems of linear equations over polynomials," *Theoretical Comp. Sci.*, vol. 39, pp. 69-88, 1985.
7. Kannan, R. and Bachem, A., "Polynomial algorithms for computing the Smith and Hermite normal forms of an integer matrix," *SIAM J. Comp.*, vol. 8, pp. 499-507, 1981.
8. MacDuffee, C. C., *Vectors and Matrices*, Math. Assoc. America, 1943.
9. Mulmuley, K., "A fast parallel algorithm to compute the rank of a matrix over an arbitrary field," *Combinatorica*, vol. 7, pp. 101-104, 1987.
10. Schwartz, J. T., "Fast probabilistic algorithms for verification of polynomial identities," *J. ACM*, vol. 27, pp. 701-717, 1980.

Fonctions symétriques et changements de bases

Annick Valibouze
LITP,Université P. et M. Curie, 4 Place Jussieu
75252 Paris Cedex 05
Unité associée au CNRS N.248 et GRECO de Calcul Formel N.60
UUCP: ... mcvax!inria!litp!avb

Abstract

This paper describes change of basis algorithms for symmetric polynomials. We consider below the three usual following bases : monomial forms, symmetric elementary and Newton polynomials. The originality consists in retaining only one representative of the orbit to make the computations. It is a crucial point if we realize that one orbit can contain commonly hundreds of terms. We implemented these algorithms in FRANZLISP with an interface MACSYMA.

INTRODUCTION

Une fonction est dite symétrique, si elle est invariante par toute permutation de ses variables.
Une difficulté apparente de manipulation des fonctions symétriques est le caractère exponentiel des groupes symétriques ($n!$ éléments).
L'algèbre des polynômes symétriques possède entre autres bases les fonctions symétriques élémentaires et les fonctions puissances.
Nous présentons ici des algorithmes de décomposition d'un polynôme symétrique dans chacune de ces deux bases qui évitent ce caractère exponentiel. Pour aboutir à cela il a fallu travailler sur des représentations contractées des polynômes, consistant à ne conserver qu'un élément par orbite.
L'ensemble de ces algorithmes, implantés en Franzlisp, constituent un sous-module de MACSYMA que j'ai nommé SYM.
Comme les coefficients d'un polynôme sont des fonctions symétriques de ses racines, ces algorithmes peuvent déboucher sur la manipulation des racines d'un polynôme et éviter l'explosion des calculs.
Les algorithmes sont suffisamment performants pour pouvoir calculer des résultants en des temps du même ordre de grandeur que la méthode classique.

1 Définitions et notations

Soient k un corps, R_n l'anneau $k[x_1, x_2, ..., x_n]$ des polynômes à coefficients sur k en les variables $x_1, x_2, ..., x_n$ et X la multivariable $(x_1, x_2, ..., x_n)$.

Pour tout élément σ de S_n (le groupe des permutations d'ordre n) et toute suite T, $(t_1, t_2, ..., t_n)$, on notera $\sigma(T)$ la suite $(t_{\sigma(1)}, t_{\sigma(2)}, \ldots, t_{\sigma(n)})$.

Un polynôme P de R_n est dit *symétrique* , si pour tout élément σ de S_n, $P(X)$ est égal à $P(\sigma(X))$.
L'algèbre des polynômes symétriques de R_n apparaît alors comme l'ensemble des invariants $R_n^{S_n}$.

Introduisons maintenant quelques notations pour les fonctions symétriques (voir [Macdonald]) : Soit I une suite finie $(i_1, i_2, \ldots, i_n)$ d'entiers positifs ou nuls. On définit le monôme X^I comme le produit $x_1^{i_1} x_2^{i_2} \ldots x_n^{i_n}$. Si $i_1 \geq i_2 \cdots \geq i_n$, I est appelée une *partition*. Cette notation des partitions est indépendante du nombre de zeros. Les *parts* de I sont les i_k non nuls. La *longueur* $lg(I)$ de I est le nombre de ses parts. Si $a_1, \ldots, a_q$ sont les parts distinctes de I avec $a_1 > a_2 > \cdots > a_q$ et $m_{a_1}, \ldots, m_{a_q}$ leur *multiplicité* (i.e. leur nombre d'occurences dans I) respective dans I, on notera également la partition I : $(a_1^{m_{a_1}} a_2^{m_{a_2}} \ldots a_q^{m_{a_q}})$. A une partition I on associe la *forme monomiale* $M_I(X)$, somme des monômes de l'orbite de X^I sous l'action de S_n :

$$M_I(X) = \sum_{\sigma \in S_n / G(I)} X^{\sigma(I)}$$

où $G(I)$ désigne le stabilisateur de I sous l'action de S_n. Si la partition I est de longueur strictement supérieure à n, le nombre de variables, on conviendra que $M_I(X)$ est nulle.
Les formes monomiales forment trivialement une base de $R_n^{S_n}$.

La forme monomiale $M_{(1^i)}(X)$ est la $i^{ième}$ *fonction symétrique élémentaire* . On la note $e_i(X)$, avec les conventions $e_0 = 1$ et $e_i = 0$ pour $i > n$. Les fonctions symétriques élémentaires forment une base de l'algèbre $R_n^{S_n}$. Si l'on considère un polynôme unitaire en x de degré n, et X l'ensemble de ses racines, le coefficient de x^i est $(-1)^i e_i(X)$.

Les formes monomiales $p_0 = n$ et $p_i(X) = \sum_{x \in X} x^i$, où i est un entier strictement positif, sont les *fonctions puissances* sur X. Elles forment également une base de l'algèbre $R_n^{S_n}$.

2 Le problème

Nous voulons réaliser les opérations élémentaires de l'algèbre $R_n^{S_n}$ et donner deux algorithmes décomposant un polynôme symétrique sur la base des fonctions symétriques élémentaires et sur celle des fonctions puissances. Confrontés à un problème d'espace, nous coderons les polynômes symétriques pour ne travailler qu'avec un seul monôme par forme monomiale (cf. paragraphe 3). Le point fondamental consiste alors en la formule permettant de calculer le produit de deux formes monomiales (lemme du produit, voir paragraphe 5). Le passage des fonctions symétriques élémentaires aux fonctions puissance, et sa réciproque, sont donnés par les formules classiques de Girard-Newton [Girard]. On les trouve également, ainsi que d'autres, dans [Macdonald, ch.2].

3 Représentation des données

Un polynôme symétrique s'écrit naturellement comme combinaison linéaire sur k de formes monomiales. Il peut donc être codé par une liste de partitions adjointes d'un coefficient élément de k. Soit I une partition. Les deux représentations, $(i_1, \ldots, i_n)$ et $(a_1^{m_{a_1}} \ldots a_q^{m_{a_q}})$ de I induisent deux codages. Avec la première on dira que le polynôme est *partitionné de type* 1. Par exemple le polynôme $p(X) = 3\, M_{(2,0,0)}(X) + 5\, M_{(1,1,0)}(X)$ est codé par la liste ((3, 2, 0, 0), (5, 1, 1, 0)). Avec la seconde, I est codé par la liste $(a_1, m_{a_1}, a_2, m_{a_2}, \ldots, a_n, m_{a_n})$. Notre polynôme p sera alors codé avec la liste ((3, 2, 1), (5, 2, 1)). Un polynôme ainsi codé sera dit *partitionné de type* 2. C'est la représentation partitionnée de type 2 qui est utilisée pour l'implémentation des algorithmes de changements de bases `ELEM` et `PUI` que nous décrirons ici.

On peut ainsi obtenir ainsi un gain de place pouvant aller jusqu'à l'exponentielle de l'ordre n du groupe symétrique.

4 Ordres

Nos algorithmes, ELEM et PUI, font intervenir des décroissances d'ordres pour, d'une part, ne jamais faire le même calcul deux fois, et d'autre part, s'assurer de leur convergence. Pour cela nous considérons deux suites finies d'entiers $u = (u_i)_{1\leq i\leq n}$ et $v = (v_i)_{1\leq i\leq n}$.

4.1 Ordre lexicographique : $<$

Il sera associé à la décomposition en les fonctions symétriques élémentaires. On dit que u est inférieur à v pour l'*ordre lexicographique*, noté ici $u < v$, s'il existe un entier k compris entre 1 et n tel que u_k soit strictement inférieur à v_k et que pour tout entier p inférieur à k on ait $u_p = v_p$. Cet ordre définit naturellement la notion de suite ordonnée lexicographiquement pour les suites produits $u \times v = ((u_p, v_p))_{1\leq p\leq n}$.
Restreint aux partitions de $\mathbf{N}^n$ il induit également un ordre total sur les formes monomiales de $R_n^{S_n}$: $M_I < M_J$ si $I < J$.

Remarque : Les plus petites formes monomiales pour cet ordre sont les fonctions symétriques élémentaires avec : $e_1 < e_2 < \cdots < e_n$.

4.2 Ordre des longueurs : $<_{lg}$

Il sera associé à la décomposition en les fonctions puissances. On dira que u est inférieur à v pour l' *ordre des longueurs* si $lg(u) < lg(v)$ ou si $lg(u) = lg(v)$ et $u < v$. On notera alors $u <_{lg} v$.

Remarque : Les plus petites formes monomiales pour cet ordre sont les fonctions puissances avec : $p_1 <_{lg} p_2 <_{lg} \cdots <_{lg} p_n <_{lg} \cdots$

5 Produit de deux formes monomiales

C'est la formule calculant ce produit qui va permettre de réaliser nos manipulations en conservant le codage, que nous avons appelé partitionné, évitant ainsi l'explosion exponentielle des développements. Soient $u = (u_i)_{1\leq i\leq n}$ et $v = (v_i)_{1\leq i\leq n}$ deux n-uplets.
Le partition engendrée par une permutation de u sera notée $P(u)$. Désignons par $u \times v$ et $u + v$ les suites $((u_i, v_i))_{1\leq i\leq n}$ et $(u_i + v_i)_{1\leq i\leq n}$ respectivement. On dira que $u \times v$ est dans l'*ordre lexicographique* sur $(\mathbf{N}^2)^n$ si pour tout $1 \leq i < n$ ou $u_i > u_{i+1}$, ou bien $u_i = u_{i+1}$ et $v_i \geq v_{i+1}$.
On définit l'action de S_n sur les n-uplets de couples de la manière suivante :

$$\begin{array}{ccc} S_n \times (\mathbf{N}^2)^n & \longrightarrow & (\mathbf{N}^2)^n \\ (\sigma\ ,\ (c_1, \ldots, c_n)) & \longmapsto & (c_{\sigma(1)}, \ldots, c_{\sigma(n)}) \end{array}$$

où pour $c_i = (a_i, b_i)$ on a $c_{\sigma(i)} = (a_{\sigma(i)}, b_{\sigma(i)})$.

Le lemme du produit : *Soit X la multivariable $(x_1, \ldots, x_n)$. Pour toute partition J et K on a :*

$$M_J(X) M_K(X) = \sum_{L \in Ad(J,K)} c_L\, M_{P(J+L)}(X) \tag{1}$$

où $Ad(J,K) = \{\sigma(K) \mid \sigma \in S_n\ ,\ J \times \sigma(K)$ est dans l'ordre lexicographique$\}$ et c_L est un entier qui peut se calculer de deux manières :

- *Soit $P(J + L) = (\lambda_1^{m_{\lambda_1}} \lambda_2^{m_{\lambda_2}} \ldots \lambda_q^{m_{\lambda_q}})$. Soit i compris entre 1 et q. Les m_{λ_i} parts λ_i de $P(J + L)$ résultent de la somme d'une sous-suite $u_i(J)$ de J et d'une sous-suite $u_i(L)$ de*

L. En désignant par $c_i(J)$ (resp. $c_i(L)$) le cardinal de l'orbite de $u_i(J)$ (resp. $u_i(L)$) sous l'action de $S_{m_{\lambda_i}}$, il vient :

$$c_L = \prod_{i=1}^{q} c_i(J) = \prod_{i=1}^{q} c_i(L). \tag{2}$$

- *La seconde égalité est donnée par le quotient de deux cardinaux de stabilisateurs :*

$$c_L = \frac{\#G(J+L)}{\#G(J \times L)}. \tag{3}$$

L'identité (2) mieux adaptée à l'algorithme ELEM nécessitera une représentation partitionnée de type 2 et la (3) proposée par Daniel LAZARD est utilisée dans le module SYM pour le produit de deux polynômes symétriques.

Remarques : K est dans l'ensemble d'admissibilité $Ad(J,K)$ puisque $J \times K$ est ordonnée lexicographiquement et $P(J+K)$ est égale à $J+K$ puisque J et K sont des partitions.

Propriété 5.1 *$P(J+L) <_{lex} J+K$ pour tout L dans $Ad(J,K)$.*

En effet, J et K étant deux partitions, on peut utiliser le lemme suivant :

Lemme : *Soient deux partitions $I = (i_1, ..., i_n)$ et $J = (j_1, ..., j_n)$, et s et t deux permutations de S_n, alors $P(I+J)$ est supérieur à $P(s(I)+t(J))$ relativement à l'ordre lexicographique.*

Démonstration par récurrence sur n : Pour $n = 1$, c'est clair. Sinon on regarde le maximum des $i_{s(k)} + j_{t(k)}$ pour k variant entre 1 et n, atteint disons pour l'indice q. De deux choses l'une : Ou bien ce maximum est strictement inférieur à $i_1 + j_1$, et alors la conclusion est assurée ; ou bien il est égal à $i_1 + j_1$, et alors $i_1 = i_{s(q)}$ et $j_1 = j_{t(q)}$. On considère alors les deux suites ordonnées $(i_2, ..., i_n)$ et $(j_2, ..., j_n)$, et leurs permutations : $i_{s(1)}, ..., i_{s(n)}$ et $j_{t(1)}, ..., j_{t(n)}$ où on a enlevé respectivement $i_{s(q)}$ et $j_{t(q)}$. Par hypothèse de récurrence on conclut.

Propriété 5.2 *Le coefficient c_K associé à K est égal à 1.*

On prend $J = (j_1, \ldots, j_n)$ et $K = (k_1, \ldots, k_n)$. Comme le coefficient associé à J est le quotient de $\#G(J+K)$ avec $\#G(J \times K)$, et que $\#G(J+K) \geq \#G(J \times K)$, pour qu'il soit égal à 1 on doit montrer l'inégalité inverse, ou ce qui revient au même, que l'égalité de deux parts de $J+K$ entraine l'égalité des deux parts de $J \times K$ qui les ont engendrées. Si on a $j_q + k_q = j_r + k_r$ avec $q < r$, ou bien $k_r = k_q$ ou bien $k_q > k_r$ puisque K est une partition. Le premier cas implique que $j_q = j_r$. Et le deuxième viendrait de $j_q < j_r$, ce qui est exclu puisque J est une partition. On voit ainsi que $(j_q, k_q) = (j_r, k_r)$.

Propriété 5.3 *J , $K <_{lex} P(J+L)$ pour toute suite L de $Ad(J,K)$.*

Ceci est totalement évident pour J, et aussi pour K par symétrie.

6 Décomposition en les symétriques élémentaires

Soit P un polynôme de $R_m^{S_m}$ de degré d. Notons p le plus petit des deux nombres d et m. Il s'agit de construire le polynôme f de p variables sur k, tel que ;

$$P(x_1, x_2, \ldots, x_m) = f(e_1(X), ..., e_p(X))$$

6.1 Principe de la décomposition

Pour réaliser cette décomposition on regarde le lemme du produit énoncé précédemment. On en déduit ainsi les formules (4) et (5) permettant de réécrire une forme monomiale avec d'autres strictement plus petites pour l'ordre lexicographique. Or on obtient exactement la règle dont on a besoin puisque toute fonction symétrique élémentaire minore pour cet ordre toute forme monomiale non symétrique élémentaire.

Soit I une partition de longueur non nulle. Pour toute décomposition de I en somme non triviale $J+K$ de partitions de longueurs non nulles, on a d'après le lemme du produit et la propriété 5.2 :

$$M_I = M_K\, M_J - \sum_{L \in Ad^*(J,K)} c_L\, M_{P(J+L)} \tag{4}$$

où $Ad^*(J,K) = Ad(J,K)\backslash\{K\}$ et dont tout élément L vérifie, d'après les propriétés 5.1 et 5.3 :

$$J <_{lex} P(J+L) <_{lex} I. \tag{5}$$

A partir de la formule (4) on voit se dessiner au moins deux stratégies. La première que nous décrivons ci-dessous est une méthode linéaire et la seconde serait une méthode dichotomique où chaque part i_p de I serait divisée comme suit :

$$j_p = k_p = \frac{i_p}{2} \qquad \text{si } i_p \text{ paire}$$

$$j_p = k_p - 1 = \lfloor \frac{i_p}{2} \rfloor \qquad \text{si } i_p \text{ impaire}$$

Dans les deux cas il est essentiel de ne décomposer chaque partition qu'une seule fois.

6.2 L'algorithme linéaire

Considérons le polynôme symétrique P que l'on désire décomposer. Pour éviter de décomposer plusieurs fois la même forme monomiale, on ordonne P dans l'ordre lexicographique décroissant sur les formes monomiales qui le constituent. Soit M_I sa forme monomiale dirigeante affectée de son coefficient c dans k. Notons n la longueur de la partition I : $(i_1, i_2, ..., i_n, 0, 0, ..., 0)$. Si M_I n'est pas déjà symétrique élémentaire, on peut l'écrire comme la somme de $J = (i_1 - 1, i_2 - 1, ..., i_n - 1)$ et de $K = (1^n)$ (i.e $M_K = e_n$). La décomposition (4) devient alors :

$$M_I = e_n M_J - \sum_{L \in Ad^*(J,K)} c_L\, M_{P(J+L)} \tag{6}$$

où le second membre, $Q(M_I)$, est considéré comme un polynôme de $\mathbf{k}[e_n][x_1, x_2, ..., x_m]^{S_m}$. On remplace P par $P1$, somme de $P - cM_I$ avec $cQ(M_I)$, que l'on ordonne comme P. Les inégalités (5) assurent que les formes monomiales de $P1$ sont strictement plus petites que M_I. Il suffit maintenant de réitérer ce processus en considérant que les coefficients sont dans l'anneau $k[e_1, \ldots, e_p]$. L'ordre lexicographique étant un bon ordre le processus s'arrête lorsque la forme monomiale de tête est une fonction symétrique élémentaire (voir la remarque sur l'ordre lexicographique).

Remarque : Dans ce cas particulier le calcul du produit d'une forme monomiale avec une fonction symétrique élémentaire est réalisable par un algorithme, `MULTELEM` (voir 6.2.2), qui

- ne commence jamais à construire une solution qui s'avèrerait ne pas appartenir à l'ensemble admissible $Ad(J,K)$ et qui serait alors éliminée en cour de calcul ;
- permet de ramener le polynôme, membre droit de (4), déjà ordonné dans l'ordre lexicographique avec I en tête qu'il sera alors aisé de retirer.

6.2.1 Algorithme ELEM

On suppose que e_i prend la valeur ei. Le symbole $\star$ entre un élément d de $\mathbf{k}[e_1, ..., e_p]$ et un polynôme symétrique exprimé sur la base des formes monomiales signifie que l'on multiplie tout les coefficient du polynôme par d.

définition : ELEM
Départ :
LIRE(P)
ORDONNER(P) dans l'ordre lexicographique décroissant.
DECOMPOSER(P)

définition : DECOMPOSER(P)
P est un polynôme symétrique trié dans l'ordre lexicographique décroissant.
M_I est sa forme monomiale de tête, et c son coefficient dans $\mathbf{k}[e_1, ..., e_p]$.
SI M_I est une fonction symétrique élémentaire
ALORS RENDRE P en remplaceant chaque M_I par $e_{lg(I)}$.
SINON
Q := DECOMP(M_I)$\star$ c
P := SOMME(Q,P-c$\star M_I$, lexico)
DECOMPOSER(P)

définition : SOMME(P1,P2,ordre)
P1 et P2 étant 2 polynômes rangés suivant l'ordre décroissant,
on ramène leur somme de nouveau rangée dans l'ordre décroissant.

définition : DECOMP(M_I)
Soient J la partition obtenue en enlevant 1 à chaque part de I,
n la longueur de I.
R := MULTELEM(M_J,n)
R' := M_I-R
RAMENER(R' + en$\star M_J$)

D'après la remarque précédente, R est ordonné dans l'ordre lexicographique décroissant. Comme pour chaque L dans $Ad^*(J,K)$ on a $P(J+L) <_{lex} I$, M_I est en tête de R, et R' est facile alors à récupérer. Les inégalités $J <_{lex} P(J+L)$ permettent de constater qu'il suffit de mettre M_J en queue de R' pour que $Q(M_I)$ soit ordonné dans l'ordre lexicographique décroissant.

6.2.2 Description de MULTELEM

Soient J et K les partitions de l'algorithme linéaire. Nous raffinons, tout d'abord, la contrainte d'admissibilité du lemme du produit, dans ce cas où $M_K = e_n$.
Etendons aux suites la notation $(a_1^{m_1} \ldots a_r^{m_r})$, réservée habituellement aux partitions, qui signifiera ici que les m_1 premiers éléments sont égaux à a_1, que les m_2 suivants sont égaux à a_2 et ainsi de suite.

Propriété 6.1 *Soit L une permutation de K où M_K est une fonction symétrique élémentaire. La suite L est admissible (i.e. dans $Ad(J,K)$) si et seulement si $P(J+L)$ est égale à $J+L$.*

En effet si $J = (a_1^{m_1} \ldots a_q^{m_q})$, comme les parts de K sont 0 ou 1, les suites admissibles sont celles du type :

$$(1^{n_1}0^{m_1-n_1}1^{n_2}0^{m_2-n_2} \ldots 1^{n_i}0^{m_i-n_i} \ldots 1^{n_q}0^{m_q-n_q}1^{n-(n_1+\cdots+n_q)}),$$

tel que n_i soit un entier naturel compris entre 0 et m_i et que $n-(n_1+\cdots+n_q)$ soit positif ou nul. Ainsi chaque part j_r+l_r d'une suite $J+L$ où L est admissible, ne peut-être qu'inférieure ou égale à la part $j_{r-1}+l_{r-1}$ qui la précède, si elle existe. Soit $P(J+L)=J+L$. Inversement on ne peut avoir $P(J+L)=J+L$ si L n'a pas la forme que nous venons de donner.

Maintenant pour réaliser le produit e_nM_J, on construit les diagrammes représentatifs [Macdonald, p.1] de toutes les partitions $P(J+L)$, où $L\in Ad(J,K)$, de la manière suivante : on rajoute n cases dans le diagramme de J sans jamais en mettre deux sur la même ligne (toutes les parts de K valent 1 ou 0) et de sorte que le diagramme final représente bien une partition qui est ici la contrainte d'admissibilité dans $Ad(J,K)$.
En remplissant d'abord les lignes représentant les plus grands exposants, on obtient les plus grandes partitions $J+L$ en premier et on peut tronquer les calculs afin de ne pas dépasser m le nombre de variables. Ceci justifie le fait que le polynôme symétrique R de DECOMP soit ordonné dans l'ordre lexicographique décroissant.
Les coefficients c_L sont alors calculés au fur et à mesure de la construction à l'aide de l'égalité (2) du lemme du produit.

6.2.3 Exemple : réduction d'une fonction puissance

Décomposons le polynôme symétrique $P(X)=p_3(X)$ en suivant l'algorithme ELEM. Pour plus de clarté nous restons sur la base des formes monomiales au lieu de prendre des représentations partitionnées de type 2 et nous évitons les troncatures dues au dépassement de cardinalité en supposant que X a au moins 3 variables. Au départ

$$P(X)=M_{(3)}(X).$$

Avec MULTELEM appliquée à $J=(2)$ et e_1 on obtient $M_{(3)}=e_1M_{(2)}-M_{(2,1)}$. Ce qui donne pour Q $-M_{(2,1)}+e_1M_{(2)}$ et P est remplacé par :

$$-M_{(2,1)}+e_1M_{(2)}.$$

La forme monomiale de tête de P est à présent $M_{(2,1)}$ et son coefficient est -1. En appliquant MULTELEM à $J=(1)$ et e_2 on obtient : $M_{(2,1)}=e_2M_{(1)}-3M_{(1,1,1)}$. Ainsi Q est égal à $3\,M_{(1,1,1)}-e_2M_{(1)}$ et P est remplacé par :

$$e_1M_{(2)}+3\,M_{(1,1,1)}-e_2M_{(1)}.$$

On continue : Comme $M_{(2)}=e_1M_{(1)}-2M_{(1,1)}$, P est remplacé par :

$$3M_{(1,1,1)}-2e_1M_{(1,1)}+(-e_2+e_1^2)M_{(1)}$$

Comme $M_{(1,1,1)}$ est une fonction symétrique élémentaire les autres formes monomiales à décomposer dans P le sont également. En remplaçant chaque M_I par $e_{lg(I)}$ dans P on obtient finalement :

$$p_3=3e_3-2e_1e_2+(-e_2+e_1^2)e_1=e_1^3-3e_1e_2+3e_3$$

6.2.4 Comparaison avec la méthode de Waring

Nous comparons ici la fonction ELEM et celle liée à la méthode de Waring [Dubrueil] écrite en Macsyma dans le livre [S,D,T] portant sur un alphabet de 3 lettres. La méthode de Waring utilise la forme développée des polynômes symétriques et non sa représentation partitionnée. Une des raisons de l'écart de temps entre les deux méthodes est le fait que même si par la méthode de Waring on trouve dès le départ le monôme de $k[e_1,\ldots,e_p]$ qui comporte la plus grande forme

monomiale du polynôme symétrique que l'on réduit, une fois construite on la développe pour la soustraire à ce polynôme. On remarque que ce monôme de $k[e_1, \ldots, e_p]$ serait celui obtenu si la fonction DECOMP de ELEM ne rendait que $e_n M_J$.
La comparaison porte sur les fonctions puissances par variation de d, la puissance :

d	WARING	ELEM
12	2 mn 37 s	24 s
15	8 mn 18 s	55 s
18	18 mn 48 s	2 mn
20	31 mn	3 mn
21	39 mn	3 mn 42

Si le cardinal de l'alphabet augmente l'écart de temps s'amplifie avec le caractère exponentiel de son groupe symétrique.
La place prise par les polynômes dans Waring empêche la résolution de certains problèmes comme celui proposé par P. Cartier (voir plus loin) faisant intervenir des orbites de cardinalité 7!.

7 Décomposition en les fonctions puissances

Soit $R = \mathbf{Z}[x_1, \ldots, x_m]$ et $P \in R^{S_m}$ de degré d. Il s'agit de construire le polynôme F de $\mathbf{Q}[x_1, x_2, ..., x_d]$ tel que :

$$P(X) = F(p_1(x), p_2(X), ..., p_d(X))$$

Dans la pratique si $p = inf(m, d)$, P s'écrit en fonction des p premières fonctions puissances. Ceci résulte de la dépendance algébrique des fonctions puissances d'ordres strictement supérieures à m en fonction des m premières.

7.1 Principe sur les formes monomiales

Le procédé, identique à celui utilisé pour les fonctions symétriques élémentaires, est fondé sur la décroissance suivant l'ordre des longueurs dont les formes monomiales minorantes sont les fonctions puissances (cf. paragraphe 4.2), et sur l'utilisation du produit de deux formes monomiales.

Soit I une partition $(i_1, \ldots, i_m)$ de longueur n non nulle. On désigne par σ_p la transposition entre le premier et le $p^{ième}$ élément. Soit r un entier quelconque compris entre 1 et n. Nous utilisons le lemme du produit avec K et J les partitions (i_r) (i.e. $M_K = p_{i_r}$) et $(i_1, \ldots, i_{r-1}, i_{r+1}, \ldots, i_m)$ respectivement, obtenant ainsi $I = P(J + \sigma_m(K))$ avec $c_{\sigma_m(K)} = m_{i_r}(I)$ et donc :

$$M_I = \frac{p_{i_r} M_J - \sum c_{\sigma_q(K)} M_{P(J+\sigma_q(K))}}{m_{i_r}(I)}, \tag{7}$$

où la sommation est étendue à tout les entiers q compris entre 1 et $m-1$ tels que $i_{q-1} > i_q$, qui est ici la contrainte d'admissibilité.
Comme pour tout entier q inférieur à $m-1$ on a $lg(I) - 1 = lg(J) = lg(P(J + \sigma_q(K)))$, la décroissance stricte suivant l'ordre des longueurs est assurée :

$$J <_{lg} P(J + \sigma_q(K)) <_{lg} I.$$

Propriété 7.1 *: Si $r = 1$ alors pour tout entier q de la sommation on a : $c_{\sigma_q(K)} = 1$.*

Preuve : Par convention 0! vaut 1. Nous utilisons l'égalité (3) du lemme du produit donnant les coefficients $c_{\sigma(K)}$. Dans le cas où $r = 1$, pour chaque permutation σ_q de la sommation on a : $P(J + \sigma_q(K)) = (i_1 + i_q, i_2, i_3, \ldots, i_{q-1}, i_{q+1}, \ldots, i_m)$. Comme la part $i_1 + i_q$ est strictement plus grande que les autres et que $i_{q-1} > i_q$, le cardinal du stabilisateur de $P(J + \sigma(K))$ est donné par :

$$\#G(P(J + \sigma_q(K))) = 1!(m_{i_1} - 1)!m_{(i_1-1)}!m_{(i_1-2)}! \ldots (m_{i_{q-1}})!(m_{i_q} - 1)!(m_{(i_q-1)})! \ldots m_0!.$$

Par ailleurs, comme $J + \sigma_q(K) = ((i_2, 0), (i_3, 0), \ldots, (i_{q-1}, 0), (i_q, i_1), (i_{q+1}, 0), \ldots, (i_m, 0))$, on conclut en constatant l'égalité de $\#G(P(J + \sigma_q(K)))$ et de $\#G(P(J \times \sigma(K)))$ qui est égal à

$$(m_{i_1} - 1)!m_{i_1-1}! \ldots (m_{i_{q-1}})!1!(m_{i_q} - 1)!(m_{(i_q-1)})! \ldots m_0!.$$

7.2 Algorithme PUI

Comme pour la réduction en les fonctions symétriques élémentaires on réalise le produit d'une forme monomiale par une fonction puissance avec un algorithme, que je nomme `MULTPUI`, adapté à ce cas particulier de produit de formes monomiales. Il est implanté dans le cas $r = 1$ pour lequel tout les coefficients sont égaux à 1 (cf. propriété 7.1) et qui permet d'obtenir aisément les partitions $P(J + L)$ admissibles rangées suivant l'ordre des longueurs décroissant.
L'algorithme `PUI` décomposant un polynôme symétrique en les fonctions puissances découle de ce qui précède et se calque sur celui donné pour les fonctions symétriques élémentaires en :

- substituant l'ordre des longueurs à l'ordre lexicographique
- substituant `MULTPUI` à `MULTELEM`
- remplaçant le test d'arrêt sur une fonction symétrique élémentaire par celui sur une fonction puissance
- remplaçant `Q := DECOMP(`M_I`)*c` par `Q := DECOMP(`M_I`)*c/`m_{i_1}.

8 Applications

Une première application est la solution d'un problème proposé par Cartier. La seconde découle d'un autre proposé par Barrucand. Nous ne donnons pas ici les méthodes utilisées pour réaliser ce type de calculs. Elles feront l'objet d'un autre article relatif aux problèmes d'élimination. Les exécutions ont été réalisées sur le VAX 780 du LITP, en utilisant `MACSYMA` et des programmes écrits en `FRANZLISP` inclus dans le module de manipulation de fonctions symétriques, `SYM`.

8.1 Problème de Cartier

On se donne le polynôme $x^7 - 7x + 3$ dont les racines sont $x_1, x_2, ..., x_7$ et on cherche le polynôme de degré 35 dont les racines sont les sommes 3 à 3 des x_i.

Ce problème fait intervenir des polynômes symétriques de degré 35 à réduire ou bien avec `ELEM` ou bien avec `PUI`. Les temps d'éxécutions sont de 2mn 30 avec `PUI` et de 27 mn avec `ELEM`.

Remarque: L'utilisation des fonctions puissances s'est avérée bien plus efficace que celle des fonctions symétriques élémentaires. Et ceci, aussi bien pour ce qui est de la complexité en temps et en espace, que pour la combinatoires liée à ce problème. Il semblerait que cette constatation soit d'ordre général.

8.2 Problème de Barrucand

Le calcul d'un discriminant dépendait du problème suivant : Etant donnés les polynômes

$$\begin{aligned} P(x) &= x^5 - e_1 x^4 + e_2 x^3 - e_3 x^2 + e_4 x - e_5 \\ Q(x) &= -55x^4 + 52e_1x^3 + (-40e_2 - 2e_1^2)x^2 + (64e_3 - 16e_1e_2 + 4e_1^3)x \\ &\quad - 64e_4 + 16e_2^2 - 8e_1^2e_2 + e_1^4, \end{aligned}$$

on cherche à calculer le produit $S = Q(x_1)Q(x_2)Q(x_3)Q(x_4)Q(x_5)$, où x_1, x_2, x_3, x_4, x_5 sont les racines de P.
On peut obtenir S comme le résultant de P et Q en x. On peut également le voir comme la cinquième fonction symétrique élémentaire en les $Q(x_1), Q(x_2), Q(x_3), Q(x_4), Q(x_5)$. Ce calcul peut se faire en passant par les fonctions puissances et nécessite 78 secondes dont 7 de "garbage collector", alors que l'utilisation du résultant de MACSYMA donne le résultat en 77 secondes dont 13 de "garbage collector".

Ce calcul permet de constater :

1. l'efficacité des fonctions puissances, même pour obtenir les fonctions symétriques élémentaires
2. l'efficacité des algorithmes de manipulations des fonctions symétriques qui n'est pas très loin de celle du résultant.

Références

[Dubreuil], P. Dubreuil, Algèbre , Cahiers scientifiques, fascicule XX, *Gauthier - Villard.*

[Girard],(1629), Invention Nouvelle en Algèbre, Amsterdam.

[Macdonald], I.G. Macdonald,(1979), Symmetric functions and Hall polynomials , *Clarendon Press*, Oxford.

[S,D,T], Y. Siret, J.H. Davenport et E. Tournier, (1986), Calcul Formel, Systèmes et algorithmes de manipulations algébriques, *Masson.*

COMPLEXITY OF STANDARD BASES IN PROJECTIVE DIMENSION ZERO

Marc Giusti
Centre de Mathématiques de l'Ecole Polytechnique
91128 Palaiseau Cedex FRANCE
bitnet : cfmagi@frpoly11 uucp : ... mcvax!inria!cmep!giusti
Unité associée au CNRS No. 169 and GRECO de Calcul Formel No. 60
(partially supported by a grant from PRC "Mathématiques et Informatique")

INTRODUCTION

The aim of this technical report is to prove that the maximal degree of elements in a standard basis, with respect to any choice of coordinates and any compatible ordering, is essentially linear in the degree of the projective variety if its dimension is zero.

This implies that the time and space complexities of constructing a standard basis in this case are essentially polynomial in the degree of the variety.

1 STABLE SUBSETS OF DIMENSION ZERO

1.1 Preliminaries

We shall use freely in this section classical results on stable subsets, as recalled in [GI]. A stable subset E of $\mathbb{N}^{n+1}$ is minimally generated by a finite subset $B(E)$. We shall denote by $D(E)$ the maximal degree of the elements of $B(E)$ and by $H(E)$ the regularity of the Hilbert function HF_E. The degree of the Hilbert polynomial is then the *dimension* $dim(E)$ of E, and the coefficient of the term of maximal degree multiplied by $dim(E)!$ is the *degree* $deg(E)$ of E.

1.2 Proposition

Let E be a stable subset of $\mathbb{N}^{n+1}$ of dimension 0. Then :

$$D(E) \leq Max((n+1)deg(E), H(E))$$

Proof: If the dimension is 0, the Hilbert function becomes constant for large values of the argument. Hence the complement of any section $E_i = E \cap \{x_i = s\}$ is constant for large s, either empty or of finite volume smaller than the degree of E. Call e the number of indices corresponding to a non empty section (*admissible* indices) . The complement of E contains any such cylinder based on E_i whose fiber is the corresponding i^{th} axis. This leads to the following decomposition of the complement of E :

1.2.1 Lemma

Let us define $G(E)$ as the complement of the union of the previously defined cylinders, i.e.

$$G(E) = \{A \in \mathbf{N}^{n+1} \mid \forall \text{ admissible } i,\ \forall w \gg 0\,, (A_0, \ldots, A_i + w, \ldots, A_n) \in E\}$$

Then $G(E)$ is a stable subset, and the complement $F(E) = G(E) \backslash E$ is finite.

Proof : Since the Hilbert function of E and $G(E)$ are equal for large values of their argument, $F(E)$ is finite.

Now let us turn to the stability of $G(E)$. By construction, $G(E)$ contains E : so if a point of $G(E)$) belongs already to E, there is no problem for any positively translated point to belong to E, hence to $G(E)$. If not, let us consider a point A in $F(E)$. As $F(E)$ is finite, for any index i the translated point $(A_0, \ldots, A_i + B_i, \ldots, A_n)$ goes outside of $F(E)$ for large B_i, hence belongs to $G(E)$.

1.2.2 Lemma

$$D(G(E)) \leq deg(E)$$

Proof : Let $A = (A_0, \ldots, A_n)$ be an element of $B(G(E))$. Fix an admissible index i ; then for every index j different from i, the point $(A_0, \ldots, A_j - 1, \ldots, A_n)$ is no longer in $G(E)$, which means that the projection of A on the coordinate hyperplane $x_i = 0$ is an element of $B(E_i)$. Moreover if v_i is the volume of the complement of E_i, the degree of this projection is upper bounded by v_i, that is :

$$A_0 + \ldots + A_n \leq A_i + v_i$$

If there is only one admissible index, the corresponding A_i is zero, and we are done. If not, summing over all indices we obtain that the degree of A is bounded above by $1/(e-1)$ times the degree $deg(E)$, and a fortiori by the degree.

1.2.3 End of the proof

We divide the basis $B(E)$ into two classes : the first ones belong to $B(G)$, the others do not. Furthermore if $A = (A_0, \ldots, A_n)$ is in the last class, there exists an index i such that the translated point $(A_0, \ldots, A_i - 1, \ldots, A_n)$ is no longer in E but stays in G, hence is in $F(E)$. Either its degree is always smaller than $(n+1)deg(E)$, and the conclusion holds using 1.2.2 ; or not, and then using again 1.2.2, we see that the maximal degree $D(E)$ is attained for an element A not in $B(G(E))$. As we have already remarked, there exists a point in $F(E)$ of degree one less, so let us compute the Hilbert function in this degree ; with the notations of [GI] :

$$HF_E(D(E) - 1) = HF_F(D(E) - 1) + HF_G(D(E) - 1)$$

The regularity of HF_G is bounded above by $(n+1)D(G(E))$, hence by $(n+1)deg(E)$ (1.2.2), itself by $D(E) - 1$. Putting all together we get :

$$HF_E(D(E) - 1) > HF_G(D(E) - 1)$$

Since $F(E)$ is finite, the Hilbert functions of E and $G(E)$ are the same, and we conclude that $H(E)$ is at least $D(E)$.

So in all cases we proved the proposition.

2 A COMPLEXITY THEOREM

Let k be a field, R the polynomial algebra $k[x_0, \ldots, x_n]$ and I a homogeneous ideal of R generated by polynomials $f_1, \ldots, f_t$ of degree at most d and of maximal size 2^M. Choose any compatible total ordering on the monomials of R. Let us consider the projective subvariety $V(I)$ of $\mathbf{P}^n$ defined by I. Call δ the modified degree of $V(I)$, i.e. the maximum of the degree of $V(I)$ and of d.

2.1 Theorem

If the dimension of $V(I)$ is zero, the maximal degree of elements of a standard basis, with respect to any compatible ordering, is smaller than $n+1$ times the modified degree of the variety.

Proof : Let us apply the proposition 1.2 to the stable subset $E(I)$ of $\mathbf{N}^{n+1}$, of dimension zero. But in this case the regularity $H(E(I))$ is smaller than $(n+1)d$ (by a result of D. Lazard [LA]).

2.2 Corollary

The time or space complexity needed to construct a standard basis in projective dimension zero is polynomial in the modified degree of the variety, δ, more precisely polynomial in $MtO(\delta^n)$.

REFERENCES

[GI] M. GIUSTI, Théorie combinatoire de la dimension d'une variété algébrique, Notes informelles de Calcul Formel VI, prépublication du Centre de Mathématiques de l'Ecole Polytechnique (1985), and Compt. Rend. du Colloque Algèbre et Algèbre effective de Rennes (1985). Augmented english version to be published in J. Symbolic Computation (1988).

[LA] D. LAZARD, Résolution des systèmes d'équations algébriques, Theoretical Computer Science 15 (1981), 77-110.

GRÖBNER BASES FOR POLYNOMIAL IDEALS OVER COMMUTATIVE REGULAR RINGS

Volker Weispfenning
Lehrstuhl für Mathematik
Universität Passau
D-8390 Passau, FRG

INTRODUCTION. The method of Gröbner basis computations introduced by Buchberger in 1965 has provided a source of algorithms for solving many basic questions concerning ideals in polynomial rings over fields (see [B2] for a survey). The success of the method has stimulated ongoing research on extensions of the method to other ground rings (and also to non-commutative polynomials, see [M], [AL], [KW]). The ring Z of integers formed a natural starting point for the study of more general classes of ground rings such as Euclidean rings and principal ideal domains (see [B1] and the articles quoted there, [KK] and [Pa]). More comprehensive axiomatic approaches to Gröbner basis constructions in $R[X_1,..X_n]$ are presented in [KN1] and [Moe]; they are bases on the respective hypotheses, that a Gröbner basis construction is available for the ideals of R, and the bases for modules of syzygies can be computed in R[X].

The ground rings studied in this paper, commutative regular rings, differ from most of the rings studied previously in two essential respects: They are in general far from being integral domains, and they are in general not Noetherian. Nevertheless, they share one feature with fields: Every element has a unique quasi-inverse. In particular, any direct product of fields is a commutative regular ring. On the other hand, they generalize Boolean rings. Accordingly, they may be construed as "compositions" of fields and Boolean algebras. This view is supported by the well-established representation of commutative regular rings as Boolean products of fields, or equivalently as rings of global sections in sheaf of fields over a Boolean space (see [Pi], [W]). This representation extends to polynomial rings over a commutative regular ring R, and in fact to most of the rings and modules arising in the algebraic geometry over R (see [SW]). In particular, any finite direct multiple $S = K[\underline{X}]^m$ of a polynomial ring over a field may be identified with the polynomial ring $R[\underline{X}]$ over the regular ring $R = K^m$.

As a consequence, the study of Gröbner bases for submodules of S (see [MM], [AK], [FSK]) can be subsumed under a slight extension of this framework. Another type of application arises from automatic theorem proving using Groebner basis techniques in polynomial rings over the two-element field 2 (see [KN2]). By extending 2 to an arbitrary Boolean ring, free parameters can be incorporated into this approach. From a more fundamental viewpoint, algebraic geometry over commutative regular rings may be regarded as the study of classical algegraic geometry within some topos (see [Lo]). All this indicates that the motivation for our study goes well beyond pure generalization.

The purpose of this paper is to present and prove the basic facts concerning the Gröbner basis method over commutative regular rings, and to indicate some of the most immediate applications. The reader is expected to have a nodding acquaintance with the Gröbner basis technique over fields. (We recommend Buchberger's survey [B2] as a reference.) The algebraic facts required about regular rings and *-rings are presented in detail in [SW]; they are outlined again in section 0. Section 1 studies the reduction relations associated with finite lists of polynomials over regular rings. Section 2 introduces and characterizes Gröbner bases, reduced and stratified reduced Gröbner bases, and provides the algorithms for the construction of these bases. Section 3 presents applications to the ideal membership problem, computing in residue rings, the dimension function of an ideal, and the module of syzygies associated with a Gröbner basis. Further applications, in particular those mentioned above, will be treated in a subsequent paper.

I am indebted to K. Madlener for stimulating conversations that started the research leading to this paper.

0. PRELIMINARIES ON REGULAR RINGS AND *-RINGS. All facts reviewed in this section are presented in detail in [SW] (comp. also [Pi]). All rings in this paper will be commutative rings with 1. A ring R is (von Neumann) regular if for every $a \in R$ there exists $b \in R$ with $a^2b = a$; $a^* = a\cdot b$ and $a^{-1} = a\cdot b^2$ are then uniquely determined by a and satisfy $a\cdot a^* = a$, $a\cdot a^{-1} = a^*$. a^* is the idempotent of a, a^{-1} the quasi-inverse of a. B(R) denotes the Boolean algebra of idempotents of R (with the operations defined by $\sim a = 1-a$, $a \sqcap b = a\cdot b$, $a \sqcup b = a + b - a\cdot b$). Any regular ring is a *-ring, i.e. a ring with an operation $a \longmapsto a^*$ associating with a the smallest idempotent $e = a^* \in B(R)$ with $e\cdot a = a$. Examples of regular rings (*-rings) are direct products R of fields (of integral domains), where a^{-1} (a^*) is the pointwise inverse of a [with $0^{-1} = 0$] (the characteristic function of a). More generally, any subring of R closed under $^{-1}$ (under *) is a regular ring (a *-ring). Conversely, any regular ring (any *-ring) can be canonically represented in this way as subdirect product of fields (of integral domains): Let Spec R be the Boolean space of prime ideals of the Boolean algebra B(R). Then the Stone representation of B(R) extends uniquely to representation of R as subdirect product of factors R_p, $p \in$ Spec R. In this representation, the support of any element of R is a clopen set equal to the support of a^* and of a^{-1}. More generally, for any Boolean combination $\varphi(a_1,\dots,a_n)$ of equations between elements of R, the set of all $p \in$ Spec R, where $\varphi(\underline{a}_p)$ holds in R_p is clopen. Thus $\varphi(\underline{a})$ defines a locally constant property of R. (If $\varphi(\underline{a}_p)$ holds in R_p then $\varphi(\underline{a}_{p'})$ holds in $R_{p'}$ for all p' in some neighborhood of $p \in$ Spec R.) More general locally constant properties are defined by (infinite) disjunctions of existence assertions about R. The most important fact about locally constant properties is the following:

COMPACTNESS PRINCIPLE. Let $\langle R,*\rangle$ be a *-ring, S an additive subgroup of R closed under multiplication with idempotents from R, let $c_1,\dots,c_m \in R$, and let $\varphi(x_1,\dots,x_n,y_1,\dots y_m)$ be a locally constant property for $\langle R,*\rangle$. If for every $p \in$ Spec R there exist $a_1^{(p)},\dots,a_n^{(p)} \in S_p$ such that $\varphi(a_1^{(p)},\dots,a_n^{(p)}, c_{1p},\dots,c_{mp}$ holds in R_p, then there exist $b_1,\dots,b_n$ such that $\varphi(b_{1p'},\dots,b_{np'}, c_{1p'},\dots,c_{mp'})$ holds in R_p for all $p' \in$ Spec R.

An (almost trivial) application is the fact that for any $a \in R$ and any ideal I of R, $a \in I$ iff for all $p \in$ Spec R, a_p is in the ideal I_p of R_p. For the present paper, our interest in *-rings arises from the following:

FACT. Let R be a regular ring and let $S = R[X_1,\ldots,X_r]$ be a polynomial ring over R. Then S is a $*$-ring with $B(S) = B(R)$ and for all $p \in \mathrm{Spec}\, S = \mathrm{Spec}\, R$, S_p is canonically isomorphic to $R_p[X_1,\ldots,X_r]$, and hence will be identified with this polynomial ring.

Similarly, any module M over a $*$-ring R has a canonical representation as subdirect product of R_p-modules $M_p = p\cdot M$, where $p \in \mathrm{Spec}\, R$.

To conclude, we indicate how (countable) regular ground rings R can be handled computationally: If R is a finite direct product of computable fields, no problem arises. In other cases, R may frequently be regarded as a regular subring of the bounded Boolean power K[B] of a computable field K by the universal countable Boolean algebra B. Then the elements of R are sums $e_1k_1 + \ldots + e_nk_n$ with e_i in B, k_i in K. In the canonical representation, the elements of R are locally constant functions from Cantor space C into K. The elements of B can be represented as disjoint unions of basic clopen subsets of $C = 2^N$ coded by finite strings of zeros and ones.

1. REDUCTION AND BOOLEAN CLOSURE. Let R be a regular ring, let $S = R[X_1,\ldots,X_r]$ be a polynomial ring over R; for any $p \in \mathrm{Spec}\, R$, we let $S_p = R_p[X_1,\ldots,X_r]$ be the canonical factor of the $*$-ring $(S,*)$ at p. T denotes the set of terms s, t, t',..., i.e. of power-products of the indeterminates X_i. Monomials are products $a\cdot t$ with $0 \neq a \in R$, $t \in T$. A monomial $a\cdot t$ occurs in a polynomial $f \in S$ if $a\cdot t$ is a summand in f; t occurs in f if $b\cdot t$ occurs in f for some $0 \neq b \in R$. A linear order $<$ on T is admissible, if $1 \leq t$, and $s < t$ implies $s\cdot t' < t\cdot t'$ for all $s,t,t' \in T$. Examples of admissible orders are the lexicographical order and the total degree order on T (see [B2]). Any admissible order on T induces a linear quasiorder $<$ on S and on all S_p : $f < g$ if there exists $t \in T$ such that t occurs in g but not in f, and for all $t < t' \in T$, t' occurs in f iff it occurs in g. HT(f), HM(f), HC(f), HI(f) denote the highest term occuring in f, the highest monomial occuring in f, the highest coefficient a of f (i.e. the coefficient of HT(f)), and the highest idempotent a^* of f, respectively. f is quasimonic if HC(f) is an idempotent.

The most important fact about admissible orders and their induced quasiorders is that both are Noetherian, i.e. do not admit infinite decreasing chains of elements. This is a wellknown consequence of Dickson's lemma (comp. [B2], [KRW]). Reduction relations on S (with respect to a fixed admissible order $<$ on T) can be defined verbatim as for polynomials over fields, with inverses replaced by quasiinverses:

Let f, g, $h \in S$, and let F be a finite list of polynomials in S. If $a\cdot t$ is a monomial occuring in g, $b\cdot s = HM(F)$, and $s \mid t$, say $t = t'\cdot s$, then g reduces to h mod f via $a\cdot t$ ($g \xrightarrow[f]{} h\ (a\cdot t)$), if $a\cdot b \neq 0$ and $h = g - a/b\cdot t'\cdot f$. g reduces to h mod f ($g \xrightarrow[f]{} h$), $a\cdot b^{-1}$, if $g \xrightarrow[f]{} (a\cdot t)$ for some monomial $a\cdot t$ occuring in g. For notational convenience, we also admit reduction by the zero-polynomial f, $g \xrightarrow[0]{} h = g$. g reduces to h mod F ($g \xrightarrow[F]{} h$), if $g \xrightarrow[f]{} h$ for some f in F. g is reducible mod F, if there exists $0 \neq f \in F$ and $h \in S$ such that $g \xrightarrow[f]{} h$. If this is not the case, g is irreducible (or in normal form) mod F. F is reduced, if $F = \{0\}$, or for all $f \in F$, $f \neq 0$, and f is quasimonic and irreducible mod $F \setminus \{f\}$. The ideal generated by f or F in S is denoted by (f) and (F), respectively. As usual, $\xrightarrow{k}$, $\xrightarrow{*}$, $<\xrightarrow{*}>$ denote the k-th power, the reflexive-transitive closure, and the reflexive-symmetric-transistive closure of the relation $\longrightarrow$, resprectively. The notation $f \downarrow g$ means that for some $h \in S$, $f \xrightarrow{*} h$ and $g \xrightarrow{*} h$.

The first fact that one likes to establish for the reduction relations $\xrightarrow[F]{}$ is that they are Noetherian. For polynomials over fields this is an easy consequence of the fact that the quasi-order < is Noetherian: Indeed, if $g \xrightarrow[f]{} h\ (a \cdot t)$, then the term t does not occur in h, and so g > h. This argument is in fact valid for infinite F as well. For a regular ground ring R that is not a field it fails, since t may still occur in h. (Notice that $a \cdot HC(f)^{-1} \cdot HC(f) = a \cdot HI(f)$, which may be different from a.) In fact, for infinite $F \subseteq S$, the reduction relation $\xrightarrow[F]{}$ may be non-Noetherian:

EXAMPLE 1.1. Let $R = K^N$ for some field K, let $S = R[X]$, $g = X$, and let $F = \{f_n = e_n X\}_{n \in N}$, where $e_n(i) = \delta_{ni}$. Then $g \xrightarrow[0]{} g' \xrightarrow[1]{} g'' \xrightarrow[2]{} g''' \xrightarrow[3]{} \ldots$, where $\xrightarrow[n]{}$ is the reduction relation mod f_n. (Notice, that in fact $(F) = (g)$.)

Nevertheless, we can show:

THEOREM 1.2. For any finite list F of non-zero polynomials in S, the reduction $\xrightarrow[F]{}$ is Noetherian.

The proof uses the following simple lemma:

LEMMA 1.3. Let $g \xrightarrow[f]{} h\ (a \cdot t)$, let $f \neq 0$, $b = HC(f)$, and let $c \in R$ with $c^* b^* = c^*$. Then (1) $c \cdot g = c \cdot h$, if $a \cdot c = 0$; (2) $c \cdot g \xrightarrow[f]{} c \cdot h$ and $h < g$, if $a \cdot c \neq 0$.

PROOF OF 1.2. Assume for a contradiction that there exists an infinite chain of reductions $g \longrightarrow g_1 \longrightarrow g_2 \longrightarrow \ldots$ mod F. Let B' be the Boolean subalgebra of B(R) generated by $HI(F) = (HI(f) : f \in F\}$. Then B' is finite and hence generated by a partition E of 1 in B(R). For any $e \in E$ and any $n \in N$, we have by 1.3, $g_{n+1} = g_n$ or $g_n \xrightarrow[F]{} g_{n+1} < g_n$. Moreover, $g_n = \sum_{e \in E} eg_n$. So by the pigeon-hole principle, some chain $e \cdot g$, $e \cdot g_1$, $e \cdot g_2, \ldots$ must contain infinitely many proper reductions, and thus form a decreasing chain with respect to the quasi-order < of S. This contradicts the fact that < is Noetherian.

As a routine consequence, we may conclude (see [H]):

COROLLARY 1.4. Let F be a finite list of polynomials in S. Then the following assertions about the reduction relation $\longrightarrow$ mod F are equivalent:

(1) $\longrightarrow$ is confluent.
(2) $\longrightarrow$ is locally confluent.
(3) $\longrightarrow$ has the Church-Rosser property.
(4) Every $g \in S$ has a unique normal form mod F.

Next we prove a very useful technical lemma (comp. [B2]).

TRANSLATION LEMMA 1.5. Let $\longrightarrow$ be the reduction relation induced by a finite list Q of polynomials in S. Let $f,g,h,h' \in S$, where $f = g + h$, $h \xrightarrow{*} h'$. Then there exist $f', g' \in S$ such that $f \xrightarrow{*} f'$, $g \xrightarrow{*} g'$ and $f' = g' + h'$.

PROOF. By induction on n, where $h \xrightarrow{n} h'$, we may assume $h \longrightarrow h'$. Let $h' = h - c/b \cdot u \cdot g$, where $c \in R$, $u \in T$, $q \in Q$, $b = HC(q)$. Let c_1 (c_2) be the coefficient of $u \cdot HT(q)$ in f (g), where c_1 (c_2) is zero, if this term does not occur in f (in g). We put $f' = f - c_1/b \cdot u \cdot q$, $g' = g - c_2/b \cdot u \cdot q$. Then $c_1 = c_2 + c$, and so $f' = g' + h'$, and by definition $f \longrightarrow f'$, $g \longrightarrow g'$.

COROLLARY 1.6. If $f - g \xrightarrow{*} 0$ mod Q, then $f \downarrow g$ mod Q.

The experience with reductions of polynomials over fields suggests that for any monomial $a \cdot t$ and any polynomial $q \in S$, $a \cdot t \cdot q \xrightarrow[q]{} 0$. Unfortunately, this is no longer true over regular rings:

EXAMPLE 1.7. Let R be a regular ring which is not a field, and let $e \neq 0, 1$ be an idempotent in R. Let $S = R[X]$ and let $q = e \cdot X + (1-e)$. Then $(1-e) \cdot q = (1-e)$ is irreducible mod q.

To avoid this kind of counterexample, we define: $q \in S$ is Boolean closed (b.c.) if $q = HI(q) \cdot q$. So for any $q \in S$, $HI(q) \cdot q$ is b.c.. We call $HI(q) \cdot q$ the Boolean closure BC(q) of q, and $(1-HI(q)) \cdot q$ the Boolean remainder BR(q) of q. So for $q \neq 0$, $BR(q) < q$ and $q = BC(q) + BR(q)$. A finite list Q of polynomials in S is Boolean closed (b.c.) if every member q of Q is b.c.. The following algorithm provides for any finite list Q in S a finite Boolean closed list BC(Q) in S such that $(Q) = (BC(Q))$. Termination is guaranteed by the fact that the quasi-order < on S is Noetherian.

Algorithm (BOOLEAN CLOSURE) Input : Q, a finite list of polynomials in S.
Output : BC(Q), a finite b.c. list of polynomials in S with (Q) = (BC(Q)).

```
BEGIN
  H := Q; H' := Ø; Q' := Ø;
  REPEAT
    REPEAT
      q:= an element of H;
      Q':= Q' U {BC(q)};
      IF BR(q) ≠ 0, THEN H':= H' U {BR(q)};
      H := H \ {q};
    UNTIL H = Ø;
    H := H'; H' := Ø;
  UNTIL H = Ø;
  BC(Q):= Q'
END
```

As an application, we can now relate reductions to ideal membership:

LEMMA 1.8. Let Q be a finite b.c. list of polynomials in S, and let $\longrightarrow$ be the reduction relation induced by Q.

(1) For all $f \in S$, $q \in Q$, $f\cdot q \xrightarrow{*} 0$.

(2) For all $f,g \in S$, $f - g \in (Q)$ iff $f \xleftrightarrow{*} g$.

PROOF. (1) Assume for a contradiction that f is minimal such that for some $q \in Q$, not $f\cdot q \xrightarrow{*} 0$. Since q is b.c., $f\cdot q = g\cdot q$, where $g = HI(q)\cdot f$. Let $a\cdot t = HM(g)$. Then $g\cdot q \xrightarrow[q]{} h\cdot q$, where $h = (g - a\cdot t) < g$. So $h\cdot q \xrightarrow{*} 0$, and hence $f\cdot q = g\cdot q \xrightarrow{*} 0$, a contradiction.
(2) "⇒": Let $f - g = a_1s_1q_1 + \dots + a_ks_kq_k$ with $a_1 \in R$, $s_1 \in T$, $q_1 \in Q$. Then by (1) $a_1s_1q_1 \xrightarrow{*} 0$, and so by 1.8 and induction on k, $f \xleftrightarrow{*} g$. "⇐": is obvious.

We close this section with a comparison of global reduction in S and pointwise reduction in the polynomial rings S_p, $p \in \text{Spec } R$.

LEMMA 1.9. Let $f \in S$ be b.c.. Then for all $p \in \text{Spec } R$ with $f_p \neq 0$, $HM(f_p) = HM(f)_p$.

PROOF. $f_p = 0$ iff $HC(f)_p = 0$.

LEMMA 1.10. Let $f,g,h \in S$, let f be b.c., and let F be a finite, b.c. list of polynomials in S.

(1) $g \xrightarrow[f]{} h\ (a\cdot t)$ iff for all $p \in \text{Spec } R$ with $f_p \neq 0$, $g_p \longrightarrow h_p\ (a_p\cdot t) \pmod{f_p}$.

(2) g is irreducible mod F iff for all $p \in \text{Spec } R$, g_p is irreducible mod F_p, where $F_p = \{f_p : f \in F\}$.

(3) F is reduced iff for all $p \in \text{Spec } R$, F_p is reduced.

(4) $g \xrightarrow{*} 0 \bmod F$ iff for all $p \in \text{Spec } R$, $g_p \xrightarrow{*} 0 \bmod F_p$.

PROOF. (1) - (3) and (4) " $\Rightarrow$ " are easy using 1.9. We show (4) " $\Leftarrow$ " by Noetherian induction on the relation $\xrightarrow[F]{}$: Let $0 \neq g \in S$ and pick $p \in \text{Spec } R$ with $g_p \neq 0$. Then there exists $h \in S$ with $g_p \longrightarrow h_p\ (a\cdot t) \xrightarrow{*} 0 \bmod F_p$. By section 0, there exists $e \in B(R)$ such that $e_p = 1$, and $g_{p'} \longrightarrow h_{p'}\ (a\cdot t) \xrightarrow{*} 0 \bmod F_{p'}$ for all $p' \in \text{Spec } R$ with $e_{p'} = 1$, and $h_{p'} = g_{p'}$ for all $p' \in \text{Spec } R$ with $e_{p'} = 0$. Put $g' = e'h + (1-e')g$. Then by (1), $g \xrightarrow[F]{} g'$, and for all $p \in \text{Spec } R$, $g'_p \xrightarrow{*} 0 \bmod F_p$. So by induction assumption, $g' \xrightarrow{*} 0 \bmod F$, and so $g \xrightarrow{*} 0 \bmod F$.

2. GRÖBNER BASES AND REDUCED GRÖBNER BASES. Among the many equivalent definitions of a Gröbner basis for polynomial ideals over fields, the following is quite natural: A finite list G of polynomials is a Gröbner basis (GB), if for every $f \in (G)$, $f \xrightarrow{*} 0 \bmod G$. So we take this as definition of a Groebner basis in S as well. G is a reduced Groebner basis, if G is reduced and a Groebner basis. For finite b.c. lists G, the equivalent definitions known for polynomials over fields are still valid:

LEMMA 2.1. Let G be a finite b.c. list of non-zero polynomials in S, and let $\longrightarrow$ be the induced reduction relation. Then the following assertions are equivalent:

(1) G is a Groebner basis.

(2) For all $0 \neq f \in (G)$, f is reducible mod G.

(3) For all $f,g \in S$, $f - g \in (G)$ implies $f \downarrow g$.

(4) $\longrightarrow$ is confluent.

PROOF. (1) $\Rightarrow$ (2) is trivial; for the converse, we may by 1.2 apply Noetherian induction on the relation $\longrightarrow$: If $0 \neq f$ is in (G), then by (2) there exists f' with $f \longrightarrow f' \in (G)$. By induction assumption, $f' \xrightarrow{*} 0$, and so $f \xrightarrow{*} 0$.

(1) $\Leftarrow$ (3): By (1), $f - g \in (G)$ implies $f - g \xrightarrow{*} 0$, and so by 1.6 $f \downarrow g$.

(3) $\Rightarrow$ (4): If $g \longleftarrow f \longrightarrow h$, then $g \xleftrightarrow{*} h$, and so by 1.8 $g - h \in (G)$, and so by (3) $g \downarrow h$. (4) $\Rightarrow$ (1): For $f \in (G)$, let f' be a normal form of f mod G. Then by 1.4, $f' = 0$.

Notice that the assumption that G is b.c. has been used essentially in 2.1. For reduced Gröbner basis this property is always true:

THEOREM 2.2. Let G be a finite list of non-zero polynomials in S. Then G is a reduced Gröbner basis iff G is reduced, b.c. and one of the equivalent properties (1) - (4) of 2.1 hold for G.

PROOF. By 2.1 it suffices to show that any reduced GB G is b.c.: Assume for a contradiction that for some $g \in G$, $BR(g) \neq 0$. Then g is irreducible mod $G \setminus (g)$, and so $BR(g) = (1-HI(g))\cdot g \in (G)$ is irreducible mod $G \setminus (g)$, and also mod (g). So $BR(g)$ is irreducible mod G contradicting 2.1.

A comparison of global Gröbner bases in S with pointwise Gröbner bases in S_p yields the following result:

THEOREM 2.3. Let G be a finite list of non-zero polynomials in S.
(1) If G is b.c., then G is a GB iff for all $p \in \text{Spec } R$, G_p is GB in S_p.
(2) G is a reduced GB iff G is b.c. and for all $p \in \text{Spec } R$, G_p is a reduced GB in S_p.

PROOF. (1) "$\Leftarrow$": Let $0 \neq f \in (G)$, and pick $p \in \text{Spec } R$ with $0 \neq f_p$. Then $f_p \in (G_p)$ and so f_p is reducible mod G_p. By 1.10 (2), this means that f is reducible mod G. "$\Rightarrow$": Let $f \in S$, $p \in \text{Spec } R$ with $0 \neq f_p \in (G_p)$. Then there exists $e \in B(R)$ such that $e\cdot f \in (G)$ and $e_p \neq 0$. Let $G' = \{g \in G : e\cdot f$ is reducible mod $(g)\}$ and let $e' = \bigsqcup \{HI(g) : g \in G'\}$. Then $e'_p = 1$, for otherwise $e\cdot(1-e')f \in (G)$ would be irreducible mod G, a contradiction. So f_p is reducible mod G_p.
(2) follows immediately from (1), 2.2 and 1.12 (3).

Reduced Gröbner bases for polynomial ideals over fields are unique (see [B2] or [KRW]). In the present situation, this is not the case: (X) and $(eX, (1-e)X)$ for $0,1 \neq e \in B(R)$ both are reduced Gröbner bases generating the same ideal. Of course, by 2.3(2), we have pointwise uniqueness:

PROPOSITION 2.4. Let G, H be reduced Gröbner bases in S generating the same ideal $I = (G) = (H)$. Then for all $p \in \text{Spec } R$, $G_p = H_p$.

In order to regain uniqueness, we define: A reduced Gröbner basis G in S is stratified, if for all $f,g \in G$, $p,p' \in \text{Spec } R$, with $f_p, g_p, f_{p'}, g_{p'} \neq 0$, $f_p \lesssim g_p$ iff $f_{p'} \lesssim g_{p'}$. This is equivalent to saying that for all $e,e' \in B(R)$ with $ef, eg, e'f, e'g \neq 0$, $e\cdot f \lesssim e\cdot g$ iff $e'\cdot f \lesssim e'\cdot g$. Since the quasi-order $<$ is a linear order on any reduced GB over a field, it follows that the same holds on any stratified reduced GB over R.

THEOREM 2.5. Let G, H be stratified, reduced Gröbner bases in S generating the same ideal $(G) = (H)$. Then $G = H$ (as sets).

PROOF. Assume for a contradiction that g is a minimal polynomial in $(G\setminus H) \cup (H\setminus G)$, and let $g \in G\setminus H$, say. By 2.4, for every $p \in \text{Spec } R$ there exists $h^{(p)} \in H$ with $g_p = (h^{(p)})_p$. By the compactness principle this implies that there exists a partition of 1 in B(R), $e_1,\ldots,e_n$, and $h_1,\ldots,h_n \in H$ such that $h_i \sim e_i \cdot h_i$, $g = e_1h_1 + \ldots + e_nh_n$, and $h_1 <\ldots< h_n$. If $n > 1$, the $g > h_1$, and so by the minimal choice of g, $h_1 \in G$, and $g_p = (h_1)_p$ for some $p \in$ Spec R. But then g is reducible mod h_1, a contradiction. So n must be 1, and hence $g = h_1 \in H$, which yields the desired contradiction.

In order to decide algorithmically, whether a given list G of polynomials in S is a GB, we need S-polynomials (comp. [B2]). Let $f,g \in S$ with $HM(f) = a\cdot s$, $HM(g) = b\cdot t$, and let $t' = u\cdot s = v\cdot t$ be the least common multiple of s and t in T. Then $SPol(f,g) = b\cdot u\cdot f + a\cdot v\cdot g$. We note that $SPol(f,g) = HI(f)\cdot SPol(f,g) = HI(g)\cdot SPol(f,g)$.

THEOREM 2.6. Let G be a finite b.c. list of polynomials in S. Then G is a Gröbner basis iff for all f,g in G with $f \neq g$, $SPol(f,g) \xrightarrow{*} 0$.

PROOF. 2.6 can be shown directly as for polynomials over fields (comp. [B2], [KRW]). We give a different proof by reduction to pointwise Gröbner bases: By 2.3, G is a GB iff for all $p \in$ Spec R, G_p is a GB in S_p iff for all $p \in$ Spec R and all $f \neq g$ in G, $SPol(f_p,g_p) = SPol(f,g)_p \xrightarrow{*} 0 \bmod G_p$ iff (by 1.12(4)) for all $f \neq g$ in G, $SPol(f,g) \xrightarrow{*} 0 \bmod G$.

Finally, we present algorithms computing Gröbner bases and stratified reduced Gröbner bases.

```
Algorithm (GRÖBNER)   Input  :  A finite b.c. list F of polynomials in S.
                      Output :  G, a b.c. Gröbner basis with (G) = (F).
BEGIN
  F' := F; H := F; H' := Ø;
  REPEAT
    REPEAT
      h := some element of H; H := H \ (h); F" := F';
      REPEAT
        f := some element of F"; F" := F" \ (f);
        g := a normal form of SPol(h,f) mod (F' ∪ H);
        IF g ≠ 0, THEN H' := H' ∪ BOOLEAN CLOSURE ({g})
      UNTIL F"= Ø
    UNTIL H = Ø;
    F' :=F' ∪ H'; H := H'; H' := Ø
  UNTIL H = Ø;
  G := F'
END
```

The fact that the algorithm is partially correct should be obvious from 2.6. Termination is more delicate and requires besides Dickson's lemma König's tree lemma ("Every finitely branching tree without infinite branches is finite.", see [Lv], IX. 2.17). We argue as follows: Let $(F',H) = (F'_1,H_1), (F'_2,H_2),\ldots$ be the sequence of pairs of lists produced by the outermost REPEAT-loop. For all $h \in H_i$, $k \in H_{i+1}$, we say (h,k) is in the relation ϱ, $h \,\varrho\, k$, if for some $f \in F_i$, $k \in BC(\{g\})$, where

SPol(h,f) $\xrightarrow{*}$ g ≠ 0 is obtained in the innermost REPEAT-loop. Then it is easy to see that the transitive closure of ϱ turns $G = \bigcup \{F'_i : i \in N\}$ into a finitely branching tree. Moreover, by definition, h ϱ k implies HI(h)·k = k, and so HT(k) < HT(h). So by Dickson's lemma, the tree has no infinite branch, and hence by König's lemma, G is finite.

The **algorithm** (STRAT-RED-GRÖBNER) outputs a stratified, reduced, b.c. Gröbner basis, when presented with a b.c. Gröbner basis F as input. Due to lack of space, we present only a sketch of the algorithm: It uses 3 auxilliary algorithms **BOOLEAN DECOMPOSITION, REDUCTION, STRATIFICATION** (BOOL DECOMP, RED, STRAT). RED replaces every polynomial f in the input list F by a quasi-monic normal form f' of f mod F∖{f}, and outputs all these normal forms. BOOL DECOMP computes from F the set A of atoms of the Boolean subalgebra of B(R) generated by {HI(f) : f ∈ F}, and replaces F by a list of blocks of polynomials generating together the same ideal as F, such that all polynomials in one block have the same highest coefficient in A. STRAT orders each of these blocks decreasingly, fills up with zeros to make their length equal, and sums up corresponding polynomials in all blocks in decreasing order.
The main algorithm (STRAT-RED-GRÖBNER) applies iteratively first BOOL DECOMP then RED to each block of polynomials, until this process becomes stationary (again by a combination of Dickson's and König's lemma). Finally, STRAT is applied to yield the desired result. The fact that the output is still a GB is established via 2.3.

3. APPLICATIONS. As in the case of polynomial rings over fields (see [B2]), the Gröbner basis technique has numerous applications in the algorithmic theory of polynomial ideals over regular rings. For lack of space, we present only some highlights:
(1) The ideal membership, inclusion and equality problem.
(2) Computing in residue class rings S/I.
(3) Computing the dimension function of an ideal.
(4) Computing generators for modules of syzygies.
As before R is a commutative regular ring and $S = R[X_1,...,X_r]$.

(1) IDEAL MEMBERSHIP, INCLUSION AND EQUALITY. The ideal membership problem for finitely generated (f.g.) ideals asks for a method to test for given f ∈ S and finite F ⊆ S, whether f ∈ (F). A solution is obtained as follows: Compute a b.c. Groebner basis G with (G) = (F) and a normal form f' of f mod G. Then f ∈ (F) iff f' = 0.
The ideal inclusion problem for f.g. ideals asks for a method to test for given finite F,F' ⊆ S, whether (F) ⊆ (F'). For a solution, compute a b.c. Gröbner basis G" of (F') and a normal form f' mod G'' for each f ∈ F. Then (F) ⊆ (F') iff f' = 0 for all f ∈ F.

(2) COMPUTING IN RESIDUE RINGS S/I. Let I be a f.g. ideal in S and let G be a b.c. Gröbner basis for I. For f ∈ S, F ⊆ S, [f] denotes the residue class of f mod I, and

$[F] = \{[g] : g \in F\}$. For every $G' \subseteq G$, we let $e(G') = 1 - \bigsqcup \{HI(g) : g \in G'\}$, and put $B(G') = \{e(G')\cdot t : t \in T,\ HT(g) \mid t$ for $g \in G'$, not $HT(g) \mid t$ for $g \in G\setminus G'\}$, $B = \bigcup \{B(G') : G' \subseteq G\}$.

THEOREM 3.1. The map $B \longrightarrow [B]$, $b \longmapsto [b]$ is bijective, and S/I is the sum of the free $e(G')\cdot R$-modules $e(G')\cdot S\ /\ e(G')\cdot I$ $(G' \subseteq G)$ having $[B(G')]$ as a basis.

PROOF. Let $0 \neq f \in S$ be in normal form mod G. Then every monomial $a\cdot t$ occuring in f is in normal form mod G. Let $G'' = \{g \in G : HT(g) \mid t\}$. Then $e(G'')\cdot a\cdot t = a\cdot t$ and $e(G'')\cdot t \in B(G'')$. So $[B]$ generates S/I. If for fixed $G' \subseteq G$, some linear combination g of monomials of the form $a\cdot e(G')\cdot t$ with $t \in B(G')$ is in I, then it reduces to 0 mod G, and so one of these monomials is reducible mod G, which is possible only if $a\cdot e(G') = 0$. So the elements of $B(G')$ are independent mod $e(G')\cdot I$.

COROLLARY 3.2. The following assertions are equivalent:
(1) S/I is finitely generated as R-module.
(2) For all $1 \leq i \leq r$, $\bigsqcup \{HI(g) : g \in G$, there exist $k_i \in N$ with $X_i^{k_i} \in HT(g)\} = 1$.
(3) $\dim_{R_p} (S_p/I_p) < \infty$ for all $p \in \mathrm{Spec}\ R$.

(3) DIMENSION FUNCTION OF AN IDEAL. Let I be a f.g. ideal in S with b.c. Gröbner basis G. We define as in [SW], $\dim I : \mathrm{Spec}\ R \longrightarrow N \cup \{\infty\}$ by $(\dim I)(p)$ = dimension of I_p in S_p (see [B2], [KW]). By [KW], the values of the function dim I can be computed easily from the values of HM(g), $g \in G$, at the points $p \in \mathrm{Spec}\ R$. In particular, we get:

THEOREM 3.3. dim I is a locally constant function which is constant on every atom of the Boolean algebra generated by $\{HI(g) : g \in G\}$.

This extends a corresponding result for prime ideals (theorem I.3.14) proved in [SW] by a different method.

(4) GENERATORS FOR MODULES OF SYZYGIES. Let $G = \{g_1,\dots g_m)$ be a b.c. Gröbner basis in S consisting of quasi-monic polynomials, and let $M = \{(h_1,\dots,h_m) \in S^m : h_1g_1 +\dots+ h_mg_m = 0\}$ be the corresponding S-module of syzygies. Let $f_{ij} = \mathrm{SPol}(g_i,g_j) = u_ig_i - u_jg_j$, where $u_i, u_j \in T$, $1 \le i<j \le m$. Since $f_{ij} \xrightarrow{*} 0$ mod G, one can compute polynomials $q_{ijk} \in S$ such that $f_{ij} = q_{ij1}g_1 +\dots+ q_{ijm}g_m$ and $HT(q_{ijk}g_k) \leq HT(f_{ij}) < HT(u_ig_i), HT(u_jg_j)$. Put $r_{ijk} = q_{ijk}$ for $k \neq i,j$, $r_{iji} = q_{iji} - u_i$, $r_{ijj} = q_{ijj} + u_j$, and put $b_{ij} = (r_{ij1},\dots,r_{ijm}) \in S^m$, $B = \{b_{ij} : 1 \leq i<j \leq m\}$.

THEOREM 3.4. B generates M as S-module.

PROOF. It is well-known (see [B2], [AL]), that for Gröbner bases G in polynomial rings over fields B generates M. So by 2.3, B_p generates M_p as S_p-module for all $p \in$ Spec R. By the facts in section 0, this implies that B generates M as S-module.

REFERENCES.

[AL] J.Apel,W.Lassner, An extension of Buchberger's algorithm and calculations in envelopping fields of Lie algebras, J.Symb.Comp., to appear.

[AK] D.Armbruster,H.Kredel, Constructing Universal Unfoldings using Gröbner Bases, J.Symb.Comp. 2 (1986), pp. 383-388.

[B1] B.Buchberger, A critical-pair/completion algorithm for finitely generated ideals in rings, in Logic and machines: Decision problems and complexity, Muenster 1983, Springer LNCS vol. 171.

[B2] B.Buchberger, Gröbner bases: An algorithmic method in polynomial ideal theory, chap. 6 in Recent trends in multidimensional system theory, N.K.Bose, Ed., Reidel Publ. Comp., 1985.

[FSK] A.Furukawa,T.Sasaki,H.Koboyashi, Gröbner bases of a module over $K[x_1,...,x_n]$ and polynomial solutions of a system of linear equations, Proc. ACM SYMSAC '86, Waterloo 1986, pp. 222-224.

[H] G.Huet, Confluent reductions: Abstract properties and applications to term rewriting systems, J. ACM 27 (1980), pp. 797-821.

[KRK] A.Kandri-Rody,D.Kapur, Algorithms for computing Gröbner bases of polynomial ideals over various Euclidean rings, in EUROSAM '84, Springer LNCS vol.174, pp. 195-205.

[KRW] A.Kandri-Rody,V.Weispfenning, Non-commutative Gröbner bases in algebras of solvable type, 1986, J.Symb.Comp., to appear.

[KN1] D.Kapur,P.Narendran, Constructing a Gröbner basis for a polynomial ideal, presented at the workshop Combinatorial Algorithms in Algebraic Structures, Otzenhausen, Sept. 1985.

[KN2] D.Kapur,P.Narendran, An equational approach to theorem proving in first-order predicate calculus, GE CRD, Schenectady, N.Y., Sept. 1985.

[KW] H.Kredel,V.Weispfenning, Computing dimension and independent sets for polynomial ideals, J.Symb.Comp., to appear.

[Lv] A.Levy, Basic Set Theory, Springer Verlag 1979.

[Lo] G.Loullis, Sheaves and boolean valued model theory, Journal of Symbolic Logic 44 (1979), pp. 153-183.

[Moe] H.M.Möller, On the computation of Gröbner bases in commutative rings, 1985, submitted.

[MM] H.M.Möller,F.Mora, New Constructive Methods in Classical Ideal Theory, J.Algebra 100 (1986), pp. 138-178.

[Mr] F.Mora, Gröbner bases for non-commutative polynomial rings, in AAECC-3, Grenoble 1985, Springer LNCS 229, pp. 353-362.

[Pa] L.Pan, On the D-bases of ideals in polynomial rings over principal ideal domains, manuscript Sept. 1985.

[Pi] R.S.Pierce, Modules over commutative regular rings, Memoirs of the AMS, vol. 70, 1967.

[SW] D.Saracino,V.Weispfenning, On algebraic curves over commutative regular rings, in Model Theory and Algebra, a memorial tribute to A.Robinson, Springer LNM vol. 489, pp. 307-383.

[W] V.Weispfenning, Model-completeness and elimination of quantifiers for subdirect products of structures, J. Algebra 36 (1975), pp. 252-277.

SOME ALGEBRAIC ALGORITHMS BASED ON HEAD TERM ELIMINATION OVER POLYNOMIAL RINGS

Tateaki Sasaki

The Institute of Physical and Chemical Research
Wako-shi, Saitama 351-01, Japan

Let F_1, ..., F_r be polynomials in $R[X_1,...,X_n]$ with $R = K[u_1,...,u_m]$. Many algebraic problems can be reduced to calculating polynomials in $K[u_1,...,u_m]$ by eliminating X_1, ..., X_n from F_1, ..., F_r. We formulate this elimination as a construction of a Gröbner basis with the total-degree order for variables X_1, ..., X_n. We apply this elimination for several typical problems and present efficient algorithms. Furthermore, some ideas which make the elimination efficient are presented, with timing data by actual implementation.

§1. Introduction

Many algebraic problems are reduced to the elimination of variables. The conventional elimination method is the leading term elimination. For polynomials F and G in main variable X, the leading term elimination is defined by the formula

$$[\mathrm{lcm}/\mathrm{lt}(F)]\cdot F - [\mathrm{lcm}/\mathrm{lt}(G)]\cdot G, \quad \mathrm{lcm} = \mathrm{LCM}(\mathrm{lt}(F),\mathrm{lt}(G)), \tag{1}$$

where lt and LCM denote the "leading term" and "least common multiple", respectively. (The resultant calculation is a successive application of the leading term elimination.)

Another kind of elimination is the head term elimination which plays an essential role in the construction of Gröbner basis of the polynomial ideal [Buch65]. (For precise definition of "head term", see §2.) For polynomials F and G with coefficients in a field, the head term elimination is defined by the formula

$$[\mathrm{lcm}/\mathrm{ht}(F)]\cdot F - [\mathrm{lcm}/\mathrm{ht}(G)]\cdot G, \quad \mathrm{lcm} = \mathrm{LCM}(\mathrm{ht}(F),\mathrm{ht}(G)), \tag{2}$$

where ht denotes the head term. Note the similarity between (1) and (2).

Although the leading term elimination is employed in many algorithms, the superiority of head term elimination is being recognized by many researches. This is because that the head term elimination is more general than the leading term elimination in that the latter can be attained by successive application of the former. Furthermore, we can choose the term order variously for head term elimination, while the term order for leading term elimination is unique when we set the main variable.

In many algorithms, the following elimination is required: given polynomials F_1, ..., F_r in $R[X_1,...,X_n]$, where $R = K[u_1,...,u_m]$, calculate polynomial(s) in $K[u_1,...,u_m]$ by eliminating X_1, ..., X_n from F_1, ..., F_r. Following Buchberger [Buch65,85], we formulate this elimination as a construction of Gröbner basis by the head term elimination. In §2, we briefly survey the construction of Gröbner basis of the polynomial ideal over polynomial ring R. In §3, we apply the head term elimination for reducing a system of algebraic equations, calculating algebraic relations, and so on. In many cases of eliminations, we are unnecessary to calculate the full set of Gröbner basis but to calculate only a subset,

(By this reason we do not call successive application of head term elimination the construction of Gröbner basis, although we formulate it in terms of the Gröbner basis.) Exploiting this, we can avoid unnecessary computation and save the time largely, which is explained in §4. The algorithms to be presented in this paper have been implemented on the Japanese algebra system GAL, and §4 shows the timing data.

§2. Head term elimination over polynomial ring

We denote the set of nonnegative integers by $\mathbb{Z}_0$ and the Cartesian product $\mathbb{Z}_0\times\cdots\times\mathbb{Z}_0$ by $\mathbb{Z}_0^r$. Let K be a field and R a ring $K[u_1,\ldots,u_m]$. We abbreviate the rings $K[u_1,\ldots,u_m]$ and $R[X_1,\ldots,X_n]$ to $K[u]$ and $R[X]$, respectively. Similarly, we abbreviate monomials $cu_1^{a_1}\cdots u_m^{a_m}$ and $CX_1^{A_1}\cdots X_n^{A_n}$, where $c\in K$ and $C\in R$, to cu^a and CX^A, respectively. The ideal generated by $F_1, \ldots, F_r$ is denoted by $(F_1,\ldots,F_r)$.

Definition 1 [order $>$ for elements of $\mathbb{Z}_0^r$]. Let $a = (a_1,\ldots,a_r)$ and $b = (b_1,\ldots,b_r)$ be elements of $\mathbb{Z}_0^r$. We define $a > b$ iff there exists an integer k such that $a_k > b_k$ and $a_i = b_i$, $i=1,\ldots,k-1$ when $k > 1$. //

Definition 2 [order $\triangleright$ for monomials in K[u] and R[X]]. Let $t_a = c_a u_1^{a_1}\cdots u_m^{a_m}$ and $t_b = c_b u_1^{b_1}\cdots u_m^{b_m}$ be monomials in K[u], where $c_a, c_b \in K$. We define the lexicographic order $\triangleright$ as $t_a \triangleright t_b$ iff $(a_1,\ldots,a_m) > (b_1,\ldots,b_m)$. We define the total-degree order $\triangleright$ as $t_a \triangleright t_b$ iff $(\Sigma a_i, a_1,\ldots,a_m) > (\Sigma b_i, b_1,\ldots,b_m)$. Similarly, we define the lexicographic order $\triangleright$ and the total-degree order $\triangleright$ for monomials T_A and T_B in R[X]. //

Definition 3 [order $\gg$ for monomials in K[u,X]]. Let $T_a = c_a u^a X^A$ and $T_b = c_b u^b X^B$ be monomials in K[u,X], where $c_a, c_b \in K$. We define $T_a \gg T_b$ iff either $X^A \triangleright X^B$ or $X^A = X^B$ and $u^a \triangleright u^b$. //

Definition 4 [head term of polynomials in K[u,X]]. The head term of a polynomial F in K[u,X], to be abbreviated to ht(F), is the highest order term of F in the sense of $\gg$. //

Definition 5 [Gröbner basis of an ideal in K[u][X]]. We regard the polynomials in K[u][X] as polynomials in K[u,X]. Then, with the definitions of order $\gg$ and head term, we can construct a Gröbner basis of a given ideal in K[u,X] according to Buchberger's procedure, which we define a Gröbner basis in K[u][X]. //

Definition 6 [reduced Gröbner basis]. A Gröbner basis $\Gamma = \{G_1,\ldots,G_s\}$ is a reduced basis if $ht(G_i) \nmid ht(G_j)$ for any elements G_i and G_j in Γ, $i\neq j$. //

It is needless to say that the head term elimination (2) is nothing but the "S-polynomial" and, when $ht(G) \mid ht(F)$, (2) defines the "M-reduction" of ht(F) by G. Furthermore, successive application of head term elimination terminates and gives a Gröbner basis of the given ideal. By the head term elimination in K[u][X], with the term order $\gg$, the variables $X_1, \ldots, X_n$ are eliminated first and resulting polynomials are in K[u]. This leads to the following well-known theorem (see, for example, [Buch85], Lemma 6.8, which gives a proof for the case of lexicographic order).

Theorem 1 [well-known]. Let $\Gamma = \{G_1,\ldots,G_s\}$ be a Gröbner basis of ideal $(F_1,\ldots,F_r)$ in K[u][X] with the term order $\gg$ (not necessarily lexicographic). Then,

$$\mathrm{Ideal}(\Gamma)\cap K[u] = \mathrm{Ideal}(\Gamma\cap K[u]). \quad // \qquad (3)$$

We note that, when applying formula (2) for $\{F_1,\dots,F_r\}$, we may choose the term order $\rhd$ and the pair $<F_i,F_j>$ arbitrarily. However, the efficiency of elimination depends on the choice strongly and we select the following choices actually.

Choice 1. We choose the total-degree order as far as possible, because this order is most desirable for efficient elimination, as is well-known empirically.

Choice 2. In applying (2) to pairs in $\{F_1,\dots,F_r\}$, we choose the lowest order pair $<F_i,F_j>$ that is not applied by the formula yet.

§3. Applications of head term elimination

In this section, we apply the head term elimination, with the total-degree order for $\{X_1,\dots,X_n\}$, to four typical algebraic calculations.

3.1. Reducing a system of algebraic equations

Consider solving a system of algebraic equations

$$\{P_1 = 0, \cdots, P_r = 0\}, \tag{4}$$

where $P_i \in K[X]$, $i=1,\dots,r$, and we assume that $(P_1,\dots,P_r)$ is zero-dimensional, i.e., the dimension of solution space is 0.

There are various ways to solve (4), and a practically very useful way is to derive equations $Q_1(X_1) = 0$ and $Q_i(X_i,X_1) = 0$, $i=2,\dots,n$, such that $ht(Q_i) = X_i^{\lambda_i}$, where λ_i is made as small as possible. We can obtain $Q_1(X_1)$ by calculating a reduced Gröbner basis with the lexcographic order $X_n \rhd \cdots \rhd X_2 \rhd X_1$. If we use the ordering $X_n, \dots, X_2 \rhd X_1$, with the total-degree order for $\{X_n,\dots,X_2\}$, then we can obtain $Q_1(X_1)$ efficiently. Similarly, we can obtain $Q_2(X_2,X_1)$ efficiently by calculating a reduced Gröbner basis with the ordering $X_n, \dots, X_3 \rhd X_2 \rhd X_1$, where the total-degree order is applied for $\{X_n,\dots,X_3\}$, and so on. It is important to note that, if $\lambda_i = 1$ for $i=2,\dots,n$, the above Gröbner basis calculation for Q_1 gives $Q_2, \dots, Q_n$ simultaneously as the following proposition assures.

Proposition 1. Let $F_i \in K[X]$, $i=1,\dots,r$, and $(F_1,\dots,F_r)$ be zero-dimensional. Let $\Gamma = \{G_1,\dots,G_s\}$ be a reduced Gröbner basis of $(F_1,\dots,F_r)$, where Γ is constructed by treating $F_1, \dots, F_r$ as elements in $R[X_2,\dots,X_n]$ with $R = K[X_1]$. Then, for each $i=1,\dots,n$, Γ contains an element of the form $X_i^{\nu_i}$ + (less order terms) which is the lowest order element among the polynomials $Q_i(X)$ such that $Q_i \in (F_1,\dots,F_r)$ and $ht(Q_i) = c_iX_i^{\lambda_i}$, $c_i \in K$.

(Proof) Since the ideal is zero-dimensional, Buchberger's theorem (see [Buch85], Method 6.9) assures that Γ contains elements of the form $X_i^{\nu_i}$ + (less order terms), $i=1,\dots,n$, and the number of such elements is only one for each i because Γ is a reduced basis. Since Γ is a Gröbner basis, the above Q_i must be reduced to 0 by successive M-reduction by elements in Γ, which means $\nu_i \le \lambda_i$. //

Corollary. If $\nu_i = 1$ for $i=2,\dots,n$ (which happens quite often actually), then Γ is composed of only elements $Q_1(X_1)$ and $Q_i = X_i - \tilde{Q}_i(X_1)$, $i=2,\dots,n$. //

Algorithm A (reducing a system of algebraic equations).

Input : Polynomials $P_1, \dots, P_r$ in $K[X]$;

Output: Polynomials $Q_1(X_1)$ and $Q_i(X_i,X_1)$, $i=2,\dots,n$, such that $Q_1,Q_i \in (P_1,\dots,P_r)$ and

$ht(Q_1) = X_1^{\nu_1}$ with as small ν_1 as possible;

Step 1: Treat $P_1, \ldots, P_r$ as elements in $R[X_2,\ldots,X_n]$, $R = K[X_1]$, and calculate a reduced Gröbner basis Γ_1 of $(P_1,\ldots,P_r)$ with the total-degree order for $\{X_2,\ldots,X_n\}$; If $\Gamma_1 = \{Q_1(X_1), X_2-\tilde{Q}_2(X_1), \ldots, X_n-\tilde{Q}_n(X_1)\}$ then return Γ_1;

Step 2: Treat $P_1, \ldots, P_r$ as elements in $R[X_3,\ldots,X_n]$, $R = K[X_1,X_2]$, and calculate a reduced Gröbner basis Γ_2 of $(P_1,\ldots,P_r)$ with the total-degree order for $\{X_3,\ldots,X_n\}$ and lexicographic order for $\{X_2,X_1\}$;

Similarly, calculate reduced Gröbner bases Γ_i, $i=3,\ldots,n$;

Return: Correct required elements from $\Gamma_1, \ldots, \Gamma_n$, and return them. //

3.2. Calculating U-resultant

Adding another equation $P_0 = u_0 + u_1X_1 + \cdots + u_nX_n$ to (4), where $u_0, \ldots, u_n$ are indeterminates, we can derive resultants which are homogeneous polynomials in $u_0, \ldots, u_n$. Among them, the lowest total-degree one is the U-resultant. The classical method of calculating the U-resultant is inefficient. In 1977, Lazard presented a practical method [Laza77], and Lazard's method has been made efficient by Kobayashi et al. by using the Gröbner basis [K&F&F86]. (This paper also presents a practical method for factoring the U-resultant into linear factors.) However, the calculation is still complicated.

Our method using the head term elimination is quite simple, yet it is efficient for small-sized problems. (For large-sized problems, the method of Kobayashi et al. will be better.) Our method calculates the U-resultant by eliminating $X_1, \ldots, X_n$ with the total-degree order for both $\{X_1,\ldots,X_n\}$ and $\{u_0,\ldots,u_n\}$ as follows.

Algorithm B (calculating U-resultant).

Input : $P_1, \ldots, P_r$ in $K[X]$ and indeterminates $u_0, u_1, \ldots, u_n$;

Output: U-resultant of $\{P_1=0, \cdots, P_r=0\}$;

Method: Treat $P_0, P_1, \ldots, P_r$ as elements in $R[X]$, $R = K[u]$, and calculate a Gröbner basis Γ of ideal $(P_0,P_1,\ldots,P_r)$ with the total-degree order for both $\{X_1,\ldots,X_n\}$ and $\{u_0,\ldots,u_n\}$;

Return: Return the lowest order element G such that $G \in \Gamma \cap K[u]$. //

3.3. Calculating algebraic relations

Let $P_1, \ldots, P_r$ be polynomials in $K[X]$, and let $\rho(u_1,\ldots,u_r)$ be a polynomial in $K[u_1,\ldots,u_r]$. If $\rho(P_1,\ldots,P_r) = 0$, then ρ is an algebraic relation of $P_1, \ldots, P_r$.

The calculation of the algebraic relations is based on the following proposition.

Proposition 2 [well-known]. $\rho(P_1,\ldots,P_r)$ is an algebraic relation of $P_1, \ldots, P_r$ $\Longleftrightarrow$ $\rho(u_1,\ldots,u_r) \in (P_1-u_1, \cdots, P_r-u_r)$, where $u_1, \ldots, u_r$, are new indeterminates.

(Proof) We give an elementary proof.

$\Longrightarrow$: Let $\rho(P_1,\ldots,P_r)$ be an algebraic relation of $P_1, \ldots, P_r$. Rewriting the polynomial ρ as $\rho((P_1-u_1)+u_1,\ldots,(P_r-u_r)+u_r)$ and expanding it around $u_1, \ldots, u_r$, we obtain $\rho(u_1,\ldots,u_r) + \Sigma C_i(P_i-u_i) = 0$, $C_i \in K[u][X]$. Hence, $\rho(u_1,\ldots,u_r) \in (P_1-u_1, \cdots, P_r-u_r)$.

$\Longleftarrow$: Let $\rho(u_1,\ldots,u_r) \in (P_1-u_1, \cdots, P_r-u_r)$, hence we can write ρ as $\rho = \Sigma C_i(P_i-u_i)$.

Substituting P_i for u_i, $i=1,\ldots,r$, in $\rho(u_1,\ldots,u_r)$, we obtain $\rho(P_1,\ldots,P_r) = 0$. //

Algebraic relations can be calculated efficiently by using the total-degree order for both $\{X_1,\ldots,X_n\}$ and $\{u_1,\ldots,u_r\}$ as follows.

Algorithm C (calculating algebraic relations).

Input : Polynomials $P_1, \ldots, P_r$ in $K[X]$;

Output: Gröbner basis of the ideal composed of algebraic relations;

Method: Treat $P_1-u_1, \cdots, P_r-u_r$ as elements in $R[X]$, $R = K[u]$, and calculate a Gröbner basis Γ of the ideal $(P_1-u_1, \cdots, P_r-u_r)$ with the total-degree order for both $\{X_1,\ldots,X_n\}$ and $\{u_1,\ldots,u_r\}$;

Return: Return $\Gamma \cap K[u]$. //

3.4. Representing a polynomial by other polynomials

Given polynomials P and $P_1, \ldots, P_r$ in $K[X]$, we want to determine whether there exists a polynomial $Q(u_1,\ldots,u_r)$ in $K[u_1,\ldots,u_r]$ such that $P = Q(P_1,\ldots,P_r)$, and we want to determine Q when exists. We see that this calculation is a special case of calculating an algebraic relation described in 3.3. Therefore, we have the following algorithm.

Algorithm D (polynomial decomposition).

Input : Polynomials P, $P_1, \ldots, P_r$ in $K[X]$;

Output: Polynomial $Q(u_1,\ldots,u_r)$, if exists, such that $P = Q(P_1,\ldots,P_r)$;

Method: Treat P, $P_1, \ldots, P_r$ as elements in $R[X]$, $R = K[u]$, and calculate a Gröbner basis Γ of $(P_1-u_1, \cdots, P_r-u_r)$ with the total-degree order for both $\{X_1,\ldots,X_n\}$ and $\{u_1,\ldots,u_r\}$. Then, $Q \leftarrow$ M-reduction of P by elements in Γ;

Return: If $Q \in K[u]$ then return Q else return NIL. //

§4. Devices for efficient elimination

Although we have formulated the variable elimination as the construction of a Gröbner basis, we need not always calculate the full set of Gröbner basis but may stop the computation when the required elimination is accomplished. For example, when we are calculating one algebraic relation, we may stop the computation when all the variables $X_1, \ldots, X_n$ are eliminated. This truncation will save the computation time drastically, as we will see from the actual timing data. Note that, with Choice 2 given in §2, the required elimination will be performed by avoiding wasteful computation as far as possible.

The second device is removal of monomial factors from the head term eliminations, which is quite easy to execute. For example, monomial factors in algebraic relations are meaningless and we can remove them. Note that, even if such a monomial factor is meaningful, we can often remove it so long as the removed factor is processed suitably. For example, suppose $P_i = \tilde{P}_i X_k^a$ in Algorithm A, then we can split the system of algebraic equations into two systems as $\{P_1=0, \cdots, P_{i-1}=0, \tilde{P}_i=0, P_{i+1}=0, \cdots, P_r=0\}$ and $\{P_1=0, \cdots, P_{i-1}=0, X_k^a=0, P_{i+1}=0, \cdots, P_r=0\}$. Hence, only if the latter system is also solved, we can remove the monomial factor X_k^a from P_i.

Performing successive head term eliminations with removal of monomial factor from each elimination does not give Gröbner basis, but the procedure terminates and we can calculate required polynomials in K[u]. In our current implementation of Gröbner basis package on the algebra system GAL, the following three modes are prepared for controlling the truncation of computation.

(T0) No truncation (default mode);

(T1) Truncate the computation when variables $X_1, \ldots, X_n$ are eliminated;

(T2) Truncate the computation when variables $X_1, \ldots, X_n$ are eliminated and a polynomial in K[u] of a specified total-degree is obtained.

Furthermore, the user can choose one of the following three modes for controlling the removal of the monomial factors.

(R0) No removal (default mode);

(R1) Monomial factors in K[u] are removed;

(R2) Monomial factors in K[u,X] are removed.

With the above-mentioned devices, algorithms B and C can be specialized as follows.

Algorithm B' (U-resultant of a known total-degree, containing no monomial factor).

Perform Algorithm B with modes (T2) and (R1). //

Algorithm C' (calculating one algebraic relation).

Perform Algorithm C with modes (T1) and (R1). //

Let us show the timing data for the above algorithms.

Problem 1 (reducing a system of algebraic equations).

$$P_1 = 2(X_4^2 + X_3^2 + X_2^2 + X_1^2) - X_1 = 0,$$
$$P_2 = 2(X_4X_3 + X_3X_2 + X_2X_1) - X_2 = 0,$$
$$P_3 = 2(X_4X_2 + X_3X_1) + X_2^2 - X_3 = 0,$$
$$P_4 = 2(X_4 + X_3 + X_2) + X_1 - 1 = 0.$$

This problem is taken from a theory of spin grass by Katsura et al.

Problem 2 (calculating a U-resultant).

$$P_1 = X_1^2 + X_2^2 - 2 = 0, \quad P_2 = X_1X_2 - 1 = 0.$$

Adding $P_0 = u_0 + u_1X_1 + u_2X_2 = 0$ to the above system, we can calculate the following polynomial as the U-resultant:

$$U(u_0,u_1,u_2) = u_0^4 - 2u_0^2u_1^2 - 4u_0^2u_1u_2 - 2u_0^2u_2^2 + u_1^4 + 4u_1^3u_2 + 6u_1^2u_2^2 + 4u_1u_2^3 + u_2^4.$$

Problem 3 (calculating an algebraic relation).

$$P_1 = (X_1^6 + X_2^6) + 522(X_1^5X_2 - X_1X_2^5) - 10005(X_1^4X_2^2 + X_1^2X_2^4),$$
$$P_2 = -(X_1^4 + X_2^4) + 228(X_1^3X_2 - X_1X_2^3) - 494X_1^2X_2^2,$$
$$P_3 = X_1X_2(X_1^2 + 11X_1X_2 - X_2^2)^5.$$

The algebraic relation of these polynomials is $P_1^2 + P_2^3 - 1728P_3 = 0$. This problem is taken from a Klein's book discussing the symmetry of regular polyhedra.

Problem 4 (polynomial decomposition).

$$P_1 = X_1^2X_2 + 2X_1^2 - 3X_1X_2 + 5X_2,$$
$$P_2 = 2X_1^3 - 4X_1^2X_2 - 3X_1X_2^2 + X_2^3,$$
$$P = -2X_1^9X_2^3 + 4X_1^8X_2^4 + 3X_1^7X_2^5 + \cdots + 4X_1^2 - 6X_1X_2 + 10X_2.$$

Given these polynomials (P is composed of 49 terms), we derive a polynomial $Q(u_1,u_2)$ such that $P = Q(P_1,P_2)$. In this example, $Q = - u_1^3u_2 + u_1u_2^2 + 2u_1 - 3u_2$.

Algorithm	Prob. 1	Prob. 2	Prob. 3	Prob. 4
A − D	607	782	121	119
B'− C'	XXXX	27	48	XXXX

Table I. Timing data (in milliseconds)

Table I shows the timing data, where the computation has been done by GAL on a FACOM-M780 computer. We see that the truncation mode is often quite effective. This effectiveness is due partly to Choice 1 given in §2 and partly to skipping the termination check in Buchberger's procedure. For Problem 2, for example, the elimination has been performed after constructing 11 S-polynomials while we must construct 16 S-polynomials for the Gröbner basis calculation. The removal mode is effective only for Problem 2 in our test. In mode (R1), we obtain the U-resultant just when X_1 and X_2 are eliminated, with successive removal of monomials u_2, u_1, and u_1. On the other hand, in mode (R0), we obtain $u_1^2u_2 \cdot U(u_0,u_1,u_2)$ just when X_1 and X_2 have been eliminated, and we have to construct 19 more S-polynomials to get $U(u_0,u_1,u_2)$. We have also solved Problem 2 by applying Algorithm B with modes (T0) and (R1) (i.e., no truncation but removal of monomial factors), and the computation time was 516 milliseconds. This shows that the removal mode is also effective considerably. Although the above test is restricted within a small number of examples which are of small-sized, we have seen that our devices were quite effective in many cases. The devices will become more effective for larger-sized problems.

Acknowledgement. The author thanks Prof. Kobayashi and Mr. Furukawa for stimulating discussions and comments.

References

[Buch65] B. Buchberger, "An algorithm for finding a basis for the residue class ring of a zero-dimensional polynomial ideal (German)", Ph.D. Thesis, Math. Inst., Univ. of Innsbruck (Austria), 1965.

[Buch85] B. Buchberger, "Gröbner bases : An algorithmic method in polynomial ideal theory", in Multidimensional Systems Theory (ed. R. Bose), Reidel, 1985, Ch.6.

[K&F&F86] H. Kobayashi, T. Fujise, and A. Furukawa, "Solving systems of algebraic equations by a general elimination method", J. Sym. Comp. Vol.5 (1988), pp. 303-320.

[Laza77] D. Lazard, "Algèbre linéaire sur $K[x_1,x_2,\ldots,x_n]$ et éllimination", Bull. Soc. Math. France, No. 105 (1977), pp. 165-190.

ALGORITHMIC DETERMINATION OF THE JACOBSON RADICAL OF MONOMIAL ALGEBRAS

Tatiana Gateva-Ivanova
Institute of Mathematics
Bulgarian Academy of Sciences
P.O. Box 373, Sofia 1090, Bulgaria

ABSTRACT. The paper considers computer algebra in a non-commutative setting. Under investigation are finitely presented associative monomial algebras and some of their recognizable properties. In [1] it was shown that for a monomial algebra A the properties of being semi-simple (in the sense of Jacobson), prime, or semi-prime are recognizable. In the paper a new interpretation of these properties is given in terms of the Ufnarovsky graph of A. This provides better algorithms for their verification. It is proved that the Jacobson radical of A is finitely generated as an ideal. It is also proved that the algebra A is semi-prime if and only if it is semi-simple in the sense of Jacobson.

1. INTRODUCTION

In this paper K denotes a fixed field of arbitrary characteristic, and the term K-algebra is used to denote an associative algebra over K.

1.1 A monomial algebra is a (finitely presented) algebra which has a presentation as $A = K\langle X\rangle /I$, where $X = \{x_1,\dots,x_n\}$ is a set of indeterminates, $K\langle X\rangle$ is the free K-algebra on it, and the ideal of relations $I = (W)$ is generated as a two-sided ideal by a finite set of monomials $W = \{w_1,\dots,w_t\}$. Throughout the paper A denotes a monomial algebra in which the generators x_i and the monomials w_j are fixed. Furthermore, we assume each w_j is of degree ≥ 2.

1.2 A monomial is called normal (mod I) if it does not contain as a subword any of the w_i's. We say a polynomial f is normal (mod I) if either $f = 0$, or if $f = \sum_{i=1}^{s} \alpha_i f_i$ with $\alpha_i \in K\setminus\{0\}$, f_i is a normal monomial for $i = 1,\dots,s$, and $f_i \neq f_j$ for $i \neq j$.

Partially supported by Contract No. 62/1987, Committee of Science, Bulgaria.

It is obvious that the set of normal (mod I) monomials in K<X> projects to a K-basis of A.

We are interested in such algebraic properties of A which are recognizable, i.e. can be established algorithmically.

It was shown in [1] that for a monomial algebra various properties such as being finite-dimensional, nil, nilpotent, semi-simple (in the sense of Jacobson), prime, semiprime are recognizable.

Here we give a new interpretation of the semi-simplicity (in the sense of Jacobson), primeness, and semi-primeness of a monomial algebra and supply better algorithms for checking them. We prove, cf. Theorem 2.19, that the Jacobson radical of A is generated as an ideal by a finite set of normal monomials. Theorem 2.25 shows that for a monomial algebra the properties of being semi-simple and semi-prime are equivalent.

We should like to thank Victor Ufnarovsky for several useful discussions and for suggesting improvements of the text.

2. JACOBSON RADICAL, SEMI-SIMPLICITY, PRIMENESS, SEMI-PRIMENESS OF MONOMIAL ALGEBRAS

We now introduce some technique which will be used in the rest of the paper.

2.1 Following Ufnarovsky, cf. [2], the monomial algebra A determines an oriented graph $U = U(A)$ by the following procedure. The vertices of U are the degree m ($m+1 = \max \{ \deg w_i \mid 1 \leq i \leq t \}$) normal monomials in K<X>. If u and v are such monomials, then there is an edge $u \to v$ from u to v if and only if there exist variables $x,y \in X$, such that $ux = yv \notin I$. We shall refer to U(A) as the Ufnarovsky graph of A.

2.2 A route of length d, $d \geq 1$, in the graph U is a sequence of edges $v_0 \to v_1 \to v_2 \to \dots \to v_{d-1} \to v_d$. A route is called simple if it contains no edge twice. A cycle is a subgraph of U such that its vertices $v_0, \dots, v_d$ satisfy $d \geq 1$, $v_0 = v_d$ and $v_0 \to v_1 \to \dots \to v_{d-1} \to v_d$ is a simple route. A route R is called cyclic if the underlying graph is a cycle, we call it the cycle associated to R . A vertex is cyclic if it belongs to a cycle.

2.3 REMARK. ([2]). There is a one to one correspondence R between the nonzero monomials of length $> m$ in A and the routes in U(A): the monomial $w = x_{i_1} \dots x_{i_s}$ is mapped to the route $R(w) = v_0 \to v_1 \to \dots \to$

where $d = s-m$, and $v_j = x_{i_{j+1}} x_{i_{j+2}} \ldots x_{i_{j+m}}$ for $0 \leqslant j \leqslant d$. Under this correspondence, a cyclic route beginning at v is represented by a word of the form $va = bv$ for suitable a and b.

2.4 EXAMPLES. Let $X = \{x_1, x_2, x_3\}$.

2.4.1 Let $A_1 = K\langle X\rangle/(W_1)$, where

$$W_1 = \{x_2^2, x_2x_3, x_3x_2, x_3^2, x_1^2x_2, x_1^2x_3, x_3x_1^2, x_3x_1x_2, x_1^3\}.$$

The graph $U(A_1)$ is given in Figure 1.

Figure 1:

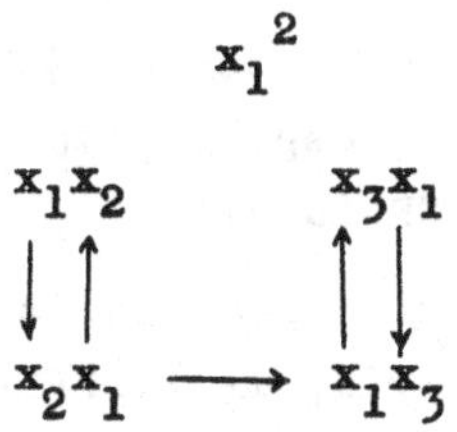

2.4.2 For $A_2 = K\langle X\rangle/(W_2)$, $W_2 = \{x_1x_2, x_1x_3, x_2x_1, x_3x_1\}$, the graph $U(A_2)$ is given in Figure 2.

Figure 2:

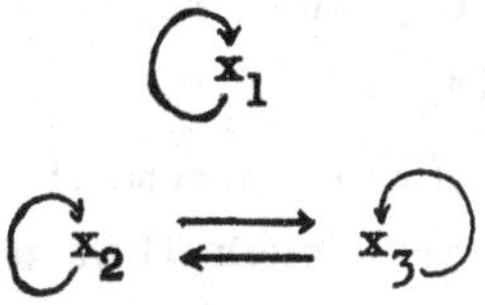

2.4.3 Figure 3 gives the graph $U(A_3)$ of the algebra $A_3 = K\langle X\rangle/(W_3)$, where $W_3 = \{x_1x_3, x_3x_1, x_2^2, x_3^2\}$.

Figure 3:

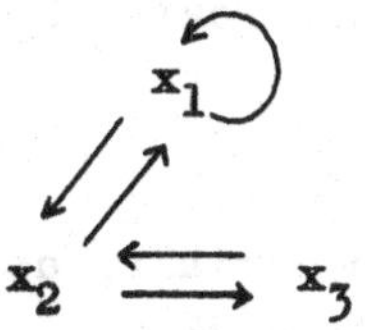

2.5 DEFINITION. A normal monomial $a \neq 1$ is called cyclic if

i) $\deg a \leqslant m$ and a is a right segment of a cyclic vertex in $U(A)$; or

ii) $\deg a > m$ and the route $R(a)$ is a subroute of a cyclic route.

2.6 DEFINITION. A normal monomial $a \neq 1$ which is not cyclic is called noncyclic.

2.7 LEMMA. A route R in U is a subroute of a cyclic route if and only if any edge in R lies on a cycle.

Applying 2.5, 2.6, 2.7 and the obvious fact that any edge in U corresponds to a normal monomial of degree m+1 one obtains:

2.8 COROLLARY. Let a be a normal monomial, deg a $\geq$ m+1. Then:

i) a is cyclic if and only if any segment of a of degree m+1 is cyclic as well;

ii) a is noncyclic if and only if there exists a noncyclic monomial of degree m+1 which is a segment of a.

2.9. If u is a monomial, j is an integer, j $\leq$ deg u, the segment (subwor consisting of the j rightmost (resp. leftmost) letters of u will be denoted (RIGHT(u,j) (resp.: LEFT(u,j)).

2.10 REMARK. Let u, v be normal monomials, deg u = m, deg v = k, and let uv $\not\equiv 0$ (mod I). Let w_j = RIGHT(u, m-j).LEFT(v, j) for $1 \leq j \leq \min(m,k)$. Then: a) If k > m (resp.: k = m) then there exists a route of length m in U connecting the vertex u with the route R(v) (resp.: with v). More precisely, if $v = v_0v_1$, deg v_0 = m then $R(uv) = u \to w_1 \to w_2 \to \dots \to w_{m-1} \to v_0 \to R(v)$ (resp.: $R(uv) = u \to w_1 \to w_2 \to \dots \to w_{m-1} \to v$).

b) If k < m then v is a proper right segment of a vertex in U and $R(uv) = u \to w_1 \to w_2 \to \dots \to w_k$.

2.11 LEMMA. Let w = uv be a cyclic monomial, deg u = m. Then there exist a monomial a (a = 1 is possible) such that uvau $\not\equiv$ 0(mod I).

PROOF. It follows from 2.5.ii that the route R(w) is a subroute of a cyclic route with associated cycle $C = v_0 \to v_1 \to \dots \to v_{d-1} \to v_0$. We can assume that $u = v_0$. Note that if v $\neq$ 1 then the length of the route R(uv) is equal to deg uv - m = deg v. Let k be an integer such that kd $\geq$ deg uv. Then by R.2.10 the route $\underbrace{C \to C \to \dots \to C}_{k \text{ times}}$ corresponds to a word of type uvau $\not\equiv$ 0(mod I).

2.12 REMARK. Let a, b, u be monomials, deg u $\geq$ m, and let au $\not\equiv$ 0(mod I) ub $\not\equiv$ 0(mod I). Then aub $\not\equiv$ 0(mod I).

2.13 LEMMA. For a normal monomial w $\neq$ 1 the following conditions are equivalent:

1) w is cyclic;

2) there exist monomials u, a, deg u = m, such that uwau $\not\equiv$ 0(mod I)

3) there exists a monomial b such that $(wb)^k \not\equiv 0$(mod I) for all k.

PROOF. (1)$\to$(2). Assume that w is cyclic. By 2.5 two cases are possible:

i) $\deg w \leqslant m$ and w is a right-hand segment of a vertex v which belongs to the cycle $C = v \to v_1 \to \dots \to v_{d-1} \to v$. Then starting from v and "going back" against the direction of C one finds, after $m - \deg w$ steps, a vertex u in C such that $uw \not\equiv 0 (\mathrm{mod}\ I)$. Obviously the monomial uw is cyclic. By Lemma 2.11 there exists a monomial a such that $uwau \not\equiv 0(\mathrm{mod}\ I)$.

ii) $\deg w > m$ and R(w) is a subroute of the cyclic route $C \to \dots \to C$, where C is a cycle. Let $v = \mathrm{LEFT}(w, m)$. Obviously v is a vertex in C, and "going back" against the direction of C, as in case (i), one finds a vertex u such that $uv \not\equiv 0(\mathrm{mod}\ I)$. Applying 2.11 again, one has $uwau \not\equiv 0(\mathrm{mod}\ I)$ for some monomial a.

We proved that (1) implies (2). The implication $(2) \longrightarrow (1)$ is obvious.

It follows from 2.12 that if the monomials u and a are as in (2) then $(wau)^k \not\equiv 0 \ (\mathrm{mod}\ I)$, hence (2) implies (3). The implication $(3) \longrightarrow (2)$ is obvious.

2.14 COROLLARY. 1) Any nontrivial segment of a cyclic monomial is cyclic as well. 2) A normal monomial containing a noncyclic segment is noncyclic as well.

2.15. The inverse image in $K\langle X\rangle$ of the Jacobson radical J of A is denoted J(A).

2.16 PROPOSITION. A normal (mod I) polynomial f belongs to J(A) if and only if f has no constant term and $ufau \equiv 0 \ (\mathrm{mod}\ I)$ for all monomials a and u such that $\deg u = m$.

PROOF. The statement follows from Proposition 14 and Theorem 16 of [1] .

2.17 COROLLARY. A normal monomial w belongs to J(A) if and only if w is noncyclic.

2.18 REMARK. The normal polynomial $f = \sum_{i=1}^{s} \alpha_i f_i$, $\alpha_i \in K \setminus \{0\}$, f_i is normal, $1 \leqslant i \leqslant s$, belongs to J(A) if and only if all the monomials f_i belong to J(A).

PROOF. For any monomials a and u the conditions

$ufau \equiv 0(\mathrm{mod}\ I)$

and

$uf_iau \equiv 0 \ (\mathrm{mod}\ I)$ for $i = 1,\dots,s$

are obviously equivalent, as $f_i \neq f_j$ for $i \neq j$. Hence, the statement follows from 2.16.

2.19 THEOREM. 1) The Jacobson radical J of a monomial algebra is the K-linear span of the set of all noncyclic monomials. 2) J is finitely generated as an ideal by the set of all noncyclic monomials of degree $\leqslant m+1$.

PROOF. The assertion (1) follows from 2.16 and 2.17.

Let J_0 be the ideal in A generated by the noncyclic monomials of degree $\leqslant m+1$. By 2.14.2 and 2.19.1 one has $J_0 \subseteq J$. Applying 2.8.ii one obtains $J \subseteq J_0$. Thus $J = J_0$ and the assertion (2) is proved.

2.20 COROLLARY. The algebra A, without unit, is radical if and only if it is finite-dimensional.

PROOF. It follows from Remark 3 of [1] that A is finite-dimensional if and only if there are no cycles in U(A) and by Theorem 2.19 this is equivalent to the condition that any element of A lies in its Jacobson radical.

An immediate consequence of Theorem 2.19 is

2.21 THEOREM. The monomial algebra A is semi-simple (in the sense of Jacobson) if and only if any normal monomial v, $1 \leqslant \deg v \leqslant m+1$, is cyclic.

2.22 COROLLARY. The monomial algebra A is semi-simple if and only if any normal monomial v, $v \neq 1$, is cyclic.

2.23. Let A_1 be the algebra defined in Ex. 2.4.1. Using Th. 2.19 one can see that the monomials x_1^2 and $x_2x_1x_3$ generate its Jacobson radical as an ideal.

2.24. It follows from Th. 2.21 that the algebras A_2 and A_3 given in Ex. 2.4.2, resp. 2.4.3, are both semi-simple.

2.25. Recall that an algebra B is prime (resp.: semi-prime) if for any $f, g \in B \setminus \{0\}$ (resp.: $f \in B \setminus \{0\}$), $fBg \neq 0$ (resp.: $fBf \neq 0$).

The following remark is essentially contained in Th. 18 of [1] .

2.26 REMARK. 1) The algebra A is prime if and only if for any normal monomials u, v there exists a monomial a such that $uav \not\equiv 0 \pmod I$. 2) The algebra A is semi-prime if and only if for any normal monomial v there exists a monomial a such that $vav \not\equiv 0 \pmod I$.

2.27 THEOREM. The monomial algebra A is semi-prime if and only if it is semi-simple in the sense of Jacobson.

PROOF. a) Assume that A is semi-prime. We shall prove that A is semi-simple. By 2.22 it is enough for this purpose to show that any monomial v, $v \neq 1$, is cyclic. Let v be a normal monomial, $v \neq 1$. It follows from 2.26.2 that v is a segment of some normal monomial w, deg $w \geq m+1$. Applying 2.26.2 again one can find a monomial a such that $waw \not\equiv 0 \pmod{I}$. It follows then from 2.12 that $(wa)^s \not\equiv 0 \pmod{I}$ for all s, hence, by 2.13, w is cyclic. By 2.14.1 v being a segment of a cyclic monomial is cyclic as well. We have proved that A is semi-simple.

b) Conversely, let A be semi-simple and let v be a normal monomial. We shall prove that $vav \not\equiv 0 \pmod{I}$ for some monomial a. The last is obvious in case that v = 1. Assume that $v \neq 1$. It follows from 2.22 that v is cyclic. By 2.13.3 there exists a normal monomial a such that $(va)^s \not\equiv 0 \pmod{I}$ for all s, which gives $vav \not\equiv 0 \pmod{I}$. Thus, by 2.26.2, A is semi-prime.

2.28 <u>THEOREM.</u> The monomial algebra A is prime if and only if the following conditions hold:

1) Any normal monomial of degree $< m$ is a right-hand segment of a vertex of U(A);

2) For any two vertices u and v of U(A) there exists a route from u to v.

<u>PROOF</u>. If A is prime then it is semi-prime and (1) follows from 2.27, 2.21 and 2.5.ii. Condition (2) can be obtained by applying 2.26.1 and 2.10.

Assume now that the conditions (1) and (2) hold. We shall prove that A is prime. By 2.26.1 it is enough to show that for any normal monomials w_1 and w_2 there exists a normal monomial a such that $w_1aw_2 \not\equiv 0 \pmod{I}$. Four different cases are to be considered: a) deg $w_i < m$, i = 1, 2; b) deg $w_1 < m$, deg $w_2 \geq m$; c) deg $w_1 \geq m$, deg $w_2 < m$; deg $w_i \geq m$, i = 1, 2. We give the details in case (a) and leave the remaining three to the reader. By (1) there exist vertices u and v of U(A) such that $u = u_1w_1$, $v = v_2w_2$. By (2) there exists a route R connecting u and v, hence, there exists a monomial b such that R = R(ubv) and $ubv \not\equiv 0 \pmod{I}$. Since $ubv = u_1w_1bv_2w_2$, it is obvious that $w_1bv_2w_2 \not\equiv 0 \pmod{I}$ and one can take $a = bv_2$.

2.29. Applying Theorems 2.21, 2.27 and 2.28 one can see that :

1) The algebra A_2 given in Example 2.4.2 is semi-prime but not prime.

2) The algebra A_3 given in Example 2.4.3 is prime (and semi-simple).

3. ALGORITHMS

We keep the preceding notation

Recall that A denotes a monomial algebra, $A = K\langle X\rangle/(W)$, where $X = \{x_1, \ldots, x_n\}$ is a set of indeterminates, $K\langle X\rangle$ is the free K-algebr on it, and $W = \{w_1, \ldots, w_t\}$ is a set of monomials in $K\langle X\rangle$, cf. 1.1. U denotes the Ufnarovsky graph of A (cf.2.1) and J is the Jacobson radical of A .

3.1. We need some more notation:

N	the set of all normal monomials of degree $\leq$ m ($m+1 = \max\{\deg w_i \mid 1 \leq i \leq t)$
$\tilde{V} = \{v_1, \ldots, v_k\}$	the set of all normal monomials of degree m (V is the set of vertices of U.)
M	the incidence matrix of the graph U (M is kxk boolean matrix with $M(i,j) = \begin{cases} 1 & \text{if there is an edge from vertex } v_i \text{ to vertex } v_j; \\ 0 & \text{otherwise.} \end{cases}$
M^k	the k-th power of M
$\tilde{M} = M \vee M^2 \vee \ldots \vee M^k$	a boolean kxk matrix, where the operation V is determined as follows: if P, Q are kxk matrices of integers, then $(P \vee Q)(i,j) = \begin{cases} 0 & \text{if } P(i,j) = Q(i,j) = 0, \\ 1 & \text{else.} \end{cases}$
T	the set of all nonciclic monomials of degree $\leq$ m (cf. 2.7)

3.2 We shall also use the following functions, defined for normal monomials w, deg w $\geq$ m.

$$\text{INDEX}(w) = \begin{cases} i & \text{if } w = v_i \in \tilde{V} \\ 0 & \text{otherwise.} \end{cases} \quad ;$$

RIGHT(w,m) and LEFT(w,m) are defined in 2.9.

3.3 REMARK. The following recursive definition can be used for finding the set T:

$$T = \left\{ w \in N \;\middle|\; \begin{array}{l} ((\deg w = m) \text{ and } (M(\text{INDEX}(w), \text{INDEX}(w)) = 0)) \text{ or} \\ ((\deg w < m) \text{ and } (\forall x \in X\ (xw \notin N) \text{ or } (xw \in T))) \end{array} \right\}$$

3.4. Note that given X and W, Program 2 of [1] finds the matrix M, hence, we have a method for finding $\widetilde{M}$ and T.

One can easily see that:

3.5. If $1 \leqslant i,j \leqslant k$ then there exists a route from vertex v_i to vertex v_j iff $\widetilde{M}(i,j) = 1$.

3.6. A normal monomial $v \neq 1$ is noncyclic iff:

a) $\deg v \leqslant m$ and $v \in T$; or

b) $\deg v > m$ and M(INDEX(RIGHT(v,m)),INDEX(LEFT(v,m))) = 0.

3.7. All normal monomials v, $v \neq 1$, are cyclic iff $T = \emptyset$ and $M(i,j) = M(j,i)$ for $1 \leqslant i \leqslant k-1$, $i+1 \leqslant j \leqslant k$.

It follows from 2.28 and 3.5 that

3.8. The algebra A is prime iff $T = \emptyset$ and $\widetilde{M}(i,j) = 1$ for $1 \leqslant i,j \leqslant k$.

3.9. Applying 2.17, 2.18 and 3.6 one can see that the following procedure decides whether a normal polynomial $f = \sum_{i=1}^{s} \alpha_i f_i \neq 0$ belongs to J.

```
BOOL PROCEDURE ELEMENT OF RADICAL ( Σ_{i=1}^{s} α_i f_i )
BEGIN
  BOOL ANSWER:= TRUE;
  FOR i=1 TO s WHILE ANSWER
  DO
    IF deg f_i ≤ m
    THEN ANSWER := (w_i ∈ T)
    ELSE ANSWER := ¬ M̃(INDEX(RIGHT(f_i,m)), INDEX(LEFT(f_i,m)))
    FI
  OD;
  ANSWER
END
```

3.10. It follows from 2.22 and 3.7 that the procedure SEMI-SIMPLICITY given bellow decides whether A is semi-simple (in the sense of Jacobson).

```
BOOL PROCEDURE SEMI-SIMPLICITY
BEGIN
  BOOL ANSWER:= (T = ∅) ;
  FOR i = 1 TO k-1 WHILE ANSWER DO
    FOR j = i+1 TO k WHILE ANSWER DO
      ANSWER:= (M̃(i,j) = M̃(j,i))
    OD
  OD;
  ANSWER
END
```

3.11. Obviously, it follows from 3.8 that the procedure PRIMENESS decides whether the algebra A is prime.

```
BOOL PROCEDURE PRIMENESS :
BEGIN
  BOOL ANSWER:= (T = ∅);
  FOR i = 1 TO k WHILE ANSWER DO
    FOR j = 1 TO k WHILE ANSWER DO
      ANSWER:= M(i,j)
    OD
  OD;
  ANSWER
END
```

REFERENCES

1. T. Gateva-Ivanova, V. Latyshev, On recognizable properties of associative algebras, J. Symb. Comp., 1988, to appear.
2. V. Ufnarovskij, A growth criterion for graphs and algebras defined by words, Mat. Zametki 31 (1982), 465-472 (in Russian); English transl.: Math. Notes 37 (1982), 238-241.

A RECURSIVE ALGORITHM FOR THE COMPUTATION OF THE HILBERT POLYNOMIAL

M.V.Kondrat'eva

Department of Mathematics & Mechanics, Moscow State University

Linin Hills, 119899 MOSCOW, U.S.S.R

E.V.Pankrat'ev

Department of Mathematics & Mechanics, Moscow State University

Linin Hills, 119899 MOSCOW, U.S.S.R

In this report we describe a new recursive algorithm for the computation of the Hilbert polynomial, which is faster than the algorithms formulated in [5]. We also give algorithms for the determination of the degree of the Hilbert polynomial and its head coefficient, which do not require any computation of the whole Hilbert polynomial. The detailed exposition of results and additional examples may be found in [6]. The algorithms are implemented in the algorithmic language REFAL.

The following notations will be used:

$R=R(m)=N_\emptyset^m$, where $N_\emptyset$ is the set of natural numbers with zero; on R the relation of partial order $\geq$ is introduced so that if $r=(r_1,\ldots,r_m) \in R$ and $s=(s_1,\ldots,s_m) \in R$, then $r \geq s$ iff for every i the inequality $r_i \geq s_i$ holds. For $r \in R$ the sum of coordinates will be designated by $|r|$.

The problem of the computation of the Hilbert polynomials arises both in commutative algebra and algebraic geometry in the description of dimension of an algebraic variety and in differential (difference) algebra by computation of differential (difference) dimension polynomials. By using Gröbner bases this problem can be reduced to the following combinatorial one ([1], p. 51).

Main problem:

For the given finite set $E=\{e_1, \ldots, e_n\}$ of points of R let $\overline{E} = \{r \in R \mid \exists\, i,\ r \geq e_i\}$, the complement of $\overline{E}$ in R be denoted by V. We have for any $s \in N$ to determine the number of points $v=(v_1, \ldots, v_m) \in V$ such that $\Sigma v_i \leq s$.

The function, describing the dependence of this number upon s, is known as the Hilbert function for the set V. For almost all natural numbers its values coincide with the values of a numerical polynomial, the so called Hilbert polynomial. We shall represent E by the rows of a matrix, denoted by the same letter, and we shall denote the corresponding Hilbert polynomial by $\omega(E,s)$, or $\omega(n,m,E,s)$, if we have to emphasize its dependence upon the dimension m of space and upon the cardinality n of E (note that n may be equal to $\emptyset$). Besides we shall use the notation $\omega(V,s)$, if we have to emphasize the dependence of the Hilbert polynomial upon the set V.

Let's mention the folloing properties of the Hilbert polynomials.

Proposition 1.

a) $\omega(n,m,E,s) = \omega(n,m,E',s)$, if matrix E' is obtained from E by a interchange of rows or columns.

b) Suppose that the p-th row of the matrix E is greater than or equal to the q-th one, i.e. $e_{pj} \geq e_{qj}$ for any j. Then $\omega(n,m,E,s) = \omega(n-1,m,E',s)$ where E' is obtained from E by deleting the p-th row.

An explicit formula for the Hilbert polynomials computation is given in [2] (see also [3]):

$$\omega(n,m,E,s) = \sum_{t\in T} (-1)^{|t|} \binom{s+m-\mu(t,E)}{m}, \tag{1}$$

where $T = \{ t=(t_1, \ldots, t_n) \mid t_i \in \{\emptyset;1\} \ \forall i\}$;

$$|t| = \sum_{i=1}^{n} t_i; \quad \mu(t,E) = \sum_{i=1}^{m} (\max_{j=1}^{n} t_j \cdot e_{ji})$$

If $n=\emptyset$, then $\omega(n,m,E,s) = \binom{s+m}{m}$

The direct computation of the Hilbert polynomial via this formula demands not less than $O(2^n.m)$ operations and may be used only for little n.

Mora and Möller have modified this algorithm by using the fact that for sufficiently big n there are in this formula identical summands with opposite signs, which mutually annihilate themself. The runtime of their algorithm is equal to $O(r^{n+1})$ [4].

In this paper a new algorithm for the computation of the Hilbert polynomials based on the joining of some vectors to the set E according to some rules is proposed.

Let's see how the Hilbert polynomial changes itself when a vector v is joined to the set E. Let $R_v = \{s \in R \mid s \geq v\}$ and $V_v = \{w \in R \mid w+v \in V\}$.

Proposition 2. The Hilbert polynomial $\omega(V,s)$ of V equals the sum of the Hilbert polynomial $\omega(V\setminus R_v,s)$ of $V\setminus R_v$ and the Hilbert polynomial $\omega(V_v,s-|v|)$ of the set V_v, the argument of which is translated on the sum of coordinates $|v|$ of the vector v.

Proof. There is one-to-one correspondence between the points of V_v, the coordinate sums of which don't exceed $s-|v|$, and the points of $V\cap R_v$, the coordinate sums of which don't exceed s.

Corollary 3. Let E be a (n x m)-matrix and v be a vector. Then $\omega(n,m,E,s)=\omega(n+1,m,E',s)+\omega(n,m,E'',s-|v|)$, where E' is obtained from E by joining the row v, and E" is obtained by subtraction the vector v from every row of E and by replacing its negative elements by zeros.

Note that E' often has some "spare" rows, which may be thrown away and the number of rows in it does not increase, but decreases. In particular, choosing the coordinates of v equal to the minimal values of entries of corresponding columns of E, we shall have a row in E', which is majorized by all other ones, so for the computation of the Hilbert polynomial we may retain in E' only the row v. The formula of the Hilbert polynomial in this case is well known. By choosing v = = (1, Ø,...,Ø), we may, when forming E, throw away all the rows of E containing a nonzero entry in the first column. Similarly they may be trown away, if v = (k, Ø, ..., Ø), where k equals the minimal nonzero value of entries in the first column of E.

By choosing the vector v in different ways we obtain different algorithms of computation of the Hilbert polynomials.

The first algorithm can be obtained by choosing v = (1, Ø,..., Ø), if E contains nonzero elements in the first column. Then E' and E" is calculated in the following way: E' consists of the row (1,Ø,...,Ø) and the rows of E, containing Ø in the first column, E" is obtained by subtraction of 1 from the nonzero entries of the first column of E.

Lemma 4. Let

$$E = \begin{bmatrix} 1 & \emptyset & \dots & \emptyset \\ \emptyset & & & \\ \vdots & & E^\circ & \\ \emptyset & & & \end{bmatrix}$$

where $e_{11} = 1$, and all other elements of the first column and the first row equal zero. Then $\omega(n,m,E,s) = \omega(n-1,m-1,E^\circ,s)$.

Proof. The mapping $(\emptyset, e_2, \dots, e_m) \dashrightarrow (e_2, \dots, e_m)$ gives us an one-to-one correspondence betweeen m-dimensional vectors that don't majorize any row of E and (m-1)-dimensional vectors that don't majorize any row of E°. The coordinate sums are invariant by this mapping.

Let's introduce operators Δ_t and Δ^{-1} on Q[x]:

$\Delta_t(p(x)) = p(x) - p(x-t)$;

Δ^{-1} is defined on the binomial coefficients by the rule:

$$\Delta^{-1}\binom{x+i}{i} = \binom{x+i+1}{i+1}$$

and is extended on the ring Q[x] by linearity. It is evident that

$$\Delta_t \cdot \Delta^{-1}(p(x)) = p(x) + p(x-1) + \dots + p(x-t+1).$$

In particular, $\Delta_1 \cdot \Delta^{-1}(p(x)) = p(x)$.

The application of the lemma 4 k times brings about the following result:

Lemma 5. Let

$$E = \begin{bmatrix} k & \emptyset & \dots & \emptyset \\ \emptyset & & & \\ \vdots & & E^{\circ} & \\ \emptyset & & & \end{bmatrix}$$

where $e_{11} = 1$, and all other elements of the first column and the first row equal zero. Then $\omega(n,m,E,s)=\Delta_k \cdot \Delta^{-1}\omega(n-1,m-1,E^{\circ},s)$.

Proof. Before the treatment of the general case we shall consider the case when E contains not more than two columns, i.e. m=1 or m=2.

If m=1, then by proposition 1 we may be assumed that E consists only of an element e, then

$$\omega(E,s) = \binom{s+1}{1} - \binom{s+1-e}{1} = e$$

Let m=2. By proposition 1 it may be assumed that the elements in the first column increase and in the second one they decrease, i.e. $e_{11} < e_{21} < \dots < e_{n1}$ and $e_{12} > e_{22} > \dots > e_{n2}$. Suppose that $e_{11} = e_{n2} = \emptyset$.

Proposition 6. Let

$$E = \begin{bmatrix} e_{11} & e_{12} \\ \vdots & \vdots \\ e_{n1} & e_{n2} \end{bmatrix}$$

where $\emptyset = e_{11} < e_{21} < \dots < e_{n1}$ and $e_{12} > e_{22} > \dots > e_{n2} = \emptyset$. Then $\deg \omega(E,s) = \emptyset$ and $\omega(E,s) = \omega(E) = \sum_{i=1} (e_{i+1,1} - e_{i1}) \cdot e_{i2}$.

Proof. The case n=1 is trivial.

Let $n > 1$. Then $\omega(E,s) = \omega(E',s) + \omega(E'',s-e_{21}) =$ $= \Delta_{(e_{21}-e_{11})}\Delta^{-1}\omega(E_1',s) + \omega(E'',s-e_{21})$, where $E_1' = (e_{12})$ and

$$E'' = \begin{bmatrix} \emptyset & e_{12} \\ \emptyset & e_{22} \\ e_{31}-e_{21} & e_{32} \\ \cdot \quad \cdot & \cdot \\ e_{n1}-e_{21} & e_{n2} \end{bmatrix}$$

It is shown above, that $\omega(E_1') = e_{12}$, hence

$$\Delta_{(e_{21}-e_{11})}\ \Delta\ \omega(E_1',s) = (e_{21} - e_{11}) \cdot e_{12}.$$

By proposition 1 the first row of E" may be deleted. It is sufficient now to use the inductive assumption.

Thus we receive the following algorithm for the Hilbert polynomial computation when m=2.

Algorithm HILBPOL2(N,E,ω)

Input: N is a natural number.

E(N,2) is a matrix of nonnegative integers.

Output: ω(x)=ω(N,2,E,x) is the Hilbert polynomial of E.

---

to sort the rows of E in ascending order of elements in the first column, deleting spare rows.

n := the number of rows in the resulting matrix E

$v_1 := e_{11}$; $v_2 := e_{n2}$

for i from 1 to n do

$e_{i1} := e_{i1} - v_1$; $e_{i2} := e_{i2} - v_2$

$$\omega(x) := \binom{x+2}{2} - \binom{x+2-v_1-v_2}{2} + \sum_{i=1}^{n-1} (e_{i+1,1} - e_{i1}) \cdot e_{i2}$$

The complexity of this algorithm is not higher, than O(N log N), as this quantity of operations is sufficient for sorting, the deleting of spare rows may be done simultaneously with sorting. The remaining actions need O(N) operations.

Note. The proposed algorithm without increasing complexity may be generelized to the case when m > 2, but matrix E contains only two nonzero columns.

The general algorithm of Hilbert polynomial computation consists of two steps. At the first one we choose the minimal value in every column of E and make the shift on the resulting vector. It will be shown below that the degree of the Hilbert polynomial of a matrix, containing zero in every column, (we shall call such matrix normed) doesn't exceed m-2. By computing Hilbert polynomial of normed matrix we shall choose the minimal nonzero element k in the first column and make the shift on the vector (k, Ø, ..., Ø).

Algorithm HILBPOL(n,m,E,ω)

Input: n is a nonnegative integer;

m is a natural number;

E is an n × m matrix of nonnegative integers.

Output: ω(x) is the Hilbert polynomial of E.

```
case
        n=Ø  ==>   ω(x) := (x+m choose m)

        n=1  ==>   ω(x) := (x+m choose m) - (x+m-|e| choose m)
   else      ==>
      v := (v_1, ..., v_m);   v_i := min_{j=1}^{n} e_ji
      for i from 1 to n do;  for j from 1 to q do
             e_ji := e_ji - v_i
      HILBPOLN(n,m,E,ω)
      ω(x) := ω(x-|v|) + (x+m choose m) - (x+m-|v| choose m)
```

The main command of this algorithm is the call of the program HILBPOLN of the Hilbert polynomial computation for a normed matrix. The algorithm HILBPOLN follows.

Algorithm HILBPOLN(n,m,E,ω)

Input: n is a nonnegative integer;
m is a natural number, m ≥ 2;
E is n x m matrix of nonnegative integers such that the first column of E contains both zero and nonzero elements.

Output: ω(x) is the Hilbert polynomial of E.

Variables: EØ is a matrix, its dimension doesn't exceed q x n;
ωØ is a polynomial;
Ns is an integer, the current value of the first coordinate;
Nr is the next value of the first coordinate.

```
if m=2
      then Hilbpol2 (n,E,ω)
else
      ω(x) := Ø
      Ns := Ø
      EØ := Ø
      for every nonzero value of e_i1 in acsending order do
            Nr := e_i1
            to append to EØ the sequence of rows
                     {(e_j2, ..., e_jm) | e_j1 = Ns}
            NØ := the number of rows in  EØ.
            HILBPOL(NØ,m-1,EØ,ωØ)
            ωØ(x) := Δ_(Nr-Ns) Δ^-1 ωØ(x)
```

```
        ω(x) := ω(x) + ωØ(x-Ns)
        Ns := Nr
    to append to EØ  the sequence of rows
    {(e_j2, ..., e_jm) | e_j1= Ns}
    to append to EØ the zero column
    HILBPOL(n,m,EØ,ωØ)
    ω(x) := ω(x) + ω(x-Ns)
```

This algorithm reduces the problem with n x m matrix to not more than n problems, each of which contains m-1 nonzero columns and not more than n rows. Thus we obtain not more than n^{m-2} of two-dimensional problems. Besides we can use other algorithms for problems which have a small number of rows.

We shall illustrate this algorithm by the example of Macauley's curves, which were used for the comparison of the algorithms in [4].

Example.

Let the ideal J be generated by the polynomials

$$f_1 = xy - z$$
$$f_i = x^{r+1-i} z^{i-2} - y^{i-1}, \quad i = 2, \ldots, r-1$$
$$f_r = y^{r-1} - xz^{r-2}$$

Let the order of variables x,y,z be such that $y > x > z$. The polynomials f_i form a Gröbner basis of J and the first term in f_i is the highest one (i = 1, ..., r). The matrix E has the form

$$E = \begin{bmatrix} 1 & 1 & 0 \\ 0 & r-1 & 0 \\ 0 & r-2 & 1 \\ . & . & . \\ 0 & 2 & r-3 \\ r-1 & 0 & 0 \end{bmatrix}$$

By algorithm HILBPOLN we get

$$\omega(r,3,E,x) = \omega_1(x) + \omega_2(x-1) + \omega_3(x-r+1), \quad \text{where}$$
$$\omega_1(x) = \omega(r-2,2,E_1,x) \,,$$
$$\omega_2(x) = \Delta_{r-2}\Delta^{-1}\omega(r-1,2,E_2,x) \,,$$
$$\omega_3(x) = \omega(r,3,E_3,x),$$

$$E_1 = \begin{bmatrix} r-1 & 0 \\ r-2 & 1 \\ . & . \\ 2 & r-3 \end{bmatrix}, \quad E_2 = \begin{bmatrix} 1 & 0 \\ r-1 & 0 \\ . & . \\ 2 & r-3 \end{bmatrix}, \quad E_3 = \begin{bmatrix} 0 & 1 & 0 \\ 0 & r-1 & 0 \\ . & . & . \\ 0 & 2 & r-3 \\ 0 & 0 & 0 \end{bmatrix}$$

By HILBPOL2 we get

$$\omega(r-2,2,E_1,x) = \binom{x+2}{2} - \binom{x+2-2}{2} + \sum_{i=\emptyset}^{r-2} 1\cdot(r-3-i) =$$

$$= \frac{(x+2)(x+1)}{2} - \frac{x(x-1)}{2} + \sum_{i=1}^{r-3} i = 2x + 1 + \frac{(r-3)(r-2)}{2}$$

In E_2 may be deleted all rows besides the first one and

$$\omega(r-1,2,E_2,x) = \binom{x+2}{2} - \binom{x+1}{2} = \binom{x+1}{1}$$

$$\omega_2(x-1) = \binom{x-1+2}{2} - \binom{x-1+2-r}{2} = \frac{(x+1)\cdot x}{2} - \frac{(x+3-r)(x+2-r)}{2} =$$

$$= (r-2)\cdot x - \frac{(r-3)(r-2)}{2}$$

E_3 contains the zero row, hence $\omega_3(x) \equiv \emptyset$.

So, we have $\omega(r,3,E,x) = 2x + 1 + \frac{(r-3)(r-2)}{2} + (r-2)\cdot x - \frac{(r-3)(r-2)}{2} =$ $= rx + 1$.

Differential dimension polynomial widely used in differential algebra is not a differentially birational invariant of an extension. However such invariants are the degree of this polynomial and its higher coefficient, the so called differential type of extension and typical differential dimension. The folllowing proposition gives an algorithm for the computation of those invariants, which doesn't require the computation of the whole polynomial.

Proposition 7.

a) $\deg \omega(n,m,E,x) \le m$;

b) $\deg \omega(n,m,E,x) = m$ iff $n=\emptyset$; in this case

$$\omega(n,m,E,x) = \binom{x+m}{m};$$

c) $\deg \omega(n,m,E,x) < \tau$ iff for any subset I, consisting of $m-\tau$ elements of the set $\{1,\ldots, m\}$, there exists a row e_I of E, such that all elements in this row in the columns having indices from I equal $\emptyset$. In particular, $\deg \omega(n,m,E,x) < m-1$ iff every column of E contains a zero element; $\omega(n,m,E,x) = \text{const}$ iff E contains a diagonal submatrix.

d) Let $n > \emptyset$ and $\deg \omega(n,m,E,x) \le \tau$. Then

$$a_\tau = \sum_{\sigma\in A(\tau,m)} \omega(n,m-\tau,E\sigma), \qquad (2)$$

where $A(\tau,m)$ is the set of combinations of m elements τ at a time, and if $\sigma=(i_1, \ldots, i_\tau) \in A(\tau,m)$, then $E\sigma$ is obtained from E by deleting the columns with indices $i_1, \ldots, i_\tau$.

Proof.

a) Formula (1) gives a representation of $\omega(n,m,E,x)$ as a sum of polynomials of degree not heigher than m.

b) If $n=0$, then $\omega(n,m,E,x)$ equals the number of monomials of m variables of degree not more than x, i.e. $\omega(n,m,E,x) = \binom{x+m}{m}$ is a polynomial of degree m. If $n > 0$, i.e. E contains some row e, then $\omega(n,m,E,x) \le \omega(1,m,e,x) = \binom{x+m}{m} - \binom{x+m-|e|}{m}$. The inequality holds for all $x > x0$. Hence $\deg \omega(n,m,E,x) \le \deg \omega(1,m,e,x) \le m-1$ ($\deg \omega(1,m,e,x) = m-1$ for any nonzero vector e).

c) We shall show that $\deg \omega(n,m,E,x) < \tau$ if for any subset I, consisting of $m-\tau$ elements of the set $\{1,\ldots, m\}$, there exists a row e_I of E such that all elements in this row in the columns with indices from I equal 0. The inverse will follow from d.

We shall use the induction on the sum Σ of entries of the matrix E. If $\Sigma=0$, then E is the zero matrix, and we can take $n=1$ and $E = e = (0, \ldots, 0)$. Then $\omega(n,m,E,x) = 0$ and $\deg \omega(n,m,E,x) = < \tau$ for any $\tau\ge 0$.

Let $\Sigma>0$. W.l.o.g. we can assume that the first column of E contains a nonzero entry. Corollary 3 with $v=(1, 0, \ldots, 0)$ implies that $\omega(n,m,E,x)$ is a sum of polynomials $\omega(n+1,m,E',x)$ and $\omega(n,m,E'',x-1)$. The inductive assumption can be applied to E"; the degree of the polynomial isn't change by the shift of argument, hence $\deg \omega(n,m,E'',x-|v|) < \tau$.

As for the degree of the first polynomial let's note that we can hold in E' only one row with nonzero element in the first column and apply lemma 4. The inductive assumption with m replaced by $m-1$ and τ replaced by $\tau-1$ can be applied to E^o.

d) We shall use induction on $k = m-\tau$.

Let $m-\tau = 1$. Suppose $v = (\min_{i=1}^{n} e_{i1}, \ldots, \min_{i=1}^{n} e_{im})$. Corollary 3 and the just obtained result imply $a_{m-1} = \sum_{j=1}^{m} \min_{i=1}^{n} e_{ij}$.

This result was obtained by different way in [3].

Let $k=m-\tau > 1$. The proof of d is essentially analogous to the proof of c and it is based on the induction on the sum Σ of entries of E. The polynomial $\omega(n,m,E,x)$ is the sum of the polynomials $\omega(n+1,m,E',x)$ and $\omega(n,m,E'',x-1)$. The inductive assumption can be applied to the matrices E' and E". The shift of argument in the second polynomial has no influence on its highest coefficient. If $1 \notin \sigma$ in (2) then $\omega(n+1,m,E'\sigma) = 0$, and $\omega(n,m,E''\sigma) = \omega(n,m,E\sigma)$. If $1 \in \sigma$ then $\omega(n,m,E\sigma) = \omega(n+1,m,E'\sigma) + \omega(n,m,E''\sigma)$ by the properties of determinants.

Example.

The computation of the difference dimension polynomial for the ring of inversive difference polynomial may be reduced to the computation of the Hilbert polynomial for the ideal generated by $x_i \cdot x_{i+m} - 1$, $i=1,\ldots,m$ in the ring $k[x_1, \ldots, x_{2m}]$. These polynomials form the Gröbner basis of the corresponding ideal and matrix E for this system is

$$\begin{bmatrix} 1 & 0 & 0 & \dots & 0 & 1 & 0 & 0 & \dots & 0 \\ 0 & 1 & 0 & \dots & 0 & 0 & 1 & 0 & \dots & 0 \\ . & . & . & \dots & . & . & . & . & \dots & . \\ 0 & 0 & 0 & \dots & 1 & 0 & 0 & 0 & \dots & 1 \end{bmatrix}$$

Direct application of (1) gives

$$\omega(m,2m,E,x) = \sum_{k=0}^{m} (-1)^k \binom{m}{k} \binom{x+2m-2k}{2m}$$

By proposition 7 the degree of this polynomial is m and its main coefficient equals 2^m. The application of the algorithm Hilbpol gives us

$$\omega(m,2m,E,x) = \sum_{k=0}^{m} (-1)^{m-k} 2^k \binom{m}{k} \binom{x+k}{k}$$

In conclusion we give an algorithm of computation of coefficients a_m, a_{m-1}, a_{m-2} of the Hilbert polynomial.

Algorithm Maincoef$(n,m,E,a_m,a_{m-1},a_{m-2})$

Input: n is a nonnegative integer;
m is a natural number;
E is a n x m matrix of nonnegative integers.

Result: a_m, a_{m-1}, a_{m-2} are integers.

---

if $n=0$

then $a_m := 1$; $a_{m-1} := 0$; $a_{m-2} := 0$

else

$a_m := 0$

$v := (v_1, \dots, v_n)$, where $v_i = \min_{j=1}^{n} e_{ji}$

$a_{m-1} := |v|$

$a_{m-2} := -\binom{|v|}{2}$

$e_{ji} := e_{ji} - v_i$, $i=1,\dots,m$; $j=1,\dots,n$

for all pairs $1 \le i < j \le m$ do

$a_{m-2} := a_{m-2} +$ Hilbert polynomial of matrix consisting of i-th and j-th columns of E

AN AFFINE POINT OF VIEW ON MINIMA FINDING
IN INTEGER LATTICES OF LOWER DIMENSIONS.

Brigitte VALLÉE
Département de Mathématiques, Université de Caen,
F-14032 Caen Cedex, France

Abstract: *We study here algorithms that determine successive minima in integer lattices of lower dimensions (n=2 or n=3). We adopt an affine point of view that leads us to a better understanding of the complexity of Gauss' algorithm and we can exhibit its worst-case input configuration. We then propose for the three dimensional case a new algorithm that constitutes the natural generalisation of Gauss' algorithm. We build in polynomial time a "minimal" basis of the lattice and we also get a new structural result - on hyperacute tetrahedra. Furthermore, our algorithm has a better computational complexity that of the LLL algorithm in the 3-dimensional case. Detailed proofs and a more thorough algorithmic discussion are given in [5]*

1. The minima of a lattice.

We consider the euclidean space $\mathbf{R}^n$. For u and v in $\mathbf{R}^n$, (u, v) is the dot product of u and v and $|v| = \sqrt{(v, v)}$ is the length of v.
A lattice L of $\mathbf{Z}^n$ is a $\mathbf{Z}$–module of rank n. L is usually given by any basis $b = (b_1, b_2, ..., b_n)$ of length $M = \max |b_i|$. There are, in a lattice, a few intrinsic objects; amongst them, a system of successive minima $(\lambda_i(L) \ / \ 1 \leq i \leq n)$ which is defined by the two properties:
(i) $\lambda_1(L)$ is a shortest non zero vector of L.
(ii) $\lambda_i(L)$ is a shortest vector of L that is linearly independent of the system $(\lambda_j(L), j < i)$.

For $n \leq 4$, this system form a basis of L, which is called *minimal*.

The Minima Finding Problem.
Find a sequence of successive minima of a lattice L of $\mathbf{Z}^n$, given by a basis b of length M.

This problem is known to be NP-hard [4]. Two approaches are thus possible:
1. *Fix* n and find an *exact* algorithm which runs in time polynomial in $\log M$.
2. Find an *approximation* algorithm which runs in time polynomial in $(n, \log M)$.

The second approach is the one usually followed: Lovasz reduced bases, obtained with the LLL algorithm [3], provide a good approximation of the successive minima.
We take here, in the lower dimensions, the first point of view. We find, directly, successive minima of a lattice, without a previous reduction of the basis. We begin with dimension 2 by analysing an affine version of Gauss' Algorithm. We give a better description of the complexity of the algorithm [2] and we find geometrical principles that we will use further for the 3–dimensional case, in the Acute Algorithm.

2. Gauss' Algorithm revisited.

We now design an affine version of Gauss' algorithm, which, given an affine basis (A, B, C) of a lattice L in $\mathbf{Z}^2$, constructs a minimal affine basis of L. As usual, we write $a = |\vec{BC}|$, $b = |\vec{AC}|$, $c = |\vec{AB}|$.

References.

1. Kolchin E.R. Differential algebra and algebraic groups. Pure and Applied Mathematics. Vol. 54. New York-London. Academic Press, XVII, 446 p. (1973)

2. Buchberger B. Ein Algorithmus zur Auffinden der Basiselemente des Restklassenringes nach einem nulldimensionalen Polynomideal. Ph. D. thesis., Univ. Innsbruch, Austria, 1965

3. Mikhalev A.V., Pankrat'ev E.V. Differential dimension polynomial of a system of differential equations. Algebra, Collect., Moskva, 1980, 57-67 (Russian) (1980)

4. Mora F., Möller H.H. The computation of the Hilbert function. Proc. EUROCAL'83. Lect. Notes Comput. Sci., 162, (1983), pp. 157-167.

5. Sit W. Well-ordering of certain numerical polynomials. Trans. Amer. Math. Soc., 212, (1975), 37-45

6. Kondrat'eva M.V., Pankrat'ev E.V., Algorithms of the computation of the characteristic Hilbert polynomials. Pakety prikladnyh programm. Analiticheskie wychislenija. Collect., Moskva, "Nauka", p. 129-146 (Russian) (1988)

Let AB be the shortest side of the triangle, J the middle point of this side. If $H(C)$ is defined by $H(C) = \{M \mid \vec{AM} = \epsilon(\vec{AC} - m\vec{AB}),\ m \in \mathbf{Z},\ \epsilon = \pm 1\}$, we replace the vertex C by a point $h(C)$ of $H(C)$ whose projection lies inside $[A, J]$. Notice that $h(C)$ is easy to calculate with the quotient of $(\vec{AB}, \vec{AC})$ with $(\vec{AB}, \vec{AB})$.

Algorithm GaussAff

Repeat

1. Permute (A, B, C) in such a way that $c^2 \le a^2 \le b^2$;
2. Replace C by $h(C)$;

until $a^2 + b^2 \ge c^2$.

The output triangle is *acute* –*i.e.* its three angles are acute–, and minimal. As long as the triangle has an obtuse angle, the *inertia* –*i.e.* the sum of the squares of the sides– is geometrically decreasing. Thus, we can characterise the worst-case configuration of the GaussAff Algorithm: it is linked to the linear recurrence $v_{n+1} = 2v_n + v_{n-1}$ that shows up in the worst-case of the modified continued fraction algorithm, as shown by Dupré [1]. This is consistent with the observation that Gauss' Algorithm is a 2–dimensional generalisation of the modified continued fraction algorithm. From there, we deduce:

Theorem 1: *Let (A, B, C) be a triangle of length M of the lattice L. Then the maximal number k of loops executed by GaussAff on the triangle ABC satisfies* $k \le \log_{1+\sqrt{2}}(M) + 1$.

3. The Acute Algorithm.

We now generalise the ideas of GaussAff Algorithm to 3 dimensions. Starting with a tetrahedron of $\mathbf{Z}^3$, we build a succession of tetrahedra whose *inertia* –*i.e.* the sum of the squares of the six edges– is gradually reduced and obtain in polynomial time a *hyperacute* tetrahedron.
A tetrahedron is called *hyperacute* iff its four faces form *acute* triangles and at least one of the vertices D is projected *inside* its opposite face.

A description of the Acute Algorithm.

Let $ABCD$ be an affine basis of a lattice L of $\mathbf{Z}^3$;

1. Find the minimal inertia face, say ABC, of the tetrahedron. Apply GaussAff to ABC, which reduces inertia; call again ABC the face so obtained, and let I be its inertia.

2. Bring D closer to that face ABC. To do so, select in the bi-lattice

$$L(D) = \{M \mid \vec{AM} = \epsilon(\vec{AD} - m\vec{AB} - n\vec{AC}),\ (m, n) \in \mathbf{Z}^2,\ \epsilon = \pm 1\}$$

a point $l(D)$ which is projected inside the triangle ABC. Note that (m, n, ϵ) are easily calculated from subdeterminants of the Gram determinant of vectors $\vec{AB}, \vec{AC}, \vec{AD}$.
Now, the tetrahedron has a vertex D which is projected *inside* its opposite *acute* face.

3. Test whether the three angles in C are acute:

if *yes*: exit; the tetrahedron is *hyperacute*.
if *no*: further construct the adjacent tetrahedron which contains the obtuse angle, say ADB and the point C'' symmetrical to C w.r.t. the middle point K of $[A, B]$. Amongst the two adjacent tetrahedra, retain the one that has smaller total inertia and return to step 1.
Note: that tetrahedron has a face whose inertia is less than $5I/6$.

The complexity of the Acute Algorithm.

Let M be the length of the input tetrahedron –*i.e.* the maximum of the lengths of its edges–. The geometrical decrease of I inside the Acute Algorithm and inside the GaussAff Algorithm allows us to show that *the total number of loops* of all executions of the GaussAff Algorithm called by the Acute Algorithm is *upper-bounded by a linear function of* $\log M$. We can next bound the lengths of the vectors themselves, and we thus obtain:

Theorem 2: *When applied to an input tetrahedron of length M, the Acute Algorithm has a worst-case complexity of at most* $O(\log M)\ \mu(M)$ *where $\mu(M)$ is the cost of a multiplication performed on two integers of absolute value less than M.*

The Output Configuration of the Acute Algorithm.

We remark that a hyperacute tetrahedron has at most one obtuse dihedral and we are then lead to the main result:

Theorem 3: *Any lattice L of $\mathbf{R}^3$ has an affine basis which is hyperacute and contains three successive minima of L. Conversely, any hyperacute tetrahedron which is an affine basis of a lattice L of $\mathbf{R}^3$ contains the first two minima of the lattice.*
If the hyperacute tetrahedron has no obtuse dihedral, then the third minimum is also an edge of that tetrahedron. Otherwise, the third minimum is an edge of the octahedron obtained by the reunion of the tetrahedron and its symmetrical w.r.t. the middle of the edge of the obtuse dihedral.

As a consequence, we obtain:

Theorem 4: *When applied on a lattice L of $\mathbf{Z}^3$ given by a basis of length M, the Acute Algorithm permits to obtain, in time polynomial in $\log M$, three successive minima of L.*

Comparing with the LLL Algorithm in 3 dimensions.

The LLL algorithm in 3 dimensions –called here L33– also operates with the Gram matrix and performs operations that resemble to some extent our Acute Algorithm. Similar complexity analyses can be conducted and we obtain worst-case complexity bounds for the two algorithms. From those bounds, our algorithm thus seems to run "faster". We have in effect conducted an experimental study [5], based on MACSYMA programs for both methods. Even though the interpretation mechanism of MACSYMA somewhat penalises our implementation of GaussAff (due to a larger number of comparisons/exchanges), simulations confirm that, on the average, our algorithm tends to exhibit better convergence properties and is, asymptotically, of a lower arithmetic complexity.

References.

[1] A. DUPRÉ: *Journal de Mathématiques* **11** (1846), pp 41-64

[2] J. LAGARIAS: Worst-Case complexity bounds for algorithms in the theory of integral quadratic forms, *Journal of Algorithms* **1** (1980), pp 142-186.

[3] A.K. LENSTRA, H.W. LENSTRA, L. LOVASZ: Factoring polynomials with rational coefficients, *Math. Annalen* **261** (1982), pp 513-534.

[4] J. STERN: Lecture Notes, National University of Singapore (1986).

[5] B. VALLÉE: Thèse de Doctorat, Université de Caen (1986).

A COMBINATORIAL AND LOGICAL APPROACH TO LINEAR-TIME COMPUTABILITY (EXTENDED ABSTRACT)

P. Scheffler and D. Seese
AdW der DDR, Karl-Weierstraß-Institut für Mathematik
Mohrenstraße 39, Berlin, DDR-1086

The paper is devoted to an investigation of the connection between the structure of graphs and the complexity of algorithmic graph problems. Especially, polynomial and linear-time computability results are regarded. Robertson and Seymour [8] defined: A <u>tree-decomposition</u> of a graph G is a pair $(T,\mathfrak{X})$ where T is a tree and $\mathfrak{X} = \{X_t: t\in V(T)\}$ is a family of subsets of V(G) with the following properties:

(1) $\bigcup_{t\in V(T)} X_t = V(G)$.

(2) For every edge e of G there exists a $t \in V(T)$ such that e has both ends in X_t.

(3) For $\{t,t',t''\} \subseteq V(T)$ holds if t' is on the path of T between t and t'' then $X_t \cap X_{t''} \subseteq X_{t'}$.

The <u>width</u> of a tree-decomposition is $\max_{t\in V(T)} |X_t| - 1$ and the <u>tree-width</u> of a graph G is the minimum width of a tree-decomposition of G. The following classes of graphs have bounded tree-width: trees, forests, k-almost trees, partial k-trees, partial k-chordal, series-parallel, Halin, k-outerplanar, bandwidth-k and cutwidth-k graphs.

A graph property is called <u>L-existential locally verifiable (L-ELV)</u> iff it is expressible in the form

$$\exists G_1\ \exists G_2 \ldots \exists G_k\ \forall a\ \ P'(G,G_1,\ldots,G_k,a) \wedge P_1(G_1)$$

where $a \in V(G)$, the G_i are (possibly mixed) subgraphs of G, P' is a locally verifiable property, i.e. $P'(G,G_1,\ldots,G_k,a)$ holds if and only if $P'(G',G_1',\ldots,G_k',a)$ holds, where G' and G_i' $(1\le i\le k)$ are the subgraphs of G, G_i respectively induced for some fixed number r by the vertex set $\{b: b\in V(G)$ and a and b have distance at most r in $G\}$ and P_1 is a boolean combination of the following atomic properties (verifiable in linear time): "G_1 is connected", "G_1 is acyclic", "$f(G_1)\le B$" and "$f(G_1)\ge B$", where B is some constant and f is a weight function on the set of subgraphs such that $f(G' \cup G'') = f(G') + f(G'') - f(G' \cap G'')$ for all subgraphs G' and G'' of a graph G.

<u>Theorem 1:</u> The following graph properties are L-ELV-properties:
G has no monochromatic triangle, G has a vertex cover of size ≤k, G has a dominating set of size ≤k, G has a partial feedback edge set of size ≤k, a minimum maximal matching has ≤k edges, G contains an independent set of size ≥k, G has a cubic subgraph, G has a maximal cut of weight ≥B, G has a Hamiltonian circuit, G has a Hamiltonian path, G has an induced path of length ≥k, G has a Steiner tree of weight ≤B, G has a spanning tree with maximum degree ≤k, G has a spanning tree with ≥d leaves. Some other properties are L-ELV-properties for fixed k and d: chromatic number ≤k, domatic number ≥k, distance d chromatic number ≤k,

chromatic index ≤k, G has a partition into ≤k cliques, disjoint paths between k pairs of vertices exist, G has a partition into ≤k perfect matchings.

Theorem 2: Let k and d be fixed natural numbers and P a L-ELV-property. Then there is an algorithm deciding P(G) in time polynomial in |V(G)| for all graphs with tree-width at most k and vertex degree at most d. Moreover, there is a linear-time algorithm if G is given with a corresponding tree-decomposition.

The degree bound on the input graphs can be omited for the subclass of clique-verifiable properties, which are defined similar as the L-ELV-properties only using cliques instead of the r-neighbourhoods. Examples of clique-verifiable properties are: G has no monochromatic triangle, G has a vertex cover of size ≤k, G has an independent set of size ≥k, G has a maximal cut of weight ≥B, G has a Steiner tree of weigth ≤B, the chromatic number of G is ≤k. The result generalizes some results from [1, 4, 5, 7, 11, 12, 13, 14] and solves some problems from [6].

Note added in proof:
Recently, theorem 2 was proved without degree bound also for the more general class of Extended Monadic Second-order (EMS-) properties [3].

References

[1] Arnborg S.: Efficient algorithms for combinatorial problems on graphs with bounded decomposability - a survey: BIT 25(1985), 2-23
[2] Arnborg S., Corneil D.G., Proskurowski A.: Complexity of finding embeddings in a k-tree: SIAM J. Alg. Disc. Meth. 8 (1987), 277-284
[3] Arnborg S., Lagergren J., Seese D.: What problems are easy for tree-decomposable graphs: to appear
[4] Bern M.W., Lawler E.L., Wong A.L.: Linear-time computations of subgraphs of decomposable graphs: J. Alg. 8 (1987), 216-235
[5] Corneil D.G., Keil J.M.: A dynamic programming approach to the dominating set problem on k-trees: SIAM J. Alg. Disc. Meth. 8 (1987), 535-543
[6] Johnson D.S.: The NP-completeness column: an ongoing guide: J. Algorithms 6 (1985), 434-451
[7] Robertson N., Seymour P.D.: Graph minors. II. Algorithmic aspects of tree-width: J. Algorithms 7 (1986), 309-322
[8] Robertson N., Seymour P.D.: Graph minors. V. Excluding a planar graph: J. Comb. Th. B 41 (1986), 92-114
[9] Scheffler P.: Dynamic programming algorithms for tree-decomposition problems: AdW der DDR, K.-Weierstraß-Inst. für Mathematik, Berlin 1986 (P-MATH-28/86)
[10] Scheffler P.: Linear-time algorithms for NP-complete problems restricted to partial k-trees: AdW der DDR, K.-Weierstraß-Inst. für Mathematik, Berlin 1987 (R-MATH-03/87)
[11] Seese D.: Tree-partite graphs and the complexity of algorithms: AdW der DDR, Karl-Weierstraß-Institut für Mathematik, Berlin 1986 (P-MATH-8/86)
[12] Takamizawa K., Nishizeki T., Saito N.: Linear-time computability of combinatorial problems on series-parallel graphs: J. ACM 29 (1982), 623-641
[13] Wimer T.V., Hedetniemi S.T.: K-teminal recursive families of graphs: Technical Report 86-May-6
[14] Wimer T.V., Hedetniemi S.T., Laskar R.: A methodology for constructing linear graph algorithms: Congr. Numer. 50 (1985), 43-60

COMPLEXITY OF COMPUTATION OF EMBEDDED RESOLUTION OF ALGEBRAIC CURVES

J.P.G. Henry and M. Merle

Centre de Mathématiques de l'Ecole Polytechnique
F-91128 Palaiseau Cedex
Unité associée au CNRS n° 169
and GRECO Calcul Formel n° 60

ABSTRACT

We study the complexity of an algorithm we gave in a former paper to compute an embedded resolution of an irreducible singular algebraic curve.

This is more complex than just finding the Puiseux expansion associated to the curve, but this computation is also more interesting because it gives not only the Puiseux pairs, and the Puiseux series but also a way to work on some other invariants of the curve (see [D] for a study of complexity of computing Puiseux Pairs, with an emphasis on the reducible case, and algebraic numbers, and [5D] for the related algorithms).

For example it will allow the mathematician to work on the mixed Hodge structure of the curve.

This complexity is shown to be polynomial in terms of the degree *d* of the polynomial of the curve.

We will try to make a study of the complexity strongly related to the real algorithm we are using; in fact this study comes after, and is motivated by, two implementations we made of resolutions of irreducible curves ([H.M]) *.

---

* Since we wrote this paper, a faster implementation of this algorithm has been done by G. Moreno.

1. THE MATHEMATICAL ALGORITHM

Resolution of singularities of complex algebraic curves is well known after the works of Newton, Puiseux, Noether, Zariski and his school (Abhyankar ...)(see [AB] for a survey on this subject, and [OZ] for a good introduction). The problem is in fact local in the neighbourhood of the singular points of the curve.

1.1. Embedded resolution of curves.

The datum is a **polynomial** F in two variables, with coefficients in $\mathbf{Z}$. One can easily reduce to the case where F is square free.

We want to compute invariants of **the germ at the origin of the function** F, such that:

- the resolution data of F, that is the minimal sequence of transformations necessary to turn F in a normal crossing function, together with the multiplicity of the exceptional divisors and their intersection diagram,
- the Puiseux pairs of each component of the germ $(F^{-1}(0), O)$ together with the intersection multiplicities, and the beginning of the Puiseux expansions of the local components,
- the order along any exceptional divisor of a given function of the two variables x and y, or a given differential form on the (x, y)-plane.

1.2. The algorithm.

The resolution algorithm is simple:

```
if the multiplicity is 1
    then stop
    else
        until the ancient multiplicity is 1 and the multiplicity is 0
            Blow-up the origin
            Translation

procedure Blow-up the origin
    until the multiplicity is 0
        Blow-up,
        Compute the equation of the exceptional divisor,
        Compute the strict transform,
        Replace the initial equation by the strict transform,
        Compute the new multiplicity.

procedure Translation
    compute the new singular points of the strict transform,
    choose a singular point,
    make a translation of coordinates to bring this singular point to the origin.
```

The program stops because multiplicity cannot increase after a blow-up and if it does not decrease, some other integer invariants must decrease.

This algorithm gives also the tree of multiplicities, and therefore the Milnor number.

1.3. An overview of the complexity.

At each step of the resolution process, the number of operations to perform is strongly related to the number of terms and to the degree of the polynomial, total transform of F. In fact, the computation of the strict transform has a complexity which depends only on the number of terms; but the complexity of the translation depends on both the number of terms and on their degrees in y.

So the first step in bounding the complexity of the resolution algorithm is to bound the degree of the polynomial. As the degree of the total transform is clearly greater than the degree of the strict transform, we will work essentially on the former. This will be the point in the next paragraph.

2. BOUNDS FOR THE DEGREE OF THE TOTAL TRANSFORM

2.1. Notations and prerequisites.

We consider a polynomial F of two variables x and y of total degree d such that F is irreducible as a power series.

In a coordinate system (x, y) where $x = 0$ is transversal to the curve C of multiplicity n at O, we consider a parametrisation of C given by:

$$y = \sum_i c_i x^{\frac{i}{n}}.$$

We define the first Puiseux exponent β_1 as the minimum of the exponents such that $c_i \neq 0$ and i not divisible by n. Let e_1 the gcd of β_1 and n. The other ones are defined by induction:

$\beta_0 = 0$ (by convention)

β_k is the first i with $c_i \neq 0$ and i not divisible by e_{k-1}. We define $e_k = gcd(\beta_k, e_{k-1})$. We define also integers $m_k = \beta_k / e_k$ and $n_k = e_{k-1}/e_k$, (we set $(n_0 = 0)$.

The kth Puiseux pair is (m_k, n_k). We stop with $e_g = 1$.

Let us recall the following formulas ([OZ]):

$$\begin{aligned}\Delta &= \sum_1^g \beta_k(e_{k-1} - e_k)\\ &= \sum_1^g (\beta_k - \beta_{k-1})e_{k-1} - \beta_g\end{aligned}$$

with the convention $\beta_0 = 0$ and where Δ is the usual notation for the multiplicity of the discriminant at O. As Δ is the intersection multiplicity at O of the curves defined by the equations F and F'_y, it is clearly bounded above by the product $d(d-1)$ because of Bézout's theorem.

2.2. Pseudo Puiseux pairs and Puiseux pairs.

Let (m_1, n_1) the first Puiseux pair of C, and $[a_0, a_1, \ldots, a_s]$ the continued fraction expansion of m_1/n_1. If we perform $a_0+a_1+\cdots+a_s$ blowing-ups centered at the intersection of the strict transform with the exceptional divisor, we get a curve with $g-1$ pairs $(m_k - m_1 n_2 n_3 \ldots n_k, n_k)$, $2 \leq k \leq g$. Notice that the multiplicity of the strict transform decreases if and only if we have just

2.2.1. Definition. - *The process of resolution of the curve is performed by three kinds of changes of variables:*

$$\text{type 1: } (x,y) = (x, xy),$$
$$\text{type 2: } (x,y) = (xy, x),$$
$$\text{type 3: } (x,y) = (x, y+a),$$

for some a in **C**.

2.2.2. Remark.

Under any of these transformations, the partial degree in x of all the monomials of the transform of a given monomial $x^u y^v$ is greater than u.

2.2.3. Remark.

Except eventually before the first blowing-up, the partial degree in x of any monomial of the transform of a polynomial G is greater than its partial degree in y.

2.2.4. Definition. - *We define the sequences $(f_0, \dots, f_b)$ and $(\gamma_0, \dots, \gamma_b)$ as follows:*
f_0 is the order of the polynomial $F(0,y)$,
$f_\ell, (1 \le \ell \le b)$ is the quotient $f_{\ell-1}/q_\ell$.
We then denote by γ_ℓ the product $p_\ell \cdot f_\ell$.

2.2.5. Definition. - *We define inductively the* **pseudo Puiseux pairs** *(pPp) of the polynomial F as follows:*

The first one (p_1, q_1) is associated to the continued fraction expansion $[h_0, h_1, \dots, h_r]$ where h_j is the length of the sequence of transformations

- beginning with the jth type 2 transformation if $j \ge 1$ and ending just before the $j+1^{st}$ type 2 transformation if $j \le r-1$, or just before a type 3 transformation if $j = r$.

- preceding the first type 2 transformation otherwise.

The first ℓ pseudo Puiseux pairs being defined, the $(\ell+1)^{th}$ is defined as follows:

We build on the same pattern a sequence $[h_0^{\ell+1}, h_1^{\ell+1}, \dots, h_{r_{\ell+1}}^{\ell+1}]$; the pair $(p_{\ell+1}, q_{\ell+1})$ is then associated to the continued fraction expansion $[h_{r_\ell}^\ell + h_0^{\ell+1}, h_1^{\ell+1}, \dots, h_{r_{\ell+1}}^{\ell+1}]$.

We stop with (p_b, q_b) when the strict transform is smooth and transversal to the exceptional divisor, that is $f_b = 1$.

2.2.6. Properties.

The set $\{f_0, \dots, f_b\}$ is equal to the set $\{e_0, \dots, e_g\}$ if $F(0,y)$ has the same order as F and equal to $\{f_0, e_0, \dots, e_g\}$ otherwise.

After the ℓth translation, that is the ℓth type 3 transformation, the equation of the strict transform F_ℓ of F has order f_ℓ and the slope of the Newton polygon of F_ℓ in the (x,y) coordinates is $-p_\ell/q_\ell$.

Notice that we are able to recover the Puiseux pairs just knowing the pPp. Given the sequence S of the continued fraction expansions associated to the pPp sequence, we reduce S applying the following process:

> each time an element of S is of type $[n]$ (resp $[0,n]$) with $n \in \mathbf{N}$ we suppress it from S changing the next member $[h_0, h_1, \dots, h_r]$ of S into $[h_0 + n, h_1, \dots, h_r]$ (resp $[0, h_0 + n, h_1, \dots, h_r]$); we will say that the associated pPp **belongs** to the next "true" Puiseux pair. Using repeatedly this process we come to the sequence of the continued fraction expansion associated to the Puiseux pair sequence.

2.3. Numerical invariants and blowing ups.

2.3.1. Proposition. - *Let (p,q) a pseudo Puiseux pair and $[h_0, h_1, \dots, h_s]$ the coefficients of the continued fraction expansion of p/q. Let $x^u y^v$ a monomial. In the appropriate chart, the total transform of this monomial under the sequence of $h_0 + h_1 + \dots + h_s$ blowing-ups is a monomial $X^{u_{s+1}} Y^{v_{s+1}}$ with*

$$u_{s+1} = pv + qu,$$
$$v_{s+1} = p_{s-1}v + q_{s-1}u,$$

where p_{s-1}/q_{s-1} is the last non exact convergent of the continued fraction expansion of p/q and X is the equation of the last exceptional divisor.

Proof : by induction on the length of the continued fraction expansion. The sequence of h_j blowing-ups induces a change:

$$y_j = x_{j+1},$$
$$x_j = x_{j+1}^{h_j} y_{j+1}.$$

One deduces the following relation on the exponents (u_{j+1}, v_{j+1}) of the transform of $x^u y^v$

$$u_{j+1} = h_j u_j + v_j,$$
$$v_{j+1} = u_j,$$

which gives:

$$u_{j+1} = h_j u_j + u_{j-1}, \quad 0 \le j \le s-1,$$
$$v_{j+1} = u_j$$

provided $u_0 = u$ and $u_{-1} = v$. One recognises the induction relation satisfied by the terms of the convergents of the continued fraction expansion of p/q:

If we note p_j/q_j the jth convergent of the continued fraction expansion of p/q, the integers p_j and q_j satisfy the relations:

$$p_{j+1} = h_j p_j + p_{j-1}, \quad 0 \le j \le s-1,$$
$$q_{j+1} = h_j q_j + q_{j-1}, \quad 0 \le j \le s-1,$$

and their sequence is initialized by:

$$p_0 = 1,$$
$$p_{-1} = 0,$$
$$q_0 = 0,$$
$$q_{-1} = 1,$$

therefore the sequences $(u_j)_j$ and $(v_j)_j$ are linear combination of the basic sequences $(p_j)_j$ and $(q_j)_j$. Thus we clearly have:

$$u_{s+1} = p_s u + q_s v$$
$$v_{s+1} = p_{s-1}u + q_{s-1}v.$$

Hence the proposition is proved.

2.3.2. Corollary. - *Let (m,n) a Puiseux pair and $[a_0, a_1, \dots, a_s]$ the coefficients of the continued fraction expansion of m/n. Let $x^u y^v$ a monomial. In the appropriate chart, the total transform of this monomial under the sequence of $a_0 + a_1 + \dots + a_s$ blowing-ups is a polynomial, the monomial of maximal degree $X^{u_{s+1}} Y^{v_{s+1}}$ of this polynomial is such that:*

where m_{s-1}/n_{s-1} is the last non exact convergent of the continued fraction expansion of m/n and X is the equation of the last exceptional divisor.

2.3.3. Remark.

The number v_{s+1} is clearly bounded above by $mv + nu$ because

$$m_{s-1} \leq m_s = m$$
$$n_{s-1} \leq n_s = n.$$

2.3.4. Proposition. - *Let (p_ℓ, q_ℓ) a pPp of the function F. When we perform the $(\ell+1)^{th}$ type 3 transformation that finishes the process attached to this pPp, the order of the total transform of F along the last exceptional divisor is $q_\ell \overline{\gamma}_\ell$, where $\overline{\gamma}_\ell$ is given by the induction formula:*

$$\overline{\gamma}_0 = f_0,$$
$$\overline{\gamma}_{\ell+1} = q_\ell \overline{\gamma}_\ell + \gamma_{\ell+1}.$$

Proof : By induction on ℓ, we know that the total transform of F after the ℓth translation has a weighted homogeneous initial form:

$$x^{q_{\ell-1}\overline{\gamma}_{\ell-1}} P_{\ell-1}(x, y),$$

where P_ℓ is a non-zero weighted homogeneous polynomial of weight $f_{\ell-1}p_\ell = f_\ell \gamma_\ell$ if x is of weight q_ℓ and y of weight p_ℓ.

We then apply the proposition 2.3.1. to any monomial of this weighted homogeneous initial form and we get the expected result: the exponent of x in the initial form of the total transform of F after the $(\ell+1)^{th}$ translation is equal to:

$$q_\ell(q_{\ell-1}\overline{\gamma}_{\ell-1} + f_\ell p_\ell) = q_\ell \overline{\gamma}_\ell.$$

2.3.5. Remark.

When the translation is associated to a "true" Puiseux pair, the order of the total transform of F along the last exceptional divisor is nothing but the quantity: $n_k \overline{\beta}_k$ where the $\overline{\beta}_k$ satisfy the following induction relation :

$$\overline{\beta}_0 = n,$$
$$\overline{\beta}_{k+1} = n_k \overline{\beta}_k + \beta_{k+1} - \beta_k$$

for $0 \leq k \leq g-1$. This shows that the set $\{\overline{\beta}_0, \ldots, \overline{\beta}_g\}$ is a subset of $\{\overline{\gamma}_0, \ldots, \overline{\gamma}_b\}$. Moreover, $\overline{\gamma}_b$ is equal to $\overline{\beta}_g$. As the $\overline{\gamma}_\ell$'s make an increasing sequence, they are bounded by $\overline{\beta}_g$, hence by Δ, for we have the formula:

$$\Delta = \sum_{k=1}^{g} (n_k - 1)\overline{\beta}_k.$$

2.3.6. Lemma. - *The degree of the total transform of F under the sequence of blowing-ups which give an embedded resolution of F is bounded by the product:*

$$2^g d m'_1 m'_2 \ldots m'_g$$

where m'_k is the maximum of the two integers $m_k - n_k m_{k-1}$ and n_k.

3. THE COMPLEXITY OF LAZINESS AND EAGERNESS

3.1. The importance of being lazy.

We have written two versions of the algorithm computing the embedded resolution of irreducible curves:

In the first version, the program had to compute at each step the full strict transform of F. Therefore it had, after each blow-up, to compute the 2 exponents of all monomial terms and then proceed to a reordering on this list, i.e to a sort.

After each translation on y, it had to expand the transforms of all monomials in y, then sum and sort all the new monomial terms. This was rather inefficient. In a later version we decided that the program should be *lazy*: at each step it was only necessary to compute the monomial terms that were needed to decide what should be the next operation. Roughly speaking, it was only necessary to compute the monomials in the initial form of the strict (or total) transform of F.

This was reached, in an automatic way, by using what is widely known as *lazy evaluation* or *call by need* (see [ASS] for a fine introduction to these points), and choosing an appropriate way of implementing data; we had to find a "good way" of writing the polynomial F and its successive transforms.

In fact, one can easily see that the chosen representation kept track of all the monomials in the initial weighted form corresponding to the first side of the Newton polygon.

This later algorithm was far more efficient than the former (cf [HM]).

We will now explain the mathematical background of this divergence, by computing the complexities of the two processes; we will show that they are far from being equivalent.

We will first give a precise definition of the main output we expect from the *lazy resolution*, and that we will call a **resolution sequence**.

3.1.1. Definition. - *Let F a polynomial in two variables x and y, irreducible as a power series in x and y. We will call S* **a resolution sequence** *of F if it is a sequence $[r_1, \ldots, r_s]$ where r_j is an element of the set $\{1, 2, 3_a; a \in \mathbf{C}\}$, and such that the jth transformation of the resolution process of F is a type 1 (resp type 2) transformation if r_j is equal to 1 (resp 2) and a translation on y of module a otherwise.*

3.2. Complexity of the LAZY resolution.

At each step of the resolution process, we only need to know the weighted quasi-homogenous polynomial, the weighted initial form of the strict transform of F.

The transformations of type 1 and 2, i.e. the blowing-ups, have little effect on this initial form: the exponents of each monomial are modified by the same linear application, which has been computed in the preceding paragraph.

In order to compute the sequence of transformations which resolve the curve, it is only necessary to compute at each step i the monomials which must be known at step $i, i+1, \ldots, s$, where s is the total number of blowing-ups and translations.

3.2.1. Proposition. - *The partial degrees of any monomial of the total transform of F which needs to be computed at any step of the resolution process are bounded above by $\overline{\beta}_g$*

Proof : To compute the weighted initial form of the total transform of F after the ℓth translation,

Because of the remarks 2.2.2 and 2.2.3 these terms can only be generated by terms the partial degrees of which are less than $\overline{\gamma}_\ell$. We know (Remark 2.3.5) that all the integers $\overline{\gamma}_\ell$ are bounded above by $\overline{\beta}_g$.

3.2.2. Proposition. - *Let G be a polynomial of total degree d'. In order to compute the orders of the total transform of G along all the exceptional divisors of F we only need to*

- *use the lazy algorithm and compute the* **resolution sequence**
- *compute, in the successive total transforms of G, only those monomials whose exponents are lower than dd'.*

Proof : If D_i is one of the exceptional divisors in the process of desingularization of F, and ν_i the exponent of D_i in the total transform of G, then the intersection multiplicity of F and G at the origin $(F \cdot G)_o$ is greater than ν_i by the projection formula. Thus, as $(F \cdot G)_o \leq dd'$ by Bézout's theorem, the proposition is proved.

3.2.3. Proposition. - *Let F be a polynomial of total degree d. The complexity of the* **lazy resolution** *of F is of order $O(d^8)$.*

Proof : 1) Let us first evaluate the complexity of each one of the elementary transformations implied by the resolution process:

Blow-ups
Translations

At each blow-up the program must

transform the exponents of each relevant monomial,
find the terms of lowest degree.

As the relevant monomials are of partial degrees bounded above by Δ, their number is less than Δ^2. Therefore the transformations on exponents is of a complexity in $O(\Delta^2)$. To find the term of lowest degrees is an operation of complexity less than the number of the terms to compare which is $O(\Delta^2)$. Finally the complexity of type 1 or type 2 operations is in $O(\Delta^2)$.

For translation we need to do 3 operations

1- Compute the root of a polynomial
2- Expand the $(y+a)^j$ in each coefficient of x^i
3- Add up the monomials in y, in each coefficient of the x^i and form again the relevant polynomial.

As we restrict ourselves for the moment to the irreducible case, the first operation is elementary, because the polynomial in y, the root of which we need, is necessarily of the form $(y-a)^k$.

As the partial degrees of all relevant monomials are at most Δ, we have to expand, in less than Δ polynomials in y, less than Δ terms, each one of degree less than Δ.

How do we do this?

In fact, it seems that the best way is to compute successive powers of $y+a$ and keep track (by some sort of what is generally known as *memo-ization*(see [HUG])) of all those until all the translations $y \longrightarrow y+a$ are finished. Let's find by recursion the complexity of such a process: Suppose that we have already computed $(y+a)^\rho$ and that we have got the result as an ordered list of coefficients, $c_0, \ldots, c_\rho$ beginning by the constant term and ending by the coefficient of y^ρ.

Then to compute the next power, we need to do the following operations:

`shift` the previous list towards right into $[0, c_0, \ldots, c_\rho]$,
`multiply` each term of the previous list by a, and put 0 at the end ($\rho+1$ multiplications),

Therefore to compute all the powers from 1 to Δ of $y+a$ is of complexity in $O(\Delta^2)$. Let us notice that such lists of coefficients are ordered.

When we want to compute the polynomial in y which is the coefficient of x^i, we first observe that, because of remark 2.2.3, there was initially only monomials in y of degree less than i; hence it is clear that to compute the new polynomial in y after translation by a we only have to proceed to (at most) $1+2+\cdots+(i+1)$ multiplications, then to (at most) $(i+\cdots+1)$ sums, which means that we have less than $O(i^2)$ operations for computing the new coefficient of x^i, knowing the powers of $y+a$. Clearly to compute all the coefficients, polynomials in y, of the x^i is in $O(\Delta^3)$.

So the cost of a translation is at worse in $O(\Delta^3)$.

2) How many blow-ups and translations do we have to perform? Let us point out that each translation (except, may be, the last one) is followed by a blow-up. We know that

$$\Delta = \sum_i e_i(e_i - 1) + e_0 - 1$$

where e_i is the multiplicity (order) of the strict transform of F at the ith blow-up. As $e_i \geq 2$ we have that the number of blow-ups is less than Δ. Thus there is less than Δ translations. It follows clearly that the total complexity of **lazy resolution** is of order $O(\Delta^4)$ which, as $\Delta \leq d^2$, is less than $O(d^8)$, as stated in the proposition.

3.2.4. Proposition. - *Let G be a polynomial of total degree d'. In order to compute the orders of the total transform of G along all the exceptional divisors of F we only need to proceed to $O(d^5 d'^3)$ elementary operations.*

Proof : We have to operate at most $O(d^2)$ translations, because of what we showed in the proof of the preceding proposition. The partial degrees of the relevant monomials are bounded above by dd' (instead of d^2 in the later proof) as stated in proposition 3.2.2. By the same arguments as in proposition 3.2.3. we find that the complexity of each translation is of order $O((dd')^3)$. Finally we find a complexity in $O(d^5 d'^3)$.

3.3. The complexity of eagerness.

The aim here, is to compute all the total (resp strict) transform at each step of the resolution process. Hence, we do a global affine process, not only a local one above a neighbourhood of the singular point, which was mimicked by the lazy process.

The lemma 2.3.6 gives an upper bound for the degrees of the total transform of F. We want to give now an example to show that this bound is essentially the best one.

3.3.1. Example.

Consider the following Puiseux expansion:

$$y = \sum_{k=1}^{g} x^{(1-1/2^k)m},$$

where m is an odd integer. This defines a curve of multiplicity 2^g, with g Puiseux pairs such that: $\beta_k - \beta_{k-1} = 2^{g-k+1}m$ for all k between 1 and g. The g integers $m'_1, \ldots, m'_g$ are all equal to m.

When we make the product of the 2^g possible Puiseux expansions, we get a polynomial F with a monomial of maximal degree, $cx^{(2^g-1)m}$, $(c \neq 0)$, hence of degree: $d = (2^g - 1)m$.

References.

[ABH] S.S. Abhyankar. *Desingularization of plane curves,* Proceedings of symposia in Pure Mathematics of the A.M.S. **40**, 1, 1-45, 1981.

[ASS] Harold Abelson and Gerald Jay Sussman with Julie Sussman. *Structure and interpretation of computers programs,* The MIT electrical engineering and computer science series, The MIT press, Mc Graw-Hill Book Company, 1985.

[D] Dominique Duval. *Rational Puiseux series,* preprint Groupe de calcul Formel de Grenoble, Institut Fourier, B.P. 74, 38402 Saint-Martin-d'Hères. France.

[2D] Claire Dicrescenzo et Dominique Duval. *Algebraic computations on algebraic numbers,* Informatique et calcul, Wiley-Masson, 1985.

[3D] Claire Dicrescenzo et Dominique Duval. *Calculs algébriques avec des nombres algébriques: exemple,* Journées de calcul formel. Luminy. Calsyf **4**, 3-28, 1985.

[5D] Jean Della Dora, Claire Dicrescenzo et Dominique Duval. *About a new method for computing in algebraic number fields,* EUROCAL 1985 volume 2, Lecture notes in computer science **204** Springer.

[HM] J.-P. Henry and M. Merle. *Puiseux Pairs, Resolutions of Curves and Lazy Evaluation,* preprint Centre de Mathématiques, Ecole Polytechnique, F-91128 Palaiseau Cedex.

[HUG] John Hughes. *Lazy memo-functions,* Functional programming languages and computer architecture. Editeur Jean-Pierre Jouannaud, Nancy, France, Lecture notes in computer science **201**,Springer, 1985.

[MIC] D. Michie. *'Memo' functions and machine learning,* Nature, **218**, 19-22, 1968.

[OZ] Oscar Zariski. *Le problème des modules pour les branches planes,* Cours donné au Centre de Mathématiques de l'Ecole Polytechnique, rédigé par F.Kméty et M.Merle. Publications du Centre de Mathématiques de l'Ecole Polytechnique, 1973, or Hermann 1986.

Polynomial Factorisation: an Exploration of Lenstra's Algorithm

J A Abbott and J H Davenport
School of Mathematical Sciences
University of Bath, Bath, Avon, England BA2 7AY

Abstract We describe various design decisions and problems encountered during the implementation of the Lenstra factoriser [Lenstra82] in REDUCE. A practical viewpoint is taken with descriptions of both successful and unsuccessful attempts at tackling some of the problems. Particular areas considered include bounding coefficients of factors, the Cantor-Zassenhaus factoriser [Cantor81], Hensel lifting, basis reduction, and trial division of polynomials. We give an empirical formula which estimates the running time of our basis reduction routine.

1. Introduction

The motivation for our study of the factorisation of polynomials over algebraic number fields stems from the general problem of symbolic integration. The factorisation package is written in REDUCE and is intended to be used directly by the user as well as indirectly through a new integration package. There have been several other factorisation algorithms proposed, for example those in [Trager76], [Wang76], and [Weinberger76], but they have been too slow for widespread use.

In the rest of this paper we shall assume some knowledge of algebraic number theory, such as the notions of algebraic integers, field conjugates, and integral bases. When discussing basis reduction we have tried to keep the notation compatible with [LLL82]. We begin by explaining our view of algebraic number fields and how we represent their elements.

1.1. Underlying Algebraic Number Code

This section describes the representations used in the Bath Algebraic Number Package — for fuller details see [Abbott86]. The basic data-structure in REDUCE is a multivariate polynomial with coefficients from some *domain.* We have introduced a new domain of algebraic numbers (called "algebraics"). This approach allows us to use many of the existing routines in REDUCE without modification (except for a couple of bug fixes, now incorporated into REDUCE 3.3). The generic algebraic number is represented by a polynomial divided by a positive integer. The "variables" in the polynomial are *algebraic kernels,* and the degree in each algebraic kernel is less than that of the respective minimal polynomial. Algebraic kernels are created under the direction of the user (or a higher level package such as an integration package [Abbott86]), and for efficiency are restricted to being algebraic integers; so in general, an algebraic number created by the user will be an algebraic kernel divided a rational integer. The responsibility for avoiding non-trivial representations of zero lies with the user of the package (e.g. mixing algebraic kernels with "hidden" algebraic relations); no attempt whatsoever is made within the package to detect such relations. However, the factoriser described in this paper may be used to verify irreducibility of a minimal polynomial, or to find its factors (when the user must decide what action is to be taken). It would be possible to write a package, using these facilities, which checked for hidden relations, but this would be much slower.

Purely for use inside the Lenstra and Cantor-Zassenhaus factorisers, we have another domain of "modular algebraics," being algebraic extensions of finite fields. The generic element of this domain is a (degree reduced) multivariate polynomial in modular algebraic kernels with integer coefficients in the range $0, \cdots, modulus-1$. As in the non-modular case, the minimal polynomial corresponding to a modular algebraic kernel is *assumed* to be irreducible over the relevant field (including any other modular algebraic kernels with which it may be mixed).

2. Coefficient Bounds

The general form of Lenstra's algorithm is to compute a factorisation in a suitable finite field, then increase the p-adic accuracy of this factorisation, and finally deduce the true non-modular factorisation. The coefficient bounds are needed to decide how far to lift the modular factorisation so that all the irreducible factors will be found. Usually the theoretical bounds are gross overestimates, though extreme cases exist where the bounds are achieved.

2.1. Denominator Bound

Let O denote the ring of integers of our algebraic number field — those elements which are zeros of a *monic* polynomial over $\mathbb{Z}$. Recall that an *integral basis* is a $\mathbb{Z}$-basis for O. In the field $\mathbb{Q}(\alpha, \beta, \gamma, \cdots)$ the basis for O used in our code is

$$basis = \{\alpha^{e_\alpha}\beta^{e_\beta}\gamma^{e_\gamma} \cdots : 0 \le e_\alpha < degree(\alpha),\ 0 \le e_\beta < degree(\beta),\ \cdots\}$$

which is not always an integral basis, but it is a $\mathbb{Q}$-basis so we must allow rational coefficients: e.g. $\frac{1}{2}(1+\sqrt{5})$ is an algebraic integer since $x^2 + x - 1 = (x + \frac{1}{2}(1+\sqrt{5}))(x + \frac{1}{2}(1-\sqrt{5}))$. We have Gauss's Lemma which says that the factors of a polynomial $f(x) \in O[x]$ may be chosen to lie in $O[x]$; so we need to know about the denominators which can appear when using our representation for elements of O. If our representations used an integral basis this problem would not arise (though many others would).

Modular techniques necessarily produce answers involving integers, as opposed to rational numbers; so some conversion must occur. There are two ways of converting p-adic numbers to rationals. One requires a bound greater than the largest allowed denominator [Wang82], the other requires an integer multiple, D, of every denominator:

$$n \bmod m \rightarrow cancel\text{-}common\text{-}factors\left[\frac{nD \bmod m}{D}\right].$$

We adopted the latter simpler scheme since we can easily find a suitable D from the discriminant of our basis [Weinberger76]. By considering the determinant for the discriminant we get the following equation (assuming $\alpha, \beta, \gamma, \cdots$ are independent):

$$discr(basis) = N_\beta N_\gamma ... (\text{discr } m_\alpha) \times N_\alpha N_\gamma ... (\text{discr } m_\beta) \times N_\alpha N_\beta ... (\text{discr } m_\gamma) \times \cdots$$

where $N_\alpha : K(\alpha) \rightarrow K$, etc. are norms and $\text{discr} : O[x] \rightarrow O$ is the polynomial discriminant. We note that in this form the discriminant is already partly factorised which aids the calculation of D above. If we computed a primitive element for the field, the basis would be simpler and so would the code to evaluate the discriminant, but finding D would be harder — of course, the discriminant itself could be used as a sufficient denominator.

The best possible denominator bound could be found by determining an integral basis for O — see, for example, the algorithm in [Ford78]. The total factorisation times in the table below suggest that it is worth computing an integral basis, largely because the lattice basis reduction becomes much faster with the smaller numbers. The times in the column headed *Integral Basis* include the time spent computing the integral basis, whereas the times in the column headed *Free I.B.* do not.

Comparison of Denominator Bounds			
Example [Lenstra82]	Estimate [Weinberger76]	Integral Basis	Free I.B.
1	5.50	5.56	5.40
2	4.08	4.70	3.94
3	9.62	9.76	9.58
4	72.8	83.8	43.6
5	2198	1211	974

In all tables, times are in seconds using REDUCE 3.2 on HLH Orion 1 (≈ VAX 750).

2.2. Numerator Bound

Now we have a denominator bound, we must see how big the numerator of any coefficient of any factor of f can be. We follow the pattern laid out in [Weinberger76]: embed everything into $\mathbb{C}$;

(i) find a bound, β, on the moduli of the roots in $\mathbb{C}$;

(ii) deduce a bound, B, on the moduli of coefficients of any factor;

(iii) compute a bound on the representation of the coefficients using the chosen $\mathbb{Q}$-basis.

For (i), we use a binary chop to calculate the least integer greater than the positive root of

$$x^m - |a_{m-1}|x^{m-1} - \cdots - |a_0|$$

as suggested in [Wang76]. Clearly this is superior to using a formula which merely estimates the positive root. Step (ii) is just a binomial expansion, but it is useful to note that for $\beta \in \mathbf{R}^+$

$$\| (x+\beta)^n \|_\infty = \binom{n}{m}\beta^{n-m} \text{ where } m = \left\lfloor \frac{n+1}{\beta+1} \right\rfloor.$$

There are also a few formulae which perform steps (i) and (ii) directly, e.g. [Mignotte81]. It is a good idea to calculate several of them and take the smallest.

Step (iii) is a bit trickier. A key fact is that the magnitude bounds apply to *all* the field conjugates of the coefficients. For clarity we restrict discussion to a simple extension $\mathbb{Q}(\alpha)$, the generalisation to multiple extensions is not hard. Consider the following system of linear equations with right hand sides bounded in absolute value:

$$\begin{bmatrix} 1 & \alpha & \alpha^2 & \cdots & \alpha^{n-1} \\ 1 & \alpha_2 & \alpha_2^2 & \cdots & \alpha_2^{n-1} \\ \cdot & \cdot & \cdot & \cdots & \cdot \\ \cdot & \cdot & \cdot & \cdots & \cdot \\ 1 & \alpha_n & \alpha_n^2 & \cdots & \alpha_n^{n-1} \end{bmatrix} \begin{bmatrix} a_0 \\ a_1 \\ \cdot \\ \cdot \\ a_{n-1} \end{bmatrix} = \begin{bmatrix} b_0 \\ b_1 \\ \cdot \\ \cdot \\ b_{n-1} \end{bmatrix}$$

where the b_i are the field conjugates of $b_0 = \sum_{i=0}^{n-1} a_i \alpha^i$. By Cramer's rule the a_i are the ratio of two determinants, the bottom one being $\sqrt{discr(basis)}$. The top determinant can be bounded in various ways. Let $B \geq \max|b_i|$, and D be the denominator bound; then Hadamard's determinant bound leads to the numerator bound

$$\frac{DBn^{1/2n} \|\alpha\|^{1/2n(n-1)}}{\sqrt{discr(basis)}}$$

where $\|\alpha\| = max\{|\alpha|, |\alpha_2|, \cdots |\alpha_n|\}$. Expanding the determinant into a sum and bounding each summand gives

$$\frac{DBn!\|m_\alpha\|_2^{n-1}}{\sqrt{discr(basis)}},$$

see [Abbott87a] which also shows that there is a misprint in [Wang76]: the following bound, attributed to Weinberger, is invalid,

$$\frac{DBn!\|\alpha\|^{n-1}}{\sqrt{discr(basis)}}.$$

2.3. Heuristic Bounds

The bounds above are very pessimistic and lead to intermediate calculations with excessively large numbers. An alternative is to use a lower "probable" bound. If the bound is too low, the "irreducible" factors found may be reducible. So we can use the following scheme:-

(i) use the "probable" bound to obtain factors;

(ii) note which factors are guaranteed to be irreducible, e.g. factors coming from a single modular factor, or by consideration of a degree set (see §3);

(iii) calculate a theoretical bound for the remaining factors which should be lower than the theoretical bound for the original polynomial;

(iv) use this bound to factorise these remaining factors.

Note that as the basis reduction takes so long it is not worth using different bounds for each of the remaining factors in step (iv) — they should all be lifted to the same modulus so that a single basis reduction will suffice for them all. Also note that the modular factorisations of the remaining factors are still valid.

Suggestions for the bound in step (i) include the height of the input polynomial (which is fooled by cases like factorising x^2-2 over $Q(\alpha)$ where $(\alpha-999)^2-2=0$), the heuristic bound in

[Wang76], and even a constant bound, on the principle that any useful factorisation would be fairly small.

3. Modular Factorisation & Hensel Lifting

This section contains notes on choosing the prime, the modular factorisation algorithm described in [Cantor81], and Hensel lifting. The polynomial to be factorised is denoted by f.

3.1. Choice of Prime

The selection of the "best" prime has to be a compromise between speedy modular factorisation and Hensel lifting, and rapid basis reduction. For a particular prime there may be many further possibilities corresponding to the various finite field extensions we can choose. For each prime we consider, we must pick a suitable extension (i.e. factorise the modular images of the minimal polynomials of the generators of the algebraic number field). A distinct degree factorisation of f in this finite field supplies enough information for us to estimate the running time of the rest of the algorithm; also it can be used to maintain a set of possible degrees of factors, perhaps even prove irreducibility.

The table below shows the effect of choosing the smallest valid prime compared with the best of the first five valid primes.

Comparison of Choices of Prime		
Example	Smallest	Best
1	5.7	5.4
2	4.1	3.8
3	12.3	10.5
4	74.6	45.0
5	2222.5	1164.4

3.2. Modular Factorisation

The Cantor-Zassenhaus algorithm is very elegant, although when we compared our implementation of CZ with the Berlekamp factoriser supplied with REDUCE, the Berlekamp algorithm [Berlekamp70] was usually faster for factorisations over a prime field. Two formulae useful in this context are:

$$(x+y)^p \equiv x^p+y^p \bmod p,$$

$$\tfrac{1}{2}(p^d-1) = \tfrac{1}{2}(p-1)(1+p+p^2+\cdots+p^{d-1}).$$

The first formula is directly applicable to finding p^{th} powers, provided p is prime. The second is useful for finding $\tfrac{1}{2}(p^d-1)^{th}$ powers. We compute tables of pth powers (similar to Berlekamp's Q-matrix) to make full use of these equations. In the CZ factoriser the result of a powering operation is always an argument to a *gcd* whose other argument is already known, so we keep the table entries reduced modulo this known polynomial. Such *gcd*s occur both in the distinct degree factorisation and in the random part of the algorithm. A small change to the distinct degree factorisation algorithm is worthwhile: instead of trying $gcd(f, x^{p^d-1}-1)$ we obtain a potentially better split by trying $gcd(f, x^{½(p^d-1)} \pm 1)$; in effect this is a step taken from the random part of the CZ algorithm.

We compute *gcd*s over a finite field by a crude PRS technique. There is no advantage to be gained by attempting to limit coefficient growth as this occurs automatically in a finite field.

We tried one unsuccessful improvement to the algorithm. >From the principle of the CZ algorithm it seemed a good idea to pick a random polynomial h, and try $gcd(f, h^{½(p^r-1)})$ for $r = 1,2, \cdots s$. However, this is less efficient because the *gcd* operation is slow compared with the powering operation, and the chance of finding a factor at the rth step is only $1/p^{s-r}$.

3.3. Hensel Lifting

We have tried four variations of lifting, and conclude that truncated quadratic lifting seems best. The four alternatives were: (i) pure linear lifting [Yun73]; (ii) pure quadratic lifting (not lifting the cofactors on the last step) [Zassenhaus69]; (iii) *fast linear* lifting, i.e. quadratic lifting until the accuracy exceeded the machine's wordsize then linear lifting (see the solution to exercise 4.6.2-22 in [Knuth81]); and (iv) *truncated quadratic* lifting, i.e. quadratic lifting until the accuracy exceeded the cube root of the lifting bound, then using one or two linear lifts to achieve the required accuracy.

Pure linear lifting suffers from having to pass through many steps to reach the necessary bound, whereas fast linear has about the same cost per step but needs far fewer steps (we neglect the time spent lifting the cofactors at the start). Pure quadratic lifting needs very few steps but each step is about four times harder than the previous one (if using classical arithmetic). Truncated quadratic lifting is very similar except that in some cases the cofactors are not lifted on the penultimate step, so two linear lifts are needed. Also in these cases pure quadratic lifting exhibits excessive overshoot which is partly avoided by the truncated lifting.

Extensive experiments showed pure linear lifting as the worst method. Fast linear and pure quadratic were roughly the same for lifts up to 10^{30}, then pure quadratic became increasingly superior. Truncated quadratic lifting was never much slower than pure quadratic and sometimes definitely faster. The table below shows the times spent Hensel lifting using the four methods described.

Comparison of Hensel Lifting Methods				
Example	Lifting Method			
[Lenstra82]	Pure Linear	Fast Linear	Pure Quadratic	Truncated Quadratic
1	5.96	2.14	1.94	2.12
2	5.96	2.00	2.54	2.62
3	16.32	4.66	5.64	4.50
4	44.46	7.68	4.94	5.20
5	1980	296	110	122
x^9-54	1375	188	68.26	43.44

We had observed that Hensel lifting consumes a significant proportion of the total factorisation time. Thus we were keen to make the lifting as quick as we could. We investigated different ways of lifting both the factors and the cofactors. Our chosen method of stepping the factors, f_i, is quite simple. Supposing the f_i are determined mod p^m and the cofactors, α_i, are accurate mod p^δ, then we step the factors by $f_i \to f_i + (\alpha_i f \bmod (f_i, p^{m+\delta}))$, reducing $f \bmod (f_i, p^{m+\delta})$ directly each time. The choice of method for stepping the cofactors is not clear. We computed the number of operations each of five methods would take for a wide selection of different inputs, but there was no obvious winner.

One surprise was that for computing $f \bmod (f_i, p^{m+\delta})$ we could employ a divide-and-conquer algorithm very similar to partial fraction decomposition [Abdali77], but in this application it requires more domain operations than the algorithm above. To see this, consider finding $f \bmod f_i$ for each i, where $f = \prod_{i=1}^{4} f_i$ with each f_i monic. Let n_i = degree f_i, N = degree f, and note that reducing a degree m polynomial modulo a monic degree n polynomial takes $max(0, (n-m+1)m)$ domain operations if all polynomials are dense. Performing $f_{12} = f \bmod f_1 f_2$ and $f_{34} = f \bmod f_3 f_4$, then finding $f_{12} \bmod f_1$, $f_{12} \bmod f_2$, $f_{34} \bmod f_3$, and $f_{34} \bmod f_4$ takes $N^2+2N-\sum n_i^2$ domain operations, which is N more than reducing $f \bmod f_i$ directly for each i.

4. Lattice Basis Reduction

Basis reduction is the key to Lenstra's factorisation algorithm. To be competitive with rival factorisers such as [Weinberger76] we need a quick routine for the basis reduction. We have already observed that choices of finite field favourable for the modular factoriser cause the basis reduction to be slow. Several versions of the basis reduction routine have been implemented, we describe both successful and unsuccessful variations. For clarity we shall assume our extension is simple, $\mathbb{Q}(\alpha)$; details about multiple extensions are in [Abbott87b].

In accordance with [LLL82] we use the following notation: b_i are the vectors defining the lattice, b_i^* are the corresponding Gram-Schmidt vectors, $\mu_{ij} = (b_j^*, b_i)$, and $d_i = \prod_{j=1}^{i} |b_j^*|^2$. The Gram-Schmidt orthogonalization is especially easy because the basis of the lattice is defined by

$$\{p^k, p^k\alpha, \cdots, p^k\alpha^{m-1}, m_\alpha, \alpha m_\alpha, \cdots, \alpha^{n-m-1}m_\alpha\}$$

(using an obvious map from polynomials in α to vectors) which is triangular so the μ_{ij} and $|b_i^*|^2$ can be found directly.

4.1. Using Rational Numbers

For simplicity, the first implemented version used the rational numbers supplied by Cambridge LISP [Fitch77]. This turned out to be hopelessly slow: the fifth example in [Lenstra82] produced a basis which took more than six hours to reduce, whereas Lenstra claimed to have completed the factorisation in under a minute. Most of the time was spent reducing the rational numbers to minimal form, i.e. integer *gcd*s. There was little consolation in the discovery that one of the factors Lenstra gave is reducible.

4.2. Trying Floating Point

We replaced all the rational numbers by floating point. No attempt to control rounding error was made. Now the basis reductions were very fast but the reduced bases were wrong. The culprit was loss of accuracy when the μ_{ij} were updated. Attempts to avoid this by recalculating the b_i^* led us to the problem of finding inner products accurately. We abandoned the use of floating point numbers because of this poor behaviour on large lattices; also numbers outside the representable range may be required, and there could be portability problems. A possibility we have yet to try is the use of high precision floating point numbers. Guidelines about the minimum accuracy necessary are given in [Schnorr85].

4.3. Using Integers

Another look at [LLL82] revealed that the d_i were sufficient denominators, so the routine need use only integer arithmetic. A new version was duly implemented (accompanied by plenty of headaches), and was between four and ten times as fast as the original. We noticed that intermediate calculations involved extremely large integers, and these were normally associated with the last few vectors.

4.4. Being Lazy

It seemed wasteful to manipulate all the μ_{ij} all the time, especially as those associated with last few vectors are not needed initially. Instead, we reduce the first few vectors, and then incorporate the next vector with the relevant μ_{ij}, and d_i. We repeat until all the vectors have been incorporated:

for $k = 1$ to n do

$(b_1, \ldots, b_{k-1}, b_k) = \mathit{reduce}(b_1, \ldots, b_k)$

NB each time round the loop $b_1, \ldots, b_{k-1}$ are already reduced.

This avoids costly arithmetic on the last few μ_{ij} and d_i while only the first few are being processed. The d_i and μ_{ij} for the first $k-1$ vectors contain enough information to allow us to find d_k and μ_{kj} without having to keep the values of the b_i^*, e.g. see algorithm R in [Kaltofen83]. Unfortunately the relevant calculations involve summing rational numbers — we know no efficient way of doing this using only integer arithmetic, but we can represent each summand as an integer and a floating point fractional part.

4.5. Leading Digits

We remembered Lehmer's algorithm for computing integer *gcds* [Knuth69] and tried to apply it to basis reduction. We scale down the large lattice to a smaller one by dividing all the coordinates of the b_i by some integer, k, and rounding. To maintain linear independence we may need to alter a single coordinate in each vector by ± 1. We then reduce the small basis, and apply the same transformation to the original basis to reduce its orthogonality defect before applying the reduction algorithm. However, this fails because the transformation matrix can have extremely large entries: e.g. when factorising the fifth example from [Lenstra82] the original basis has numbers with 104 digits, the reduced basis has numbers with 35 digits, the transformation matrix taking the reduced basis to the original basis has numbers with 69 digits (hardly surprising since the reduced basis is nearly orthogonal), but its inverse has entries with 241 digits. We explain why this is a problem. Let L be the large lattice, S the smaller lattice, so $L = kS+\varepsilon$ where k is the scale factor and ε is a matrix with entries not exceeding k in absolute value. Say S reduces to the matrix R, i.e. $R = US$ for some unimodular U. The hope was that UL would have smaller orthogonality defect than L, but this is unlikely as $UL = kUS+U\varepsilon = kR+U\varepsilon$, and $U\varepsilon$ may have entries far larger than any entry in kR or even L.

4.6. Preprocess

A totally different approach was to preprocess the basis to reduce its orthogonality defect by a few quick and simple transformations before using the full power of Lovász's algorithm. Just by looking at the original bases one can see many obvious reductions. Hence the idea of writing a routine to simulate crudely the reduction algorithm. In essence the new algorithm pretended that $b_i^* = b_i$, hoping that normally this would not be too inaccurate. A crude analysis of this algorithm yielded an atrocious worst case complexity. Nevertheless it reduced the bases produced during factorisation of the five examples in [Lenstra82] about thirty times as quickly as the original routine, and the final orthogonality defects were only slightly greater than those of the fully reduced bases. The bad news is that occasionally it can be very slow. What happens is that the reduction of one pair of vectors causes another pair to become unreduced, and this can repeat several thousand times. We tried a hybrid using the lazy integer version to complete the reduction when more than, say, 100 repeats had occurred. This hybrid showed good overall performance although the time taken varied erratically with the size of the input basis.

4.7. Using blocks

We noticed that the program often swapped the same pair of vectors several times in succession. Each time there is a swap the μ_{ij} must be updated. We can achieve this more efficiently by calculating the two linear combinations directly from the pair of vectors and only then updating the μ_{ij} and d_i. The benefits were instant: this routine was as fast as the preprocess version and showed no signs of erratic variation. [Schönhage84] discusses a similar approach allowing block reduction of several vectors before updating all the μ_{ij}.

Extensive timings using this reduction code have yielded an empirical formula for the complexity: $time \approx (n^7\delta^5 d^{10})^{1/3}$ where $n = log(Hensel\ bound)$, δ is the finite field extension degree, and d is the dimension of the lattice (i.e. the original extension degree). The formula was derived by a trivariate regression of $log(time)$ against $log(n)$, $log(d)$, and $log(\delta)$.

4.8. Comparison

Below is a table of times taken by five of the versions mentioned above. The headings *Rational, Integer, Lazy, Preprocess,* and *Block* refer respectively to the original implementation using LISP rational numbers, the version using the d_i as denominators, the lazy version, the preprocessor assisted by the lazy routine, and the version which uses blocks of two vectors. The bases used for testing all come from trying to factorise Lenstra's five examples (using Weinberger's estimate for the denominators); the sixth basis is an alternative basis produced from Lenstra's fifth example.

Comparison of Basis Reduction Routines

Basis	Time taken for the Reduction (seconds)				
	Rational	Integer	Lazy	Preprocess	Block
1	2.96	0.30	0.32	0.22	0.30
2	13.14	1.46	1.16	0.80	0.78
3	63.38	5.02	3.56	2.52	3.08
4	2356	248	107	27.02	55.22
5	18714	5428	2801	2291	1449
6	23356	4881	2362	1214	1423

5. Combinatorial Phase

In common with other similar (exponential time) factorisation algorithms there is a combinatorial juggling act as the last step. Some tricks allowing rapid detection of wrong combinations are described in [Abbott85]. These tricks can be applied to factorisation over algebraic number fields. However, division of algebraic numbers is a highly expensive operation; and, particularly in large degree extensions, the same tricks turn out to be slower than performing a polynomial trial division. Thus, we need a new idea. There are two candidates: polynomial division, checking each coefficient of the quotient as it is produced; and performing the division modulo a large prime, cf [Langemyr87]. For the former, we know bounds on the sizes of coefficients of true factors of the polynomial so if the bounds are exceeded in the quotient then the division must fail. In this way trial divisions doomed to failure can be stopped after only two or three coefficients have been found. Both the denominators and numerators of each coefficient can be checked — the denominator bound is totally fixed whereas the numerator bound varies with the degree of the quotient being formed.

Even before attempting the polynomial division, the program can make a few rapid checks. If a degree set has been found then it can be used to ignore combinations which would give factors of impossible degree. Also lifting a little too far and checking the coefficients of putative factors will cut down the number of trial divisions, but this is not applicable if a heuristic lifting bound has been used. Great care should be exercised when lifting further than necessary as the basis reduction becomes significantly slower.

6. Conclusion

The Lenstra algorithm is applicable to problems of a useful size with a bearable running time — it is certainly much faster than the algorithm in [Trager76]. There should be a definite improvement in speed when a procedure for picking good finite fields has been found. There remains plenty of scope for further investigation: e.g. better bounds on coefficients, and faster basis reduction. It is pleasing to find that the Cantor-Zassenhaus factoriser and the Lenstra factoriser are both elegant and practical.

References

[Abbott85] J A Abbott, R J Bradford and J H Davenport, "A Remark on Factorisation," *SIGSAM Bulletin* **19**(2)(May 1985).

[Abbott86] J A Abbott, R J Bradford and J H Davenport, "The Bath Algebraic Number Package," *Proc. ACM SYMSAC 86,* ed. B W Char (1986).

[Abbott87a] J A Abbott and J H Davenport, "A Remark on a Paper by Wang: another surprising property of 42," *submitted to Math. Comp.*

[Abbott87b] J A Abbott, *Ph.D. thesis,* in preparation.

[Abdali77] S K Abdali, B F Caviness and A Pridor, "Modular Polynomial Arithmetic in Partial Fraction Decomposition," *Proc. 1977 MACSYMA Users' Conference,* NASA (Jul 1977).

[Berlekamp70] E R Berlekamp, "Factoring Polynomials over Large Finite Fields," *Math. Comp.* **24**(111) (Jul 1970).

[Cantor81] D G Cantor and H Zassenhaus, "A New Algorithm for Factoring Polynomials over Finite Fields," *Math. Comp.* **36**(154) (Apr 1981).

[Fitch77] J P Fitch and A C Norman, "Implementing LISP in a High-Level Language," *Software — Practice and Experience 7,* (1977).

[Ford78] D J Ford, "On the Computation of the Maximal Order in a Dedekind Domain," *Ph.D. thesis,* Ohio State University (1978).

[Kaltofen83] E Kaltofen, "On the Complexity of Finding Short Vectors in Integer Lattices," *Proc. EUROCAL 83,* Springer LNCS #162 (1983).

[Knuth81] D E Knuth, "Seminumerical Algorithms," Addison-Wesley (1981) 2nd ed.

[Langemyr87] L Langemyr and S McCallum, "The Computation of Polynomial Greatest Common Divisors Over an Algebraic Number Field," To appear in Proc. EUROCAL 87.

[Lenstra82] A K Lenstra, "Lattices and Factorisation of Polynomials over Algebraic Number Fields," *Proc. EUROCAM 82,* LNCS 144 pp.32-39 (1982).

[LLL82] A K Lenstra, H W Lenstra and L Lovász, "Factoring Polynomials with Rational Coefficients," *Math. Ann.,* (261) (1982).

[Mignotte76] M Mignotte, "Some Problems about Polynomials," *Proc. ACM SYMSAC,* ed R D Jenks, (1976).

[Mignotte81] M Mignotte, "Some Inequalities About Univariate Polynomials," *Proc. SYMSAC 81,* (1981).

[Schönhage84] A Schönhage, "Factorisation of Univariate Integer Polynomials by Diophantine Approximation and an Improved Basis Reduction Algorithm," *Preprint for Proc. ICALP 1984,* (1984).

[Schnorr85] C P Schnorr, "A More Efficient Algorithm for Lattice Basis Reduction," *Preprint,* Universität Frankfurt (Oct 1985).

[Trager76] B M Trager, "Algebraic Factoring and Rational Function Integration," *Proc. ACM SYMSAC,* ed. R D Jenks, (1976).

[Wang76] P S Wang, "Factoring Multivariate Polynomials over Algebraic Number Fields," *Math. Comp.,* **30**(134) (Apr 1976).

[Wang82] P S Wang, M J T Guy and J H Davenport, "P-adic Reconstruction of Rational Numbers," *SIGSAM Bulletin,* (2) (May 1982).

[Weinberger76] P J Weinberger and L P Rothschild, "Factoring Polynomials over Algebraic Number Fields," *ACM ToMS,* **2**(4) (Dec 1976).

[Yun73] D Y Y Yun, "The Hensel Lemma in Algebraic Manipulation," *Ph.D. thesis,* MIT (1973).

[Zassenhaus69] H Zassenhaus, "On Hensel Factorisation I," *J.N.T.,* **1** (1969).

A Matrix-Approach for Proving Inequalities

A. Ferscha
Institut für Statistik und Informatik
Universität Wien
A-1080 Wien

ABSTRACT

For special inequalities $p \leq q$, where p, q are algebraic expressions such that for p and q *corresponding matrices* P, Q can be given, proofs can be performed by manipulating the rows of Q, such that the manipulation yields P. The paper gives an Ω with $P = \Omega(Q)$ where Ω can be seen as an *algorithm* in the classical sense because $\Omega = \omega \circ \omega \circ \ldots \omega$ and $\omega$is a manipulation of colums of some matrix. Two special manipulations $\omega^{<}$ and $\omega^{>}$ are presented as $<$*-ordering* and $>$*-ordering functions.* Furtheron it is shown how P, Q are to be chosen to be *corresponding* to p, q, i.e. mappings φ are given such that $p = \varphi(P)$ and $q = \varphi(Q)$ by example. For those φ-s it is shown that $p = \varphi(P) \leq \varphi(\Omega(P)) = \varphi(Q) = q$. Although p, q need to be very special, a lot of capabilities of the introduced φ, ω exist, for example it can be proven algorithmically that $\sqrt[n]{a_1 \, a_2 \ldots a_n} \leq \frac{1}{n} \sum_{i=1}^{n} a_i$. Hence the perspectives of the method are, that improvements of the given φ, ω could give algorithms for a wider range of inequalities (for example polynomials) to be implemented in some Computer Algebra systems.

INTRODUCTION

Although there is a well established "Theory of Inequalities" (see [1], [2], [3],[4]) the manners and methods in this field are manyfold and vary not only between, but also within the special classes of inequalities. To access inequalities to Computer Algebra the need for uniformity arises and systematic methods are appreciated for developing algorithms. A first approach in treating *fundamental inequalities* ([1], [2]) from an algorithmic point of view gives Kovacec in [5], showing that inequalities of Hölder- or Minkowski-type are all of the form:

$$p(x_1, x_2, \ldots x_n) \leq q(x_1, x_2, \ldots x_n) \qquad \text{(i)}$$

with $(x_1, x_2, \ldots x_n)$ being an element of $S \subseteq \mathbb{R}^{+n}$ and p, q being algebraic expressions. In many cases there is a $(\bar{x}_1, \bar{x}_2, \ldots \bar{x}_n)$ for which (i) reduces to

$$p(\bar{x}_1, \bar{x}_2, \ldots \bar{x}_n) = q(\bar{x}_1, \bar{x}_2, \ldots \bar{x}_n)$$

or in other words: There is a $\tau : \mathbb{R}^n \mapsto \mathbb{R}$ and points $(y_1, y_2, \ldots y_n), (z_1, z_2, \ldots z_n) \in \mathbb{R}^n$ such that

$$\tau(y_1, y_2, \ldots y_n) = p(x_1, x_2, \ldots x_n) \leq q(x_1, x_2, \ldots x_n) = \tau(z_1, z_2, \ldots z_n). \qquad \text{(ii)}$$

An algorithm now, for proving (i) would be:

1) Find some τ, and $\vec{y}, \vec{z} \in \mathbb{R}^n$ so that (ii) holds.

2) Find some $G \subseteq \mathbb{R}^n$ with $\vec{y}, \vec{z} \in G$ where with $f : G \mapsto G$ for every $\vec{t} \in G$ the inequalitiy $\tau(\vec{t}) \leq \tau(f(\vec{t}))$ holds, and there is some $m \in \mathbb{N}$ such that $f^m(\vec{y}) = \vec{z}$.

One possible extension [5] of the idea is to interpret $\vec{y}, \vec{z}$ as matrices, where results are given based on the Banach Fixedpointlemma. This paper is to show a pragmatical method for Computer Algebra, where no Fixedpoint considerations are necessary.

In the following we will prove special inequalities $p \leq q$ by manipulating *corresponding matrices* to the expressions p, q. In short: P is a *corresponding matrix* to p if $p = \varphi(P)$; so is Q for q if $q = \varphi(Q)$ for a special φ. Constructing Q out of P - or just showing, that Q is constructable out of P - is a proof for $p \leq q$. The **PRELIMINARIES** are to introduce necessary

DEFINITIONS of Ω and φ and to give a **PROOF** for $\varphi(P) \leq \varphi(Q)$. Some examples showing the **CAPABILITIES OF THE METHOD** will follow. **EXTENSIONS** to the method are (appreciated and) possible!

PRELIMINARIES

DEFINITION 1 Let $M^{n,m}$ be the set of all $(n \times m)$-matrices $A = (a_{i,j})$ with $a_{i,j} \geq 0$ for $1 \leq i \leq n$, $1 \leq j \leq m$. We call

$$\omega^{<}_{i,j} : M^{n,m} \mapsto M^{n,m}$$

a *<-ordering function* so that for a $A \in M^{n,m}$ with columns $A = [\vec{a_1}, \vec{a_2}, \ldots \vec{a_i}, \ldots \vec{a_j}, \ldots \vec{a_m}]$

$$\omega^{<}_{i,j}(A) := [\vec{a_1}, \vec{a_2}, \ldots \min(\vec{a_i}, \vec{a_j}), \ldots \max(\vec{a_i}, \vec{a_j}), \ldots \vec{a_m}], \tag{1.1}$$

where $\min(\vec{a_i}, \vec{a_j})$ represents a new column $\vec{a_i}$ with elements $a_{k,i} := \min(a_{k,i}, a_{k,j})$ for $1 \leq k \leq n$, and $\max(\vec{a_i}, \vec{a_j})$ stands for a column $\vec{a_j}$ with elements $a_{k,i} := \max(a_{k,i}, a_{k,j})$ for $1 \leq k \leq n$. (Analogously $\omega^{>}_{i,j} : M^{n,m} \mapsto M^{n,m}$ with

$$\omega^{>}_{i,j}(A) := [\vec{a_1}, \vec{a_2}, \ldots \max(\vec{a_i}, \vec{a_j}), \ldots \min(\vec{a_i}, \vec{a_j}), \ldots \vec{a_m}], \tag{1.2}$$

is a *>-ordering function.*)

The following example shows the application of *ordering functions* on some $A \in M^{n,m}$

EXAMPLE

$$A := \begin{bmatrix} 2 & 2 & 6 & 3 \\ 8 & 1 & 2 & 5 \\ 0 & 9 & 1 & 7 \end{bmatrix} \in M^{3,4}$$

$$\omega^{<}_{3,4}(A) := \begin{bmatrix} 2 & 2 & 3 & 6 \\ 8 & 1 & 2 & 5 \\ 0 & 9 & 1 & 7 \end{bmatrix} \in M^{3,4} \qquad \omega^{>}_{1,3}(A) := \begin{bmatrix} 6 & 2 & 2 & 3 \\ 8 & 1 & 2 & 5 \\ 1 & 9 & 0 & 7 \end{bmatrix} \in M^{3,4}$$

DEFINITION 2 We call two matrices $A, B \in M^{n,m}$ *algorithmically comparable* $A \sim B$ if there is a finite series

$$\Omega^{<}_{N} = \omega^{<}_{i_N, j_N} \circ \omega^{<}_{i_{N-1}, j_{N-1}} \circ \cdots \circ \omega^{<}_{i_1, j_1} \qquad i_k, j_k, N \in N \tag{2.1}$$

of *<-ordering functions* such that

$$B = \Omega^{<}_{N}(A),$$

in this case B is *algorithmically before* A and we write

$$B \prec A. \tag{2.2}$$

In the case that there is a finite series of *>-ordering functions* with $\Omega^{>}_{N} = \omega^{>}_{i_N, j_N} \circ \omega^{>}_{i_{N-1}, j_{N-1}} \circ \cdots \circ \omega^{>}_{i_1, j_1}$ $i_k, j_k, N \in N$ so that $B = \Omega^{>}_{N}(A)$, we call B *algorithmically after* A: $B \succ A$. (In the Example above $B := \omega^{<}_{3,4}(A) \prec A$ and $C := \omega^{>}_{3,1}(A) \succ A$.)

REMARK: To be *algorithmically comparable* the two matrices $A, B \in M^{n,m}$ necessarily need to have the same sets of rows. So if $\{a_{i,1}, a_{i,2}, \ldots a_{i,m}\}$ is the set of elements of row i of matrix A then for $A \sim B$ The following needs to hold:

$$\{a_{i,1}, a_{i,2}, \ldots a_{i,m}\} = \{b_{i,1}, b_{i,2}, \ldots b_{i,m}\} \qquad 1 \leq i \leq n \tag{2.3}$$

DEFINITION 3 The mapping

$$\varphi\colon M^{n,m} \mapsto \mathbf{R}$$

is called $<$-*supporting* if

$$\varphi(A) < \varphi(\omega^{<}_{i,j}(A)) \qquad \forall A \in M^{n,m}, \ \forall \omega^{<}_{i,j}, \text{ with } i < j \tag{3.1}$$

and is called $>$-*supporting* if

$$\varphi(A) > \varphi(\omega^{<}_{i,j}(A)) \qquad \forall A \in M^{n,m}, \ \forall \omega^{<}_{i,j}, \text{ with } i < j \tag{3.2}$$

where in any case:

$$A = \omega^{<}_{i,j}(A) \Rightarrow \varphi(A) = \varphi(\omega^{<}_{i,j}(A)) \quad \forall A \in M^{n,m} \tag{3.3}$$

After the definitions above we have to give the following theorem which is very essential for proving special inequalities by manipulating the columns of corresponding matrices:

THEOREM 1 Let $A, B \in M^{n,m}$ be two matrices with $B \prec A$ and φ be $<$-*supporting*, then the following Inequality holds:

$$\varphi(A) < \varphi(B) \tag{4}$$

PROOF The Proof is simple. From **DEFINITION 3** it is easy to see that if φ is $<$-*supporting* and $A \in M^{n,m}$ the inequality

$$\varphi(A) < \varphi(\Omega^{<}_{N}(A)) \tag{5}$$

holds, just because $\varphi(A) < \varphi(\omega^{<}_{i_1,j_1}(A)) < \varphi(\omega^{<}_{i_2,j_2} \circ \omega^{<}_{i_1,j_1}(A)) < \varphi(\omega^{<}_{i_N,j_N} \circ \cdots \circ \omega^{<}_{i_2,j_2} \circ \omega^{<}_{i_1,j_1}(A)) = \varphi(\Omega^{<}_{N}(A))$. On the other hand $B \prec A$ implies that B can be constructed from A by applying a finite series of $<$-*ordering* functions on A as in (2.1). From (2.2) and (5) we see that (4) holds ∎

Finally let us give two examples of usefull mappings φ.

LEMMA 1: [5] For any $A \in M^{n,m}$

$$\varphi_{*,+}(A) := \sum_{j=1}^{m} \prod_{i=1}^{n} a_{i,j} \qquad \text{is } < -\textit{supporting}, \tag{6.1}$$

$$\varphi_{+,*}(A) := \prod_{j=1}^{m} \sum_{i=1}^{n} a_{i,j} \qquad \text{is } > -\textit{supporting}. \tag{6.2}$$

PROOF Let us first consider (6.1). As we see in (1.1) $\omega^{<}_{i,j}$ only influences the columns i and j. On the other hand (6.1) does not exert any influence on the ordering of columns. Therefore we can restrict our considerations to the case of a matrix with two columns $A = [\vec{a_1}, \vec{a_2}]$ and simply have to show that for $B = \omega^{<}_{1,2}(A) = [\vec{b_1}, \vec{b_2}]$ and A, B having different sets of columns

$$\varphi_{*,+}(A) < \varphi_{*,+}(B) = \varphi_{*,+}(\Omega^{<}_{1}(A)). \tag{7.1}$$

(In the case of $A = B$, but also if only A, B have the same sets of columns, we would have $\varphi_{*,+}(B) = \varphi_{*,+}(A)$ because of the column-ordering-independence of (6.1).)

To prove (6.1) we first need the following: If $0 \le a < b$ and $0 \le \tilde{a} < \tilde{b}$ then

$$a * \tilde{b} + \tilde{a} * b < a * \tilde{a} + b * \tilde{b}. \tag{8.1}$$

Let now A, B, a, b, $\tilde{a}$, $\tilde{b}$, where a, b, $\tilde{a}$, $\tilde{b}$ meet (8.1) be as follows:

$$A = \begin{bmatrix} a_{1,1} & a_{1,2} \\ a_{2,1} & a_{2,2} \\ \vdots & \vdots \\ a_{n,1} & a_{n,2} \end{bmatrix} \qquad B = \begin{bmatrix} \min(a_{1,1}, a_{1,2}) & \max(a_{1,1}, a_{1,2}) \\ \min(a_{2,1}, a_{2,2}) & \max(a_{2,1}, a_{2,2}) \\ \vdots & \vdots \\ \min(a_{n,1}, a_{n,2}) & \max(a_{n,1}, a_{n,2}) \end{bmatrix} = \begin{bmatrix} b_{1,1} & b_{1,2} \\ b_{2,1} & b_{2,2} \\ \vdots & \vdots \\ b_{n,1} & b_{n,2} \end{bmatrix}$$

$$a := \prod_k a_{k,1} \quad \textit{for } k \textit{ where } a_{k,1} < a_{k,2} \qquad b := \prod_k a_{k,2} \quad \textit{for } k \textit{ where } a_{k,1} < a_{k,2}$$

$$\tilde{a} := \prod_k a_{k,2} \quad \textit{for } k \textit{ where } a_{k,1} \geq a_{k,2} \qquad \tilde{b} := \prod_k a_{k,1} \quad \textit{for } k \textit{ where } a_{k,1} \geq a_{k,2}$$

We can write (7.1) because of (8.1) as $a*\tilde{b}+\tilde{a}*b = \prod_{k<} a_{k,1} * \prod_{k\geq} a_{k,1} + \prod_{k<} a_{k,2} * \prod_{k\geq} a_{k,2} = \prod_{i=1}^{n} a_{i,1} + \prod_{i=1}^{n} a_{i,2} = \varphi_{*,+}(A) < \varphi_{*,+}(B) = \prod_{i=1}^{n} b_{i,1} + \prod_{i=1}^{n} b_{i,2} = \prod_{k<} a_{k,1} * \prod_{k<} a_{k,2} + \prod_{k\geq} a_{k,1} * \prod_{k\geq} a_{k,2} = a * \tilde{a} + b * \tilde{b}$ ∎

To prove (6.2) we have to show that

$$\varphi_{+,*}(A) > \varphi_{+,*}(B) = \varphi_{+,*}(\Omega_1^{<}(A)), \tag{7.2}$$

and do need the inequality: for $0 \leq a < b$ and $0 \leq \tilde{a} < \tilde{b}$:

$$(a + \tilde{b}) * (\tilde{a} + b) > (a + \tilde{a}) * (b + \tilde{b}), \tag{8.2}$$

which is just the same as (8.1). We define A, B as above but:

$$a := \sum_k a_{k,1} \quad \textit{for } k \textit{ where } a_{k,1} < a_{k,2} \qquad b := \sum_k a_{k,2} \quad \textit{for } k \textit{ where } a_{k,1} < a_{k,2}$$

$$\tilde{a} := \sum_k a_{k,2} \quad \textit{for } k \textit{ where } a_{k,1} \geq a_{k,2} \qquad \tilde{b} := \sum_k a_{k,1} \quad \textit{for } k \textit{ where } a_{k,1} \geq a_{k,2}$$

and can analogously write (7.2) because of (8.2) as $(a + \tilde{b}) * (\tilde{a} + b) = (\sum_{k<} a_{k,1} + \sum_{k\geq} a_{k,1}) * (\sum_{k\geq} a_{k,2} + \sum_{k<} a_{k,2}) = \sum_{i=1}^{n} a_{i,1} * \sum_{i=1}^{n} a_{i,2} = \varphi_{+,*}(A) > \varphi_{+,*}(B) = (a + \tilde{a}) * (b + \tilde{b}) = (\sum_{k<} a_{k,1} + \sum_{k\geq} a_{k,2}) * (\sum_{k<} a_{k,2} + \sum_{k\geq} a_{k,1}) = \sum_{i=1}^{n} b_{i,1} * \sum_{i=1}^{n} b_{i,2}$ ∎

It is not neccessary to consequently give definitions for all possible combinations of ordering functions with φ-like functions based on $\omega_{i,j}^{>}$, so for example to give a theorem on a relation like (4). Of course there is a kind of duality which we don't want to present explicitely and will therefore - when showing some capabilities - restrict to $\varphi_{*,+}, \varphi_{+,*}$ under $\omega^{<}$.

CAPABILITIES OF THE METHOD

Before showing some special capabilities of the method we want to demonstrate the idea proving a simple, but nevertheless important elementary inequality upon $(a, b, c \geq 0)$ [3]:

$$(b+c)\,(c+a)\,(a+b) \geq 8\,abc \tag{9}$$

Without loss of generality let us assume $a \leq b \leq c$ and let us write (9) as:

$$a^2c + ab^2 + 2\,abc + ac^2 + a^2b + b^2c + bc^2 \geq 8\,abc \tag{9.1}$$

If we now assume that the lefthandside $l = a^2c + ab^2 + 2\,abc + ac^2 + a^2b + b^2c + bc^2$ as well as the righthandside $r = 8\,abc$ in (9) are $\varphi_{*,+}$ - *values* of some matrices L and R

$$l = \varphi_{*,+}(L) \qquad r = \varphi_{*,+}(R) \tag{10}$$

then we can prove $\varphi_{*,+}(R) \leq \varphi_{*,+}(L)$ and simultaneously (9) by showing that

$$L \prec R,$$

i. e. finding some $\Omega_N^<$ such that $L = \Omega_N^<(R)$. To comply with (10) we choose L, R in a way that their columns represent single terms of l and r, but we also have to meet $L \sim R$ as remarked ((2.3)) - the latter is the reason for permutating some columnelements in R:

$$L = \begin{bmatrix} a & a & a & a & a & a & b & b \\ a & b & b & b & c & a & b & c \\ c & b & c & c & c & b & c & c \end{bmatrix} \in M^{3,8} \qquad R = \begin{bmatrix} a & a & a & a & a & a & b & b \\ b & b & b & b & c & c & a & a \\ c & c & c & c & b & b & c & c \end{bmatrix} \in M^{3,8}$$

Now L can be constructed out of R by applying a sequence of $<$-*ordering* functions:

$$R = \begin{bmatrix} a & a & a & a & a & a & b & b \\ b & b & b & b & c & c & a & a \\ c & c & c & c & b & b & c & c \end{bmatrix} \quad \mapsto \omega_{6,8}^< \mapsto \quad \begin{bmatrix} a & a & a & a & a & a & b & b \\ b & b & b & b & c & a & a & c \\ c & c & c & c & b & b & c & c \end{bmatrix} \quad \mapsto$$

$$\omega_{1,7}^< \mapsto \begin{bmatrix} a & a & a & a & a & a & b & b \\ a & b & b & b & c & a & b & c \\ c & c & c & c & b & b & c & c \end{bmatrix} \quad \mapsto \omega_{2,5}^< \mapsto \quad \begin{bmatrix} a & a & a & a & a & a & b & b \\ a & b & b & b & c & a & b & c \\ c & b & c & c & c & b & c & c \end{bmatrix} = L$$

In short:

$$L = \omega_{2,5}^< \circ \omega_{1,7}^< \circ \omega_{6,8}^<(R) = \Omega_3^<(R) \tag{11}$$

Hence $L \prec R$ and as $\varphi_{*,+}$ is $<$-*supporting* (9) holds because of (4) ∎

Having shown that (9) holds is not the only result of (11)! It is easy to see, that $X_1 = \omega_{6,8}^<(R) = \Omega_1^<(R)$ also proves an inequality (where X_1 is a matrix corresponding to a lefthand-side), namely:

$$a^2b + 6\,abc + bc^2 \geq 8\,abc,$$

or $b\,(a^2 + c^2) \geq 2\,abc$ or even

$$a^2 + c^2 \geq 2\,ac$$

which can also be proven by choosing corresponding matrices $\bar{L}, \bar{R}$:

$$\bar{L} = \begin{bmatrix} a & c \\ a & c \end{bmatrix} \qquad \bar{R} = \begin{bmatrix} a & c \\ c & a \end{bmatrix}$$

Furthermore $X_2 = \omega_{1,7}^< \circ \omega_{6,8}^<(R) = \Omega_2^<(R)$ as well as $X_3 = \omega_{5,6}^< \circ \omega_{2,5}^< \circ \omega_{4,7}^<(R) = \Omega_3^<(R)$ prove inequalities. To summarize: any finite series $\Omega_N^<$ of $<$-*ordering* functions applied on R might give a new inequality. Hence we can use the tool to create new inequalities, knowing that just a lot of them are unimportant - but some of them might be interesting. The following example is to give a taste of those capabilities:

EXAMPLE Let (without loss of generality) $a_1 \le a_2 \le a_3 \cdots \le a_n$ be n positive reals and assume $A, B \in M^{n,n}$ with $B \prec A$:

$$A := \begin{bmatrix} a_1 & a_2 & a_3 & \dots & a_n \\ a_n & a_1 & a_2 & \dots & a_{n-1} \\ a_{n-1} & a_n & a_1 & \dots & a_{n-2} \\ \vdots & \vdots & \vdots & \ddots & \vdots \\ a_2 & a_3 & a_4 & \dots & a_1 \end{bmatrix} \qquad B := \begin{bmatrix} a_1 & a_2 & a_3 & \dots & a_n \\ a_1 & a_2 & a_3 & \dots & a_n \\ a_1 & a_2 & a_3 & \dots & a_n \\ \vdots & \vdots & \vdots & \ddots & \vdots \\ a_1 & a_2 & a_3 & \dots & a_n \end{bmatrix}$$

Applying $\varphi_{*,+}$ shows that

$$\varphi_{*,+}(A) = n(\prod_{i=1}^{n} a_i) \le \sum_{i=1}^{n} a_i^n = \varphi_{*,+}(B)$$

which might also be proven with $\varphi_{+,*}$. If we assume $a_1 \le a_2 \le a_3 \cdots \le a_n$ and ϕ monotonically increasing, further the elements of A, B be $b_1 = \phi(a_1), b_2 = \phi(a_2), \dots b_n = \phi(a_n)$, then

$$n(\prod_{i=1}^{n} \phi(a_i)) \le \sum_{i=1}^{n} \phi(a_i)^n$$

holds and it might for example cause ($\phi = \sqrt[n]{\ }$):

$$n \sqrt[n]{a_1 \, a_2 \, a_3 \dots a_n} \le (\sqrt[n]{a_1})^n + (\sqrt[n]{a_2})^n + \dots + (\sqrt[n]{a_n})^n,$$

or (with $C = n$) better known as the relation between the *geometric* and *arithmetic mean* of a set $\{a_1, a_2, \dots a_n\}$ with elements $a_i \ge 0$:

$$\sqrt[n]{a_1 \, a_2 \, a_3 \dots a_n} \le \frac{1}{n} \sum_{i=1}^{n} a_i$$

For another variation allow $a = a_1 \le a_2 = a_3 = a_4 \cdots = a_n = b$ and A, B as above to obtain

$$\varphi_{*,+}(A) = nab^{n-1} \le a^n + (n-1)b^n = \varphi_{*,+}(B)$$

but also

$$\varphi_{+,*}(A) = (a + (n-1)b)^n \ge na(nb)^{n-1} = \varphi_{+,*}(B).$$

New inequalities can also be generated by adding or cancelling rows in A, B (so that $A \sim B$ still holds). For example if we define C as a matrix containing A m times and D analogously for B:

$$C := \begin{bmatrix} A \\ A \\ \vdots \\ A \end{bmatrix} \in M^{nm,n} \qquad D := \begin{bmatrix} B \\ B \\ \vdots \\ B \end{bmatrix} \in M^{nm,n}$$

we could obtain ($C \sim D$):

$$\varphi_{*,+}(C) = n(\prod_{i=1}^{n} a_i^m) \le m \sum_{i=1}^{n} a_i^n = \varphi_{*,+}(D).$$

EXTENSIONS

We remarked when proving **LEMMA 1** that (6.1) does not matter any ordering of columns. Generally we can call some φ like in **DEFINITION 3** to be *column-ordering-independent* if

$$\varphi([\vec{a_1},\vec{a_2},\dots\vec{a_i},\dots\vec{a_j},\dots\vec{a_m}]) = \varphi([\vec{a_1},\vec{a_2},\dots\vec{a_j},\dots\vec{a_i},\dots\vec{a_m}]) \qquad \forall i,j \tag{12a}$$

and *row-ordering-independent* if

$$\varphi([\vec{a_1},\vec{a_2},\dots\vec{a_i},\dots\vec{a_m}]) = \varphi([\pi(\vec{a_1}),\pi(\vec{a_2}),\dots\pi(\vec{a_i}),\dots\pi(\vec{a_m})]) \qquad \forall i,j \tag{12b}$$

where $\pi(\vec{a_i})$ is a permutation of the elements of $\vec{a_i}$.

It is not necessary to show that $\varphi_{*,+}$ meets $(12a)$ and $(12b)$. Hence with

DEFINITION 4 We call the matrix $\underline{A} \in M^{n,m}$ the (*algorithmic*) $<$-*bound* if

$$\underline{A} \prec A \qquad \forall A \in M^{n,m}$$

The $>$-*bound* we denote with $\overline{A} \in M^{n,m}$.

we are now able to give a theorem that causes a restriction as well as an extension to **THEOREM 1**:

THEOREM 2 Let $A, \underline{A} \in M^{n,m}$ be two matrices. If φ is $<$-*supporting* and *row-ordering-independent* then

$$\varphi(A) < \varphi(\underline{A}) \qquad \forall A \neq \underline{A} \tag{13}$$

holds (**PROOF** omitted).

From (13) we learn that one side of an inequality needs to correspond to an $\varphi(\underline{A})$ of some $\underline{A} \in M^{n,m}$ - and this is the restriction -, but also that the problem of finding a sequence $\Omega_N^<$ or $\Omega_N^>$ to prove the inequality reduces to the problem to show that $A \sim \underline{A}$, - and this is an improvement for $\varphi_{*,+}$, just because it allows permutations of elements within columns! In this case there might be a lot of matrices A_i such that $\varphi_{*,+}(A_i) = \varphi_{*,+}(A)$. For example

$$\varphi_{*,+}\begin{bmatrix} a_1 & a_1 & a_1 \\ a_2 & a_2 & a_2 \\ a_3 & a_3 & a_3 \end{bmatrix} = \varphi_{*,+}\begin{bmatrix} a_1 & a_2 & a_3 \\ a_2 & a_1 & a_1 \\ a_3 & a_3 & a_2 \end{bmatrix} = \varphi_{*,+}\begin{bmatrix} a_1 & a_2 & a_3 \\ a_3 & a_1 & a_2 \\ a_2 & a_3 & a_1 \end{bmatrix}$$

In this case, if we want to show $A \sim B$ we have to choose the right A_i, i.e. those one that meets (2.3). This problem in terms of graph theory (for an introduction covering algorithms see [6], [7]): Let $R_A = \{R_{A,1}, R_{A,2}, \dots R_{A,n}\}$ be a partition of the set of elements $\{a_{1,1},\dots a_{n,n}\}$ of some matrix $A := (a_{i,j})$ such that $R_{A,i}$ is the set of elements of row i. Also let $C_A = \{C_{A,1}, C_{A,2}, \dots C_{A,n}\}$ be a partition of $\{a_{1,1},\dots a_{n,n}\}$ where $C_{A,i}$ is the set of elements of column i. To show $A \sim B$ is to pick one element out of $C_{A,i}$ for every column i such that the picked elements form $R_{B,i}$ and repeat picking elements of the (reduced) $C_{A,i}$-s until all $R_{B,i}$-s are built. We can define a bipartite graph $G = (N, S)$ with nodes $N = C_A \cup R_B$ and arcs S between every $a \in C_{A,i}$ and one $b \in R_{B,i}$ for all i. So if G exists for some A then $A \sim B$:

$$A := \begin{bmatrix} a_1 & a_3 & a_2 \\ a_2 & a_1 & a_3 \\ a_3 & a_2 & a_1 \end{bmatrix} \qquad B := \begin{bmatrix} a_1 & a_2 & a_3 \\ a_1 & a_2 & a_3 \\ a_1 & a_2 & a_3 \end{bmatrix}$$

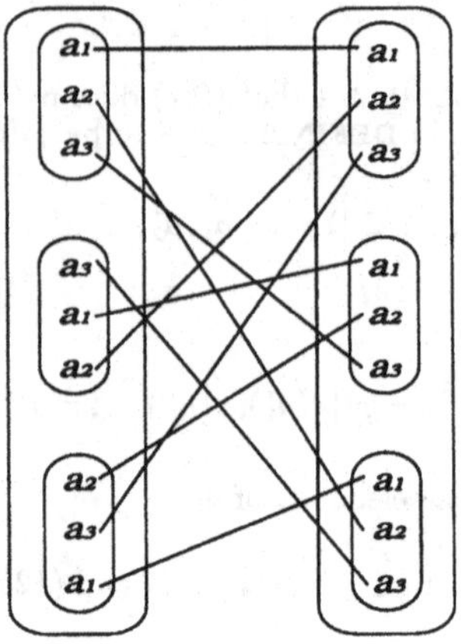

Note: A, B are to prove $3\ a_1 a_2 a_3 \leq a_1^3 + a_2^3 + a_3^3$, a special case of $n(\prod_{i=1}^n a_i) \leq \sum_{i=1}^n a_i^n$ from above!

CONCLUSION AND PERSPECTIVES

The algorithmic approach shows that proving, or at least generating inequalities systematically is easy work. It could eventually be done with conventional combinatorial methods (i.e. finding Ω) combined with Computer Algebra methods (i.e. termmanipulation, matrixmanipulation). Although the capabilities of the method are still restricted to very special classes of inequalities, perspectives widen with every new φ like in **DEFINITION 3**. So finding functions φ, either $<$-***supporting*** or $>$-***supporting***, will enlarge the capabilities! For general inequalities involving polynomials at least some improved φ seem to exist. On the other hand, an improvement of the *ordering functions*-concept could help to find a way form P to Q (for example quantifying the improvement of one step ω towards the target). Even in defining the corresponding matrices improvements seem to be possible (with some help by matching algorithms).

REFERENCES

[1]
Hardy, G., Littlewood, J., Polya, G.: *Inequalities*, 2nd Edition, Cambridge University Press, Cambridge, 1952.

[2]
Mitrinovic, D.: *Analytic Inequalities*, Die Grundlehren der mathematischen Wissenschaften, Bd. 165, Springer-Verlag, Berlin, 1970.

[3]
Mitrinovic, D.: *Elementary Inequalities*, P. Noordhoff Ltd., Groningen, 1964.

[4]
Bottema, O., Djordjevic, R., Janic, R., Mitrinovic, D. and Vasic, P.: *Geometric Inequalities*, Wolters-Noordhoff Publishing, Groningen, 1969.

[5]
Kovacec, A.: *Eine Methode zum Nachweis von Ungleichungen auf einheitlicher, algorithmischer Grundlage*, Dissertation, Universität Wien, Wien, 1980.

[6]
Lawler, E.: *Combinatorial Optimization: Networks and Matroids*, Holt, Rinehart and Winston, New York, 1976.

[7]
Christofides, N.: *Graph Theory: An Algorithmic Approach*, Academic Press, London, 1975.

USING AUTOMATIC PROGRAM SYNTHESIZER AS A PROBLEM SOLVER: SOME INTERESTING EXPERIMENTS

Pavol Návrat, Ľudovít Molnár, Vladimír Vojtek
Department of Computer Science and Engineering
Slovak Technical University
Mlynská dolina, 812 19 Bratislava, Czechoslovakia

1. INTRODUCTION

The task of the automatic program synthesis essentialy means the automatic solving of a specified problem in such a way that the solution is comprehensible to the computer, i.e., it is a program. We have elaborated such an automatic program synthesizer called TRINS which is described in [8] in detail. Now an interesting question arises: is such an automatic synthesizer able to generate solutions not only for computers, but for other mechanisms as well?

Intuitively, it is clear that in order to synthesize a program automatically, the system must possess some more general problem solving capabilities which are independent of the target language (i.e. computer) chosen. On the other hand, however, we are far from claiming an automatic program synthesizer to be some sort of a general problem solver. In our investigations we have found the possibilities of generating automatically the solutions of problems of a wider class and for a wider class of mechanisms that originally intended. In this paper we are describing our effort to use the synthesizer to generate simple plans for some robot-like mechanisms.

The control system of a cognitive robot is a complex hierarchical system consisting of several control levels. The subject of our investigation is the strategic level which includes the system of automatic planning and the control of its activity. Thus we shall not deal with the more concrete levels where much more detailed planning of the robot activity takes place. We are fully aware of the fact that robot programming is very troublesome and that ceratin principal difficulties are to be mastered on these levels.

2. SYSTEM TRINS

System TRINS was originally developed for the automatic synthesis of operations on abstract data types (cf. [7], [3], [5]). However, having updated the knowledge base, we can use it for other tasks as well.

TRINS operates according to the transformation method (similarly to, e.g., [1]). The specification is expressed in a language based on a predicate calculus. Other symbols can be included depending on the problem environment. We adopt the usual choice to reserve the symbol "compute" to denote the output part and the symbol "where" to denote the input part of the problem specification.

In our planning environment, the output part corresponds to the goal situation and the input part to the initial situation.

The specification will be written in the form:

COMPUTE: goal situation
WHERE: initial situation

It must be stressed that while the problem specification expresses completely the substance of the problem, it does not say a word about an algorithm of its solution. However, it typically includes several pieces of knowledge which will be useful in forming the solution. The algorithm will be derived in the process of transformation.

The transformation rules will be written in the form:

(goal to be satisfied) $\longrightarrow$ (operator which can do it)
$\Longrightarrow$ (changes in the situation) IF (applicability condition).

The transformation rules are organized in the knowledge base. Its organisation depends on the form of the transformation rules, methods of accessing them and the possibilities of modifying them. For a problem environment with a relatively small number of rules, the tree appears to be the appropriate structure for the representation of the base, because it allows to express a hierarchy and bindings among individual transformation rules which provides for a more efficient access.

The synthesis process consists of a sequence of particular steps. In each step, there is a set of applicable rules determinated for a chosen goal (subgoal). (The output part of the problem specification is set as the first goal). One rule is selected from among them (according to certain criteria of heuristic nature, or by means of interaction) and this rule is applied to the chosen goal. Then the next goal is chosen, etc. The synthesized program is formed by a composition of satisfied goals.

The synthesis process is controlled by a control algorithm. Its core consists of the algorithm A^{**} of the heuristic search of solution using ints [7], [2], [4], which employs global and local heuristic information from the given problem environment, combined with a man-machine interaction to increase the efficiency of the synthesis process. This algorithm is also responsible for resolving possible conflicts arising in the process of satisfying the subgoals. We shall return to this subject later.

System TRINS is implemented in LISP.The system is supported by a set of graphical operations which were used to display a process of transformation. We have conducted several experiments with it, solving problems from diverse environments. In the following section we will focus our attention to synthesizing plans for a robot-like mechanism.

3. THE SYNTHESIS OF PLANS

3.1. The synthesis of plans for solving problems from the blocks world

The notion "blocks world" denotes a model of an environment which comprises solely equally shaped blocks, a table and a manipulating mechanism (a robot). The robot is able to move one free block at a time. A block can be moved on top of another one or on the table (if there is room on it). We have defined the following operators: (cf. [10])

1. (MOV1 X Y Z) - move an object X from an object Y on the top of an object Z.
2. (MOV2 X Y) - move an object X from the table on the top of an object Y.
3. (MOV3 X Y) - remove an object X from the top of an object Y and put it on the table.

The problem environment as well as the problem specification are described in a language based on a predicate calculus with the symbols ON, ONT, CLR included:

1. (ON X Y) - an object X is on an object Y.
2. (ONT X) - an object X is on the table.
3. (CLR X) - there is no o bject on an object.

Example 1. Let us have the initial and the goal situations specified as follows:

COMPUTE: (ON A B) * (ON B C) * (ON C D) * (ON D E)
WHERE: (ONT A) * (ONT B) * (ONT D) * (ON C A) *
(ON E D) * (CLR C) * (CLR B) * (CLR E)

The knowledge base must also contain transformation rules describing operators MOV1, MOV2, and MOV3.

TR1:

(ON X Z)⟶(MOV1 X Y Z)

⟹(ON X Z) * (CLR Y) * (CLR X)

IF (ON X Y) * (CLR X) * (CLR Z)

TR2:

(ON X Y)⟶(MOV2 X Y)

⟹(ON X Y) * (CLR X) IF (ONT X) * (CLR X) * (CLR Y)

TR3:

(ONT X)⟶(MOV3 X Y)

⟹(ONT X) * (CLR X) * (CLR Y) IF (ON X Y) * (CLR X)

The algorithm A^{**} requires the definition of the evaluating function FST for selecting the next state for expansion as well as other heuristic functions FOP (for selecting the next operator to apply), L and T (for identifying local minimum of the function FST).

For instance, let us define the heuristic part of the evaluation function FST as a sum of two functions. It should be noted, however, that there are methods, to discover certain heuristics even automatically. The first feature describes the overall distance of the blocks in a given situation, from their positions in the goal situation, and is given by the sum of particular distances of the blocks. The second feature describes the "preparedness" of the blocks to take their respective positions and is given by the sum of the numbers of the blocks positioned on top of each block which has not been put in its goal position yet.

The synthesis begins with the goal

(ON A B) * (ON B C) * (ON C D) * (ON D E)

This goal cannot be solved immediately (there is no appropriate transformation rule) and, therefore, must be decomposed. Note that any goal in this problem environment is expressed by a conjunction (the goal decomposition is represented by an AND-node in the search graph).

All four subgoals are evaluated and the one with the minimal value of FST becomes a candidate for the expansion. In this case, both subgoals

(ON D E)

and

(ON C D)

receive minimal evaluation so we arbitrarily resolve to choose the former.

In the knowledge base there are two transformation rules TR1 and TR2 the left hand sides of which match the subgoal. However, having evaluated their applicability conditions under the current state of description given by the WHERE-part of the problem specification, we find none of them directly applicable. To apply TR2, condition

(ONT D) * (CLR D) * (CLR E)

must be satisfied. To apply TR1, condition

(CLR D) * (CLR E)

as well as

(ON D Y)

must be both satisfied with Y not determined yet. Therefore, we shall try to satisfy the applicability condition of TR2. We apply TR3 to the initial goal. Then we can apply TR2. Continuing the transformation of the problem specification in a similar manner we eventually reach the desired goal. The sequence of transformation rules applied is TR3, TR2, TR1, TR2, TR2 which corresponds to this sequence of operators:

(MOV3 E D)(MOV 2 D E)(MOV C A D)(MOV2 B C)(MOV2 A B)

It is the synthesized plan. Note that for the given problem this plan is better than the plan synthesized by the known plan formation system WARPLAN:

(MOV3 E D)(MOV2 D E)(MOV2 C A)(MOV2 C D)(MOV2 B C)(MOV2 A B)

Our system was able to synthesize plans for various other interesting problems. For example the problem

COMPUTE: (ON A B) * (ON B C)

WHERE: (ON C A) * (ONT A) * (ONT B) * (CLR C) * (CLR B)

which causes difficulties to such systems as HACKER [6] could be solved successfully.

The reason for the success lies partly in the fact that our system provides a framework for a more efficient and effective use of heuristic information.

3.2. The synthesis of navigation plans

Another problem environment for which our system has synthesized solutions is the navigation of a robot which aims to reach the desired situation. The activity of the robot comprises, besides selfrelocation also relocation of other objects from the scene. There are several predicates defined which allow to describe a scene (cf.,e.g., [10]), for example, that

- there is a door between room R1 and room R2,
- there is a robot near the object X,

- the door D is open, etc.

The activity of the robot can be described by a set of operators, such as

- go to the object X,
- enter the room R2 through the door D from the room R1,
- push the object X close to the object Y, etc.

The operators are defined by the following transformation rules:

TR21:(NEARTOR X)⟶(GOTO X)
⟹(NEARTOR X) IF (INROOMR R) * (INROOM X R)

TR22: (INROOMR R2)⟶(ENTERIN D R1 R2)
⟹(INROOMR R2)
IF (INROOMR R1) * (NEARTOR D) * (DOOROPEN D)*
(CONNECTS D R1 R2)

TR23: (NEARTO X Y)⟶(PUSH X Y)
⟹(NEARTO X Y)
IF (INROOMR R) * (INROOM Y R) * (NEARTOR X)

TR24:(INROOM X R2)⟶(ENTERWITH X D R1 R2)
⟹(INROOM X R2)
IF (NEARTOR X) * (NEARTO X D) * (DOOROPEN D) *
(CONNECTS D R1 R2)

The heuristic part of the evaluation function should express the following features:

- for all objects which are to be relocated; - their distances d, from their respective goal positions,
- for all objects which are to be relocated; - their distances r, from the robot,
- the distance v of the robot from its goal position.

During the relocation of an object in its goal posititon the distance of the robot from other objects, or its distance from its goal position can be increased. To ensure that the value of the heuristic function decreases even when the robot departs from its goal position or the r_i´s increase, the d_i´s contribute to the value of the heuristic function with their third power and r_i´s with their second power. The heuristic function can be formulated as

h = sum(i=1,k, (power(d_i,3) + power(r_i,2))) + v

where k is the number of the objects which are to be relocated.

As an example of a problem solved by our synthesizer, let us consider the following specification:

COMPUTE: (NEARTO A C) * (NEARTO B C) * (NEARTOR C)
WHERE: (PUSHABLE B) * (PUSHABLE A) *

(CONNECTS D R1 R2) * (CONNECTS D R2 R1)*
(INROOM A R1) * (INROOM B 31)*
(INROOM C R2) * (INROOM D R1)*
(INROOM D R2) * (DOOROPEN D) * (NEARTOR B) * (ONFLOORR)

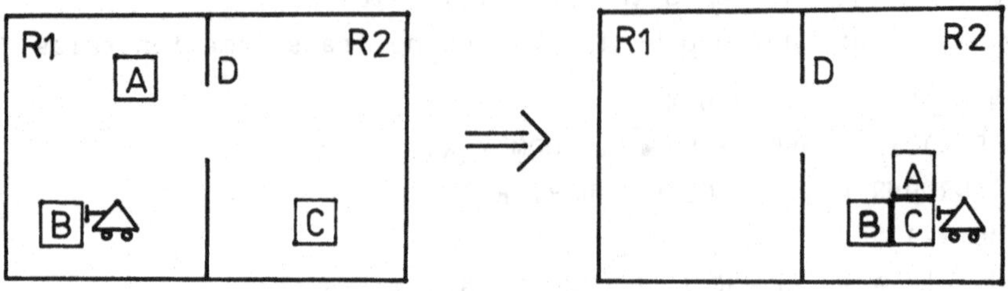

Fig.1.

The task of the robot is to gather all the objects A, B and C tohe-ther along with itself in one room. Our system synthesized the following plan for it:

(PUSH B D)(ENTERWITH B D R1 R2)(PUSH B C)(GOTO D)
(ENTERIN D R2 R1)(GOTO A)(PUSH A D)
(ENTERWITH A D R1 R2)(PUSH A C)(GOTO C)

It should be noted that despite the apparent simplicity of the presented problem, similar problems are altogether not easy, if at all sol vable, for dedicated plan formation systems.

3.3. The synthesis of robot´s plans for machine assembly

A characteristic feature of the considered assembly of machine parts configurations is inserting of objects one into another. The problem domain is described by a set of operators, such as:

1. (TAKEOUT X Y Z) - take out an object X from a hole Y in a body Z.
2. (INSERT X Y Z) - insert an object X into a hole Y in a body Z.
3. (PUT X) - put an object X on a working table.
4. (TAKEFROM X) - take an object X from a working table.

The specification language is extended by reserved words allowing to describe various situations, such as:

(TRANSVERSE X Y) - in an object X is transversely inserted an object Y.
(HOLE Y Z) - in an object Z is a hole Y.
(ONT X) - an object X is on a working table.
(EMPTYH) - the hand is empty.

(EMPTY Y) - there is no object in a hole Y.

(ALONE X) - there is no object in an object X.

(NOTALONE X) - there is some object in an object X.

(FREE X) - an object X is free to be gasped.

(INSERTED X Y) - an object X is in a hole Y.

(HOLDS X) - the hand holds an object X.

The knowledge base can be extended by following transformation rules:

(ALONE Z)⟶(TAKEOUT X Y Z).
⟹(ALONE Z) * (HOLDS X)
IF (TRANSVERSE X Z) * (EMPTYH) * (FREE X)

(TRANSVERSE X Z)⟶(INSERT X Y Z)
⟹(TRANSVERSE X Z) * (EMPTYH)
IF (EMPTY Y) * (ALONE X) * (HOLDS X)

(ONT X)⟶(PUT X)
⟹(ONT X) * (EMPTYH)
IF (HOLDS X)

(HOLDS X)⟶(TAKEFROM X)
⟹(HOLDS X)
IF (ONT X) * (EMPTYH)

Heuristic function can be for this class of problems formulated as:
$h = sum(i=1,k,(x_i + y_i)) + m$

where k is the number of objects. The elements x_i, y_i and m are computed as follows:

IF the i-th object is in (or on) an object as required for the goal configuration
THEN x_i:=0 and y_i:=0
ELSIF there are p other objects in the i-th object
THEN x_i:=r and $y_i := r^{2p}-1$;

where r is the number of all holes in all objects,

IF the hand is empty
THEN m:=0
ELSE m:=1;

Consider the example problem depicted on Fig.2.

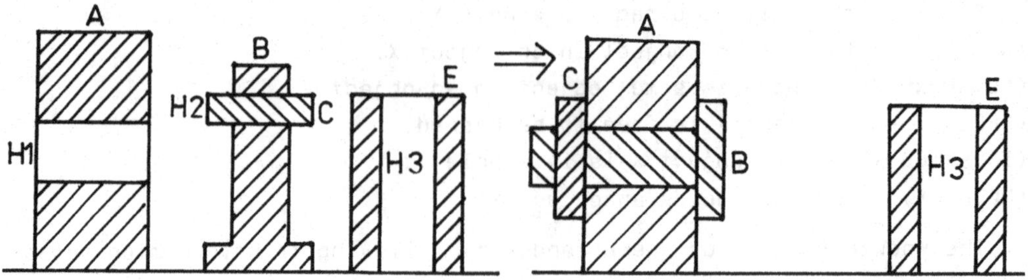

Fig. 2.

It can be specified as follows:

COMPUTE: (TRANSVERSE A B) * (TRANSVERSE B C)
WHERE: (HOLE H1 A) * (HOLE H2 B) * (HOLE H3 E) *
(ONT A) * (ONT B) * (ONT E) *
(EMPTYH) * (EMPTY H1) * (EMPTY H3) *
(ALONE A) * (NOTALONE H2) * (ALONE C) * (ALONE E) *
(FREE A) * (FREE B) * (FREE C) * (FREE E) *
(TRANSVERSE B C) * (INSERTED C H2)

The plan of robot´s activity synthesized from this specification by applying the transformation rules is the following:

(TAKEOUT C H2 B)(PUT C)(TAKEFROM B)
(INSERT B H1 A)(TAKEFROM C)(INSERT C H2 B)

4. CONCLUSIONS

We have tried to show how our automatic synthesizing system TRINS can be used to generate solutions of problems from diverse problem environments. More specifically, we have presented examples of solving the problems concerning

- the reconfiguration of the scene from the blocks word, and
- the robot navigation.
- the assembly plan for a robot.

Together with the examples of the automatic program synthesis described elsewhere (cf. [8]) all the examples from these diverse environments show that our system has been fairly successful in accomplishing the difficult task of automatically generating problem solutitons. Its relative efficiency and efficacy relies on quite a sophisticated use of heuristic information in the form of local and global heuristics. Besides that, problem-dependent knowledge is represented by transformation rules, which is for the above mentioned problem environments quite suitable. However, this is certainly not the case for any problem environment. It would be advantageous to provide alternative forms of knowledge representation.

As far as the comparison with other automatic planners is concerned, we have already made some remarks in section 3. From a more general point of view, however, we must stress that originally our system TRINS was an automatic program synthesizer. It relied on problem specific knowledge (on programming) in the form of heuristic functions and transformation rules, and roughly speaking, by stripping off the problem specific knowledge we obtained an "essential"-TRINS which preserved its general problem solving capability. Of course, for some other problem domain the related problem specific knowledge is to be provided. In our case, we have chosen the domain of the automatic plan formation. We greatly benefit from the possibilities of expressing the specific problem solving knowledge which are embedded in TRINS. So we could, in fact, manage to avoid arising conflicts in the problem solving process. Our (global) heuristic function has local minima in those "conflicting" states. Our (local) heuristic function helps to recognize them, and therefore, according to the algorithm A^{**}, they will be avoided. Of course, it seems fruitful to follow also other paths of investigation of this problem. For example, there could be devised so called macro-operators which ought to aid in overcoming the surroundings of local extremes of the heuristic function. More generally, we must consider a heuristic search guided by a non-linear heuristic function.

Recognizing the central importance of the proper knowledge representation formalisms, the directions for future work include seeking a wide-spectrum but still uniform formalism allowing to express expertise knowledge not only as transformation rules and not forcing to express expert domain-specific search-control knowledge exclusively as a (heuristic) function.

Even further, we do not consider the transformation itself (i.e.deduction) to be the only method useful here. There have been investigations to find the possibilities of using analogy or induction to synthesize programs automatically.

The control algorithm of our system would cause our system to behave as

- a general problem solver (when it has none or only little search-control knowledge), employing any of the universal but weak methods (e.g., the heuristic or even the blind search).

- a domain specific problem solver (an expert) (when it has enough search-control knowledge), employing the expertise knowledge to reduce substantially the search.

It is our determination that the devised framework of global and local heuristics used in the control algorithm reflects more universal properties of the problem solving process. For example, hints as a form of local heuristics which are used to overcome local minima and subsequent local maxima of the heuristic function (i.e. of the global heuristics) represent additional heuristic knowledge useful where the global heuristics is inadequate. Some analogy to logic programming and to prolog in particular could be traced. Prolog is considered impure because of its cut and fail primitives. On the other hand, however, it is quite reasonable to consider the rationale behind them to be precisely the form of local heuristic information, i.e., that inserting a cut symbol in a clause represents some hint. The fact that more adequate forms of expressing the control knowledge in a future logic programming language ought to be sought is another matter.

REFERENCES

[1] MANNA Z., WALDINGER R.: Synthesis: Dreams → Programs. Technical Note 156. Menlo Park. SRI International 1977, 96pp.

[2] MOLNÁR Ľ., NÁVRAT P., VOJTEK V.: A New Algorithm for Heuristic Search Which Uses Hints. In: Proc. Int. Symposium System Analysis and Simulation, Part I, A.Sydow (ed.), Berlin, Akademie Verlag 1985, pp. 177-181.

[3] MOLNÁR Ľ., NÁVRAT P.: System of Automatic Synthesis of Programs with Abstract Data Types. In: Tagungsbuch 30. Int. Wiss. Kolloquium TH Ilmenau, Vortragsreihe Informatik, 1985, pp. 155-158.

[4] MOLNÁR Ľ., NÁVRAT P., VOJTEK V.: Heuristic Search with Global and Local Heuristics. Computers and Artificial Intelligence, Vol.5, 1986, No.5, pp. 417-427.

[5] MOLNÁR Ľ., NÁVRAT P., VOJTEK V.: Automatic Program Synthesis with Abstract Data Types. Computers and Artificial Intelligence, Vol.6, 1987, No.1, pp. 51-58.

[6] SUSSMAN G.J.: A Computational Model of Skill Acquisition. Ph.D. Thesis. M.I.T. Cambridge 1973. 199 pp.

[7] VOJTEK V.: Contribution to the Automatic Program Synthesis. Ph.D. Thesis, Slovak Technical University, Bratislava 1985. 99 pp.

[8] VOJTEK V., MOLNÁR Ľ., NÁVRAT P.: Automatic Program Synthesis Using Heuristics and Interaction. Computers and Artificial Intelligence, Vol. 5, 1986, No.5, pp. 429-442.

[9] WARREN D.H.D.: WARPLAN: A System for Generating Plans. Res. Report 76, Edinburgh, University of Edinburgh 1974, 23 pp.

[10] HRIVÍK P.: Automatic regressive planning of robot´s activity. Ph.D.Thesis, Slovak Academy of Sciences, Bratislava 1983. 101 pp.

STRONG SPLITTING RULES IN AUTOMATED THEOREM PROVING

Matthias B a a z
Institut für Algebra und Diskrete Mathematik
Technische Universität Wien
A-1040 Wien, Wiedner Hauptstraße 8-10

Alexander L e i t s c h
Dept.of Computer and Information Sciences, University of Delaware
103 Smith Hall, Newark, DELAWARE 19716 USA

Most problem reduction methods in automated theorem proving are based on equivalence transformations of formulas in predicate logic [Bl]. Unfortunately, these methods often prove to be too weak. Thus, the authors introduce a method to split a theorem into lemmas, which is based on proof-theoretic observations: A logical theorem is considered to be a consequence of stronger theorems, which possess a simpler quantificational structure. Some different methods are shown to construct such stronger theorems ([BaLe2],pp 289). These methods can be applied to complex quantificational formulas, where the classical methods fail. It can be formulated for general formulas in predicate logic, as well as for set of clauses.

Splitting for set of clauses:

Let $\Gamma = \Gamma_0 \cup \{C\}$ be a set of clauses, where $C = C_1 \cup C_2$ s.t. $C_1 \cap C_2 = \phi$ and var $(C_1) \cap \text{var}(C_2) = \{x\}$ for a variable x. Define $\Gamma_1 = \Gamma_0 \cup \{C_{1d}^{\,x}\}$, $\Gamma_2 = \Gamma_0 \cup \{C_2\}$, where d is a constant symbol not occuring in Γ.

Motivation:

If $\Gamma = \Gamma_0 \cup \{C\}$ s.t. $C = C_1 \cup C_2$ and $\text{var}(C_1) \cap \text{var}(C_2) = \phi$ then Γ is unsatisfiable iff $\Gamma_0 \cup \{C_1\}$ and $\Gamma_1 \cup \{C_2\}$ are both unsatisfiable. Instead of refuting Γ itself, we refute $\Gamma_0 \cup \{C_1\}$, $\Gamma_0 \cup \{C_2\}$. If there is no clause which decomposes into variable disjoint sets, this property can be "enforced" by introduction of a constant. The price we pay is, that we only have:

Theorem: $\Gamma_0 \cup \{C^x_{1d}\}$ and $\Gamma_0 \cup \{C_2\}$ unsatisfiable ➡ Γ unsatisfiable.

Example:

$\Gamma_0 = \{\{P(x),Q(x,y)\},\{\sim Q(u,f(v)),P(u)\},\{\sim P(w),Q(w,w)\}\}$

$C = \{\sim P(z),\sim Q(x,f(z))\}$, $C_1 = \{\sim P(z)\}$, $C_2 = \{\sim Q(x,f(z))\}$,

$C^z_{1d} = \{\sim P(d)\}$, $\Gamma_1 = \Gamma_0 \cup \{\sim P(d)\}$, $\Gamma_2 = \Gamma_0 \cup \{\sim Q(x,f(z))\}$.

(Note, that the clause $C' = \{\sim P(d),\sim Q(x,f(z))\}$ is not a substitution instance of C and C does not subsume C'.)

If (like in the example) not only Γ but in addition Γ_1 and Γ_2 an unsatisfiable, splitting is called effective.

Considerable effort is spent on the development of criteria for effective splitting: First, a characterization of effective splitting using Herbrands universe is given ([BaLe2], theorem 4.1,pp 293). Later on, some syntactical criteria for sets of clauses are discussed, by which the efficiency of splitting can be determined; some of these criteria especially apply to clause sets, which can be subjected to the set-of-support strategy.

Finally, the positive effect of splitting on the search expenses of a theorem prover is investigated:

Theorem ([BaLe2],pp 305) There exists a sequence of sets of clauses such that

1. without splitting an exponential number of linear deductions is necessary,
2. with splitting the problem becomes linear.

Applications to PROLOG are obvious.

REFERENCES

[BaLe1] Baaz M., Leitsch A.: Eine Methode zur automatischen Problemreduktion, Informatik-Fachberichte 106(1985), pp.154-163.

[BaLe2] Baaz M., Leitsch A.: Die Anwendung starker Reduktionsregeln im automatischen Beweisen, Proceedings of the Austrian Academy of Science II, Vol.194(1985), pp.287-307.

[Bl] Bledsoe W.W.: Splitting and reduction heuristics in automated theorem proving, Artifical Intell.2(1971), pp.57-78.

TOWARDS A REFINED CLASSIFICATION OF GEOMETRIC SEARCH AND COMPUTATION PROBLEMS

Thomas Fischer
Karl-Weierstraß-Institut für Mathematik
Akademie der Wissenschaften der DDR
PF 1304, Berlin 1086, DDR

1. Introduction

Geometric computation problems may be classified according to several aspects. For instance, a classification can be made with respect to the geometric objects involved, e.g. points, boxes, spheres, polytopes, or to the relations to be computed such as intersection, incidence, inclusion, proximity, etc. From the viewpoint of complexity theory one is interested in discriminating geometric problems according to the computational effort necessary for their algorithmic solution.

In the context of algorithm design it appears desirable to have the possibility of comparing problems of different kind. One would like to have a measure that allows to say that one problem is in a specified sense more difficult than another or to identify problems that are essentially equivalent, respectively. A complexity theory that is not able to discriminate between apparently different problems would be useless. This is meant when speaking of the intrinsic complexity of a computation problem. It can be defined as the minimum time that is necessary in some specified model of computation for correctly solving the problem in question in the worst case.

For almost all problems of practical relevance it has been proved impossible to determine the intrinsic complexity of a problem exactly. Therefore, one tries to derive upper and lower bounds on this function, and the complexity theorist's "dream" is to find such bounds matching at least in their order of magnitude. Unfortunately, until now the latter has been proved to be a very difficult task for almost all interesting problems in the area of computational geometry. In fact, matching upper and lower bounds are known only when restricting problems to planar instances.

In the following section we summarize the known results in this area and give a survey on the classes of problems that have been identified so far. Then we show that a further discrimination can be obtained when considering geometric problems in spaces of arbitrary dimension.

This is shown here by proving an $\Omega(d\ n \log n)$ lower bound for the problem of computing all intersecting pairs within a collection of n boxes in R^d, which separates this problem from strictly easier ones such as Lexicographic Sorting and others. Due to the limited space in this volume we restrict ourselves to the area of intersection problems only. However, especially the idea of constructing mappings into cell complexes used in Theorem 4.1 below has been proved to be a powerful proof technique also in other areas of computational geometry. Some of these results are listed in Section 4.3, for further details we refer to forthcoming papers [F1] [F2].

2. Classification of Geometric Problems

When restricting attention to planar instances geometric problems can be roughly classified into three groups of problems:

The lowest level of complexity is represented typically by various search problems which can be shown to satisfy $O(\log n)$ time bounds. Note that in some cases this is only a measure for the overhead activity leading to the items of the accessed set (usually, a sequence of comparisons), while the activity of assembling the query responses may require additional time proportional to its actual size.

A second group is formed by some problems that are known to satisfy linear time bounds. A familiar example is the computation of a smallest circle that encloses a given set of n points [Meg]. Other problems satisfy linear time bounds only under certain restrictions on the input, e.g. the construction of the intersection of two convex polygons.

Most of the nontrivial lower bounds that are presently known, however, are worst case bounds of the order of magnitude $\Omega(n \log n)$, where n is an integer associated with the input size of the corresponding problem. The list of problems that belong to this class is large and it includes decision and computation problems of rather distinct nature, e.g. set operations, convex hull problems, proximity problems, many intersection problems, etc.

This classification does not imply that all the problems within the same class are computationally equivalent. In the contrary, with regard to the diversity of presently known upper bounds this seems even unlikely. Moreover, there is also another observation that indicates that each of these classes is indeed finer structurized.

Most nontrivial bounds have been obtained by using the technique of transformation. It applies when a problem of unknown difficulty can be shown to be at least as hard as another problem of known difficulty. Some of the known relationships between selected problems satisfying an $\Omega(n \log n)$ time bound are shown in Figure 1. Two problems are linked by an edge if one of them can be shown to be more or at least as difficult than the other one placed below. Thus in order to refine the classification introduced above we have to identify edges that represent the proper relation.

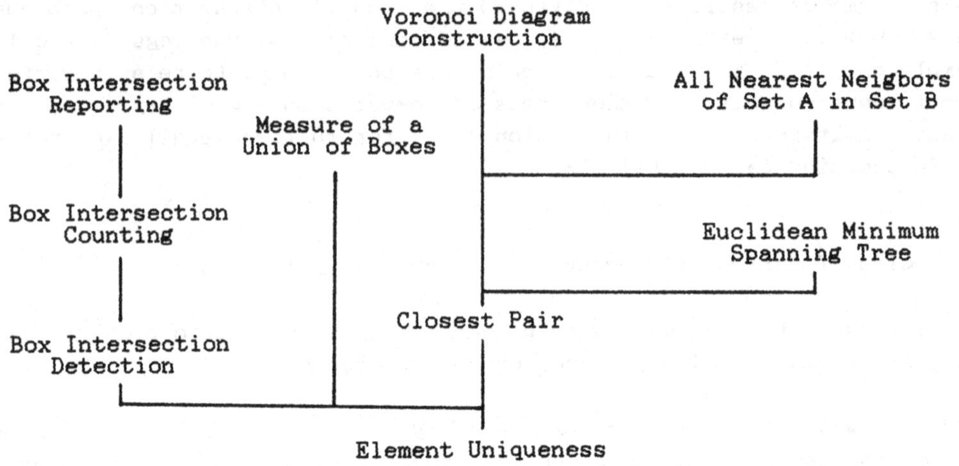

Fig. 1: The graph representing the partial order of computation problems that satisfy an $\Omega(n \log n)$ time bound in the planar case.

Of course, this transformation method prerequires the existence of at least one fundamental problem for which a nontrivial lower bound can be proved directly. Such computational prototypes for the class of problems satisfying an $\Omega(n \log n)$ bound under the algebraic decision tree model are e.g. Sorting, Extreme Points, and Element Uniqueness. They all share the fundamental trait that their lower bounds are derived from the cardinality of the set of permutations of n letters. For a survey of known results in this area we refer to Preparata and Shamos [PS] and Lee and Preparata [LP].

The lower bounds for the problems mentioned above have been usually derived in the planar or even unidimensional case. Though they remain also valid for the higher dimensional generalizations of the considered problems, it is natural to seek for possible improvements in that case. In fact, it is known that the dimension of the considered space is an essential parameter in the worst case analysis of many fundamental algorithms in this area. Typically it appears as an exponent of n or log n in such bounds.

Surprisingly, rather little is known on lower bounds depending on both parameters - the input size n and the dimension d. With respect to the class of $\Omega(n \log n)$ problems Lueker has shown that for computing the ranks for a set of n points in R^d $O(d\, n \log n)$ coordinate comparisons are necessary and sufficient [L]. Another computation problem that requires at least $\Omega(d\, n \log n)$ comparisons is the intersection reporting problem for a set of n d-dimensional boxes ([FW], below we prove two sharper versions of this result).

On the other hand there are many problems which prove to be much easier when considering them in d-space. The best known example is Sorting, it requires only n log n + d n comparisons to lexicographically sort a list of n d-dimensional vectors [StW]. Thus the class of Ω(n log n) problems decomposes into at least two subclasses: While many problems can be shown to have a solution in less than O(n log n + d n) time (further examples are given below), there are some problems that are proper multidimensional, i.e. their complexity can be shown to grow proportionally with the dimension of the considered space.

Thus further research activities should be directed on a refinement of the known classification of geometric problems according to their dependence on d. In particular, the following two questions arise:

(1) Which other problems can be shown to belong to one of the two subclasses mentioned above?

(2) Are there further subclasses characterized by quite another types of dependence on d ?

In this paper we concentrate ourselves on the area of intersection problems. Furthermore, we present a result that partially answers the second of the above stated problems.

3. Linear Problems

Throughout this paper we deal with problems of the following type:

Intersection Search Problems: Given n geometric objects stored in a data base, find all k objects that intersect with a given query object.

Intersection Decision (Counting, Reporting) Problems: Given a set of n geometric objects, decide whether there is a pair of objects having a nonempty intersection (count or find all k such pairs, resp.).

In the simplest case the objects may be assumed to be points in the d-dimensional Euclidean space. The detection of a nonempty intersection then reduces to a test for the identity of two points, and the intersection search problem becomes a problem known as Exact Match Search which can be solved with only log n + d ternary comparisons [M].

The corresponding intersection decision problem is known as Element Uniqueness Problem. It is one of the fundamental problems that are often used as a computational prototype for the class of Ω(n log n) problems. One might suspect that in order to improve upon existing lower bounds it would suffice to derive such bound for computational prototypes and then to carry it over by extending the technique of transformation to multidimensional problems. However, this hope is destroyed by the observation that the most frequently occurring prototype is provably not much harder than Lexicographic Sorting.

Theorem 3.1 The d-dimensional Element Uniqueness Problem can be solved in O(n log n + d n) time and O(d n) space.

Proof. The problem can be solved by first lexicographically sorting the n points given as input and then scanning the sorted list for a pair of immediately succeeding points that are identical. While the first step can be done with n log n + d n comparisons [StW], the following scanning requires only O(d n) time.

Element Uniqueness is a typical example of a linear computation problem. This denotation refers to the fact that problems of this kind are associated with a linear ordering of the input set. They can therefore be solved in the same way by determining this ordering and then scanning the ordered sequence according to the relation in question. Further instances are other computation problems on sets such as Set Equality, Set Inclusion, and the set operations.

4. Lower Bounds in Axis-Parallel Geometry

4.1. Search Problems. We first recall the known results on search problems in this area. The simplest problem representing a higher level of complexity is the Point Enclosure Problem. It requires to efficiently determine all boxes from a set of n d-dimensional boxes that enclose a given query point. As shown in a previous paper in the worst case d log n - O(d log d) comparisons are necessary for finding this subset [FW].

More familiar is the orthogonal d-Range Search Problem that refers to the subset of points that belong to a d-dimensional axis-parallel box (i.e. a d-range) given as query object. This problem is known to satisfy a lower bound of 2 d log n - O(d log d) comparisons, see [BM] and [S]. Of course, since boxes are more complex objects than points this bound carries over to the Box (d-Range) Intersection Search Problem.

Finally, we only mention here that for all three problems there are time optimal (but very space consuming) algorithms that achieve these lower bounds, for details we refer to [FW]. Thus essentially all search problems that are reasonably definable in an axis-parallel geometry are proper multidimensional compared with the linear problems mentioned above.

4.2. Intersection Reporting Problems. We now turn to the computationally more involved Box Intersection Reporting Problem. It is the natural d-dimensional generalization of the more familiar Rectangle

Intersection Reporting Problem that has become well known because of its importance for VLSI layout verification applications.
In the following we present a lower bound that is satisfied even in the case of a sparse set of boxes. A class of box configurations is said to be sparse, if for each configuration C_n in this class the number of intersecting pairs of boxes is linear in n, where n is the size of C_n.

We consider a configuration C_n constructed as follows: Let C_{dm} be a set of boxes in d-space,

$$C_{dm} = \{B_1, \ldots, B_{dm}\},$$

where m is an integer, $m \geq 1$, and

$$B_1 = [0,1] \times [0,m] \times \ldots \times [0,m],$$

$$B_2 = [1,2] \times [0,m] \times \ldots \times [0,m],$$

$$\vdots$$

$$B_m = [m-1,m] \times [0,m] \times \ldots \times [0,m],$$

$$B_{m+1} = [0,m] \times [0,1] \times \ldots \times [0,m],$$

$$\vdots$$

$$B_{dm} = [0,m] \times [0,m] \times \ldots \times [m-1,m].$$

Define further C_{n-dm} as a set of boxes that are supposed to be degenerated into points,

$$C_{n-dm} = \{x_1, \ldots, x_{n-dm} \;/\; x_i \in [o,m]^d, \;\; 1 \leq i \leq n - dm\},$$

where n is chosen such that $n = (d\,m)^2$. Then consider the configuration

$$C_n = C_{dm} \cup C_{n-dm}.$$

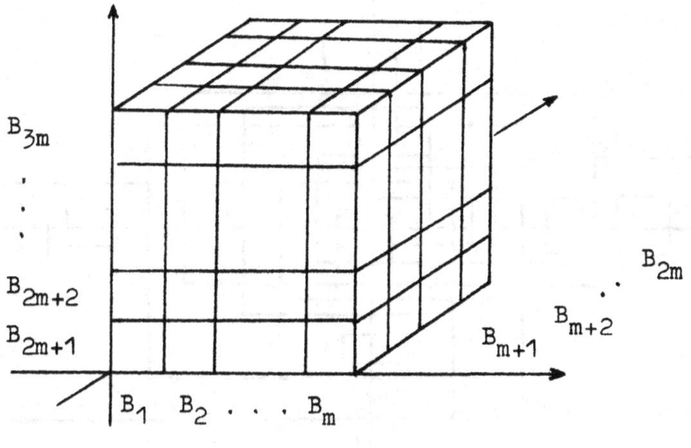

Fig. 2: C_{3m}

Obviously, the union of C_{dm} is a d-dimensional cube that consists of m^d cells, each of which is the intersection of exactly d boxes, see Fig. 2. Furthermore, the sets of boxes defining a cell are distinct for distinct cells.

Of course, there are $(m^d)^{n-dm}$ distinct mappings of the points in C_{n-dm} into the set of cells of the d-cube $[0,m]^d$ defined by C_{dm}. Since distinct cells represent distinct sets of boxes, any such mapping corresponds to a combination of intersecting pairs that differs from all combinations induced by other mappings. Hence, for the class of configurations defined above there are $(m^d)^{n-dm}$ distinct outputs possible, which yields for the number of comparisons, T(n,d), the lower bound

$$T(n,d) \geq (n-dm)\ d \log m$$

$$= (d/2)\ n \log n - O(d \sqrt{n} \log n).$$

The number of intersecting pairs of boxes in C_{dm} is apparently less than d^2m^2. Since any box in C_{n-dm} intersects with d rectangles in C_{dm}, the total number of pairs of intersecting boxes does not exceed $d(n-dm) + d^2m^2 = O(dn)$. Thus we have proved the following result:

Theorem 4.1 Any comparison based algorithm for the Box Intersection Reporting Problem requires $\Omega(d\ n \log n)$ comparisons in the worst case even for sparse configurations.

Though the result of Theorem 4.1 is sufficient for the task of separating the Box Intersection Reporting problem from the class of linear problems studied in the preceding section, it would be interesting to know whether it is possible to further improve that bound. The following result shows that this is indeed possible when considering a more dense configuration of boxes that is illustrated in Fig. 2 for the planar case.

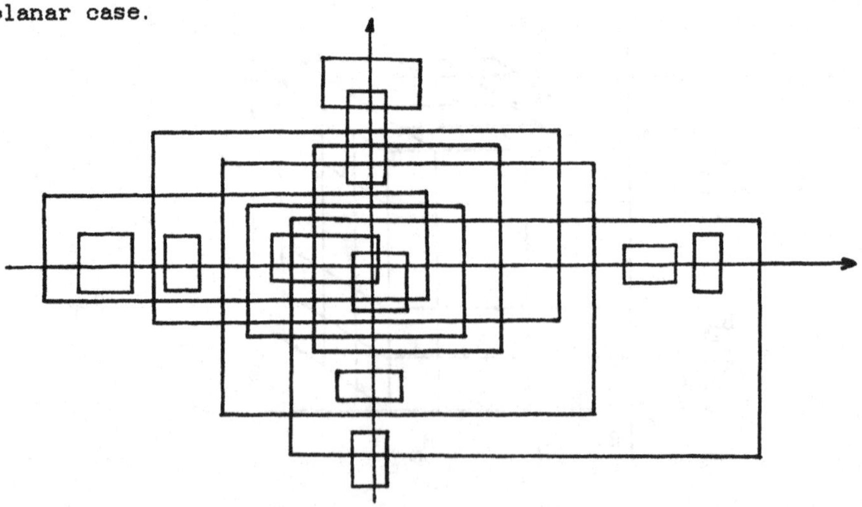

Fig. 3: Planar rectangle configuration.

Theorem 4.2 For any rational $\varepsilon > 0$, it is possible to construct a class of box configurations such that for sufficiently large n any comparison based algorithm for the intersection reporting problem has to perform at least $2(1-\varepsilon)\, d\, n \log n - O(d\, n \log(d/\varepsilon))$ comparisons in the worst case.

Sketch of proof. Consider a configuration of d-dimensional boxes such that exactly m boxes are distributed to each of the 2d half axes only, while another 2 d m boxes are placed so that they intersect each of the d coordinate axes (i.e. they enclose the origin of the coordinate system as shown in Fig. 3 for m = 2). Thus the entire collection of boxes consists of n = 4dm boxes distributed in the described way.

Let N(m,d) denote the maximum number of distinct combinations of pairs of intersecting boxes that can be obtained by configurations satisfying these assumptions. Then by induction one can derive the recurrence

$$N(m+1,d) \geq (m+1)^{4d^2} N(m,d),$$

which implies the lower bound

$$N(m,d) \geq (m!)^{4d^2}$$

when using the trivial estimate $N(1,d) \geq 1$. This gives the lower bound $d\, n \log n - O(d\, n \log d)$ proved in [FW].

When using another distribution of the boxes to the two groups it is possible to further improve this bound. In fact, assumming a distribution e.g. in proportion ε to $1-\varepsilon$ (where $1-\varepsilon$ corresponds to the contribution to the group centered around the origin), we are able to create more distinct configuration than before.

Suppose that 2md boxes are distributed to the half axes as before but let kmd boxes belong to the center group, $k > 2$. Then we have $n = (1-\varepsilon)n + \varepsilon n = kmd + 2md$, i.e. $\varepsilon = 2md / n$. It is not very difficult to see that by inserting kd new boxes into a given box configuration (i.e. into its center group) altogether

$$((m+1)^{2d})^{kd}$$

distinct new configurations can be created. This can be seen by counting the possibilities for varying one side of a new box independently from all other sides. Thus one obtains the recurrence

$$N(m+1,d) \geq (m+1)^{2kd^2} N(m,d),$$

which implies the lower bound

$$T(n,d) \geq 2\, k\, d^2\, m\, (\log m - \log e)$$

$$= 2\, (1-\varepsilon)\, n\, d\, (\log(\varepsilon n/(2d)) - \log e)$$

$$= 2\, (1-\varepsilon)\, d\, n \log n - O(d\, n \log(d/\varepsilon)).$$

The following result shows that there remains only a small gap between the lower bound obtained in Theorem 4.2 and the theoretically best possible bound that could be derived by such a combinatorial argument.

Theorem 4.3 By means of counting the number of distinct outputs no better lower bound than $2\ d\ n \log n\ -\ O(d\ n \log d)$ can be obtained.

We omit the proof here because it is not necessary for the purpose of the present investigations (it can be derived from a recent result of Gyarfas, Lehel and Tuza, [GLT]). However, the following remarks are necessary at this point:

The lower bounds in the preceding sections refer only to the number of comparisons that have to be performed in the algorithmic solution of the considered problems. For the total running time we have take into account also the size of the output of these algorithms. For dense configurations of rectangles this number can be of order $O(n^2)$. This underlines the importance of the weaker bound obtained in the proof of Theorem 3.1 for a family of sparse configurations. Efficient algorithms for the intersection reporting problem have been found in papers by Six and Wood [SW] and Edelsbrunner [E]. They both satisfy worst case bounds depending on polylogarithmic expressions. On the other hand, however, for the case of planar configurations typically occurring in VLSI applications even linear expected behavior has been observed [BHH]. Hence, one should expect that the rather pessimistic worst case bounds could be improved when restricting attention to classes of structurally "simple" configurations.

4.3 Related Results. Now we shortly summarize which other of the problems mentioned in the graph shown in Fig. 1 can be proved to satisfy an $\Omega(d\ n \log n)$ worst case bound. First of all one should be interested whether it is possible to obtain bounds of this order of magnitude also for the corresponding counting and detection problems, resp. In fact, recently we succeeded in proving such bound for the Box Intersection Counting Problem in the model of algebraic computation trees [F1]. This bound is obtained by a combination of the idea of constructing a large number of distinct mappings used in Theorem 4.1 with the fundamental result of Ben-Or [B]. Moreover, similar arguments apply to various computation problems concerning a collection of n boxes in R^d including e.g. the Measure-of-Union problem (also known as Klee's rectangle problem), which requires to efficiently compute the measure of a union of such collection [F1].

Thus almost all computation problems in the geometry of boxes can be shown to be harder than Sorting. On might suspect therefore that the intrinsic complexity of a computation problem is essentially determined by the complexity of the geometric objects involved. However, this would be a two simple explanation. In fact, there are also problems involving only n points in R^d that can be shown to satisfy an $\Omega(d\ n \log n)$ lower bound. Known examples are the computation of the ranks of n points (the so-called ECDF problem [L] and the computation of all nearest neighbors of the points of a set A in another set B [F2]. These examples show that the complexity of computation problems depends also on the complexity of the relation to be computed.

We now turn our attention to the second question posed in the introduction: Are there problems characterized by a principally different algorithmic behavior with respect to d ? This seems unlikely in an axis-parallel geometry because there are only 2 d directions for which a variation of the hyperplanes defining the boundary of geometric objects is possible. Therefore we give up this restriction and consider now objects that are defined as intersection of half-spaces in arbitrary orientation, i.e. polyhedra.

5. Lower Bounds in Polyhedral Geometry

We first mention that there are some problems known in this area which do not represent a higher level of complexity, e.g. the Polyhedral Membership Problem [YR] [MP] and the Half-Space Retrieval Problem. They both satisfy $\Omega(d \log n)$ lower bounds under the linear decision tree model.

In general, however, region search proves to be more difficult in polyhedral geometry than the corresponding d-range search in the axis-parallel case. This can be seen for the case of a query region assumed as an arbitrary convex d-polytope having a large number of facets. The following bound relates the worst case complexity of region search to both the number of data records and the complexity of the considered query region.

Theorem 5.1 If the query regions are d-polytopes with at most f_{d-1} facets, then any decision tree for the corresponding region search problem must have depth $f_{d-1} \log (n/f_{d-1} + 1)$.

Sketch of proof. Consider an arbitrary d-polytope as shown in Fig. 4 below. Choose a point in the interior and assume straight lines drawn from this point through the facets of the polytope. Then suppose that n points $(n > f_{d-1})$ are distributed to the straight lines such that about n/f_{d-1} points correspond to each line.

Obviously, if we translate a facet into the direction of the corresponding straight line, there are certain limits such that the number f_{d-1} of facets of the given polytope does not change if traslation is restricted to be within these limits.

Suppose that the n/f_{d-1} points corresponding to each line are distributed on a small interval between these limits. Hence, by translating the facets independently, from the given polytope we can construct a family of combinatorially equivalent polytopes having f_{d-1} facets each.

When translating a facet the response to the corresponding region search problem must change whenever passing one of the points distributed to the interval. Thus there are $(n/f_{d-1}+1)$ distinct responses that can be created in this way. Since any hyperplane defining a facet of the polytope may be translated independently, there are

$$(n/f_{d-1} + 1)^{f_{d-1}}$$

distinct query regions altogether. This implies the bound of the theorem.

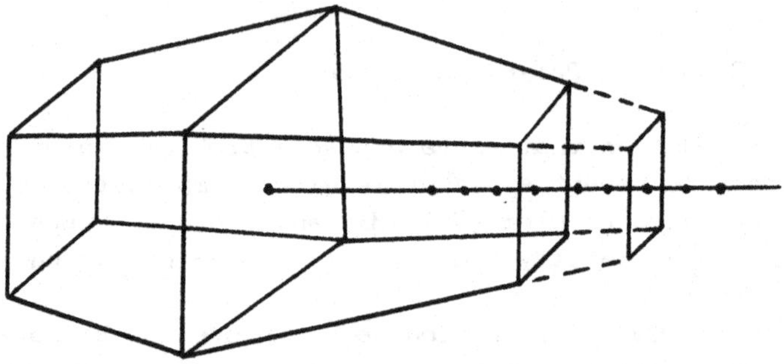

Fig. 4

For the case of the d-dimensional generalization of the octahedron we have $f_{d-1} = 2^d$, see [G], which gives an $\Omega(2^d \log n)$ bound. This shows that region search in polyhedral geometry is indeed a more difficult problem than in the axis-parallel case studied in the preceding section.

6. Concluding Remarks

Of course, the idea of the proof of Theorem 3.2 can be also used to obtain on $\Omega(2^d n \log n)$ lower bound to the corresponding intersection reporting problem. We omit the details here and conclude with some open problems:

1. What is the simplest computation problem in the class of problems satisfying an $\Omega(d n \log n)$ bound?

2. Which other problems belong to this class? Especially, to which subclass belongs the Box Intersection Decision Problem (which is apparently easier than the corresponding reporting problem) ?

3. What is the intrinsic complexity of the Line Intersection Decision (Reporting) Problem in d-space? Obviously, it is also in an intermediate position between the d-dimensional Element Uniqueness and the Box Intersection Problem.

References

[B] M. Ben-Or: Lower bounds for algebraic computation trees, Proc. 15th STOC (1983), 80-86

[BHH] J.L. Bentley, D. Haken, R. Hon: Fast geometric algorithms for VLSI tasks, Proc. Comput. Conf. 1981, 88-92

[BM] J.L. Bentley, H.A. Maurer: Efficient worst case data strutures for range searching, Acta Inform. 13 (1980), 155-168

[E] H. Edelsbrunner: A new approach to rectangle intersections, Intern. J. Comp. Math., 13 (1983), 209-229

[F1] Th. Fischer: Lower bounds in the geometry of boxes (to appear)

[F2] Th. Fischer: Voronoi diagrams and lower bounds (to appear)

[FW] Th. Fischer, K. Wolfrum: On the inherent complexity of geometric problems in d-space, Proc. MFCS'86, LNCS 233, 315-324

[G] B. Gruenbaum: Convex Polytopes, Wiley Interscience 1967

[GLT] A. Gyarfas, J. Lehl, Zs. Tuza: How many atoms can be defined by boxes? Combinatorica 5 (1985), 193-204

[LP] D.T. Lee, F.P. Preparata: Computational geometry - a survey, IEEE C-33 (1984), 1072-1101

[L] G. Luecker: A data structure for orthogonal range queries, Proc. 19th FOCS, 1978, 28-34

[M] K. Mehlhorn: Data Structures and Algorithms, Vol. 3, Springer 1984

[Meg] N. Megiddo: Linear time algorithm for linear programming in R^3 and related problems, SIAM J. Comp. 12(1983), 759-776

[MP] J. Moravek, P. Pudlak: New lower bound for polyhedral membership problem with an application to linear programming, MFCS'84, LNCS 176, 416-424

[PS] F.P. Preparata, M.I. Shamos: Computational Geometry - an Introduction, Springer 1985

[S] J.B. Saxe: On the number of range queries in k-space, DAM 1 (1979), 217-225

[SW] H.W. Six, D. Wood: Counting and reporting intersections of d-ranges, IEEE C-31 (1982), 181-187

[StW] L.J. Stockmeyer, C.K. Wong: On the number of comparisons to find the intersection of two relations, SIAM J. Comp., 8 (1979), 388-404

[YR] A.C. Yao, R.L. Rivest: On the polyhedral decision problem, SIAM Journal on Computing 9 (1980), 343-347

Matrix Padé Fractions

George Labahn and Stan Cabay †

Department of Computing Science, University of Alberta
Edmonton, Alberta, Canada T6G 2H1

† Supported in part by NSERC #A8035

ABSTRACT

For matrix power series with coefficients over a field, the notion of a matrix power series remainder sequence and its corresponding cofactor sequence are introduced and developed. An algorithm for constructing these sequences is presented.

It is shown that the cofactor sequence yields directly a sequence of Padé fractions for a matrix power series represented as a quotient $B(z)^{-1}A(z)$. When $B(z)^{-1}A(z)$ is normal, the complexity of the algorithm for computing a Padé fraction of type (m,n) is $O(p^3(m+n)^2)$, where p is the order of the matrices A(z) and B(z).

For power series which are abnormal, for a given (m,n), Padé fractions may not exist. However, it is shown that a generalized notion of Padé fraction, the Padé form, introduced in this paper does always exist and can be computed by the algorithm. In the abnormal case, the algorithm can reach a complexity of $O(p^3(m+n)^3)$, depending on the nature of the abnormalities. In the special case of a scalar power series, however, the algorithm complexity is $O((m+n)^2)$, even in the abnormal case.

1. Introduction.

Let

$$A(z) = \sum_{i=0}^{\infty} a_i z^i, \tag{1.1}$$

where a_i, i = 0, ... , is a p x p matrix with coefficients from a field K, be a formal power series. Loosely speaking, a matrix Padé approximant of A(z) is an expression of the form $U(z)\cdot V(z)^{-1}$, or $V(z)^{-1}\cdot U(z)$, where U(z) and V(z) are matrix polynomials of degree at most m and n, respectively, whose expansion agrees with A(z) up to and including the term z^{m+n}.

The definition of a Padé approximant can be made more formal in a variety of ways. For example, Rissanen [9] restricts V(z) to be a scalar polynomial and allows U(z) to be a p x q matrix. Typically, however, U(z) and V(z) are p x p polynomial matrices, and V(z) is further restricted by the condition that the constant term, V(0), is invertible (c.f., Bose and Basu[1], Bultheel[3], and Starkand[12]). In this paper, we call such approximants matrix Padé fractions, which is consistent with the scalar (p=1) case (c.f., Gragg[7]).

Definition 2.1: The pair of matrix polynomials (U(z),V(z)) is defined to be a **Right Matrix Padé Form** (RMPFo) of type (m,n) for the pair (A(z),B(z)) if

I. $\partial(U(z)) \leq m, \ \partial(V(z)) \leq n,$

II. $$A(z)\cdot V(z) \ + \ B(z)\cdot U(z) \ = \ z^{m+n+1}W(z) \tag{2.3}$$

where W(z) is a formal matrix power series, and

III. The columns of $V(z)$ are linearly independent over the field K. ■

The matrix polynomials U(z), V(z), and W(z) are usually called the right numerator, denominator, and residual (all of type (m,n)), respectively. Note that when B(z) = -I, Definition 2.1 corresponds to the definition of Padé form for a single matrix power series A(z) given in Labahn[8].

For ease of discussion, we use the following notation. For any matrix polynomial

$$U(z) = u_0 + u_1 z + \cdots + u_k z^k , \tag{2.5}$$

we write U (i.e., the same symbol but without the z variable) to mean the p(k+1) by p vector of matrix coefficients

$$U = \left[u_0, u_1, \cdots, u_k\right]^t, \tag{2.6}$$

where the transpose is at the symbolic level.

Let

$$S_{m,n} = \left[\begin{array}{cccc|cccc} a_0 & & & & b_0 & & & \\ & \cdot & & & & \cdot & & \\ \cdot & & \cdot & & \cdot & & \cdot & b_0 \\ \cdot & & & a_0 & \cdot & & & \cdot \\ & & & \cdot & \cdot & & & \cdot \\ & & & \cdot & & & & \cdot \\ a_{m+n} & \cdot & \cdot & a_m & b_{m+n} & \cdot & \cdot & b_n \end{array}\right] \tag{2.7}$$

denote a Sylvester matrix for A(z) and B(z) of type (m,n). Then equation (2.3) can be written as

$$S_{m,n} \cdot \begin{bmatrix} V \\ U \end{bmatrix} = 0. \tag{2.8}$$

Theorem 2.2: (Existence of Matrix Padé Forms) For any pair of power series (A(z), B(z)) and any pair of nonzero integers (m,n), there exists a RMPFo of type (m,n).

One case when the RMPFo is unique is given by

Definition 2.3. A pair (U(z),V(z)) of p x p matrix polynomials is said to be a **Right Matrix Padé Fraction** (RMPFr) of type (m,n) for the pair (A(z),B(z)) if

I. (U(z),V(z)) is a RMPFo of type (m,n) for (A(z),B(z)), and

II. The constant term, V(0), of the denominator is an invertible matrix. ■

For a particular m and n, however, matrix Padé fractions need not exist. Therefore, in this paper, we introduce the notion of a matrix Padé form, in which the condition of invertibility of V(0) is relaxed. The definition is a generalization of a similar one given for the scalar case (c.f., Gragg[7]). It is shown that matrix Padé forms always exist, but that they may not be unique. In general, matrix Padé forms need not have an invertible denominator, V(z). However, for m and n given, by obtaining a basis for all the Padé forms, we are also able to construct a matrix Padé form with an invertible V(z), in the case that one does exist.

In the one dimensional case, some algorithms that calculate Padé approximants for normal power series (Gragg[7]) include the ϵ-algorithm of Wynn, the η-algorithm of Bauer, and the Q-D algorithm of Rutishauser. Algorithms that are successful in the degenerate non-normal case are given by Brent et al[2], Bultheel[4], Rissanen[11], and Cabay and Choi[5].

The matrix case parallels the scalar situation in that most algorithms are restricted to normal power series. Algorithms that require the normality condition include those of Bultheel[3], Bose and Basu[1], Starkand[12], and Rissanen[10]. An algorithm that calculates Padé approximants in a non-normal case is given by Labahn[8]. However, in this algorithm there are still strict conditions that need to be satisfied by the power series before Padé approximants can be calculated.

The primary contribution of this paper is an algorithm, MPADE, for computing matrix Padé forms for a matrix power series. Central to the development of MPADE are the notions of a matrix power series remainder sequence and the corresponding cofactor sequence, which are introduced in section 4. These are generalizations of notions developed by Cabay and Kossowski[6] for power series over an integral domain. The cofactor sequence computed by MPADE yields a sequence of matrix Padé fractions along a specific off-diagonal path of the Padé table for A(z).

Unlike other algorithms, there are no restrictions placed on the power series in order that MPADE succeed. For normal power series, the complexity of MPADE is $O(p^3 \cdot (m+n)^2)$ operations in K. This is the same complexity as some of the algorithms proposed by Bultheel[3], Bose and Basu [1], Starkand [12], and Rissanen[10]. In the abnormal case, the complexity of the algorithm can reach $O(p^3 \cdot (m+n)^3)$ operations in K, depending on the nature of the abnormalities.

2. Matrix Padé Forms and Matrix Padé Fractions.

Let A(z) and B(z) be formal power series

$$A(z) = \sum_{i=0}^{\infty} a_i z^i \ , \ B(z) = \sum_{i=0}^{\infty} b_i z^i \qquad (2.1)$$

with coefficients from the ring of p x p matrices over some field K. Throughout this paper it is assumed that the leading coefficient, b_0, of B(z) is an invertible matrix. For non-negative integers m and n, let

$$U(z) = \sum_{i=0}^{m} u_i z^i \ , \ V(z) = \sum_{i=0}^{n} v_i z^i \qquad (2.2)$$

denote p x p matrix polynomials.

Condition II ensures that the denominator, V(z), is an invertible matrix polynomial.

The problem with Padé fractions, as mentioned in the previously , is that they do not always xist. However, let

$$T_{m,n} = \left[\begin{array}{cccc|cccc} a_0 & & & & b_0 & & & \\ \cdot & \cdot & & & \cdot & \cdot & & \\ \cdot & & \cdot & & & & & b_0 \\ \cdot & & & a_0 & \cdot & & & \cdot \\ & & & \cdot & & & & \cdot \\ & & & \cdot & & & & \\ a_{m+n-1} & \cdot & \cdot & a_m & b_{m+n-1} & \cdot & \cdot & b_n \end{array}\right] \tag{2.9}$$

nd define

$$d_{m,n} = \begin{cases} 1, & m = 0\,, n = 0, \\ det(T_{m,n}), & otherwise. \end{cases} \tag{2.10}$$

Then, a sufficient condition for the existence of a RMPFr is given by

Theorem 2.4. If $d_{m,n} \neq 0$, then every RMPFo of type (m,n) is an RMPFr of type (m,n). In addition, a RMPFr of type (m,n) is unique up to multiplication on the right by a nonsingular p x p matrix having coefficients from the field K. ∎

In the next section we also require

Theorem 2.5. Let A(z) and B(z) be given by (2.1). If m and n are positive integers such that $d_{m,n} \neq 0$, then RMPFo's (P(z),Q(z)) of type (m-1,n-1) for (A(z),B(z)) are unique up to multiplication of P(z) and Q(z) on the right by a nonsingular matrix from K. In addition, the leading term R(0) of the residual in condition II for RMPFo's,

$$A(z)\cdot Q(z) + B(z)\cdot P(z) = z^{m+n-1}R(z)\,, \tag{2.11}$$

is a nonsingular matrix. ∎

3. Matrix Power Series Remainder Sequences.

We define a **Right Matrix Padé Table** for (A(z), B(z)) to be any infinite two-dimensional collection of RMPFo's of type (m,n) for (A(z),B(z)) with m = 0, 1, ... and n = 0, 1, ... It is assumed that there is precisely one entry (i.e., one RMPFo) assigned to each position in the table. From Theorem 2.2, it follows that a right matrix Padé table exists for any given (A(z), B(z)). However, the table is not unique, because RMPFo's are not unique. This is unlike the definition of a Padé table for scalar power series (c.f. Gragg[7]), since here a Padé table consists of a collection of Padé fractions, which are unique.

A matrix power series pair (A(z),B(z)) is said to be **normal** (c.f., Bultheel[3]) if $d_{m,n} \neq 0$ for all m,n. For normal power series, it follows from Theorem 3.2 that every entry in the right matrix Padé table is a RMPFr. Consequently, from condition II in Definition 2.3 of RMPFr's, a right matrix

Padé table for normal power series may be made unique by insisting that the constant term, V(0), in the denominator of any Padé fraction be the identity matrix.

Following the convention used in the scalar case (c.f., Gragg[7]), we also define

$$(U(z), V(z)) = (z^m I, 0) \quad \text{for} \quad m \geq -1, \; n = -1, \tag{3.1}$$

and

$$(U(z), V(z)) = (0, z^n I) \quad \text{for} \quad m = -1, \; n \geq 0. \tag{3.2}$$

A right matrix Padé table appended with (3.1) and (3.2) is called an **extended right matrix Padé table** (c.f., Gragg[7]). The use of an extended table is strictly for initialization purposes. The entries given by (3.1) and (3.2) are not right matrix Padé forms (indeed the (-1,-1) entry is not even a matrix polynomial). However they do satisfy property II of Definition 2.1. For example, for $m \geq -1$ and $n = -1$, we have that

$$A(z)V(z) + B(z)U(z) = z^{m+n+1}W(z) \tag{3.3}$$

with

$$W(z) = B(z); \tag{3.4}$$

while, for $m = -1$ and $n \geq 0$, we have (3.3) with

$$W(z) = A(z). \tag{3.5}$$

Given the power series (2.1) and any non-negative integers m and n, we introduce a sequence of points

$$(m_0,n_0), \; (m_1,n_1), \; (m_2,n_2), \; \cdots \tag{3.6}$$

in the extended right matrix Padé table by setting

$$(m_0,n_0) = \begin{cases} (m-n-1,-1) & \text{for} \quad m \geq n \\ (-1,n-m-1) & \text{for} \quad m < n \end{cases} \tag{3.7}$$

and

$$(m_{i+1},n_{i+1}) = (m_i + s_i \,, n_i + s_i \,), \; i = 0, 1, 2, \; \cdots \,, \tag{3.8}$$

where $s_i \geq 1$. Observe that

$$m_i - n_i = m - n \,, \; i = 0, 1, 2, \; \cdots \,, \tag{3.9}$$

and consequently the sequence (3.6) lies along the m-n off-diagonal path of the extended right matrix Padé table. In (3.8), the s_i are selected so that

$$d_{m_{i+1},n_{i+1}} \neq 0 \tag{3.10}$$

and

$$d_{(m_i+j),(n_i+j)} = 0\ , \tag{3.11}$$

or $j = 1, 2, \cdots, s_i - 1$.

For i = 1, 2, ..., let $(U_i(z), V_i(z))$ be the unique RMPFr (c.f., Theorem 3.2) of type (m_i, n_i) for A(z),B(z)). Thus $\left[V_i\ ,\ U_i\right]^t$ satisfies

$$S_{m_i,n_i} \cdot \begin{bmatrix} V_i \\ U_i \end{bmatrix} = 0 \tag{3.12}$$

and, according to (2.3), there exists a matrix power series $W_i(z)$ such that

$$A(z) \cdot V_i(z) + B(z) \cdot U_i(z) = z^{m_i + n_i + 1} W_i(z). \tag{3.13}$$

Generalizing the notions of Cabay and Kossowski[6], we introduce

Definition 3.1. The sequence

$$\left\{W_i(z)\right\},\ i = 1, 2, \cdots, \tag{3.14}$$

is called the **Power Series Remainder Sequence** for the pair (A(z),B(z)). The sequence of pairs

$$\left\{(U_i(z), V_i(z))\right\},\ i = 1, 2, \cdots, \tag{3.15}$$

is called the corresponding **cofactor sequence.** The integer pairs $\{(m_i,\ n_i)\}$ are called **nonsingular nodes along the m - n off-diagonal path of the extended right matrix Padé table for (A(z),B(z)).** ■

We note that each term of a power series remainder sequence is unique up to multiplication on the right by a nonsingular matrix. This is also true for each term of the corresponding cofactor sequence.

Initially, when $m \geq n$, observe that $m_1 = m - n$ and $n_1 = 0$ (i.e., $s_0 = 1$), because in (3.2) the nonsingularity of b_0 implies that $d_{(m-n),0} \neq 0$. Thus, $V_1(z)$ is some arbitrary nonsingular matrix from K and, using (3.12), $U_1(z)$ can be obtained by solving

$$\begin{bmatrix} b_0 & & \\ \cdot & \cdot & \\ b_{m_1} & \cdots & b_0 \end{bmatrix} U_1 = - \begin{bmatrix} a_0 \\ \vdots \\ a_{m_1} \end{bmatrix} V_1\ . \tag{3.16}$$

That is, $U_1(z)$ can be obtained by multiplying the first m_1+1 terms of the quotient power series $B^{-1}(z) \cdot A(z)$ on the right by $-V_1(z)$.

Initially, when $m < n$, depending on a_0 there are two cases to consider. The simple case, when $det(a_0) \neq 0$, yields

$$d_{0,(n-m)} = \det \begin{bmatrix} a_0 & & \\ & \cdot & \\ & & \cdot \\ a_{n-m-1} & \cdot\ \cdot & a_0 \end{bmatrix} \neq 0 . \tag{3.17}$$

Thus, $s_0 = 1$, $m_1 = 0$, and $n_1 = n-m$. Then, the RMPFr $(U_1(z),V_1(z))$ of type (m_1,n_1) is determined by setting $U_1(z)$ to be an arbitrary nonsingular matrix from K and then solving

$$\begin{bmatrix} a_0 & & \\ \cdot & \cdot & \\ \cdot & & \cdot \\ a_{n_1} & \cdot\ \cdot & a_0 \end{bmatrix} V_1 = -\begin{bmatrix} b_0 \\ \cdot \\ \cdot \\ b_{n_1} \end{bmatrix} U_1 . \tag{3.18}$$

That is, when $m < n$ and $det(a_0) \neq 0$, $V_1(z)$ can be obtained by multiplying the first n_1+1 terms of the quotient power series $A^{-1}(z)\cdot B(z)$ on the right by $-U_1(z)$.

When $m < n$ and $det(a_0) = 0$, we must first determine the smallest positive integer s_0 (i.e., the smallest $m_1 = m_0+s_0$ and $n_1 = n_0+s_0$) so that $d_{m_1,n_1} \neq 0$. Notice that we must have $s_0 \geq n-m+1$. Once s_0 has been obtained, then $(U_1(z), V_1(z))$ is obtained by solving

$$S_{m_1,n_1} \cdot \begin{bmatrix} V_1 \\ U_1 \end{bmatrix} = 0 . \tag{3.19}$$

In section 5, we give an algorithm which computes a RMPFo of type (m,n) for (A(z),B(z)) by performing a sequence of the above types of initializations (albeit, each for different power series).

When the power series pair (A(z),B(z)) is normal, only the initializations corresponding to (3.16) and (3.18) are required. Thus, for normal power series $s_i = 1$ for all $i \geq 1$, and the algorithm reduces to a sequence of truncated power series divisions.

There are also some non-normal power series that share this property. For each pair of integers m and n, let $r_{m,n}$ be the rank of the matrix $T_{m,n}$. Then normality is equivalent to

$$r_{m,n} = (m+n)\cdot p \tag{3.20}$$

for all m and n. A matrix power series pair (A(z), B(z)) is said to be **nearly-normal** (c.f. Labahn[8]) if, for all integers m and n,

$$r_{m,n} = k_{m,n}\cdot p \tag{3.21}$$

for some integer $k_{m,n}$. Clearly, every normal power series is also a nearly-normal power series. In addition, all scalar power series are nearly-normal.

For a nearly-normal power series pair (A(z),B(z)) it is easy to see that when a_0 is singular, then $a_0 = 0$. This follows from the observation that the rank of a_0 is just $r_{0,1}$, which, if it is not p, must be zero. Also, if $a_0 = \cdots = a_{k-1} = 0$ and $a_k \neq 0$, then a_k must be a nonsingular matrix for similar reasons. When $k > m$ this implies that there are no nonsingular nodes along the m - n off-diagonal path before and including the node (m,n). Otherwise, when $k \leq m$, the initialization (3.19) becomes

$$\begin{bmatrix} 0 & & & \vert & b_0 & & \\ & & & \vert & & \cdot & \\ a_{m_1} & & & \vert & \cdot & & \cdot \\ \cdot & \cdot & & \vert & \cdot & & b_0 \\ \cdot & & \cdot & \vert & \cdot & & \\ a_{m_1+n_1} & \cdot\ \cdot & a_{m_1} & \vert & b_{m_1+n_1} & \cdot\ \cdot & b_{n_1} \end{bmatrix} \cdot \begin{bmatrix} V_1 \\ \\ U_1 \end{bmatrix} = 0, \tag{3.22}$$

where $s_0 = k+1$, $m_1 = k$ and $n_1 = n-m+k$. Consequently the RMPFr $(U_1(z),V_1(z))$ of type (m_1,n_1) is obtained from (3.22) first by setting $U_1(z) = z^{m_1} \cdot U$, where U is any nonsingular matrix from K. Then, $V_1(z)$ is obtained by multiplying the first n_1+1 terms of the quotient power series $(z^{-m_1} \cdot A(z))^{-1} \cdot B(z)$ on the right by -U. Thus, also for nearly-normal power series (and therefore also for all scalar power series), all initializations reduce to truncated power series divisions.

Corresponding to the Power Series Remainder Sequence, we introduce

Definition 3.2. The sequence

$$\left\{ (\, P_i(z), Q_i(z) \,) \right\}, \quad i = 1, 2, \cdots , \tag{3.23}$$

where $(P_i(z),Q_i(z))$ is the $(m_i-1,\, n_i-1)$ entry in the extended matrix Padé table for (A(z),B(z)), is called a **predecessor sequence** of the power series remainder sequence. ■

Theorem 3.3: For i = 1, 2, ... , the predecessors $(P_i(z),Q_i(z))$ are unique up to right multiplication by a nonsingular matrix from K. In addition, the leading term of the residual, $R_i(0)$, is nonsingular. ■

The main result of this section is

Theorem 3.4. For any positive integer k, (k - 1, k) is a nonsingular node in the Padé table for $(W_i(z),R_i(z))$ if and only if $(m_i+k,\, n_i+k)$ is a nonsingular node in the Padé table for (A(z), B(z)). ■

Theorem 3.4 allows us to calculate nonsingular nodes of a pair of power series by calculating nonsingular nodes of the residual pair of power series. This gives us an iterative method of calculating nonsingular nodes.

Theorem 3.5: The cofactor and predecessor sequences for (A(z),B(z)) satisfies

$$\begin{bmatrix} U_{i+1}(z) & P_{i+1}(z) \\ V_{i+1}(z) & Q_{i+1}(z) \end{bmatrix} = \begin{bmatrix} U_i(z) & P_i(z) \\ V_i(z) & Q_i(z) \end{bmatrix} \cdot \begin{bmatrix} I & 0 \\ 0 & z^2 \cdot I \end{bmatrix} \cdot \begin{bmatrix} V'(z) & Q'(z) \\ U'(z) & P'(z) \end{bmatrix} \tag{3.25}$$

where $(U'(z),V'(z))$ is the RMPFr of type $(s_i-1,\, s_i)$ for $(W_i(z),R_i(z))$, and where $(P'(z),Q'(z))$ is its predecessor. ■

4. The Algorithm:

Given non-negative integers m and n, the algorithm MPADE below makes use of Theorem 3.5 to compute the cofactor and predecessor sequences (3.10) and (3.18), respectively. Thus, intermediate results available from MPADE include those RMPFr's $(U_i(z),V_i(z))$ for (A(z),B(z)) at all the nonsingular nodes (m_i, n_i), i=1,2, ... , k-1, smaller than (m,n), along the off-diagonal path $m_i - n_i$ = m-n. The output gives results associated with the final node (m_k,n_k). If (m,n) is also a nonsingular node, then the output $(U_k(z),V_k(z))$ is a RMPFr of type (m,n) for (A(z),B(z)), and $(P_k(z),Q_k(z))$ is a RMPFo of type (m-1,n-1). If (m,n) is a singular node, then the output $(U_k(z),V_k(z))$ is simply a RMPFo of type (m,n) for (A(z),B(z)), and now $(P_k(z),Q_k(z))$ is set to be the RMPFr of type (m_{k-1}, n_{k-1}).

Note that, when (m,n) is not a nonsingular node, a simple modification of MPADE allows the computation of all RMPFo's of type (m,n) for (A(z),B(z)). It is only necessary to arrange to compute q columns of $[V'_k, U'_k]$, rather than p, in order to form a basis for the solution space of the equation in step 3.1 of MPADE. From this basis, it is then possible to construct a p x p matrix V(z), and a corresponding U(z), for which (U(z),V(z)) is a RMPFo of type (m,n) for (A(z),B(z)) and has the property that V(z) is an invertible matrix, assuming such a RMPFo exists. This enhancement is not included in MPADE primarily to simplify the presentation of the algorithm.

ALGORITHM (MPADE):

Step 1: # Initialization #

If $m \geq n$

then set

1.1) $i \leftarrow 1$

1.2) $s_0 \leftarrow m - n$

1.3) $$\begin{bmatrix} m_1 \\ n_1 \end{bmatrix} = \begin{bmatrix} s_0 \\ 0 \end{bmatrix}$$

1.4) $$\begin{bmatrix} U_1(z) & P_1(z) \\ V_1(z) & Q_1(z) \end{bmatrix} = \begin{bmatrix} -B(z)^{-1} \cdot A(z) \bmod z^{s_0+1} & z^{s_0-1} \cdot I \\ I & 0 \end{bmatrix}$$

else set

1.5) $i \leftarrow 0$

1.6) $$\begin{bmatrix} m_0 \\ n_0 \end{bmatrix} = \begin{bmatrix} m-n \\ 0 \end{bmatrix}$$

1.7) $$\begin{bmatrix} U_0(z) & P_0(z) \\ V_0(z) & Q_0(z) \end{bmatrix} = \begin{bmatrix} 0 & z^{m-n-1} \cdot I \\ I & 0 \end{bmatrix}$$

Step 2: # Search for next nonsingular node #

2.1) $s_i \leftarrow 0$

2.2) d ← 0

2.3) Do while $n_i+s_i < n$ and d = 0

2.4) Set $s_i \leftarrow s_i + 1$

2.5) Compute the residual $W_i(z)$ such that

$$(A(z)\cdot V_i(z) + B(z)\cdot U_i(z)) \bmod z^{m_i+n_i+2s_i+1} = z^{m_i+n_i+1}\cdot W_i(z)$$

2.6) Compute the residual $R_i(z)$ such that

$$(A(z)\cdot Q_i(z) + B(z)\cdot P_i(z)) \bmod z^{m_i+n_i+2s_i-1} = z^{m_i+n_i-1}\cdot R_i(z)$$

2.7) Compute

$$d = det(T'_{(s_i-1),s_i}).$$

determined from the power series $W_i(z)$ and $R_i(z)$

2.8) End do

tep 3: # Compute RMPFr for residuals #

3.1) Solve

$$S'_{(s_i-1),s_i}\cdot\begin{bmatrix}V'\\U'\end{bmatrix} = 0,$$

where S' is the Sylvester matrix determined from $W_i(z)$ and $R_i(z)$

tep 4: # Compute predecessor for residuals #

4.1) If $s_i > 1$ and $d \neq 0$,

then solve

$$S'_{(s_i-2),(s_i-1)}\cdot\begin{bmatrix}Q'\\P'\end{bmatrix} = 0\,,$$

where again S' is the Sylvester matrix determined from $W_i(z)$ and $R_i(z)$

else set

4.2) $$\begin{bmatrix}Q'(z)\\P'(z)\end{bmatrix} = \begin{bmatrix}I\\0\end{bmatrix}$$

tep 5: # Advance along off-diagonal for Padé fractions #

5.1) $m_{i+1} \leftarrow m_i + s_i$

5.2) $n_{i+1} \leftarrow n_i + s_i$

5.3) $$\begin{bmatrix}U_{i+1}(z) & P_{i+1}(z)\\V_{i+1}(z) & Q_{i+1}(z)\end{bmatrix} \leftarrow \begin{bmatrix}U_i(z) & P_i(z)\\V_i(z) & Q_i(z)\end{bmatrix}\cdot\begin{bmatrix}I & 0\\0 & z^2\cdot I\end{bmatrix}\cdot\begin{bmatrix}V'(z) & Q'(z)\\U'(z) & P'(z)\end{bmatrix}$$

5.4) i ← i + 1

tep 6: # termination test #

If $n_i < n$

then go to step 2

Else $k \leftarrow i$

$$\text{return}(\begin{bmatrix} U_k(z) & P_k(z) \\ V_k(z) & Q_k(z) \end{bmatrix})$$

5. Complexity of MPADE Algorithm

In assessing the costs of MPADE, it is assumed that classical algorithms are used for the multiplication of polynomials. Only the more costly steps are considered. For these steps, Table 5.1 below provides crude upper bounds on the number of multiplications in K performed during the i-th pass through MPADE.

Step	Bound on Number of Multiplications
2.5	$2p^3(m_i+n_i+2)s_i$
2.6	$2p^3(m_i+n_i+2)s_i$
2.7	$8p^3(s_i-1)^3/3$
3.1	$2p^3s_i^2$
4.1	$2p^3s_i^2$
5.3	$4p^3(m_i+n_i+2)(s_i+1)$

Table 5.1 : Bounds on Operations per Step

In step 2.7 of MPADE, it is assumed that the Gaussian elimination method is used to obtain the LU decomposition of $T_{(s_i-1),s_i}$. In addition, it is assumed that Gaussian elimination is accompanied with bordering techniques. Thus, as s_i increases by 1 in step 2.4, the results of the previous pass through the **while** loop are used to achieve the current LU decomposition. The bound for step 2.7 in Table 5.1 assumes we do not take any advantage of the special nature of $T_{(s_i-1),s_i}$.

For step 3.1, it is assumed that the LU decomposition of $T_{(s_i-1),s_i}$ from step 2.7, is used to simplify the triangulation of $S_{(s_i-1),s_i}$. The solution $[V', U']$ is obtained finally by solving this triangularized $S_{(s_i-1),s_i}$. Similar observations apply to step 4.1.

An upper bound for the number of multiplications in K required by MPADE is obtained by summing the costs in Table 5.1 for i=0,1, ... ,k. We use the fact that

$$\sum_{i=0}^{k} s_i = m \text{ , if } m \geq n \text{ , and } \sum_{i=0}^{k} s_i = n \text{ , if } m < n \text{ .} \tag{5.8}$$

In addition,

$$\sum_{i=0}^{k} m_i^{\alpha} s_i^{\beta} \leq m^{\alpha+\beta} \text{ and } \sum_{i=0}^{k} n_i^{\alpha} s_i^{\beta} \leq n^{\alpha+\beta} \text{ .} \tag{5.9}$$

Then, step 2.7 has a complexity of $O(p^3(m+n)^3)$ and the remaining steps a complexity of $O(p^3(m+n)^2)$, at worst. When (A(z),B(z)) is normal (i.e., $s_i = 1$ for all i), then the cost of step 2.7 is ero, since $T_{(s_i-1),s_i}$ is already in triangular form, and the complexity of MPADE then reduces to)$(p^3(m+n)^2)$. This is also the case when (A(z),B(z)) is nearly-normal. In this case s_i is often larger han one, but the matrix $T_{(s_i-1),s_i}$ is always in triangular form and so again the complexity is)$(p^3(m+n)^2)$. In particular, in the scalar case the complexity of MPADE is $O((m+n)^2)$.

When the power series is neither normal nor nearly-normal, MPADE still provides significant avings even in the case where most intermediate nodes are singular. For example, if only the midlle node (n/2,n/2) along the main diagonal is nonsingular, then MPADE has a complexity of $(n/2)^3/3 + 8(n/2)^3/3 = 2n^3/3$. This is a saving of a factor of 4 over the simple use of Gaussian elimnation. Algorithms requiring normality, on the other hand, break down when even one intermediate ode is singular.

References

1. N.K. Bose and S. Basu, "Theory and Recursive Computation of 1-D Matrix Padé Approximants," *IEEE Trans. on Circuits and Systems*, **4** pp. 323-325 (1980).

2. R. Brent, F.G. Gustavson, and D.Y.Y. Yun, "Fast Solution of Toeplitz Systems of Equations and Computation of Padé Approximants," *J. of Algorithms*, **1** pp. 259-295 (1980).

3. A. Bultheel, "Recursive Algorithms for the Matrix Padé Table," *Math. of Computation*, **35** pp. 875-892 (1980).

4. A. Bultheel, "Recursive Algorithms for Nonnormal Padé Tables," *SIAM J. Appl. Math*, **39** pp. 106-118 (1980).

5. S. Cabay and D.K. Choi, "Algebraic Computations of Scaled Padé Fractions," *SIAM J. of Computation*, **15** pp. 243-270 (1986).

6. S. Cabay and P. Kossowski, "Power Series Remainder Sequences and Padé Fractions over Integral Domains," *J. of Symbolic Computation*, (to appear).

7. W.B. Gragg, "The Padé Table and its Relation to Certain Algorithms of Numerical Analysis," *SIAM Rev.*, **14** pp. 1-61 (1972).

8. G. Labahn, *Matrix Padé Approximants*, M.Sc. Thesis, Dep't of Computing Science, University of Alberta, Edmonton, Canada (1986).

9. J. Rissanen, "Recursive Evaluation of Padé Approximants for Matrix Sequences," *IBM J. Res. Develop.*, pp. 401-406 (1972).

10. J. Rissanen, "Algorithms for Triangular Decomposition of Block Hankel and Toeplitz Matrices with Application to Factoring Positive Matrix Polynomials," *Math. Comp.*, **27** pp. 147-154 (1973).

11. J. Rissanen, "Solution of Linear Equations with Hankel and Toeplitz Matrices," *Numer. Math*, **22** pp. 361-366 (1974).

12. Y. Starkand, "Explicit Formulas for Matrix-valued Padé Approximants," *J. of Comp. and Appl. Math*, **5** pp. 63-65 (1979).

COMPUTATION OF GENERALIZED PADÉ APPROXIMANTS

G. Németh and Magda Zimányi
Central Research Institute for Physics of the Hungarian Academy of Sciences
Budapest 114, P.O.B. 49, Hungary, H-1525

The traditional framework of the Padé method assumes that the power series (Taylor series) represents the function in a certain sense and that the coefficients of the power series are known. The Padé approximants can be computed by solving a linear system of equations. It is well known that the power series has roughly speaking, an extrapolatory character and therefore it seems that the Padé approximants almost surely inherit this property. If we use other series having an interpolatory character, e.g. the series expansion in terms of orthogonal polynomials in the interval considered instead of the power series, we shall call the approximants derived in an analogous way to the Padé method *generalized Padé approximants* [1]. It is shown that for some classes of functions the generalized Padé approximants give better results than the classical Padé approximants.

The REDUCE computer algebra system [2] is used for calculating the coefficients. If the original series has rational coefficients, the approximants have rational coefficients, too.

Let us suppose that $f(x)$ has the expansion

$$f(x) = \sum_{k=0}^{\infty} c_k F_k(x)$$

where $F_k(x)$ is a polynomial of order n (corresponding to the interval of the approximation). The rational fraction $P_m(x)/Q_n(x)$ is called the generalized Padé approximant if the following relation is satisfied

$$Q_n(x)f(x) - P_m(x) = O(F_{n+m+1}(x))$$

where the notation O indicates a function for which the series in $F_k(x)$ begins with the $n+m+1$-th term. We consider approximations in the interval $(0,\infty)$, therefore the choice $F_k(x) = L_k^{(\alpha)}(x)$ is appropriate where $L_k^{(\alpha)}(x)$ denotes the Laguerre polynomial $(\alpha > -1)$.

Two classes of functions are considered. The function

$$\varphi_\mu(x) = \mu \int_0^1 (1-u)^{\mu-1} e^{-xu}\, du, \qquad \mu > 0, \qquad 0 \le x < \infty$$

is an integral function, however, its Padé approximants do not converge geometrically in $(0,\infty)$ to $\varphi_\mu(x)$. It has been shown [3] that the generalized Padé approximants converge in the interval $(0,\infty)$ to $\varphi_\mu(x)$ uniformly with a geometric rate of convergence.

The function

$$\psi_\alpha(x) = \frac{x^{1-\alpha}}{\Gamma(1-\alpha)} \int_0^\infty t^{-\alpha} \frac{1}{1+t} e^{-xt}\, dt, \qquad 0 < x < \infty$$

has algebraic singularity at $x = 0$, and has essential singularity at $x = \infty$, and its Padé approximants do not converge geometrically in $(0, \infty)$ to $\psi_\alpha(x)$. It is shown [4] that the generalized Padé approximants converge with a geometric rate of convergence in $0 < x \leq \infty$ to $\psi_\alpha(x)$ excluding a small neighbourhood of the point $x = 0$. At the point 0 for some values of α an algebraic rate of convergence exists.

Coefficients of the generalized Padé approximants can be obtained by solving a system of linear equations. As the coefficients of the series in Laguerre polynomials are rational numbers, it is possible to obtain the approximants exactly by using the computer algebra system REDUCE. Results of our calculations are given for several examples. It can be seen from the results that the generalized Padé approximation can be a useful method for obtaining rational approximants of a function.

References

[1] Cheney,E.W. *Introduction to Approximation Theory,* McGraw-Hill, New York, 1966.

[2] Hearn,A.C. *REDUCE 3.2 Manual,* The RAND Corporation, Santa Monica, 1983.

[3] Németh,G. *Notes on Some Estimations in Rational Approximations I.* Periodica Mathematica Hungarica, Vol.19(1), 1988, pp.1-17.

[4] Németh,G. *Generalized Padé Approximations to Special Functions: Singular Functions* (in Hungarian), Alk. Mat. Lapok, Vol.13, 1987-88, pp.97-125.

A CRITICAL PAIR CRITERION FOR COMPLETION MODULO A CONGRUENCE[1]

Leo Bachmair
Department of Computer Science
SUNY at Stony Brook
Stony Brook, NY 11794, U.S.A.

Nachum Dershowitz
Department of Computer Science
University of Illinois
Urbana, IL 61801, U.S.A.

Extended Abstract

Rewrite systems are collections of directed equations (rules) used to compute by repeatedly replacing subterms in a given formula until a simplest form possible (normal form) is obtained. Many formula manipulation systems such as REDUCE or MACSYMA use equations for simplification in this manner. Canonical (i.e., terminating Church-Rosser) rewrite systems have the property that all equal terms (and only equal terms) simplify to an identical normal form. Deciding validity in theories for which canonical systems are known is thus easy and reasonably efficient. A number of canonical systems have been derived with the Knuth-Bendix completion procedure [5]. Unfortunately, the Knuth-Bendix procedure can not be applied to axioms such as commutativity that induce non-terminating rewrite sequences. There are also some practical limitations in its handling of associativity, as pointed out by Peterson and Stickel [6]. Associativity and commutativity are typical equations that are more naturally viewed as "structural" axioms (defining a congruence relation on terms) rather than as "simplifiers" (defining a reduction relation).

Given a set of axioms A and a rewrite system R, we denote by R_A the corresponding relation of *rewriting modulo A*, defined as the application of rules in R via A-matching. For example, if A consists of the associativity and commutativity axioms for addition, and the rewrite system R consists of a single rule $f(x,x) \to x$, then $f(x+y, y+x)$ can not be rewritten by R, whereas it can be rewritten to $x+y$ (and $y+x$) by R_A. Extensions of the Knuth-Bendix procedure to rewriting modulo a congruence have been described by Peterson and Stickel [6] (for sets A of associativity and commutativity axioms), by Jouannaud and Kirchner [3], and Bachmair and Dershowitz [1]. The fundamental operations in these procedures are A-matching and A-unification.

Two terms s and t are said to be *A-unifiable* if and only if there is a substitution σ (called an *A-unifier*), such that $s\sigma$ and $t\sigma$ are equivalent with respect to A. The above-mentioned completion procedures apply to theories for which complete sets of A-unifiers can be computed. If $u \to v$ and $s \to t$ are rules and p is a non-variable position in u, such that u/p and s are A-unifiable with a complete set of unifiers Σ, then the "rewriting ambiguity" $v\sigma \leftarrow_R u\sigma \to_{R_A} u\sigma[t\sigma]$ determines an *A-critical pair* $v\sigma = u\sigma[t\sigma]$, for each σ in Σ.

Completion augments a given rewrite system R by so-called "extended" rules[2] and then systematically computes A-critical pairs to check whether the two terms $v\sigma$ and $u\sigma[t\sigma]$ reduce to an identical normal form. If the test is successful for all critical pairs, then the rewrite system is canonical (in the sense of defining normal forms that are unique up to equivalence in A). Otherwise, the offending equations have to be turned into rules and critical pair computation continues with the new rules. (It is possible that endless new rules are generated and completion does not terminate. Completion may even fail, when an equation can not be oriented into a rule.) The most expensive part of completion is the reduction of terms to normal form. Critical pairs in which both terms reduce to identical normal forms are redundant. Various techniques, called *critical pair criteria*, have been proposed for standard completion for detecting redundancies more efficiently than by normalization of terms (see Bachmair and Dershowitz [2] for an overview). We sketch a similar technique for rewriting modulo a congruence.

[1]This research was supported in part by the National Science Foundation under grant DCR 85-13417.

[2]In the case of associative-commutative completion, extended rules are of the form $f(f(s,t),x) \to f(u,x)$, where x is a new variable and $f(s,t) \to u$ is a (non-extended) rule for which f is an associative-commutative operator.

A rewrite step $s \to_R t$ is said to be *blocked* (with respect to A) if it is by application of some rule with a substitution σ, such that no term $x\sigma$ can be rewritten by R_A. For example, if R contains rules $x+0 \to 0$ and $-(x+y) \to -x + -y$, then the rewrite step $-((x+0)+y) \to_R -(x+0) + -y$ is not blocked, because the term $x+0$, which is substituted for x, can be rewritten by R_A. Non-blocked rewrite steps can be replaced by a sequence of *blocked* steps. For instance, the above rewrite step can be replaced by $-((x+0)+y) \to_{R_A} -(x+y) \to_{R_A} -x + -y \leftarrow_{R_A} -(x+0) + -y$.

We say that an A-critical pair is blocked if both rewrite steps in the corresponding rewriting ambiguity $v\sigma \leftarrow_R u\sigma \to_{R_A} u\sigma[t\sigma]$ are blocked. All non-blocked A-critical pairs that are obtained from non-extended rules are redundant and can be disregarded by completion. The restriction to non-extended rules is crucial, though in the case of associative-commutative completion blocking can also be applied in a restricted form to extended rules. Specifically, A-critical pairs obtained by applying an extended rule $f(s,x) \to f(t,x)$ with a substitution σ, such that $y\sigma$ can be rewritten by R_A, for some variable y in s, can also be disregarded. It may be necessary, on the other hand, to instantiate the "extension variable" x by a term that can be rewritten.

For example, consider the set of two rules $a+b \to c$ and $(a+a)+(b+b) \to d$. The only A-critical pairs are those involving the extended rules $(a+b)+x \to c+x$ and $((a+a)+(b+b))+x \to d+x$, and require that the extension variable x be instantiated by a term that can be rewritten. But even though all A-critical pairs are non-blocked, the system is not canonical, as the term $(a+b)+(a+b)$ has two different normal-forms, $c+c$ and d, that are not equivalent with respect to associativity and commutativity.

Experiments that we have run using the associative-commutative completion procedure implemented in the rewrite rule laboratory RRL [4] indicate the usefulness of blocking. Results are summarized in Table 1 (the last column of the table refers to the ratio b/t of the respective times needed to obtain a canonical system with and without blocking).

Table 1. Associative-Commutative Completion

	STANDARD	BLOCKING			
	Critical pairs	Critical pairs	Blocked	Redundant	b/t
Abelian groups	86	95	50	45	0.75
Associative-commutative rings	248	317	151	166	0.83
Modules	416	462	271	191	0.68
Lattices	151	151	63	88	0.38
Boolean rings	99	109	38	71	0.50
Non-deterministic machines	284	284	131	153	0.45

References

[1] L. Bachmair and N. Dershowitz. Completion for rewriting modulo a congruence. To appear in *Theor. Comput. Sci.* (1988).

[2] L. Bachmair and N. Dershowitz. Critical pair criteria for completion. *J. Sym. Comput.* (1988) **6**:1-18.

[3] J.-P. Jouannaud and H. Kirchner. Completion of a set of rules modulo a set of equations. *SIAM J. Comput.* (1986) **15**:1155–1194.

[4] D. Kapur, D.R. Musser, and P. Narendran. Only prime superpositions need be considered in the Knuth-Bendix completion procedure. *J. Sym. Comput.* (1988) **6**:19-36.

[5] D. Knuth and P. Bendix. *Simple word problems in universal algebras.* In *Computational Problems in Abstract Algebra*, ed. J. Leech, Pergamon Press, Oxford, 1970, pp. 263–297.

[6] G. Peterson and M. Stickel. Complete sets of reductions for some equational theories. *J. ACM* (1981) **28**:233–264.

SHORTEST PATHS OF A DISC INSIDE A POLYGONAL REGION

Günter Werner
Mathematics Sektion, Friedrich-Schiller-University
Jena 6900
GDR

Research in robotics has stimulated considerable interest in finding algorithms for solving collision avoidance problems; this in turn has created interest in computational complexity of motion planning problems. An overview about the variety of these problems, the motivations, applications, methods for solutions, complexity results and references is given in Whitesides'85.

We examine the following problem: Given two polygonal nonintersecting chains C1, C2, the walls in the plane and a nonnegative real d. Let further the chains be connected by straight line segments a and b, respectively in such a way, that the chains C1, C2 and the segments a and b form together a simple polygon P. The first task is to decide whether there is a path of a disc with the diameter d without colliding the walls through the polygon entering a and exiting b. If such a path exists, the second task is to compute it.

There are different techniques to solve this problem. O'Dunlaing, Sharir and Yap'83 obtain an O(nlogn) algorithm. They use the notions of the Voronoi diagram and of the configuration space. Let de denote the Euclidean distance in the plane. Let further C1 and C2 be chains of a simple polygon as posed earlier. The internal distance Di of the chains C1 and C2 with respect to P is defined to be

$$Di(C1,C2) = \min(de(x,y) : x \in C1,\ y \in C2,\ \overline{xy} \text{ inside } P).$$

Obviously, the decision task has to be answered negatively if and only if the internal distance of C1 and C2 is smaller than d.

There are at least two problems to measure the internal distance. The first, C1 and C2 are infinite point sets, and then it is to check whether a line segment lies inside P.

Let rVk be the reflex vertices of the chain Ck (k=1,2) and xok the set of all orthogonal projections of x on Cl (l=1,2, k≠l).

$$D1(C1,C2) = \min(de(x,y) : x \in rV1,\ y \in rV2,\ \overline{xy} \text{ inside } P)$$

$$D2(C1,C2) = \min(de(x,y) : x \in rVk,\ y \in xok,\ \overline{xy} \text{ inside } P,\ k=1,2)$$

Lemma 1: Dk(C1,C2) = min(D1(C1,C2),D2(C1,C2))

Chains Ck′ are defined in the following way: Each edge of Ck is translated in the normal direction with the length d into the interior of P. Each reflex vertex is translated twice, its translations will be connected by an edge. Intersection points together with the tranlations of reflex vertices of Ck form an approximation Ck′ of the configuration space. As intersections of line segments the Ck′ are computable in time O(nlogn).

Lemma 2: If D1 (C1,C2) $<$ d, then there are reflex vertices u,v in C1′ and C2′ respectively, with de(u,v) $<$ d.

Lemma 2 is a sieve for all these vertices x,y in C1 and C2, for which $\overline{xy}$ is in P and which may be nearer to one another than d.
By Bentley, Shamos′76 all the pairs u,v with de(u,v) $<$ d can be reported in time O(nlogn).

Lemma 3: It is decidable in time O(nlogn), whether D1(C1,C2) is smaller than d.

To each edge e of Ck denotes re the rectangle with the sides e and d. REk is the set of all rectangles re with e $\in$ Ck.

Lemma 4: D2(C1,C2) $<$ d if and only if Ck intersects REl, k$\neq$l.

Theorem: Di(C1,C2) $<$ d is decidable in time O(nlogn).

In practice the dominating parts of the decision task are all of the kind to solve line intersection problems.
Once the decision is answered the computation of the shortest path is done with terms as point visibility and point location in time O(tn), where t is the number of turns.

References:

Whitesides, S.H., "Computational Geometry and Motion Planning", in: Toussaint, G.T. (ed.), Computational Geometry (North Holland, 1985)

O′Dunlaing, C., Sharir, M., Yap, C.K., "Retraction: A new approach to motion planning", 15th ACM STOC (1983), 207-220

Bentley, J.L., Shamos, M.I., "Divide-and-Conquer in Multidimensional Space", Proc. 8th ACM STOC (1976), 220-230

Rabin's width of a complete proof and the width of a semialgebraic set

Tomás Recio*& Luis M. Pardo*

Dept. de Matematicas
Facultad de Ciencias (Universidad de Cantabria)
39006 Santander, SPAIN

1 Introduction

Let $p_1(\underline{X}),\ldots,p_m(\underline{X})$ be polynomials in $\mathbf{R}[X_1,\ldots,X_n]=\mathbf{R}[\underline{X}]$. According to Rabin [Ra] — as generalized by Jaromczyk[J] — an $r\times k$ array of polynomials in n variables

$$\begin{matrix} p_{1,1}, & \cdots & p_{1,k} \\ \cdots & \cdots & \cdots \\ p_{r,1}, & \cdots & p_{r,k} \end{matrix}$$

is said to be a complete proof of $\{p_1\geq 0,\ldots,p_m\geq 0\}$ in $\mathbf{R}^n$ if and only if

$$\{p_1\geq 0,\ldots,p_m\geq 0\}=\{p_{1,1}\geq 0,\ldots p_{1,k}\geq 0\}\cup\ldots\cup\{p_{r,1}\geq 0,\ldots,p_{r,k}\geq 0\}$$

This number k is, by definition, the *width of this complete proof*.

Both Rabin and Jaromczyk apply these concepts to furnish lower bounds for the complexity of some geometrical algorithms in the *Algebraic Decision Tree Model*. Their main technical result consists in showing that under some conditions on the collection $\{p_1,\ldots,p_m\}$ the width of any complete proof must be bigger than m and under the same conditions the cost of any algebraic decision tree that solves the *Membership Problem to the semialgebraic set* $\{p_1\geq 0,\ldots,p_m\geq 0\}$.

With respect to the obtaining of lower bounds it seems, however, that the more recent estimates of Ben-Or [B] are better fitted in many cases than the ones found by bounding the width, as observed in [B].

On the other hand the study of sets defined in $\mathbf{R}^n$ by means of systems of polynomial equalities and inequalities (semialgebraic sets) has been considerably developed by real algebraic geometers in the past few years and now a quantitative theory of semialgebraic sets is available after the work of L. Bröcker ([Br1], [Br2], [Br3]) (*c.f.* also [A-B]).

In this paper we introduce a new quantitative invariant for semialgebraic sets closely related to the concept of width, showing its relation with the s and t *invariants* of Bröcker and therefore stating an upper bound for the width of a semialgebraic set in $\mathbf{R}^n$. From here we develop the concept of width of the congruence class of a semialgebraic set A and we observe that this is a

*Partially supported by CAICYT 2280/83

lower bound for the cost of any algebraic decision tree that solves the membership problem for A. In fact, the use of the width as lower bound of complexity in [Ra] is a consequence of its coincidence with the width of the congruence class.

Considering the upper bounds of the width obtained from the relation with the s and t invariants of Bröcker we can give a theoretical explanation for the above mentioned remark of Ben-Or concerning the better behaviour of his lower bound. Nevertheless, several examples are computed in order to compare the performance of both techniques as lower bounds.

Moreover, within this more general framework, we are able to extend the results of Rabin and Jaromczyk computing the width of a simultaneous non-negative system of polynomials under weaker hypotheses than theirs. In particular we obtain the conclusions of [A-B] for regular systems of parameters ([A-B] §3 Prop. 3.1).

Our analysis can also be applied to construct decision problems with very small Ben-Or lower bound and such that the bound given by the width is strictly better.

In the following we shall work over the field of real numbers $\mathbf{R}$ but all the results equally hold for any real closed field R.

2 Width of basic closed semialgebraic sets

Let F be a semialgebraic subset of $\mathbf{R}^n$ and Y its Zariski closure (*c.f.* [B-C-R] for definitions and basic properties). For every proper closed semialgebraic set A in $\mathbf{R}^n$ we define the ***width of A in F***, $w'(A,F)$, as the minimal non-negative integer r such that:

$$A \cap F = \{\underline{x} \in \mathbf{R}^n | f_{1,1}(x) \geq 0, \ldots, f_{1,r}(x) \geq 0\} \cap F \cup \ldots \cup \{\underline{x} \in \mathbf{R}^n | f_{s,1}(x) \geq 0, \ldots, f_{s,r}(x) \geq 0\} \cap F$$

for some $f_{i,j} \in \mathbf{R}[\underline{X}]$.

We set $w'(\emptyset, F) = -1, w'(\mathbf{R}^n, F) = 0$. By considering strict polynomial inequalities we can also introduce the concept of ***width of an open semialgebraic subset B of*** $\mathbf{R}^n$ ***in F***, $w(B,F)$.

Now we define $w'(F) = \max\{w'(A,F) | A$ is a closed semialgebraic subset of $\mathbf{R}^n\}$ and, for every $k \in \mathbf{N}, w'(k) = max\{w'(F) | F$ semialgebraic set, $dim(F) = k\}$ (*c.f.* [B-C-R] for the concept of dimension of a semialgebraic set). In a similar way we can define $w(F)$ and $w(k)$.

On the other hand L. Bröcker has introduced in [Br1] and [Br2] some quantitative invariants (s, s', t and t') measuring the complexity of semialgebraic sets and obtained some upper bounds for them.

We can slightly modify Bröcker's definitions in order to have invariants with respect to a semialgebraic set F. Thus, for a proper basic closed semialgebraic set A in $\mathbf{R}^n$ (*c.f.* [Br2]) we define the ***s'-invariant of A in F***, $s'(A,F)$, as the minimal non-negative integer r such that

$$A \cap F = \{f_1 \geq 0, \ldots, f_r \geq 0\} \cap F$$

for some $\{f_1, \ldots, f_r\}$ in $\mathbf{R}[X]$. Again we set $s'(\emptyset, F) = -1$ and $s'(\mathbf{R}^n, F) = 1$. Following a similar procedure we can also define for semialgebraic subsets of $\mathbf{R}^n \supseteq B, C, D$; where B is basic open, C is just open and D is just closed, the invariants $s(B,F)$, $t(C,F)$ and $t'(D,F)$, (for example $t'(D,F)$ will be the minimum non-negative integer r such that $D \cap F = \{f_{1,1} \geq 0, \ldots, f_{1,s}\} \cap F \cup \ldots \cup \{f_{r,1} \geq 0, \ldots, f_{r,s}\} \cap F$). The following propositions are immediate from the definitions and Bröcker's results.

Proposition 2.1 *For every $k \in \mathbf{N}$ both $w(k)$ and $w'(k)$ are finite. Moreover, we have:*
i) $w(k) \leq s(k) \leq 2s(k-2)$ where $s(k) = k$ for $k = 1,2,3$
ii) $w'(k) \leq s'(k) \leq (1/2)k(k+1)$.□

Proposition 2.2 *Let A, F be semialgebraic subsets of $\mathbf{R}^n$. Then if A is closed $w'(A,F) \leq t(\mathbf{R}^n - A, F)$, if, moreover, A is basic $w'(A,F) \leq t(\mathbf{R}^n - A, F) \leq s'(A,F)$.*□

An analogy to Proposition 2.2 can be stated for open semialgebraic sets. Notice that the inequalities in Proposition 2.2 are not in general equalities. For instance, if we consider the basic closed semialgebraic subset of $\mathbf{R}^2$ defined as

$$A = \{xy^2 \geq 0, x+1 \geq 0, y \geq 0\}$$

it can be easily shown that $w'(A,\mathbf{R}^2) \leq 2$. Now, after some computations, we obtain $t(\mathbf{R}^2 - A, \mathbf{R}^2) = s'(A,\mathbf{R}^2) = 3$. Therefore, it seems to be of some interest to look for sufficient conditions on semialgebraic sets A, F and with A basic closed such that $w'(A,F) = s'(A,F)$.

For an algebraic set V in $\mathbf{R}^n$ let us denote as $Cent(V)$ the set of its central points. Next, we state the following technical

Lemma 2.3 *Let V be an algebraic subset of $\mathbf{R}^n$; F a semialgebraic set, open in V; V irreducible in F and $\{p_1,\ldots,p_m\}$ a collection of polynomials in n variables. Let us assume that*

$$Cent(V) \supseteq Clos(\{p_1>0,\ldots,p_m>0\}\cap F)\cap F = \{p_1 \geq 0,\ldots,p_m\geq 0\}\cap F \neq \emptyset \qquad (1)$$

where $Clos(M)$ denotes the closure of M in $\mathbf{R}^n$ with repect to the euclidean topology.

Then for every closed semialgebraic subset A of $\mathbf{R}^n$ the following two statements are equivalent.

i) $A\cap F = \{p_1\geq 0,\ldots,p_m\geq 0\}\cap F$

ii) $\{p_1\geq 0,\ldots,p_m\geq 0\}\cap F \supseteq A\cap F$ and there is $q \in \mathbf{R}[\underline{X}]$ non-identically zero on F such that $A\cap F - \{q=0\} \supseteq \{p_1\geq 0,\ldots,p_m \geq 0\}\cap F - \{q=0\}$.□

Let v, F and $\{p_1,\ldots,p_m\}$ as in the above Lemma and consider the following notations $v_k = \{p_k = 0,\ldots,p_m=0\}\cap V$, $F_k = V_k \cap F$ for $k=1,\ldots,m$ and set $V_{m+1} = V$ and $F_{m+1} = F$.

Theorem 2.4 *With the above notations, if for every $k=1,\ldots,m$ V_k is irreducible in F_k and the following hypothesis hold:*

i) $Cent(V_{k+1}) \supseteq Clos(\{p_1>0,\ldots,p_k>0\}\cap F_{k+1})\cap F_{k+1} = \{p_1\geq 0,\ldots,p_k\geq 0\}\cap F_{k+1}\neq\emptyset$

ii) $Clos(\{p_k<0\}\cap F_{k+1})\cap F_{k+1} = \{p_k \leq 0\}\cap F_{k+1} \neq \emptyset$

then $w'(\{p_1\geq 0,\ldots,p_m\geq 0\},F) = s'(\{p_1\geq 0,\ldots,p_m \geq 0\},F) = m$

Proof: By induction on m. The case $m=1$ is obvious. Let us assume the Theorem to be true for the case $m-1$. Then by *reductio ad absurdum*, let us suppose that

$$\{p_1\geq 0,\ldots,p_m\geq 0\}\cap F = \{q_{1,1}\geq 0,\ldots,q_{m-1,1}\geq 0\}\cap F \cup \ldots \cup \{q_{1,r}\geq 0,\ldots,q_{m-1,r}\geq 0\}\cap F$$

For every $j\in\{1,\ldots,r\}$ let $s(j)\leq m-1$ be such that $q_{i,j}$ is not identically zero on F_m if and only if $i \leq s(j)$. Define the polynomial $q = q_{1,1}\cdots q_{s(1),1}\cdots q_{1,r}\cdots q_{s(r),r}\in \mathbf{R}[\underline{X}]$. We have that q is not identically zero on F_m since V_m is irreducible in F_m. Therefore,

$$\begin{aligned}\{p_1\geq 0,\ldots,p_{m-1}\geq 0\}\cap F_m - \{q=0\} &= (\{q_{1,1}>0,\ldots,q_{s(1),1}>0\}\cap F_m - \{q=0\})\cup\ldots\\ &\cup\ (\{q_{1,r}>0,\ldots,q_{s(r),r}>0\}\cap F_m - \{q=0\}).\end{aligned}$$

Now, let $r'\leq r$ be such that $\{q_{1,j}>0,\ldots,q_{s(j),j}>0\}\cap F_m - \{q=0\}\neq\emptyset$ iff $j\leq r'$. Then we have

$$\begin{aligned}\{p_1\geq 0,\ldots,p_{m-1}\geq 0\}\cap F_m - \{q=0\} &= (\{q_{1,1}>0,\ldots,q_{s(1),1}>0\}\cap F_m - \{q=0\})\cup\ldots\\ &\cup\ (\{q_{1,r'}>0,\ldots,q_{s(r'),r'}>0\}\cap F_m - \{q=0\})\end{aligned}$$

From Lemma 2.3 we conclude

$$\begin{aligned}\{p_1 \geq 0, \ldots, p_{m-1} \geq 0\} \cap F_m &= \{q_{1,1} > 0, \ldots, q_{s(1),1} > 0\} \cap F_m \cup \ldots \\ &\cup \{q_{1,r'} > 0, \ldots, q_{s(r'),r'} > 0\} \cap F_m\end{aligned}$$

and by the induction hypothesis $s(j) = m-1$ for some $j \leq r'$. However, since $\{q_{1,j} > 0, \ldots, q_{m-1,j} > 0\} \cap F_m - \{q = 0\} \neq \emptyset$ there will be $y_0 \in \{p_m < 0\} \cap \{q_{1,j} > 0, \ldots, q_{m-1,j} > 0\} \cap F - \{q = 0\} \neq \emptyset$. Contradiction. □

Considering $\mathbf{R}[V]$ the ring of polynomial functions on an irreducible algebraic set V, and considering $\{f_1, \ldots, f_r\}$ in $\mathbf{R}[V]$ a regular system of parameters on a simple point $p \in V$, there is an open neighborhood of p, U_p, such that this collection of polynomials verifies the conditions stated in the above Theorem when $F = U_p$. Thus, we can conclude 2) of Prop 3.1 in [A-B] as an immediate Corollary of 2.4.

On the other hand, in 2.4 we need weaker hypothesis than those of [J]. In fact the "inclusion property" introduced by Jaromczyk is equivalent to be "irreducible and central" in $\mathbf{R}^n$ and we have not needed the "sign independent condition" in our development. For instance, let us consider the polynomials in $\mathbf{R}[X,Y]$, $\{p_1(X,Y) = X - 1/2, p_2(X,Y) = X^3 - X^2 - Y^2\}$. This collection of polynomials does not verify either the "strong sign independent condition" or the "inclusion property" for $\{p_2 = 0\}$ in $\mathbf{R}^2$, but this collection verifies our hypothesis and then we can conclude $w'(\{p_1 \geq 0, p_2 \geq 0\}, \mathbf{R}^2\} = s'(\{p1 \geq 0, p_2 \geq 0\}, \mathbf{R}^2) = 2$.

3 Width of congruence classes

In §2 we have studied the width of a basic closed semialgebraic set under some conditions. In order to get lower bounds for the complexity of the Membership Problem to a semialgebraic set A we are going to develop the concept of width of the congruence class of a semialgebraic set.

The *congruence relation* of two semialgebraic sets that appears undercovered in Lemma 2.3 can be defined as follows:

Given A, B, F three semialgebraic subsets $\mathbf{R}^n$ we say that A and B are *congruent in* F if and only if $dim(A \triangle B) \cap F < dim F$.

When the Zariski closure of F in $\mathbf{R}^n$ is irreducible, this property is equivalent to the existence of $q \in \mathbf{R}[\underline{X}]$, non-identically zero on F, such that

$$A \cap F - \{q = 0\} = B \cap F - \{q = 0\}$$

Thus Lemma 2.3 states that if F is open in its Zariski closure, and this is irreducible, then for every finite collection of polynomials $\{p_1, \ldots, p_m\}$ verifying equation (1) and for every closed semialgebraic subset A of $\mathbf{R}^n$ we have $A \cap F = \{p_1 \geq 0, \ldots, p_m \geq 0\} \cap F$ if and only if $\{p_1 \geq 0, \ldots, p_m \geq 0\} \cap F \supseteq A \cap F$, with A and $\{p_1, \ldots, p_m \geq 0\}$ congruent in F.

Now, we are interested in developing the concept of *width of the congruence class of a semialgebraic set A of* $\mathbf{R}^n$ *in* F as

$$w_{\text{con}}(A, F) = \min\{w'(B, F) \mid B \text{ closed}, A \text{ and } B \text{ congruent in } F\}$$

We set $w_{\text{con}}(\emptyset, F) = -1$ and $w_{\text{con}}(\mathbf{R}^n, F) = 0$. Let us remark that

$$w_{\text{con}}(A, F) = \min\{w(B, F) \mid B \text{ open}, A \text{ and } B \text{ congruent in } F\}$$

and that no hypothesis on the topological condition of A has been assumed for our definition. As in §2, we can also introduce here the concept $w_{\text{con}}(F) = \max\{w_{\text{con}}(A, F) \mid A \text{ semialgebraic}\}$, and, similarly, for every non-negative integer k the invariant $w_{\text{con}}(k) = \max\{w_{\text{con}}(F) \mid dim F = k\}$. We have the following immediate Propositions:

Proposition 3.1 *For every $k \in \mathbf{N}$ the invariant $w_{con}(k)$ is finite and we have:*
i) $w_{con}(k) \leq w'(k)$
ii) $w_{con}(k) \leq w(k)$. □

Proposition 3.2 *Let A, F be semialgebraic subsets of $\mathbf{R}^n$. Then, if A is closed $w_{con}(A,F) \leq w'(A,F) \leq t(\mathbf{R}^n - A, F)$. If, moreover, A is basic $w_{con}(A,F) \leq w'(A,F) \leq t(\mathbf{R}^n - A, F) \leq s'(A,F)$.* □

A similar result can be stated for open semialgebraic subsets of $\mathbf{R}^n$.

On the other hand, considering the semialgebraic subset A of $\mathbf{R}^2$ given as $\{X - 1/2 \geq 0, X^3 - Y^2 - X^2 \geq 0\}$ we know that $w'(A, \mathbf{R}^2) = 2$; but $w_{\text{con}}(A, \mathbf{R}^2) = 1$ since A is congruent tö $X^3 - Y^2 - X^2 > 0\}$ in $\mathbf{R}^2$. In particular, we see that in the above Propositions the inequalities are not in general equalities.

In the following result we are going to introduce sufficient conditions on a finite collection of polynomials $\{p_1, \ldots, p_m\}$ which grant that the width of the congruence class of $A = \{p_1 \geq 0, \ldots, p_m \geq 0\}$ is equal to the width of A and also equal to m. In particular, we also get sufficient conditions to grant that the width of a basic open semialgebraic set $\{p_1 > 0, \ldots, p_m > 0\}$ is exactly equal to m.

Let V be an algebraic subset of $\mathbf{R}^n$, F a semialgebraic subset, open in V and V irreducible in F. Let us consider $\{p_1, \ldots, p_m\}$ in $\mathbf{R}[V]$ and the notations introduced in §2.

Theorem 3.3 *If for every $k = 1, \ldots, m$ the algebraic set V_k is irreducible in F_k and the following properties hold:*
i) For every $h \in \mathbf{R}[\underline{X}]$ if $h(F_k) = \{0\} \Rightarrow h = 0$ in $\mathbf{R}[V]/(p_k, \ldots, p_m)$
ii) $Cent(V_{k+1}) \supseteq \{p_1 > 0, \ldots, p_k > 0\} \cap F_{k+1} \neq \emptyset$
$Cent(V_{k+1}) \supseteq \{p_k \leq 0\} \cap F_{k+1} = Clos(\{p_k < 0\} \cap F_{k+1}) \cap F_{k+1} \neq \emptyset$
Then,

$$\begin{aligned} w_{con}(\{p_1 \geq 0, \ldots, p_m \geq 0\}, F) = w_{con}(\{p_1 > 0, \ldots, p_m > 0\}, F) = \\ w'(\{p_1 \geq 0, \ldots, p_m \geq 0\}, F) = w(\{p_1 > 0, \ldots, p_m > 0\}, F) = \\ s'(\{p_1 \geq 0, \ldots, p_m \geq 0\}, F) = s(\{p_1 > 0, \ldots, p_m > 0\}, F) = m \; \square \end{aligned}$$

Proof: By induction on m. The case $m = 1$ is trivial and let us suppose the above claim holds for $m - 1$. By *reductio ad absurdum* let us assume there is $h \in \mathbf{R}[\underline{X}]$ non-identically zero on F such that

$$\begin{aligned} \{p_1 \geq 0, \ldots, p_m \geq 0\} \cap F - \{h = 0\} = \{q_{1,1} \geq 0, \ldots, q_{m-1,1} \geq 0\} \cap F - \{h = 0\} \cup \ldots \\ \ldots \cup \{q_{1,r} \geq 0, \ldots, q_{m-1,r} \geq 0\} \cap F - \{h = 0\} \end{aligned}$$

Since $\mathbf{R}[V]$ is a noetherian ring there is $r \in \mathbf{N}$ such that $h = g p_m{}^r$ in $\mathbf{R}[V]$ and $g \in \mathbf{R}[\underline{X}]$ non-identically zero on F_m. Thus we have:

$$\begin{aligned} \{p_1 \geq 0, \ldots, p_{m-1} \geq 0\} \cap F_m - \{g = 0\} = \{q_{1,1} \geq 0, \ldots, q_{m-1,1} \geq 0\} \cap F_m - \{g = 0\} \cup \ldots \\ \cup \{q_{1,r} \geq 0, \ldots, q_{m-1,r} \geq 0\} \cap F_m - \{g = 0\} \end{aligned}$$

For each $j \in \{1, \ldots, r\}$ let $s(j) \leq m - 1$ be such that $q_{i,j}$ is not identically zero on F_m if and only if $i \leq s(j)$ and let us define $q = q_{1,1} \cdots q_{s(1),1} \cdots q_{1,r} \cdots q_{s(r),r} \in \mathbf{R}[\underline{X}]$. As in Theorem 2.4 q is not identically zero on F_m since V_m is irreducible in F_m. Let us also consider $r' \leq r$ such that

$\{q_{1,j} > 0, \ldots, q_{s(j),j}) > 0\} \cap F_m - \{q.g = 0\} \neq \emptyset$ iff $j \leq r'$. The existence of r' is guaranteed by the irreducibility of V_m in F_m. Then we have

$$\{p_1 \geq 0, \ldots, p_{m-1} \geq 0\} \cap F_m - \{q.g = 0\} = \{q_{1,1} > 0, \ldots, q_{s(1),1} > 0\} \cap F_m - \{q.g = 0\} \cup \ldots$$
$$\ldots \cup \{q_{1,r'} > 0, \ldots, q_{s(r'),r'} > 0\} \cap F_m - \{q.g = 0\}$$

By the induction hypothesis we conclude $s(j) = m - 1$ for some $j \leq r'$.

On the other hand, since $\{q_{1,j} > 0, \ldots, q_{m-1,j} > 0\} \cap F_m - \{q.g = 0\} \neq \emptyset$ there will be $y_0 \in \{p_m < 0\} \cap \{q_{1,j}, \ldots, q_{m-1,j} > 0\} \cap F - \{qg = o\} \neq \emptyset$. Contradiction. □

Let us consider now V an irreducible algebraic subset of $\mathbf{R}^n$ and $f_1, \ldots, f_m \in \mathbf{R}[V]$ a regular system of parameters of V in a simple point $p \in V$. A very similar argument to the above but in a context of germs of algebraic sets allows us to conclude that

$$w\text{con}(\{f_1 > 0, \ldots, f_m > 0\}, V) = m$$

which, in particular, implies 1) of Prop. 3.1 in [A-B].

4 Width vs. connected components

In [Ra] Rabin introduced the algebraic decision tree model to analyze certain decision problems. This model considers the algorithms as ternary trees in which each non-leaf node represents the evaluation of the sign of a polynomial. Thus an algebraic decision tree can be understood as a description of a semialgebraic set A and the kind of problems studied by this model can be translated as dealing with the membership to some semialgebraic set.

Given an algebraic decision tree T solving a membership problem to a closed semialgebraic set A, Rabin defines the cost of T (or the height of T) as the maximum number of tests performed for any input. In fact, an algebraic decision tree gives a collection of conjunctions and disjunctions of polynomial equalities and inequalities describing A. Now we have the following:

Proposition 4.1 *Let A be a semialgebraic subset of $\mathbf{R}^n$. Then, for every algebraic decision tree T that solves the membership problem to A we have*

$$w_{con}(A, \mathbf{R}^n) \leq cost(T) \; \square$$

Proof: Considering T as an algebraic decision tree that solves the membership problem of A in $\mathbf{R}^n$, let $\{\Gamma_1, \ldots, \Gamma_s\}$ be the collection of all the paths in T such that they lead to an affirmative answer.

For every path $\Gamma = \{sign(h_1) = \theta_1, \ldots, sign(h_r) = \theta_r\}$ let Γ^* be the semialgebraic subset of $\mathbf{R}^n$ given by relaxation of the strict sign conditions occuring in Γ (*i.e.* $> \to \geq, < \to \leq$). Now let A_1 be the semialgebraic closed subset of $\mathbf{R}^n$ given as:

$$A_1 = \bigcup_{i=1,\ldots,s} \Gamma_i^*$$

We have that A_1 is congruent to A in $\mathbf{R}^n$ and that the width of A_1 is smaller or equal than the maximum of the lengths of the paths $\Gamma_1, \ldots, \Gamma_s$. Thus

$$w_{\text{con}}(A, \mathbf{R}^n) \leq \max\{leng(\Gamma_i) \mid i = 1, \ldots, s\} \leq cost(T) \square$$

However, Ben-Or [B] has developed another lower bound of $cost(T)$, essentially giving $cost(T) \in \Omega(\log(\#(A)) - n)$, where $\#(A)$ is the number of connected components of A. In many cases it has been shown that Ben-Or's lower bound is more accurate than the width in order to study $cost(T)$.

The reason is now clear since there is no $M \in \mathbf{N}$ such that for every closed semialgebraic subset A of $\mathbf{R}^n$ $\log(\#(A)) \leq M$; on the other hand it is clear from our considerations that $w_{\text{con}}(A, \mathbf{R}^n) \leq n$ for every semialgebraic subset A of $\mathbf{R}^n$ (*c.f.* [Br2]).

Therefore in decision problems of high complexity Ben-Or's lower bound will be always better.

For instance, consider a decision problem depending on a parameter $n \in \mathbf{N}$ such that the nth problem is solved by a closed semialgebraic set A_n in $\mathbf{R}^n$ such that $\log(\#(A_n)) = n^4$. In this case clearly the width will not be useful.

Nevertheless, there are many decision problems with bounded number of connected components of A_n. For example, to decide whether n real numbers are given or not in order. This decision problem can be associated to the semialgebraic set $A_n = \{\underline{x} \in \mathbf{R}^n \mid x_1 < \ldots < x_n\}$. In this example $\#(A_n) = 1$ and Ben-Or's result simply states $cost(T_n) \in \Omega(-n)$ for every T_n solving the membership to A_n. Since, from our results, it can be immediately shown that $w(A_n, \mathbf{R}^n) = n - 1$ and, then, for every n, $n - 1 \leq cost(T_n)$, this yields clearly a better lower bound.

Finally, we remark that is is easy to deduce, using Tarski principle, that both techniques to compute lower bounds can be similarly applied to the case of a model of computation operating only within a given real closed field by changing suitably the definition of connectedness. For example, in the field of real algebraic numbers, an assumption which seems to be more realistic that working with reals. Moreover, we observe that the lower bounds thus obtained do not depend on the real closed field considered.□

References

[A-B] C. Andradas, E. Becker, *On the Real Stability Index of Rings and its Application to Real Algebraic Geometery*, Publ. Sem. Logique et Algèbre Réelle, Univ. Paris VII, 1984-85.

[B] M. Ben-Or, *Lower bounds for Algebraic Computation Trees*, Proc. 15th ACM Annual Symp. on Theory of Comp., pp 80-86, 1983

[B-C-R] J. Bochnak, M. Coste, M. F. Roy, *Géometrie Algebrique Réelle*, forthcoming book in Springer.

[Br1] L. Bröcker, *Minimale Erzeugung von Positivbereichen*, Geometriae Dedicata **16**, pp 335-350, 1984

[Br2] L. Bröcker, *Spaces of Orderings and Semialgebraic Sets*, Quadratic and Hermitian Forms CMS Conference Proc. 4, pp 231-248, 1984

[Br3] L. Bröcker, *Description of Semialgebraic Sets by few polynomials*, ICPAM Summer School on Real Algebraic Geometry, Nice 1985

[J] J. W. Jaromczyk, *An Extention of Rabin's Complete Proof Concept*, Springer LNCS 118, pp 321-326, 1981

[Ra] M. O. Rabin, *Proving Simultaneous Positivity of Linear Forms*, J. of Computer and System Sciences 6, pp 639-650, 1972

Practical Aspects of Symbolic Integration over $Q(x)$

David M. Gillies Bruce W. Char*
Department of Computer Science
University of Waterloo
Waterloo, Canada N2L 3G1

Introduction

y restricting the domain of application of the Risch differential equation [4] $y' + fy = g$ for nown functions f and g, it has been shown by Kaltofen that the Risch *third possibility* cannot ccur [3]. This allows the construction of a specialized degree-bounding algorithm for solving the isch equation over $Q(x)$. A Hermitian-like algorithm has been proposed by Davenport [2]. Its pplication is not restricted to $Q(x)$ but it is useful and interesting to study the performance of e algorithm solely in this domain. Both of these algorithms have been implemented in Maple 4.1] at the University of Waterloo. Timings of both algorithms were compared to determine their rengths and weaknesses. We observed glaring differences in performance. This analysis indicates e necessity of a poly-algorithmic approach to the integration problem over $Q(x)$. An efficient way f solving the Risch equation in the base field $Q(x)$ will lead to improvement of general integration gorithms because of the recursive nature of the Risch procedure.

In order to choose between the two methods a decision procedure must be discovered which n quickly determine which algorithm to use from the known portions of the Risch equation. We esent here some analysis which will aid in an inexpensive choice of the algorithm.

Consider the Risch equation

$$y' + \frac{P}{q_1^{a_1} \cdots q_n^{a_m}} y = \frac{R}{q_1^{b_1} \cdots q_n^{b_n}}$$

here $q_1 \cdots q_n$ are relatively prime square-free polynomials. One quantity, which can be obtained minimal expense, we define as

$$D = \sum_{i=1}^{n} (b_i - a_i)$$

he value of n in this summation is also a useful quantity in predicting run-time complexity. nalysis has been done on anticipating coefficient growth in the Hermitian method but we do not ve an accurate and inexpensive way of estimating the coefficient growth *a priori*. If coefficient owth is detected (eg. if coefficients increase by 10^6, say) early in the Hermitian algorithm then would be advisable to switch algorithms in mid-stream discarding any partial solution already mputed.

*This work was supported in part by grant A5471 of the Natural Sciences and Engineering Research Council of nada.

Entries in table 1 show some of the running times and their relation to the above quantities. Timings are in seconds on a VAX 11/785 machine. The data in table 1 is representative of larger test batches run through the two programs. When both D and n are small, the degree bounding method was usually better but the Hermitian approach did result in several faster timings in the case where the estimated bound was loose. When D is large and n small, the Hermitian method is the clear choice of the algorithm to use. Whereas, when D is small and n large, the opposite is true. In the case where both quantities are large, it is difficult to make the proper choice of algorithm as there are instances in which each method is *much* better than the other. Table 2 summarizes the heuristic used to decide upon the proper algorithm.

	problem	D	n	Hermitian	Degree Bounding
1	$\int (x^{101}+1)e^x dx$	0	0	0.89	1.90
2	$\int \frac{x^{101}+1}{x+1} e^x dx$	0	0	30.95	9.89
3	$\int \frac{x^2-x-1}{x^2(x+1)^2} e^x dx$	small	small	10.35	2.62
4	$\int \frac{x^4+2x^3-7x^2-16x-6}{x^2(x+1)^2(x+2)^2(x+3)^2} e^x dx$	small	small	45.07	6.82
5	$\int \frac{x-100}{x^{101}} e^x dx$	large	small	2.46	56.21
6	$\int \frac{x-99}{(x+2)^{102}} e^x dx$	large	small	10.61	465.91
7	$\int \frac{1-\frac{1}{100}x^{100}}{x^{101}} e^{\frac{1}{100}x^{100}} dx$	large	small	2.30	58.54
8	$\int \frac{x^2(42x^{45}-252x^{44}+\cdots-3x^3)}{(x-1)^2(x-2)^3(x-3)^4} e^{x^{42}} dx$	small	large	37.76	6.80
9	$\int \frac{x^4-4x^3-29x^2-34x-6}{x^2(x+1)^3(x+2)^4(x+3)^5} e^x dx$	small	large	48.92	8.09
10	$\int \frac{x^{99}(x^6+100x^5+1415x^4+\cdots+12000)}{(x+1)^2(x+2)^3\cdots(x+5)^6} e^x dx$	small	large	399.85	88.19

TABLE 1: Timings in seconds.

D	n	method used
small	*small*	Degree Bounding
large	*small*	Hermitian
small	*large*	Degree Bounding
large	*large*	?

TABLE 2: Heuristic.

References

[1] B. CHAR, K. GEDDES, G. GONNET, AND S. WATT, *Maple User's Guide*. Watcom Publication Waterloo, Ontario, 1985.

[2] J. DAVENPORT, *The Risch differential equation problem*, SIAM J. Comput., (1986), pp. 903 918.

[3] E. KALTOFEN, *A note on the Risch differential equation*, Proc. Eursam '84, (Nov. 1986 pp. 359–366.

[4] R. H. RISCH, *The problem of integration in finite terms*, Trans. AMS, (1969), pp. 167–189.

Integration: Solving the Risch Differential Equation

By John Abbott
School of Mathematical Sciences
University of Bath, Bath, England BA2 7AY

Abstract We describe the first *complete* implementation of Davenport's algorithm [Davenport86] for the solution of the Risch differential equation. Our code forms part of a new integration package written in REDUCE which operates over algebraic number fields.

1. Introduction

[Risch69] introduced the term *Risch differential equation* (=RDE) to denote the equation $y'+fy=g$ which comes from trying to solve $\int ge^F$ where $f=F'$. This is a fundamental part of any system for symbolic integration, and before Davenport's algorithm, solutions were found by clearing denominators (e.g. [Fitch81], [Davenport82], [Davenport85]), bounding the degree of the numerator, substituting into the RDE and solving the resulting system of equations for the coefficients. The process of solving these equations may in turn require the solution of a *parametric* RDE: $y'+fy=\sum c_i g_i$, where the c_i are unknowns.

However, Davenport's algorithm constructs the solution directly in a style akin to Hermite's integration algorithm. Davenport's algorithm may generate a recursive RDE problem but g is always known fully — the parametric RDE does not occur. However, Davenport's method only works if f is *weakly normalised;* and this can easily be enforced.

2. Outline of the Algorithm

We assume $f, g \in C(x, \theta_1, \theta_2, \cdots, \theta_t)$, an elementary extension of the constants C: each θ_i is a log or exp and is transcendental over $C(x, \theta_1, \cdots, \theta_{i-1})$. We denote the derivation by $' = \frac{d}{dx}$. We regard f, g, and the solution as rational functions in θ_t with coefficients in $C(x, \theta_1, \cdots, \theta_{t-1})$. A vital observation is that: $p \in C(x, \theta_1, \cdots, \theta_{t-1})[\theta_t]$ is square-free as a polynomial in θ_t iff $gcd(p, p') = 1$, unless $\theta_t \mid p$ and θ_t is an exponential. In the base case $(\theta_0 = x)$ this is just the well-known test for square-freeness.

Note that any irreducible factor of $den(y)$ (the denominator of the solution) must divide $den(f)$ or $den(g)$. The first step is to find that part of the solution whose denominator does not divide $den(f)$. Let p be the product of those factors of $den(g)$ dividing $den(f)$, and $q=den(g)/p$ — we ignore the factor θ_t if it is an exp. Let $\prod q_i^i$ be the square-free decomposition of q. Using partial fractions we may write g as $g_{rest} + b_i/q_i^i$ where degree $b_i <$ degree q_i^i and $q_i \nmid den(g_{rest})$. The solution must contain a term α/q_i^{i-1} and no term with a denominator a higher power of q_i. Substituting this term into the RDE gives:

$$\frac{d}{dx}\left[\frac{\alpha}{q_i^{l-1}}\right] + \frac{\alpha}{q_i^{l-1}} f = \frac{(1-l)q_i'\alpha}{q_i^l} + \frac{\alpha' + \alpha f}{q_i^{l-1}}$$

so we compute α so that the q_i^{-l} term of g eliminated (recall $gcd(q_i, q_i')=1$). We repeat this until q_i no longer divides $den(g)$; if we obtain a term in g of the form $non\text{-}zero/q_i$ then the original equation is insoluble. Similarly the other q_j are removed from $den(g)$, or we prove the original equation insoluble.

Now every irreducible factor of $den(g)$ divides $den(f)$. If $den(g) \mid den(f)$ then the solution must be a polynomial. If not we can ensure $den(g) \mid den(f)$. Let p be a square-free polynomial which occurs to a higher power in $den(g)$ than in $den(f)$. We succesively eliminate terms with most negative power of p: either fy dominates; or both y' and fy dominate — weak normalisation precludes cancellation. a.pp Now the solution must be a polynomial (including θ_t^{-1} terms for exponential θ_t). We treat the base, log, and exp cases separately. We use degree (f) for a rational function f to mean the degree of the numerator minus the degree of the denominator, e.g. degree $(1/\theta_t) = -1$. degree 0 is undefined.

Base Case This is the simplest case as all coefficients are constants. We equate coefficients of the terms of highest degree on either side of the equation. There are three cases according as y', fy, or both have highest degree — weak normalisation of f precludes any possible cancellation when both terms dominate.

Logarithmic Case Coefficients are no longer constant. Again we equate highest degree terms, and get three cases. If y' dominates the leading coefficient of the answer is found by integration, and the constant of integration comes from the second highest terms. When both dominate the leading coefficient of the answer is the solution to another RDE. We have to be careful to avoid RDEs allowing homogeneous solutions, but this can be detected easily and cured by a simple transformation.

Exponential Case This case is different because the derivative of a polynomial has the *same* degree as the polynomial itself, and also the polynomial includes θ_t^{-1} terms. We calculate the θ_t and θ_t^{-1} polynomial parts of the answer by almost identical methods. If degree $f > 0$ then we obtain the leading term by a simple division. Otherwise degree $f \leq 0$ and we get another RDE to solve — weak normalisation of f preventing cancellation. Finally the θ_t-free part of the solution comes from either a RDE or an integration.

3. Practical Considerations & Performance

Davenport's algorithm relies only on simple operations (e.g. polynomial arithmetic and gcd, and partial fractions) *except for* a need to test whether a given log or exp already lies in a particular field — i.e. the Structure Theorem [Rothstein79]. The Structure Theorem tells us that the only relations between logs/exps are the obvious ones. So we use *gcd*s to test if a log exists in a given field; and for an exp we search for a linear relation amongst arguments to exps.

Davenport describes two decompositions of rational functions, which are convenient for the proofs but are cumbersome in implementation. We prefer to decompose g (and sometimes f)

into just two parts: that being eliminated, and the rest. So we need only update two parts as the value of g changes — it is more costly to update many parts.

Two small modifications appear to be computationally worthwhile. Firstly, when eliminating factors in $den(g)$ not in $den(f)$ we eliminated one factor at a time; we could reduce powers instead by updating the square-free decomposition after each elimination. Secondly, in the exp case we know the highest power of θ_t^{-1} in the solution, so we could transform the RDE such that its solution is a polynomial only in θ_t, and then divide this polynomial by a power of θ_t.

These routines are implemented in REDUCE which includes a Risch-Norman package, so we have been able to compare the two implementations. We have not found any examples where the existing integrator finds the answer sooner than our integrator, and usually our code is several times faster. Sometimes our integrator produced an integral in less time than the differentiation code could reverse the process.

4. Conclusion

Davenport's algorithm is easier to implement than previous algorithms, and the parametric RDE is unnecessary. The code is succint and efficient. We have to ensure weak normalisation, but this involves only simple transformations. The Structure Theorem code is a vital part of any algebra system.

References

[Davenport82] J H Davenport, "The Parallel Risch Algorithm (I)," *Proc. EUROCAM 82,* LNCS 144(1982), also

[Davenport85] J H Davenport, and B M Trager, "On the Parallel Risch Algorithm (II)," *ACM ToMS,* 11(4) (Dec 1985).

[Davenport86] J H Davenport, "The Risch Differential Equation Problem," *SIAM J Comp.,* 15(4) (Nov 1986).

[Fitch81] J P Fitch, "User Based Integration Software," *Proc. 1981 ACM SYMSAC(Snowbird),* (1981).

[Risch69] R H Risch, "The Problem of Integration in Finite Terms," *Trans. AMS,* #139 (1969).

[Rothstein79] M Rothstein, and B F Caviness, "A Structure Theorem for Exponential and Primitive Functions," *SIAM J Comp.,* 8(3) (Aug 1979)

COMPUTATION AND SIMPLIFICATION IN LIE FIELDS

J.Apel, W.Lassner
Naturwissenschaftlich-Theoretisches Zentrum
Karl-Marx-Universität, Sektion Mathematik
DDR-7010 Leipzig

1. Introduction

Computations in non-commutative quotient fields (Lie fields) of enveloping algebras of Lie algebras are of interest for physicists in connection with spectrum generating algebraical methods as well as in constructive ways in representation theory of Lie algebras.

Lie fields are well defined from algebraical point of view but no really computation is possible without turning right fractions into left fractions. We presented algorithms solving this problem in [1] and [2].

In section 2 of the present paper we show how these algorithms may be applied to perform the field operations +, -, *, and $()^{-1}$.

The question of simplification arises in a natural way in computation in domains whose elements are equivalence classes of objects. In section 2 we present a method for zero equivalence simplification. This method allows a decision about the equality of two left fractions in Lie fields.

A definition of a canonical representant of a left fraction independent of the algorithm for its computation is given in section 3. An important property of this canonical element is that it has minimal degree among all representants of the left fraction. Therefore, it is not only of principal interest but also important for the efficience of subsequent computations. A canonical simplifier for Lie fields using Gröbner bases of left and right ideals in enveloping algebras of Lie algebras is presented in the sections 4 and 5.

We present an antiisomorphism between enveloping algebras U(L) and U(L′) such that Gröbner basis and syzygy computations of right ideals may be transformed into those of left ideals. This mapping is especially of interest for practical programming since only one subroutine is needed for the treatment of left and right ideals.

2. Computation and Zero Equivalence Simplification in Lie Fields

Let L be a finite dimensional Lie algebra over a field K and U(L) the enveloping algebra of L. The Poincaré-Birkhoff-Witt theorem ensures that for a basis $\{x_1,\ldots,x_n\}$ of L the elements $x_1^{i_1}x_2^{i_2}\ldots x_n^{i_n}$ $(i_j \in \mathbb{N})$ form a vector space basis of U(L).
The Ore-condition is satisfied in U(L), i.e. for all y_1, $y_2 \in U(L)$ there exist z_1, $z_2 \in U(L)$ and w_1, $w_2 \in U(L)$ not both equal to zero, such that $y_1z_1=y_2z_2$ and $w_1y_1=w_2y_2$. That ensures the existence of a quotient skew field D(L) of U(L), the Lie field of U(L).
The skew field D(L) consists of equivalence classes of pairs $(y,z) \in U(L)\times U(L)$, $y\neq 0$. The equivalence relation $\sim$ is defined by $(y_1,z_1) \sim (y_2,z_2)$ iff there exists a pair $(w_1,w_2) \in U(L)\times U(L)\backslash\{(0,0)\}$ such that $w_2y_1=w_1y_2$ and $w_2z_1=w_1z_2$. The equivalence classes are called left fractions and the class which contains (y,z) is denoted by $y^{-1}z$. The abbreviations z or y^{-1} are used if y=1 or z=1, respectively.
In D(L) the operations +, -, •, $()^{-1}$ are defined by

$+ \quad : \quad y_1^{-1}z_1 + y_2^{-1}z_2 = (w_1y_1)^{-1}\bullet(w_1z_1 + w_2z_2)$

where $(w_1,w_2) \in U(L)\times U(L)\backslash\{(0,0)\}$ such that $w_1y_1=w_2y_2$.
The left fraction $1^{-1}\bullet 0 = 0$ is the zero-element of +.

$- \quad : \quad -(y^{-1}z) = y^{-1}\bullet(-z)$

$\bullet \quad : \quad y_1^{-1}z_1 \bullet y_2^{-1}z_2 = (w_2y_1)^{-1}\bullet(w_1z_2)$

where $(w_1,w_2) \in U(L)\times U(L)\backslash\{(0,0)\}$ such that $w_2z_1=w_1y_2$.
The left fraction $1^{-1}\bullet 1 = 1$ is the unit-element of •.

$()^{-1} \quad : \quad (y^{-1}z)^{-1}=z^{-1}y$ for $z \neq 0$.

We emphazise that the so defined operations are independent of the choosen representants of the equivalence classes.
An expression zy^{-1} is called right fraction. Note that in general neither (z,y) nor (y,z) are representants of the element zy^{-1} of D(L).
Clearly, the pairs (y,z) and (wy,wz) are equivalent for any non-zero element w of U(L). Analogous to the commutative case we introduce extension and cancellation. The pair (wy,wz) is said to be got by extension of (y,z). If there exist a non-zero element u and elements y′ and z′ of U(L) such that y=uy′ and z=uz′ then (y′,z′) is said to be obtained by cancellation of (y,z).
Note, that in difference to the commutative case there may exist different representants of the same left fraction for which no further cancellation is possible.
Up to here the Lie field D(L) of the enveloping algebra U(L) is complete defined from the algebraical point of view. But it is always

necessary to determine pairs (w_1,w_2) fulfilling Ore conditions for practical calculations in D(L).

The first general algorithm to solve Ore conditions in enveloping algebras U(L) of Lie algebras L was given in [1]. The algorithms in [1] and [2] extend the methods of Gröbner bases and Buchberger's algorithm [3] to enveloping algebras. Furthermore, a general method was given for finding a left fraction $c^{-1}d$ to a given right fraction ab^{-1} such that $ab^{-1}=c^{-1}d$. Both a method for the computation of one representant of $c^{-1}d$ and a method for the computation of a generating set for all representants of $c^{-1}d$ have been presented. The latter appears as a special case (m=2) of the following problem also solved in [1] and [2].

PROBLEM : Given a finite basis $(g_1,\ldots,g_m)$ of a left ideal J of U(L).

Find a finite set $\{(h_{11},\ldots,h_{1m}),\ldots,(h_{s1},\ldots,h_{sm})\}$, $h_{ij} \in U(L)$, $1 \le i \le s$, $1 \le j \le m$, that generates the U(L)-module of the syzygies of $g_1,\ldots,g_m$, i.e. $\sum_{j=1}^{m} h_{ij}g_j = 0$, $1 \le i \le s$, and any syzygy $(h_1,\ldots,h_m)$ is an U(L)-combination

$$(h_1,\ldots,h_m) = \sum_{i=1}^{s} k_i(h_{i1},\ldots,h_{im}),\ k_1,\ldots,k_s \in U(L).$$

Since the elements of D(L) are equivalence classes and the operations are defined using representants we are led to the problem of algebraic simplification. We answer the problem of zero equivalence simplification and construct a canonical simplifier in D(L). For the definition and review of the algebraic simplification problem we refer to [4].

Zero equivalence simplification means to decide whether a pair (a,b) represents the class zero. Since any representant of the zero class has the form (y,0) where $y \neq 0$ one has to decide whether b equals to zero. But this is easy because the zero element of U(L) has an unique representation according to the Poincaré-Birkhoff-Witt basis. With other words the problem reduces to the problem of zero equivalence simplification in U(L) which is solved. Zero equivalence simplification arises from the fundamental task of a Computer-Algebra-System (CAS) to decide whether two expressions represent the same element, i.e. the same equivalence class of objects. This equality problem for two elements reduces to zero equivalence simplification of the difference when a subtraction is defined. However, the subtraction of two fractions requires to solve an Ore-condition. This is in general a nontrivial task as illustrated by the following example.

Example 1: Let W_2 be the Weyl algebra over the field of complex numbers $\mathbb{C}$ generated by the elements q, p satisfying the commutation relation pq - qp = 1. Decide the question whether the equality $(qp+2)^{-1}*(p^2) = (q^2)^{-1}*(qp-1)$ is satisfied. This problem cannot be solved by simple (non-commutative) multiplications in W_2 as one could argue from the experiences in commutative algebra but requires a solution of the Ore-condition $(p^2)*z_1 = (qp+2)*z_2$.
The algorithm given in [1] transforms the left fraction $(qp+2)^{-1}*(p^2)$ into the equivalent right fraction pq^{-1}. The equality of a left fraction to a right fraction $a^{-1}b = cd^{-1}$, however, is solvable by the multiplication ac = bd.
So the above equality can be easy proved by verifying the equation $(qp-1)*q = q^2*p$.
The zero-equivalence simplification of $(qp+2)^{-1}*(p^2)-(q^2)^{-1}*(qp-1)$ is more convinient for a computer which has to deal with various kinds of expressions. This requires the solution of the Ore-condition $z_1(qp+2) = z_2q^2$. That yields $z_1=q$ and $z_2=p$. The equation $(q(qp+2))^{-1}(qp^2-p(qp-1)) = 0$ is easy to verify.

3. The Problem of Canonical Simplification in D(L)

Canonical simplification [4] in the quotient field $K(x_1,\ldots,x_n)$ of the polynomial ring $K[x_1,\ldots,x_n]$ can be solved by cancellation of an fraction to its normal form. An important property of quotient fields $K(x_1,\ldots,x_n)$ is the existence of a representant r for any element e of $K(x_1,\ldots,x_n)$ such that any other representant of e is an extension of r. r is uniquely determined up to extension by a non-zero element of K. It is known that any representant of a fraction $b/a \in K(x_1,\ldots,x_n)$ can be transformed to its normal form by many algorithms as GCD or factorization implemented in modern CAS.
Now we observe various difficulties in the non-commutative quotient field D(L).

i) First, the factor decomposition of elements of U(L) is not unique.
ii) Second, there will not exist a representant r of a fraction from which all other representants could be obtained by extension.

The following example illustrates the two propositions for the Weyl algebra.

Example 2: $p*(qp-1)$ and $q*p*p$ are two prime factor decompositions of qp^2. In both representants $(qp+2, p^2)$ and $(q^2, qp-1)$ of the left fraction $(qp+2)^{-1}*(p^2)=(q^2)^{-1}*(qp-1)$ no terms can be cancelled. So a representant r as required in ii) does not exist for the left fraction.

In commutative polynomial rings cancellation of a fraction leads to a unique representant of the equivalence class. So complete cancellation may be used as canonical simplificator. Example 2 demonstrated that this idea transfered to the non-commutative case will in general not produce canonical representants.

Let us discuss the situation more detailed. We recall that left fractions are equivalence classes of pairs. The union of the set of representants of the left fraction $y^{-1}z$ with $\{(0,0)\}$ forms an U(L)-left module $M(y^{-1}z)$. $M_D(y^{-1}z) = \{a \mid \exists\ b : (a,b) \in M(y^{-1}z)\}$ and $M_N(y^{-1}z) = \{b \mid \exists\ a : (a,b) \in M(y^{-1}z)\}$ are left ideals of U(L).

From now we will write module and ideal instead of left module and left ideal. Our canonical simplifier makes use of a constructive method for finding a generating set and a distinguished element of the module $M(y^{-1}z)$. Let $<$ be a noetherian ordering on the elements of the Poincaré-Birkhoff-Witt basis B which is compatible with the multiplication, i.e. for all $s,t,u \in B$ it is satisfied $s<t \iff LBE(us)<LBE(ut)$. LBE(f) denotes the maximal (leading) basis element appearing in f with non-zero coefficient. This ordering can be extended to a partial ordering on U(L) also denoted by $<$ using $y_1 < y_2 \iff LBE(y_1) < LBE(y_2)$. We recall that for a so defined ordering $<$ the existence of Gröbner bases of ideals with respect to $<$ is ensured.

Definition : Let $<$ be a partial ordering in U(L) as described above. The *canonical representant* $c=(c_1,c_2)$ of the left fraction $y^{-1}z$ with respect to $<$ is the element of $M(y^{-1}z)$ for which c_1 is the minimal (with respect to $<$) non-zero element of $M_D(y^{-1}z)$ normalized to leading coefficient 1.

The next lemma justifies this definition.

Lemma 1: Let $y^{-1}z$ be a left fraction and $<$ an ordering as above.

(i) $M_D(y^{-1}z)$ contains one and only one minimal (w.r.t. $<$) non-zero element c_1 with leading coefficient 1.

(ii) There exists only one element c_2 such that $(c_1,c_2) \in M(y^{-1}z)$.

Proof : (i) There exist minimal non-zero elements inside any subset of U(L) which includes at least one non-zero element since $<$ is noetherian and the zero element is incomparable to any element of U(L). The difference of two incomparable elements of U(L) with the same leading coefficient is zero or lower than these both elements.

Therefore, a module includes only one minimal non-zero element with leading coefficient 1. If there would be two different minimal elements then their difference would be also element of the module and lower than the both elements. That contradicts to the assumption.

(ii) Assume there would be two elements c_2^1 and c_2^2 such that $(c_1, c_2^i) \in M(y^{-1}z)$ for i=1,2. Then also $(0, c_2^1 - c_2^2) \in M(y^{-1}z)$. But the only element of $M(y^{-1}z)$ with the first component equal to zero is (0,0) and it follows $c_2^1 = c_2^2$.

■

Lemma 2: Any Gröbner basis of the ideal $M_D(y^{-1}z)$ with respect to $<$ contains an element c'_1 such that $\exists\, k \in K : c_1 = k \cdot c'_1$ where c_1 is the minimal element from lemma 1.

The proof is trivial since c_1 must be reducible with respect to any Gröbner basis of $M_D(y^{-1}z)$.

4. Subalgorithms for the Canonical Simplifier

Let $F_L = (f_1, \ldots, f_m)$ and $G_L = (g_1, \ldots, g_l)$ be two bases of a left ideal J_L of U(L). Let X_L and Y_L be two transformation matrices such that $G_L^T = X_L F_L^T$ and $F_L^T = Y_L G_L^T$. Assume R_L to be a matrix whose rows generate the module of left syzygies of G_L. Then the rows of the matrix

$$Q_L = \begin{bmatrix} I_m - Y_L X_L \\ R_L X_L \end{bmatrix}, \; I_m \text{ is the } m \times m \text{ - unit matrix,}$$

generate the module of left syzygies of F_L according to theorem 2 in [2]. So we have the relations

$$G_L^T = X_L F_L^T \;,\; F_L^T = Y_L G_L^T \;,\; 0 = R_L G_L^T \;, 0 = Q_L F_L^T \qquad (*).$$

Now we introduce the analogue matrices X_R, Y_R, R_R, and Q_R for two bases F_R and G_R of a right ideal J_R of U(L), such that the equations

$$G_R = F_R X_R^T \;,\; F_R = G_R Y_R^T \;,\; 0 = G_R R_R^T \;,\; 0 = F_R Q_R^T \qquad (**)$$

are satisfied. The rows of R_R resp. of $Q_R = \begin{bmatrix} I_m - X_R^T Y_R^T \\ X_R^T R_R^T \end{bmatrix}$ generate the right syzygy modules of G_R resp. of F_R.

In [1] an algorithm SYZYGY was given which starting with F_L produces G_L, X_L, Y_L, R_L, and Q_L as described above. In particular G_L is a left Gröbner basis of the left ideal J_L generated by F_L. The leading coefficients of the elements of G_L are normalized to 1. Let us denote this algorithm by LEFT_SYZYGY(F_L;G_L,X_L,Y_L,R_L,Q_L) now. If we are not interested in syzygies but only in the Gröbner basis and the transformation matrix X_L then we may skip some parts (these are the steps 2. , 8.7.1. , 8.7.2.1. , and 8.8. in [1]) of the algorithm and denote this reduced version by LEFT_GROBNER_BASIS(F_L;G_L,X_L). If one is interested to get only one syzygy of F_L one may insert an interrupt condition into LEFT_SYZYGY. The resulting algorithm is denoted by RIGHT_LEFT_FRACTION(F_L;G_L,X_L,Y_L,R_L,Q_L) since it is used to transform right fractions into left fractions. The matrices X_L, Y_L, R_L, and Q_L calculated up to the interrupt satisfy also the equations (*). However, in general G_L is no longer a left Gröbner basis and the rows of R_L resp. Q_L generate only a submodule of the left syzygy module of G_L resp. F_L.

The analogue algorithm for transforming left fractions into right fractions we denote by LEFT_RIGHT_FRACTION(F_R;G_R,X_R,Y_R,R_R,Q_R).

While LEFT_SYZYGY, LEFT_GROBNER_BASIS, and RIGHT_LEFT_FRACTION are based on the computation of Gröbner bases of left ideals LEFT_RIGHT_FRACTION requires the computation of Gröbner bases of right ideals. It would be easy to formulate a new algorithm for that purpose. However, the calculations for right ideals can be transfered also to that of left ideals. The idea is to find a transformation φ such that roughly spoken

LEFT_RIGHT_FRACTION(F_R;G_R,X_R,Y_R,R_R,Q_R) =
RIGHT_LEFT_FRACTION($\varphi(F_R)$;$\varphi(G_R)$,$\varphi(X_R)$,$\varphi(Y_R)$,$\varphi(R_R)$,$\varphi(Q_R)$)

where RIGHT_LEFT_FRACTION is performed in an enveloping algebra $U(L')$ with respect to an ordering $<'$. Indeed, such a transformation exists. Let L' be the Lie algebra with the basis $\{\tilde{x}_1,\dots,\tilde{x}_n\}$ and the structure constants $\tilde{c}^i_{jk}=-c^i_{jk}$, i.e.

$$[x_j,x_k] = \sum_i \tilde{c}^i_{jk}\tilde{x}_i = -\sum_i c^i_{jk}\tilde{x}_i$$

where $[x_j,x_k]=\sum_i c^i_{jk}x_i$ are the commutators in L.

We assume the elements of $U(L')$ to be represented with respect to the basis $\{\tilde{x}_n^{i_n}\dots\tilde{x}_1^{i_1} \mid i_j \in \mathbb{N}\}$. φ is defined to be the linear mapping $\varphi : U(L) \rightarrow U(L')$ with $\varphi(x_1^{i_1}\dots x_n^{i_n}) = \tilde{x}_n^{i_n}\dots\tilde{x}_1^{i_1}$. We extend φ to matrices $A=(a_{ij})$, $a_{ij} \in U(L)$ resp. to subsets $M \subseteq U(L)$ by $\varphi(A)=(\varphi(a_{ij}))$ resp. $\varphi(M)=\{\varphi(m) \mid m \in M\}$.

It is easy to verify the properties

(i) φ is an one to one mapping, i.e. there exists an inverse mapping φ^{-1} of φ

(ii) $\varphi(ab) = \varphi(b)\ \varphi(a)$, $\varphi^{-1}(ab) = \varphi^{-1}(b)\ \varphi^{-1}(a)$ (***)

(iii) $\varphi(AB)^T = \varphi(B)^T\ \varphi(A)^T$, $\varphi^{-1}(AB)^T = \varphi^{-1}(B)^T\ \varphi^{-1}(A)^T$.

Since Gröbner bases depend on an ordering as described in the previous section it is necessary to introduce such an ordering $<'$ in $U(L')$. This will be done in the natural way by

$$\forall\ f,g \in U(L') : f <' g \leftrightarrow \varphi^{-1}(f) < \varphi^{-1}(g).$$

In particular $LBE(\varphi(f))=\varphi(LBE(f))$ is satisfied. The ordering $<'$ is also noetherian and compatible with the multiplication. If $<$ is the total degree ordering in $U(L)$ with $x_{i_1} < x_{i_2} < \ldots < x_{i_n}$ where $(i_1,i_2,\ldots,i_n)$ is a permutation of $(1,2,\ldots,n)$ then $<'$ is the total degree ordering in $U(L')$ with $\tilde{x}_{i_1} <' \tilde{x}_{i_2} <' \ldots <' \tilde{x}_{i_n}$.

We write LId(F) resp. RId(F) for the left ideal resp. right ideal generated by F.

Lemma 3: Let L' be the Lie algebra, φ the mapping, and $<'$ the ordering introduced above. Then the following assertions are satisfied.

(i) J is a right ideal in $U(L)$ generated by $F=(f_1,\ldots,f_m)$ if and only if $\varphi(J)$ is a left ideal of $U(L')$ generated by $\varphi(F)$.

(ii) F is a right Gröbner basis (w.r.t. $<$) of J if and only if $\varphi(F)$ is a left Gröbner basis (w.r.t. $<'$) of $\varphi(J)$.

(iii) S generates the right syzygy module of F if and only if $\varphi(S)$ generates the left syzygy module of $\varphi(F)$.

Proof : (i) ($\Rightarrow$) That $\varphi(J)$ is left ideal of $U(L')$ is easy to verify by (***) and linearity of φ.

Let $g \in J$. Then $\varphi(g) \in \varphi(J)$. $g=\sum_{i=1}^{m} f_i a_i$, $a_i \in U(L)$ since F generates J. $\varphi(g)=\sum_{i=1}^{m} \varphi(a_i)\ \varphi(f_i)$ follows by (***) and linearity.

Therefore, $\varphi(g) \in LId(\varphi(F))$ and $\varphi(J) \subseteq LId(\varphi(F))$.

That $LId(\varphi(F)) \subseteq \varphi(J)$ is trivial, since $\varphi(F) \subseteq \varphi(J)$ by definition of the extension of φ to sets. In conclusion $\varphi(J)=LId(\varphi(F))$.

(ii) ($\Rightarrow$) Let $g \in \varphi(J)$. $\varphi^{-1}(g)$ has a right normal representation $\varphi^{-1}(g)=\sum_{i=1}^{m} f_i h_i$ with $LBE(f_i h_i) \leqslant LBE(\varphi^{-1}(g))$ since F is right Gröbner basis of J and $\varphi^{-1}(g) \in J$. Application of φ to this representation provides a left normal representation $g=\sum_{i=1}^{m} \varphi(h_i)\varphi(f_i)$ with $LBE(\varphi(h_i)\varphi(f_i)) \leqslant' LBE(g)$ of g. In conclusion $\varphi(F)$ is a left Gröbner basis.

(iii) ($\Rightarrow$) If s is right syzygy of F then $Fs^T=0$ and by (***) it follows that $\varphi(s)\ \varphi(F)^T=0$. Therefore, $\varphi(S)$ generates a submodule of the left syzygy module of $\varphi(F)$. Let r be a left syzygy of $\varphi(F)$, i.e. $r\varphi(F)^T=0$. Applying φ^{-1} to this equation it follows by (***) that $\varphi^{-1}(r)$ is right syzygy of F. Therefore, $\varphi^{-1}(r)$ is right U(L)-combination of S. Applying φ to this combination a representation of r as left U(L′)-combination of $\varphi(S)$ is obtained. That completes the proof that $\varphi(S)$ generates the left syzygy module of $\varphi(F)$.

(i) (ii) (iii) ($\Leftarrow$) These proofs are analogous to the first three using the properties of φ^{-1} instead of these of φ.

∎

We summarize the assertion of lemma 3 by saying that the diagram.

$$
\begin{array}{ccc}
F_R \subset U(L) & \longrightarrow & F_L \subset U(L') \\
\Big\downarrow & & \Big\downarrow \\
G_R, X_R, Y_R, R_R, Q_R & \longleftarrow & G_L, X_L, Y_L, R_L, Q_L
\end{array}
$$

LEFT_RIGHT_FRACTION (w.r.t. $<$)   RIGHT_LEFT_FRACTION (w.r.t. $<'$)

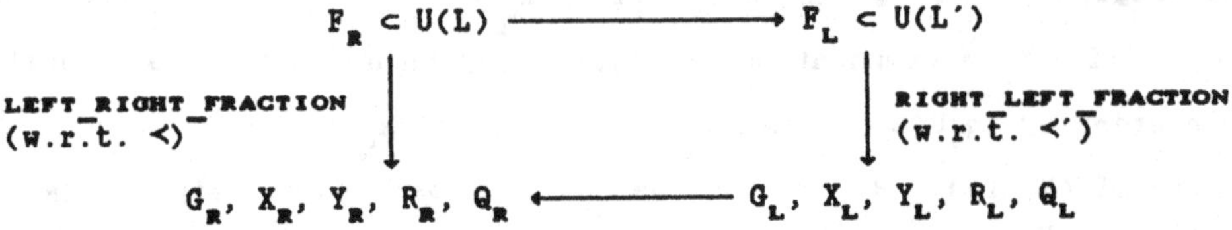

is commutative.

That the matrices at the left hand side of the diagram satisfy the equations (**) follows immediately by property (***) and the fact that the matrices at the right hand side fulfil the equations (*).

Therefore, only one subroutine for Gröbner basis and syzygy calculations is needed for the practical programming of the canonical simplifier defined in the next section. This enables memory efficient programming.

5. An Algorithm for Canonical Simplification

First we give a sketch and second an algorithm for the construction of the canonical representant c of the left fraction $y^{-1}z$ defined in section 3.

Starting from the representant (y,z) of the left fraction $y^{-1}z$ one can construct an equivalent right fraction uw^{-1} by application of LEFT_RIGHT_FRACTION. We apply LEFT_SYZYGY on the right fraction uw^{-1}. The result is a generating set $((y_1,z_1),\ldots,(y_r,z_r))$ of the module $M(y^{-1}z)$. $M_D(y^{-1}z)$ and $M_N(y^{-1}z)$ are generated by $F_D = (y_1,\ldots,y_r)$ and $F_N = (z_1,\ldots,z_r)$, respectively.

By use of LEFT_GROBNER_BASIS we construct a Gröbner basis G_D of the left ideal M_D and a matrix X_D such that $G_D^T = X_D F_D^T$. Now we choose the minimal element c_1 of G_D. c_1 may be represented as an U(L)-combination

of F_D by $c_1 = xF_D^T$ where x is a row of X_D. Last the numerator c_2 is computed by $c_2 = xF_N^T$. The element $c=(c_1,c_2)$ is the canonical representant (w.r.t. $<$) of the left fraction $y^{-1}z$.

ALGORITHM CANONICAL_REPRESENTANT($y^{-1}z$; $S_<(y^{-1}z)$)

1. $F_R:=(y\ ,\ z)$
2. call LEFT_RIGHT_FRACTION($F_R;G_R,X_R,Y_R,R_R,Q_R$)
3. $F_L:=q$ (q is the only nonzero row of Q_R)
4. call LEFT_SYZYGY($F_L;G_L,X_L,Y_L,R_L,Q_L$)
5. $F_D:=(1\ 0)\cdot Q_L^T$; $F_N:=(0\ 1)\cdot Q_L^T$
6. call LEFT_GROBNER_BASIS($F_D;G_D,X_D$)
7. $c_1:=$ the minimal (w.r.t $<$) element of G_D
8. x := the row of X_D such that $c_1=xF_D^T$
9. $c_2:=xF_N^T$
10. $S_<(y^{-1}z):=(c_1,c_2)$

6. Concluding Remarks

The canonical simplifier presented is rather time consuming since it makes use of three Gröbner basis computations. Furthermore, the result of the algorithm is more than only a canonical element. In fact, the rows of the matrix $X_D\cdot(\ F_D^T\ \ F_N^T\ )$ form a left Gröbner basis of the module $M(y^{-1}z)$. This property follows from the special feature of $M(y^{-1}z)$ that (0,0) is the only element of it which has zero as first component. Up to now we do not know a more efficient way for the computation of the canonical representant. Perhaps, there are also other elements in the equivalence class of a left fraction which may be choosen as canonical representant and can be computed faster. However, the choice of our canonical representant is such that it has minimal degree among all representants of the fraction. Therefore, this canonical representant provides big advantages for subsequent calculations. The way of finding better canonical simplifiers should be to look for faster methods to compute our canonical representant which is defined independent of our algorithm for its computation.

We refered to the algorithm presented in [1] as subalgorithm of the canonical simplifier. This algorithm does not use reduced Gröbner bases. Of course, it is also possible to use the algorithm for reduced Gröbner bases presented in [2] which is more effective for large bases and produces the same canonical representant for the same fixed ordering. However, we want to emphasize that reduced Gröbner bases are not necessary for our algorithm and that any Gröbner basis is suitable for our purpose.

7. References

1. Apel J., Lassner W., An algorithm for calculations in enveloping fields of Lie algebras. Proc.Int.Conf. on Computer Algebra and its Applications in Theoretical Physics, JINR, D11-85-791, Dubna, 1985.
2. Apel J., Lassner W., Computation of reduced Gröbner bases and syzygies in enveloping algebras., Full paper contribution at the SYMSAC 86, Waterloo, 1986.
3. Buchberger B., An algorithm for finding a basis for the residue class ring of a zero-dimensional polynomial ideal (in German). Ph.D. Thesis, Univ. Insbruck, 1965.
4. Buchberger B., Collins G.E., Loos R., Computer algebra. Springer, Wien New York, 1982.

A PACKAGE FOR THE ANALYTIC INVESTIGATION AND EXACT SOLUTION OF DIFFERENTIAL EQUATIONS

Thomas Wolf
Friedrich Schiller Universität Jena
Rechenzentrum, Humboldt Str. 2, GDR 6900 Jena

To find exact solutions of differential equations (DEs) one usually performs a variety of tests and ansaetze as e.g. the symmetry ansatz, separation ansatz, ansaetze for first integrals. When carrying through these ideas one has to solve overdetermined systems of partial DEs (PDEs). Whereas the formulation of the overdetermined system for each ansatz is achieved by different programs we use one program to investigate the overdetermined systems. This program CRACKSTAR contains three basic modules for the decoupling of DEs, for the integration of exact DEs and for separation. They are explained by examples. With these new abilities it becomes possible for CRACKSTAR to solve systems of DEs which are more difficult, i.e. the ansatz to be carried through may be less restrictive. An example shows that in this way not only point symmetries but also more general dynamical symmetries of ordinary DEs (ODEs) can be determined. For the investigation of Lie-symmetries of systems of PDEs an interval strategy is applied which reduces the effort for CRACKSTAR considerably. Further two new concepts, an integral ansatz for ODEs and a generalized separation ansatz for PDEs are described as well as the programs which generate overdetermined systems for each ansatz.

1. Motivation and concept

Many approaches and ansaetze are known for solving DEs analytically. A number of ansaetze are generalized by the invariance of a DE with respect to point transformations, i.e. the existence of Lie-symmetries. A number of ODEs without symmetries can be solved by other simple ansaetze. For example

$$y''+2yy'+f(x)(y'+y^2)=g(x) \qquad \text{(Kamke 1951 , eq. 6.37)}$$

has no Lie-symmetry for general f,g. But with

$$u(x)=y'+y^2 \qquad (1)$$

we get $u'+f(x)u=g(x)$, i.e. the original DE is reduced to a special Riccati DE and a linear first order DE. An investigation of a considerable part of the nonlinear second order ODEs in Kamke (1951) showed (Wolf,Pohle 1987) that most of them have Lie-symmetries but some of them can only be solved by special techniques. In a comparison of different ansaetze by Char (1981) it also becomes obvious that every ansatz covers only a limited range of equations. All that can be done so far is to perform a large variety of tests to look for useful properties which enable an analytical solution. Furthermore every test should be as general as possible. The properties discussed in this paper are dynamical symmetries of ODEs, Lie-symmetries of systems of PDEs, ansaetze such as (1) and a generalized separation ansatz for PDEs. The proposed strategy of performing all these investigations is similar to making measurements in physics where various physical quantities are measured implicitly by transforming them to one quantity which can be measured with high precision, e.g. a frequency. Analogously to measure a frequency we test the integrability of an overdetermind system of DEs or even solve this system. Therefore any property of interest can be investigated if a program transforms it into an equivalent set of overdetermined PDEs. The overdetermination is a direct consequence of the restrictive character of the property which need not necessarily be fulfilled. The program package CRACKSTAR is then carrying out the substantial algebraic work. The advantage is that the set of properties can be enlarged relatively easily by writing new corresponding transformation programs. Further, optimizing CRACKSTAR is beneficial to the investigation of all interesting properties. The more efficient this program is, the less overdetermined the system to be solved may be. This means that the property in question, which is equivalent to the system, may be less restrictive and therefore the chances of finding it realized increase. Example 3) shows, how increasing generality of the investigated symmetries (point-, contact-, dynamical symmetries) implies less and less overdetermined systems, i.e. less and more complicated equations for functions of more variables. The following restrictions exist for applying this concept:

1. The property in question must be expressible as an equation, i.e. the demand could not be that a function should only take integer values.
2. The overdetermined system should consist of PDEs which are polynomials of the free functions and their derivatives with coefficients that depend arbitrarily but explicitly on the variables.

In section 2. an overview of CRACKSTAR is given. The following sections describe programs which translate the condition for dynamical symmetries of ODEs, Lie-symmetries of systems of PDEs, ansaetze such as (1) and a generalized separation ansatz into an overdetermined system of PDEs and give examples.

2. The treatment of overdetermined systems

The package CRACKSTAR mainly consists of three modules for decoupling, integration and separation. An overview is given in table 1.

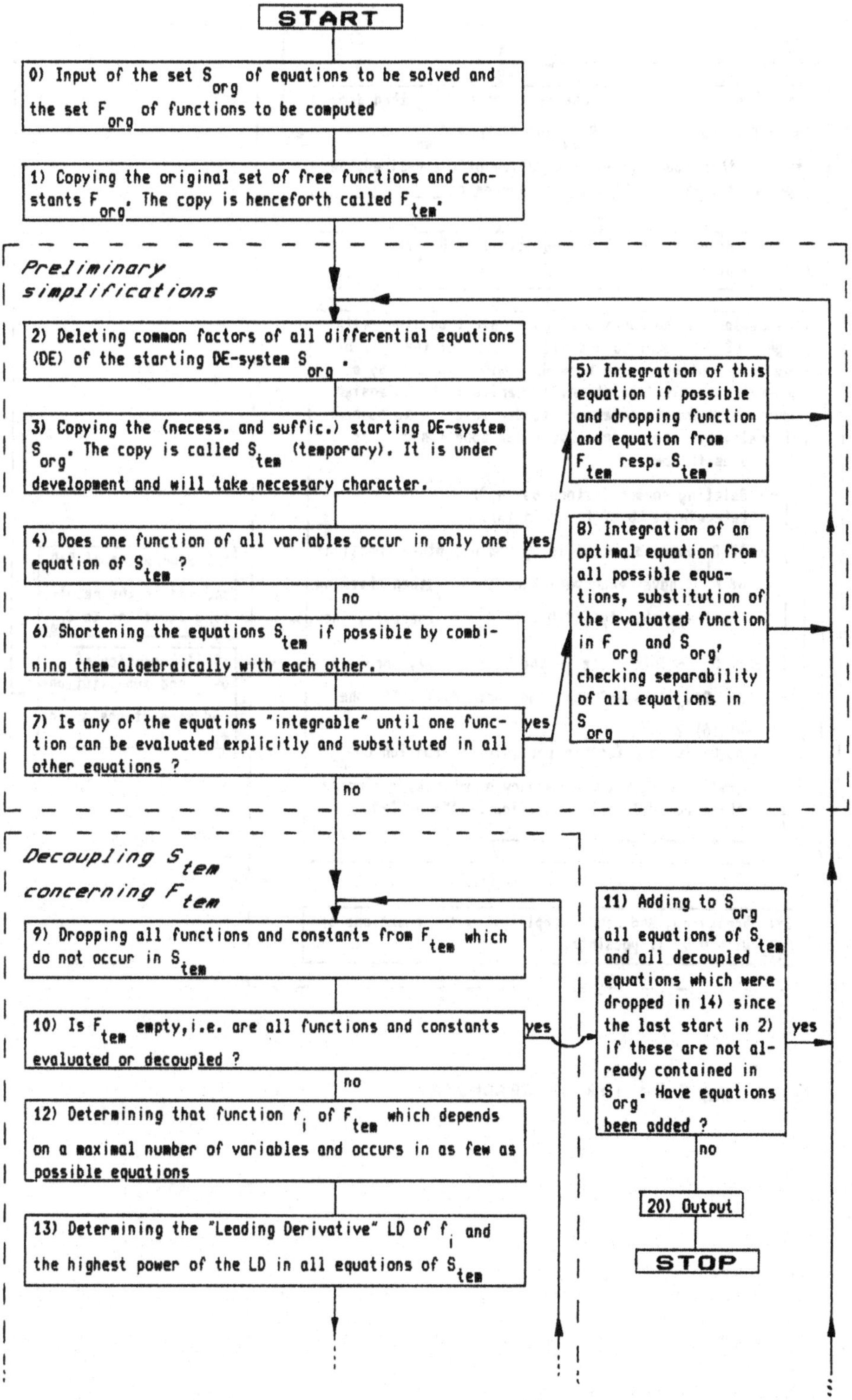
START
0) Input of the set S org of equations to be solved and the set F org of functions to be computed
1) Copying the original set of free functions and constants F org. The copy is henceforth called F tem.
Preliminary simplifications
2) Deleting common factors of all differential equations (DE) of the starting DE-system S org.
3) Copying the (necess. and suffic.) starting DE-system S org. The copy is called S tem (temporary). It is under development and will take necessary character.
4) Does one function of all variables occur in only one equation of S tem ?
yes
no
5) Integration of this equation if possible and dropping function and equation from F tem resp. S tem.
6) Shortening the equations S tem if possible by combining them algebraically with each other.
7) Is any of the equations "integrable" until one function can be evaluated explicitly and substituted in all other equations ?
yes
no
8) Integration of an optimal equation from all possible equations, substitution of the evaluated function in F org and S org, checking separability of all equations in S org
Decoupling S tem concerning F tem
9) Dropping all functions and constants from F tem which do not occur in S tem
10) Is F tem empty, i.e. are all functions and constants evaluated or decoupled ?
yes
no
11) Adding to S org all equations of S tem and all decoupled equations which were dropped in 14) since the last start in 2) if these are not already contained in S org. Have equations been added ?
yes
no
12) Determining that function f i of F tem which depends on a maximal number of variables and occurs in as few as possible equations
13) Determining the "Leading Derivative" LD of f i and the highest power of the LD in all equations of S tem
20) Output
STOP

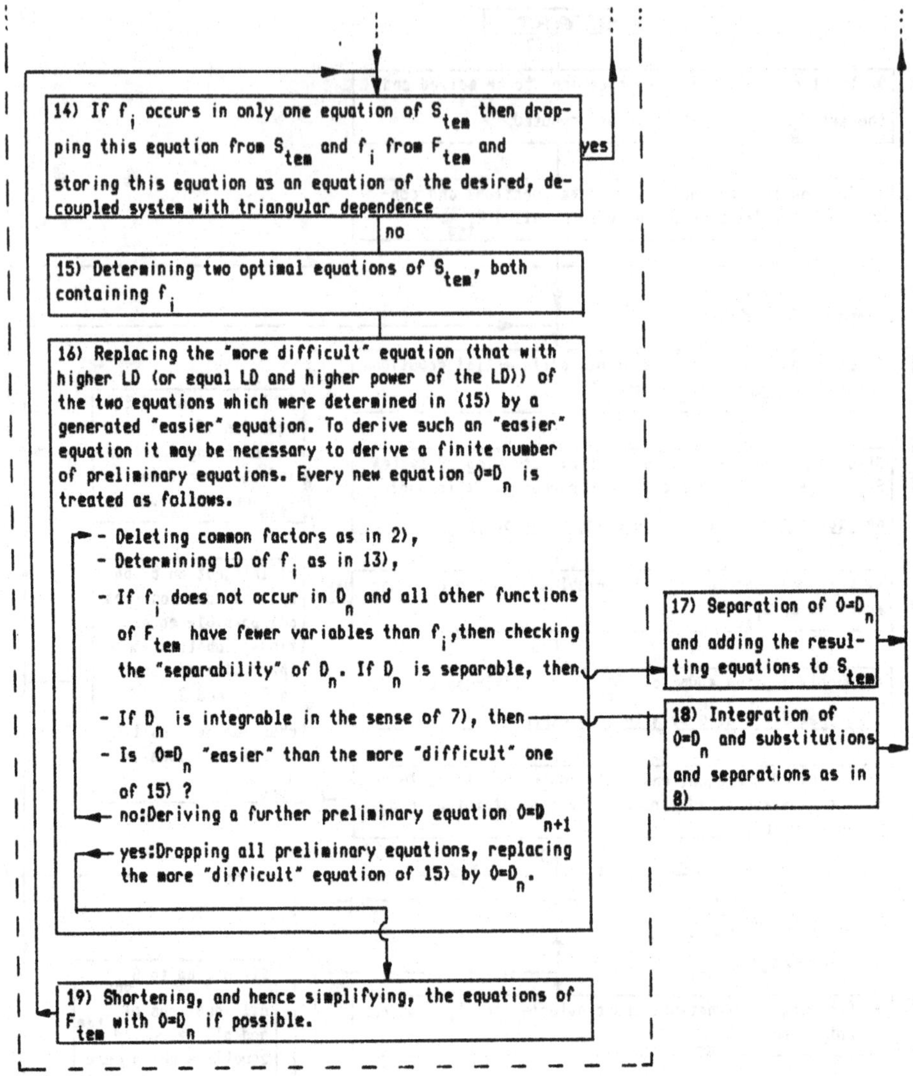
14) If f_i occurs in only one equation of S_{tem} then dropping this equation from S_{tem} and f_i from F_{tem} and storing this equation as an equation of the desired, decoupled system with triangular dependence
yes
no
15) Determining two optimal equations of S_{tem}, both containing f_i
16) Replacing the "more difficult" equation (that with higher LD (or equal LD and higher power of the LD)) of the two equations which were determined in (15) by a generated "easier" equation. To derive such an "easier" equation it may be necessary to derive a finite number of preliminary equations. Every new equation $0=D_n$ is treated as follows.
- Deleting common factors as in 2),
- Determining LD of f_i as in 13),
- If f_i does not occur in D_n and all other functions of F_{tem} have fewer variables than f_i, then checking the "separability" of D_n. If D_n is separable, then
- If D_n is integrable in the sense of 7), then
- Is $0=D_n$ "easier" than the more "difficult" one of 15) ?
no: Deriving a further preliminary equation $0=D_{n+1}$
yes: Dropping all preliminary equations, replacing the more "difficult" equation of 15) by $0=D_n$.
17) Separation of $0=D_n$ and adding the resulting equations to S_{tem}
18) Integration of $0=D_n$ and substitutions and separations as in 8)
19) Shortening, and hence simplifying, the equations of F_{tem} with $0=D_n$ if possible.

Table 1. Overview of CRACKSTAR

Notes to single boxes

0) The input also consists of flags for swiching on/off
- the test in 2), not to divide by factors containing free functions or constants which are to be determined,
- the test concerning integrability including integration,
- the simplification test in 6) and 19)

and of flags for fixing
- the maximal length of equations to be integrated,
- the maximal number of restarts, which have to be decited in box 11),
- the maximal length of expressions of the output.

1) In the course of computation, evaluated functions will be removed from F_{tem} and new functions and constants arising from integration will be added to F_{tem}. Intermediary results for the functions F_{org} (which are originally to be computed) are stored.

2) Deleting in every equation all common factors of all terms which neither contain free functions or their derivatives nor constants which are to be determined.

6) A fourth module, which is sometimes useful, provides an algebraic combination of two equations, so that an equivalent equation which is as short as possible can replace the longer of both equations. In box 19) this new equation must in addition not be more "difficult" (in the sense of box 16) than the longer equation to be replaced.

7) If more than one equation is integrable until one function f_i (of all variables occuring in this equation) can be eliminated explicitly,then
- integration of that equation which is as short as possible and is to be integrated as few as possible,
- substitution of f_i in all equations S_{org} and in the preliminary results for the original functions F_{org},
- checking separability of the new equations of S_{org}, if the expression, which is replaced for f_i, only contains free functions of fewer variables than f_i,
- adding to F_{tem} the new constants and functions of interation.

11) If all equations are solved resp. decoupled, i.e. the remaining equations S_{tem} have a triangular dependence of the functions F_{tem}, it may be useful, to add S_{tem} to S_{org} and restart.

20) The output consists of the results for the functions F_{org} and if any equations are not solved, then in addition the decoupled (necessary) equations S_{tem} and the sufficient equations S_{org} containing the functions F_{tem}.

The decoupling module

This module is applicable for equations which are polynomials of unknown constants and of functions and their derivatives. It corresponds to box 16) in table 1. The idea is to eliminate the highest powers of the highest derivatives of one unknown function and hence the highest derivatives themselves. This is achieved by differentiation and algebraic combination of two equations and the replacement of one of both by the new equation. Repeating this procedure, the differential order of the function can be decreased successively until an equation results which does not contain it. This equation can be a contradiction, the identity or a decoupled equation containing still other functions. An important point of the implemented algorithm is that the number of equations to be stored does not increase steadily because new equations replace the former ones, and in this way storage requirements do not grow so fast.

Example 1). To decouple the nonlinear heat equation

$0=f_{,t}-f_{,xx}-f-f^3 \quad =: D_1$ and the Korteweg de Vries equation

$0=f_{,t}+f_{,xxx}-6ff_{,x} \quad =: D_2$, concerning $f(x,t)$, the first steps are:

$D_1 := D_1-D_2$, $D_3 := D_{1,t}+D_{2,xxx}$, $D_3 := D_3+D_{2,xx}$, $D_3 := D_3-6fD_{2,x}$,

$D_2 := D_3+(3f^2-6f_{,x}+1)D_2$.

After further eleven steps an algebraic equation of degree 25 results. The procedure stops with an algebraic equation of ninth degree after further 42 steps when an identity is obtained. During execution it had been divided by $f_{,x}$ and therefore in addition the case $f_{,x}=0$ $(\longrightarrow f_{,t}=0)$ has to be considered which is a common solution of both original equations. The solution f=const. includes the real roots of the algebraic equations and is therefore the general solution.

A more detailed description of the decoupling module including possible outcomes is given in Wolf (1985). In CRACKSTAR this module is applied only until an equation is generated that is better than the worse of the two equations combined, and not until one decoupled equation without the function in question results (see table 1.). As a consequence the equations generated do not enlarge so rapitly.

The integration module

The efficiency of the decoupling module increases considerably if equations as e.g.

$$0 = 2xyf_1f_{1,x}+yf_1^{\,2}+\sin(x)f_{2,xy}+\cos(x)f_{2,y} =: D \qquad (2)$$

for $f_1(x), f_2(x,y)$ can be integrated to obtain

$$f_2=-xy^2f_1^{\,2}/(2\sin(x))+c_1(x)+c_2(y)/\sin(x) \qquad (3)$$

and to replace f_2 in all other equations. The module is able to integrate all exact DEs which are polynomials of the unknown functions and their derivatives. Furthermore those variables which occur explicitly in any term must occur as a power, or otherwise they must occur in that term also as a variable of an unknown function.

The method is to integrate by parts, starting with those terms which contain the highest derivatives of an unknown function. This derivative must occur linearly, if integration is to be possible. Denoting with I the x-integral of (2), the procedure provides

$$I:=0, \quad I:=I+\int(2xyf_1)d(f_1)= xyf_1^{\,2} \;, \quad D:=D - \frac{\partial}{\partial x}(xyf_1^{\,2})$$

$$I:=I+\int\sin(x)\; d(f_{2,y}) \;, \; D:=D - \frac{\partial}{\partial x}(\sin(x)f_{2,y}) \equiv 0 \;\longrightarrow \text{end.}$$

After x- and y-integration, functions of integration are added. Further preliminary and intermediary tests to realize non-integrability more readily are described in Wolf (1986a). Integration is only performed if the equation contains one function linearly which is a function of the maximal number of variables in that equation and if the equation is integrable until this function can be eliminated explicitly.

As a direct consequence of the integration ability, the separation of equations becomes necessary.

The separation module

A result of integration and the following replacement of, for example f_2 by (3) is that single variables occur only explicitly in some equations and no longer as variables of unknown functions. These equations divide into more equations. If possible, the module brings the equation 0=D in the form

$$D= \sum_i^n C_i(x^{a_i})*D_i(f_k(y^{b_i}),y^{b_i})$$

with linearly independent explicit expressions C_i and differential expressions D_i which do not depend on the only explicitly occuring variables x^{a_i}. The equation 0=D is then replaced by n equations $0=D_i$. This module is also useful for programs which establish the original overdetermined system.

compatibility and cooperation of the modules

Every module can handle equations that are generated by the other modules. The integration of an exact DE provides linearly occuring functions and constants of integration. Therefore after elimination and replacement of, for example f_2 by (3), the new functions c_1,c_2 and their derivatives occur, as f_2 did, only polynomially in all other equations. This is the restriction on equations for being manipulable by the decoupling module. With a more general integration module the new functions would occur as arguments of elementary functions and the decoupling module could not handle them. In order to decouple functions which occur as, for example, x in $F(x)=\sin(x)+x^3+3$, it would be necessary to be familiar with the inverse function of F. Therefore the treatment of DEs which are polynomials of the free functions and their derivatives can be carried through by comparatively simple means.

The frame of CRACKSTAR as depiced in table 1. has been developed in the course of symmetry investigations of a considerable part of the nonlinear second and higher order ODEs in Kamke (1951). This meant not only the establishment of an order of integrations, separations and decoupling, but also a "fine tuning" of priorities in the boxes 4,15 on which the efficiency and storage requirements are sensitively dependent.

For example, assuming the case that more than one equation is integrable in box 4), should one then integrate a short equation, that is often to be integrated with many new functions of integration, or should one integrate a longer equation but less often?

Whereas the integration of short equations and the subsequent replacement of one function f is very useful, this need not be the case if the equation is large. Then it could be that removing the highest derivatives of f by decoupling provides a less rapid increase of the length of the equation than replacing f directly after integration. It is necessary to come to an ad hoc decision without comparing the effort involved in both methods, because on the one hand it may be a lengthy procedure to decouple f with many durations of the boxes 14,15,16, and on the other hand integration and replacement of f by a large expression may exhaust the available storage. To decide such a question interactively is also not practicable. The reason is that comparing hundreds of equations or very large equations on the screen is not effective and furthermore such decisions have to be made very often. If, for example, the symmetry investigation shall be applicable not only as a main procedure but as a substep in more general investigations, then a totally automatic solution of overdetermined systems is necessary.

Some programs which translate special properties in overdetermined DEs are described below.

3. Determining dynamical symmetries of ODEs and Lie-symmetries of PDEs

The concept of symmetries of DEs and their use for reducing the order of the DEs has been developed by Sophus Lie (1912). Not only can these infinitesimal symmetries be used for solving DEs; they are also closely related to conservation laws which play an important role in physics.

The Lie-symmetries are defined as a forminvariance of the DE against infinitesimal transformations

$$\bar{x} = x + \xi(x,y) \;, \; \bar{y} = y + \eta(x,y) \;. \qquad (4)$$

In the case of PDEs, x and ξ are vectors and in the case of systems of DEs, y and h are vectors. The symmetry condition then takes the form of linear, coupled PDEs for ξ and η, which are overdetermined unless it is an ODE of the first order which is to be investigated conserning symmetries. In this case ξ and η have to be further restricted (see Char (1981) for corresponding investigations). For ODEs, the symmetry ansatz

(4) can easily be generalized to generators ξ,η which also contain derivatives of y. For such dynamical symmetries, overdetermined systems can also be derived, here with the program LIEORD (Wolf, Pohle 1987).

Example 2).

$$0=y''+(f+\frac{f'^2}{f^2}-\frac{f''}{2f})y+(y-\frac{3f'}{2f})y'-y^3-\frac{f'}{2f}y^2 \quad , \qquad f(x) \text{ free,}$$

Kamke (1951) eq. 6.35

LIEORD and CRACKSTAR provide

$\xi(x,y)=f^{-1/2}$, $\eta(x,y)=yf'f^{-3/2}/2$ $\longrightarrow$ 1 symmetry.

The example shows that the ODE may even contain free functions.

Example 3).

$$y^{(4)}=2y^{(3)}+5y''-6y'-y'^3+3y'^2y''-10y''\sin^2(y)+10y'\sin^2(y)-5y'^2\sin(2y)$$
$$+4\sin(2y)\sin^2(y)-2k^2x^2(y''+y'-\sin(2y)) \qquad (5)$$

(Dittes 1986)

For all three ansaetze for ξ,η, no symmetry exists.

$\xi(x,y),\eta(x,y)$		15 equations	with	105 terms
$\xi(x,y,y'),\eta(x,y,y')$	w.l.o.g. $\xi=0$ $\longrightarrow$	9 "	"	119 "
$\xi(x,y,y',y''),\eta(x,y,y',y'')$	w.l.o.g. $\xi=0$ $\longrightarrow$	5 "	"	465 "

The effort to investigate more general ansaetze is increasing strongly although ξ can be set to zero without loss of generality for higher symmetries. As well as having fewer equations, their length grows, and functions of more variables are to be determined from them. For the last ansatz, the decoupling (the cyclus of boxes 14,15,16 in table 1.) has to be performed for a considerable time, until the integrable equation $\partial^8\eta/(\partial y'')^8=0$ results.

Whereas symmetry investigations of ODEs become difficult if the ODE is of high order (n) and only few but large equations result, for example for $\eta=\eta(x,y,y',..,y^{(n-2)})$, this is not the case for PDEs. For PDEs the ansatz (4) provides a large set of DEs,because in the symmetry condition the coefficients of all the different products of all powers of all possible partial derivatives of y can be individually set to zero. Amoung these equations are always simple integrable equations, so that the storage rquirements decrease rapitly. To reduce the effort of formulating the full symmetry condition,an interval strategy is applied by the program LIEPAR. At first a preliminary small set of first order equations for ξ, η is derived which are solved as far as possible with CRACKSTAR. On the basis of the preliminary results for ξ,η the full symmetry condition is derived for the remaining free functions in ξ,η. For systems of PDEs, the preliminary symmetry conditions are first investigated individually for each original equation before the full condition is investigated for again each original equation.

Example 4). For the Kadomtsev-Petviasvili equation

$0=(4u,_t+6uu,_x+u,_{xxx}),_x+3u,_{yy}$, $u=u(t,x,y)$, the full condition formulated at once contains 178 terms and implies 56 different equations. Having the preliminary condition solved (4 equ., each with one term) the full condition shrinks to 70 terms and implies 23 different equations.

The ansatz of the next section is applicable for ODEs without simple symmetries.

4. An integral ansatz

We suppose, that for the ODE

$$0 = \omega(x,y,..,y^{(n)}) \tag{6}$$

we are given n-k first integrals $\varphi_i=c_i$ (c_i=const.), i.e. functions $\varphi_i(x,y,y',..,y^{(n-k)})$, whose constancy is guaranteed by (6)

$$\varphi_i=c_i \quad \longleftrightarrow \quad 0=\frac{d\varphi_i}{dx}=\varphi_i,_x+\varphi_i,_y y'+...+\varphi_i,_{y^{(n-2)}}y^{(n-1)}+\varphi_i,_{y^{(n-1)}}\omega .$$

By elimination of $y^{(k+1)},..,y^{(n)}$ in (6) with the first integrals, a functional z results with

$$y^{(k)} = z(x,y,..,y^{(k-1)},c_1,..,c_{n-k}) \quad , \quad c_i=\text{const.} \tag{7}$$

To find z without the knowledge of first integrals, we replace in (6)

$y^{(k)}$ by z ,

$y^{(k+1)}$ by $\frac{dz}{dx} = z,_x+z,_y y'+...+z,_{y^{(k-2)}}y^{(k-1)}+z,_{y^{(k-1)}}z$

and regard $x,y,..,y^{(k-1)}$ as independent variables, to obtain for z

$$0 = \omega(x,y,..,y^{(k-1)},z,\frac{dz}{dx},..,\frac{d^{n-k}z}{dx^{n-k}}) \tag{8}$$

This equation does not need to be solved in general, but the solution for z must only contain n-k constants c_i To solve the relatively complicated PDE (8), one can make as general as possible an ansatz for z, so that nevertheless (7) can be solved, e.g. for k=1

$$z=a(x)*b(y) \tag{9a}$$
$$z=a(x)y+b(x) \tag{9b}$$
$$z=a(x)y^{\alpha}+b(x)y \tag{9c}$$
$$z=a(x)y^2+b(x)y+c(x) \tag{9d}$$

Though equ. (8) is nonlinear for z, in contrast to the symmetry concept, which provides linear equations, it is possible to solve ODEs with an ansatz of this kind, which do not have, for example, Lie-symmetries. About 75% of the nonlinear second order ODEs listed in Kamke (1951) have at least one Lie-symmetry. Nearly the half of the remaining equations allow ansaetze of the kind described in this section.

In the corresponding program DECOMP one of the standard ansaetze (9) or any other ansatz can be chosen which is then inserted in (8). After that the separation module formulates the overdetermined system for the free functions in the ansatz.

Example 5). For the equation

$$0=y''+y'(ny+2/x)+2ny^2/x+y(mx^2-2/x^2)+px \quad , \; m,n,p \quad \text{const.} \qquad (10)$$

(Schmutzer 1983) the program DECOMP yields that the ansatz $z=a(x)y+b(x)$ is possible only for $p-2m/n=0$. Then a,b read $a=-2/x$, $b=x^2m/n$, i.e. solutions of the linear DE $0=y'+2y/x+x^2m/n$ are special solutions of (10). The symmetry concept is not applicable here because (10) does not have Lie-symmetries.

5. A generalized separation ansatz

The ansatz of this section is applicable to finding special solutions for PDEs 0=D that are polynomials of the unknown functions and their derivatives. It consists of writing the unknown functions as a sum of products of functions of fewer variables, e.g. for f(x,y)

$$f(x,y)= \sum_i^m a_i(x)*b_i(y). \qquad (11)$$

The technique is to replace f in the PDE 0=D by (1). It is then possible to regard $0=D$, $0=\frac{\partial D}{\partial y}, \ldots$, $0=\frac{\partial^m D}{\partial y^m}$ as an overdetermined system for a_i and to decouple them with CRACKSTAR. In this way one equation not containing any a_i, and containing x only explicitly, is generated. With the separation module all coefficients of all different explicit x-expressions can be set to zero to obtain an overdetermined system of DEs for the b_i. This system can again be treated with CRACKSTAR.

Example 6). For the PDE

$$0=ff,_x+xf,_{xy}-e^y/y =: D \qquad (12)$$

the ansaetze $f=a(x)*b(y)$ and $f=a(x)b(y)+c(x)$ lead to contradictions (Wolf 1986b). For

$$f=a(x)b(y)+c(y) \qquad (13)$$

the system $0=D$, $0=D,_y$ can be decoupled concerning a(x) to obtain one equation without a(x) which is separable concerning x and provides two equations, one of which is already c-decoupled:

$$0=(b'(y-1)-b''y)(b''-2b'^2).$$

The first solution $b=k\int e^y/y \, dy$ gives with (13) in (12) the contradiction $a'=0$. With the second solution $b=1/(y+n)$, n=const., inserted in (13),(12), the system $0=D$, $0=D,_x$ can be decoupled concerning c(y) to obtain one equation containing a(x) and explicitly y. y-separation provides $a'=1$, i.e. $a=x$ and hence $c=e^y(1+n/y)$. The obtained special solution of (12) is $f=(x+ye^y)/(y+n)$, n=const.

The probability of the existence of solutions (11) with finite m for a given arbitrary PDE is not extraordinarily high in practice. It is more promising to investigate the ansatz (11) as a substep, e.g. for finding Lie-symmetries of first order ODEs or for an ansatz like (9a) for z. Already simple ansaetze for Lie-symmetries, such as $\xi=a(x)*b(y)$, $\eta=0$; $\xi=0$, $\eta=a(x)*b(y)$; $\xi=\xi(x)$, $\eta=\eta(y)$ are remarkable successful (Char 1981).

6. Concluding remarks

To find exact solutions of DEs, a number of tests and ansaetze which are as general as possible, have to be performed. The concept is to use programs which translate the desired property in an overdetermined system of PDEs and to use one program to investigate this system. This program package CRACKSTAR as well as the translation programs are written in FORMAC 85 (Denk 1986). In contrast to a comparable program SIMPSYS of F. Schwarz (Schwarz 1986), CRACKSTAR seems to be more generally applicable and to contain modules which are more elaborated. With powerful programs for the automatic solution of overdetermined systems of DEs it becomes possible to couple tests, for example to determine the symmetries of equation (8) for z.

The properties discussed (symmetries, ansaetze of sec. 4.,5.) can only serve as examples though they are the approaches most frequently applied. Though the proposed strategy allows a variety of general ansaetze to be investigated, it must be mentioned that nevertheless the chances of finding exact solutions of difficult nonlinear DEs such as for example (5) is rather small.

Acknowledgement

I am grateful to H.J.Pohle for many discussions and many valuable hints.

References

Char,B. (1981). SYMSAC '81 proceedings, p.44
Denk,W. (1986). FORMAC 85 users manual, FSU Jena
Dittes,F. (1986). privat communication
Kamke,E. (1951). Differentialgleichungen Lösungsmethoden und Lösungen, Akad. Verlagsges. Geest & Portig K.G., Leipzig
Lie,S. (1912). Vorlesungen über Differentialgleichungen mit bekannten infinitesimalen Transformationen, Leipzig
Schmutzer,E. (1983). Unified field theory of more than 4 dimensions including exact solutions, Erice proceedings, Singapore
Schwarz,F. (1986). Symmetries of differential equations: From Sophus Lie to computer algebra, preprint
Wolf,T. (1985). J.Comput.Phys. 60, 437
Wolf,T. (1986a). Analytic integration of expressions containing arbitrary functions, preprint
Wolf,T. (1986b). Exact solutions of partial differential equations by a generalized separation ansatz, preprint
Wolf,T. and Pohle,H.J. (1987). preprint, to be published in COMPUTING

An Algorithm for the Integration of Elementary Functions

MANUEL BRONSTEIN

Department of Mathematics
University of California
Berkeley, CA 94720[1]

Abstract. Trager (1984) recently gave a new algorithm for the indefinite integration of algebraic functions. His approach was "rational" in the sense that the only algebraic extension computed is the smallest one necessary to express the answer. We outline a generalization of this approach that allows us to integrate mixed elementary functions. Using only rational techniques, we are able to normalize the integrand, and to check a necessary condition for elementary integrability.

INTRODUCTION

Intuitively, an *elementary function* is a function that can be obtained from the rational functions by repeatedly adjoining logarithms, exponentials, and algebraic functions. The *problem of integration in finite terms* is to decide in a finite number of steps whether a given elementary function has an elementary indefinite integral, and to compute it explicitely if it exists. This problem was studied extensively during the last century, but the difficulties of the algebraic function case caused Hardy (1916) to state that "there is reason to suppose that no such method can be given". This conjecture was eventually disproved by Risch (1970), who outlined a theoretical solution to the problem. While the transcendental cases of Risch's algorithm have been implemented (Moses 1971), his proof in the general case (Risch 1968, 1969a) did not present a practical algorithm that could be used to solve integration problems. More recently, Davenport (1981) published an algorithm for the integration of purely algebraic functions for which he has a partial implementation. Finally, Trager (1984) gave a "rational" algorithm for the integration of purely algebraic functions which appears easier to implement. In this paper, we outline a generalization of Trager's approach which we use to integrate mixed elementary functions.

The main theorem.

We use the concepts of differential fields and elementary extensions as defined by Rosenlicht (1972). In addition, if L is a differential extension field of K, and $\theta \in L$, we say that θ is an *elementary monomial over* K if θ is transcendental over K, is either a logarithm or an exponential of an element of K, and $K(\theta)$ and K have the same subfield of constants.

Let k be a differential field of characteristic 0, containing an element x such that $x' = 1$, and K be its constant field. The *problem of integration if finite terms on* k is, given $f \in k$, to determine in a finite number of steps whether there exists an elementary extension L of k and $g \in L$ such that

[1] Current address: IBM Research Division, T.J.Watson Research Center, Yorktown Heights, NY 10598.

$f = g'$, and to find them if they exist. A closely related problem is the ***Risch differential equation problem on k:*** given $f, g_1, \dots, g_m \in k$, to find in a finite number of steps $h_1, \dots, h_r \in k$ and a set Δ of $m+r$ linear equations over K such that the equation $z' + fz = \sum_{i=1}^{m} c_i g_i$ holds for $z \in k$ and $c_1, \dots, c_m \in K$ if and only if $z = \sum_{j=1}^{r} d_j h_j$ where $d_1, \dots, d_r \in K$ and $(c_i, d_j)_{ij}$ satisfy Δ.

Risch (1969b) proved that if both problems are solvable on k, then they are solvable on $k(\theta)$ where θ is an elementary monomial over k. With his solution to the Risch d.e. problem on $K(x)$ (1969b), he had an algorithm for the integration of purely transcendental elementary functions. Since there are algorithms for solving both problems on any finite algebraic extension of $K(x)$ (Davenport 1981, Davenport 1984, Trager 1984), the following theorem, by providing the algebraic step of the Risch algorithm, yields a complete algorithm for the integration of arbitrary elementary functions:

THEOREM. *Let k be a differential field of characteristic 0, containing an element x such that $x' = 1$, and let θ be an elementary monomial over k. If both the problem of integration in finite terms and of the Risch d.e. are solvable on any finite algebraic extension of k, then they are solvable on any finite algebraic extension of $k(\theta)$.*

Function fields and places.

Consider the integrand

$$(E) \qquad f = \frac{\log(x) + \sqrt{\log(x) + \sqrt{\log(x)}}}{1 + \log(x)}.$$

We view f as an element of the field $k(\theta, y)$ where $k = \mathbb{Q}(x)$, $\theta = \log(x)$ is an elementary monomial over k, and y is algebraic over $k(\theta)$ and satisfies $y^4 - 2\theta y^2 + \theta^2 - \theta = 0$ (we have $f = \dfrac{y+\theta}{1+\theta}$). Such an extension field of k is called an ***algebraic function field of one variable over k.***

In general, let our integrand f be an element of $L = k(\theta, y)$ where k is a differential field of characteristic 0, θ is an elementary monomial over k, and y is algebraic over $k(\theta)$. Let also K be the constant field of k, and suppose that both the integration and Risch d.e. problems are solvable on k.

Since L is an algebraic function field of one variable over k, we can use the machinery of Chevalley (1951), in particular the notions of places (points of the associated Riemann surface), and divisors. We write ν_p for the order function at a k-place p of L. We recall that each k-place p of L either lies above (i.e. contains) a monic ireducible polynomial $P \in k[\theta]$, in which case it is called a ***finite*** place, or above θ^{-1}, in which case it is called an ***infinite*** place.

We define a k-place p of L to be ***special*** if either

(1) p is infinite, or

(2) $\theta = \exp(\eta)$ for $\eta \in k$, and p lies above 0.

A non-special place is called ***normal.*** $f \in L$ is called ***normal*** if $\nu_p(f) \geq -r_p$ at all the normal places p where r_p is the ramification index of p above $k(\theta)$.

Bases and canonical forms.

$L = k(\theta, y)$ is a finite dimensional vector space over $k(\theta)$, so let $n = [L : k(\theta)]$ be its dimension. Let $\overline{k[\theta]}^L$ be the integral closure of $k[\theta]$ in L (i.e. the set of elements of L which do not have any finite pole). Among the various bases for L over $k(\theta)$, we compute a basis $w_1, \ldots, w_n$ which is also a set of generators for $\overline{k[\theta]}^L$ over $k[\theta]$. Such a basis is called an *integral basis* and can be computed by the algorithm in (Trager 1984). We also use that algorithm to make $w_1, \ldots, w_n$ normal over $\theta = \infty$.

Since $w_1, \ldots, w_n$ is a basis for L over $k(\theta)$, any $g \in k(\theta, y)$ can be written uniquely as

$$g = \sum_{i=1}^{n} \frac{P_i}{Q_i} w_i = \sum_{i=1}^{n} \frac{C_i w_i}{E}$$

where E is monic and $(E, C_1, \ldots, C_n) = (1)$. We can also uniquely write $E = \theta^b \bar{E}$, where $b \geq 0$ and $(\bar{E}, \theta) = (1)$. Thus, g can be written uniquely as

$$g = \sum_{i=1}^{n} \frac{C_i w_i}{\bar{E}\theta^b} = \sum_{i=1}^{n} \frac{B_i w_i}{\theta^b} + \sum_{i=1}^{n} \frac{D_i w_i}{\bar{E}}$$

where $(\bar{E}, D_1, \ldots, D_n) = (1)$, and $(\theta, B_1, \ldots, B_n) = (1)$ if $b > 0$.

We define the finite special part of g to be:

$$g^- = \begin{cases} 0, & \text{if } \theta = \log(\eta) \text{ for } \eta \in k \\ \sum_{i=1}^{n} \frac{B_i w_i}{\theta^b}, & \text{if } \theta = \exp(\eta) \text{ for } \eta \in k. \end{cases}$$

In both cases, let the *canonical representation of* g be

$$g = g^- + \sum_{i=1}^{n} \frac{A_i w_i}{D}$$

where D is monic, $(D, A_1, \ldots, A_n) = (1)$, and $(D, \theta) = (1)$ in the exponential case. Such a representation is unique, and is useful for testing whether an integrand is normal. We have (Bronstein 1987):

PROPOSITION. *$f \in k(\theta, y)$ is normal if and only if D is squarefree in the canonical representation of f.*

The next step of the algorithm is to compute the canonical representations of $w_1', \ldots, w_n'$. We find that $(w_i')^- = 0$ for $i = 1 \ldots n$, so putting the representations over a common denominator, we find $E, M_{ij} \in k[\theta]$ such that

$$w_i' = \sum_{j=1}^{n} \frac{M_{ij} w_j}{E}.$$

Furthermore, E must be squarefree and, in the exponential case, prime with θ.

The Hermite Reduction

This step of the algorithm reduces an integrand to a normal one. Let $f \in L$, and let

$$f = f^- + \sum_{i=1}^{n} \frac{A_i w_i}{D}$$

be its canonical representation. Let $D = D_1 {D_2}^2 \cdots {D_{m+1}}^{m+1}$ be a squarefree factorization of D. If $m = 0$ then D is squarefree, so f is normal. Otherwise, let $V = D_{m+1}$ and $U = \frac{D}{V^{m+1}}$. Let E and the M_{ij}'s be as in the previous section. We multiply D and the A_i's by $\frac{E}{\gcd(E, UV)}$ to ensure that $E | UV$. Following Trager (1984), we then ask whether there exist $B_1, \ldots, B_n, C_1, \ldots, C_n \in k[\theta]$ such that

$$\int \sum_{i=1}^{n} \frac{A_i w_i}{UV^{m+1}} = \sum_{i=1}^{n} \frac{B_i w_i}{V^m} + \int \sum_{i=1}^{n} \frac{C_i w_i}{UV^m}.$$

Differentiating both sides, equating the coeefficients of each w_i, clearing denominators and reducing modulo V, we get the following linear system for $B_1, \ldots, B_n$:

$$(S) \qquad \begin{cases} A_1 \equiv -mUV'B_1 + \frac{UV}{E} \sum_{j=1}^{n} M_{j1} B_j & \pmod V \\ A_2 \equiv -mUV'B_2 + \frac{UV}{E} \sum_{j=1}^{n} M_{j2} B_j & \pmod V \\ \quad \cdots & \\ A_n \equiv -mUV'B_n + \frac{UV}{E} \sum_{j=1}^{n} M_{jn} B_j & \pmod V \end{cases}$$

We can prove that (S) always has a solution when $m > 0$ and the conditions on E, U, and V are satisfied. After solving (S) for the B_i's, the C_i's are given by

$$C_i = \frac{A_i - \frac{UV}{E} \sum_{j=1}^{n} M_{ji} B_j + mUV'B_i}{V} - UB_i'$$

We repeat this process until D is squarefree. At that point, we have $g, h \in L$ such that $f = f^- + g' + h$, so our new integrand is $f^- + h$ which is normal.

For the example (E) above, we find $(w_1, w_2, w_3, w_4) = (1, y, y^2, y^3)$, and the canonical representation of f is

$$f = \frac{\theta w_1 + w_2}{\theta + 1}$$

so f is already normal.

The Residues and the Resultant Criterion

Assume now that our integrand $f \in L$ is normal. At every normal k-place p of L, we can define a function τ_p from the normal elements of L into $\bar{k}$ which verifies the following property of residues: for $u \in \bar{K}L$, $\tau_p(\frac{u'}{u}) = \nu_p(u)$. We can then prove the following generalization of Rothstein's criterion for non-integrability (1977, Theorem 1):

THEOREM. *Suppose that y is integral over $k[\theta]$. Let $f \in k(\theta, y)$ and let $f = f^- + \sum_{i=1}^n \frac{A_i w_i}{D}$ be its canonical representation. Write $\sum_{i=1}^n A_i w_i = \frac{G}{H}$, where $G \in k[\theta, y]$ and $H \in k[\theta]$. Let $F \in k[\theta, y]$ be the defining polynomial for $k(\theta, y)$, and z be an indeterminate over k. Define $R \in k[z]$ by:*

$$R(z) = resultant_\theta(prim_z(resultant_y(G - zHD', F)), D)$$

where $prim_z$ stands for taking the primitive part with respect to z. If f is normal, then the roots of R are nonzero rational multiples of the $\tau_p(f)$ at all the normal k-places of $k(\theta, y)$ where f has a pole. Furthermore, if f has an elementary integral over k, then $R(z)$ has constant roots, i.e. $R(z) = \alpha P(z)$ where $\alpha \in k$ and $P \in k[z]$ is monic and has constant coefficients.

This criterion is checked by the algorithm once the Hermite reduction is completed.

For example (E) above, we find

$$g(\theta, y) = \frac{G(\theta, z)}{H(\theta)} = \frac{\theta + y}{1}$$
$$R(z) = resultant_\theta(prim_z(resultant_y(G - zHD', F)), D)$$
$$= z^4 + 4xz^3 + 8x^2z^2 + 8x^3z + 5x^4$$

so (E) does not have an elementary integral.

Integrating Normal Elements

From now on, we have a normal integrand $f \in L$ which satisfies the above criterion. From Liouville's theorem (Risch 1969b, Rosenlicht 1972), f has an elementary integral over k if and only if there exist $v \in L$, $u_i \in \bar{K}L$, and $c_i \in \bar{K}$ such that

$$(*) \qquad f = v' + \sum_{i=1}^n c_i \frac{u_i'}{u_i}.$$

v is called the algebraic part of the integral, and the sum is called the logarithmic part. We outline in the next sections how to find them, or prove that they do not exist.

The Logarithmic Part.

We can use the τ_p's in order to find a divisor for each u_i of equation $(*)$. Indeed, we can prove that if $(*)$ has a solution, then $\tau_p(f) = \sum_{i=1}^n c_i \nu_p(u_i)$ at any normal k-place p of L. Those can be computed at each pole of p f by the formula $\tau_p(f) = r_p value_p(f \frac{P}{P'})$ where r_p is the ramification index of p above $k(\theta)$ and p lies above $P \in k[\theta]$. Also, we can compute $\tau_p(f) = \sum_{i=1}^n c_i \nu_p(u_i)$ at any special place p by recursively integrating one coefficient of the Puiseux expansion of f at p (Risch 1968, Bronstein 1987). We then compute a basis $r_1, \ldots, r_l$ for the vector space generated by the $\tau_p(f)$'s over $\mathbf{Q}$.

Expressing each $\tau_p(f)$ as a linear combination of the r_i's with rational coefficients, we get l divisors $D_1, \ldots, D_l$ such that, if $(*)$ has a solution, then it has one where the divisor of each u_i is a multiple of D_i, and where $c_i = \frac{r_i}{h}$, h a positive integer (Risch 1968, Davenport 1981, Bronstein 1987). The problem is now to find whether a multiple of each D_i is the divisor of a function (called the *problem of points of finite order*, which has been solved by Trager (1984)[2].

The Algebraic Part.

This part of the algorithm is executed after the logarithmic part when θ is a logarithm, but before the logarithmic part when θ is an exponential. Since f is normal, v' must also be normal. This implies that v has no poles at the normal places, hence that there exist $G_1, \dots, G_n \in k[\theta]$ such that either $v = \sum_{i=1}^n G_i w_i$ (in the logarithmic case), or $v = \theta^{-m} \sum_{i=1}^n G_i w_i$ (in the exponential case). From the orders of f at the special places, we can compute m and an upper bound B_i on the degree of each G_i. We then set $G_i = \sum_{j=1}^{B_i} c_{ij}\theta^j$ where the c_{ij}'s are finitely many undetermined coefficients in k.

We compute the principal parts of the Puiseux expansions of v at all the special places. Their coefficients are linear combinations of the c_{ij}'s with coefficients in $\bar{k}$. We differentiate termwise and equate the the resulting series with the Puiseux expansions of f. After recursively integrating one element of $\bar{k}$ (or solving one Risch d.e. over $\bar{k}$ in the exponential case) for each coefficient, we obtain a linear system over $\bar{k}$ for the c_{ij}'s. If that system has no solution, then f has no elementary integral, otherwise we find v and the u_j's which are algebraic over k (Risch 1968, Bronstein 1987).

The nth-root Case.

In the special case where $y^n \in k(\theta)$, we can differentiate the expression $v = \sum_{i=1}^n G_i w_i$ (or $v = \theta^{-m} \sum_{i=1}^n G_i w_i$), and plug it into equation (*). Equating the coefficients of each w_i on both sides, we get one Risch d.e. over $k(\theta)$ for each G_i. Such an equation can be solved recursively with the transcendental algorithms of (Risch 1969b, Rothstein 1977, Davenport 1986). In this case, computations of Puiseux expansions are not required.

Solving Risch d.e.'s

Our approach is essentially the same as in Risch (1968): we first get a lower bound for $\nu_p(z)$ at all the finite k-places p of L. This gives us a polynomial $D \in k[\theta]$ such that zD must be integral over $k[\theta]$ and satisfy a different Risch d.e. That second equation is solved by expanding its coefficients (f and the g_i's) at the infinite places, as well as expanding zD, which is of the form $\sum_{i=1}^n G_i w_i$ where $G_i \in k[\theta]$ is of bounded degree, and matching the principal part of the expansions as in the previous section.

Acknowledgements

I would like to thank Jean Della Dora, Claire Dicrescenzo, Daniel Lazard and Evelyne Tournier for providing me with the facilities needed to type and print this paper, as well as Barry Trager for his numerous suggestions on this research.

[2] That solution is now implemented in the Scratchpad II computer algebra system (August 1988).

References

Bronstein, M. (1987), "Integration of Elementary Functions," Ph.D. thesis, Dpt. of Mathematics, Univ. of California, Berkeley.

Chevalley, C. (1951), "Algebraic Functions of One Variable," Math. Surveys Number VI, American Mathematical Society, New York.

Davenport, J.H. (1981), "On the Integration of Algebraic Functions," Lecture Notes in Computer Science No. 102, Springer-verlag, New York.

Davenport, J.H. (1984), *Intégration Algorithmique des Fonctions Elémentairement Transcendantes sur une Courbe Algébrique*, Annales de l'Institut Fourier **34, fasc.2**, 271–276.

Davenport, J.H. (1986), *The Risch Differential Equation Problem*, SIAM Journal on Computing **15, No.4**, 903–916. Also Technical Report 83-4, Dpt. of Computer and Information Sciences, Univ. of Delaware.

Hardy, G.H. (1916), "The Integration of Functions of a Single Variable," Cambridge U. Press, Cambridge, England.

Moses, J. (1971), *Symbolic Integration: the Stormy Decade*, Communications of the ACM **14**, 548–560.

Risch, R. (1968), *On the Integration of Elementary Functions which are built up using Algebraic Operations*,, Report SP-2801/002/00. System Development Corp., Santa Monica, CA.

Risch, R. (1969a), *Further Results on Elementary Functions*, Report RC-2402. IBM Corp., Yorktown Heights, NY.

Risch, R. (1969b), *The Problem of Integration in Finite Terms*, Transactions American Mathematical Society **139**, 167–189.

Risch, R. (1970), *The Solution of the Problem of Integration in Finite Terms*, Bulletin American Mathematical Society **76**, 605–608.

Rosenlicht, M. (1972), *Integration in Finite Terms*, American Mathematical Monthly **79**, 963–972.

Rothstein, M. (1977), *A New Algorithm for the Integration of Exponential and Logarithmic Functions*, in "Proceedings 1977 MACSYMA Users Conference," NASA Pub. CP-2012, pp. 263–274.

Trager, B. (1984), "Integration of Algebraic Functions," Ph.D. thesis, Dpt. of EECS, Massachusetts Institute of Technology.

Index of Authors

Abbott, J. A. 391, 465
Abramov, S. A. 45
Aoki, Y. 134
Apel, J. 468
Baaz, M. 424
Bachmair, L. 452
Bartels, R. H. 64
Böffgen, R. 48
Bowyer, A. 244
Brackx, F. 208
Bradford, R. J. 315
Bronstein, M. 491
Buchmann, J. 54
Burge, W. H. 138
Cabay, S. 438
Char, B. W. 64, 463
Cowell, R. G. 71
Creutzburg, R. 161
Czapor, S. R. 260
Davenport, J. H. 244, 391
Dershowitz, N. 452
Drska, D. 178
Dulyan, L. S. 172
Eltekov, V. A. 216
Fedorova, R. N. 1, 174
Ferscha, A. 403
Fischer, T. 426
Fitch, J. P. 95, 202
Folz, H. G. iii
Fukui, Y. 163
Gateva-Ivanova, T. 355
Gerdt, V. P. 1, 81, 93, 206
Gerez, S. H. 149
Gianni, P. 293
Gillies, D. M. 463
Giusti, M. 333
Goldman, V. V. 120
Govorun, N. N. 1
Grigor'ev, D. Yu. 11
Gurin, N. I. 116
Hall, R. G. 95
Henry, J. P. G. 381
van den Heuvel, P. 120
van Hulzen, J. A. 120
Ilyin, V. A. 225
Jahn, K.-U. 204
Kalkbrener, M. 282
Kaltofen, E. 317
Katkov, V. L. 179
Kazasov, C. 132
Klimov, D. M. 97
Kondratéva, M. V. 365
Kornyak, V. V. 174
Kostov, N. A. 206
Kostova, Z. T. 206
Kredel, H. 270
Krishnamoorthy, M. S. 317
Kryukov, A. P. 225, 233
Kusche, K. 246
Kutzler, B. 246
Labahn, G. 438
Langemyr, L. 50, 298
Lassner, W. 468
Leitsch, A. 424
Liska, R. 178
MacCallum, M. A. H. 34
Martin, B. 311
Mayr, H. 246
McCallum, S. 298
Merle, M. 381
Milne, P. S. 244
Molnár, Ľ. 412
Mulder, R. 149
Mutrie, M. P. W. 64
Návrat, P. 412
Németh, G. 450
Nisheva, M. 52
Paczynski, J. 107
Padget, J. A. 244
Pankratév, E. V. 365
Pardo, L. M. 456
Pearce, P. D. 202
Perlt, H. 176
Pfister, G. 311
Pohst, M. 54
Popov, M. D. 179
Ranft, J. 176
Recio, T. 456
Reichel, H. iii
Reichert, M. A. 48
Rodionov, A. Ya. 192, 233
Roider, B. 258
Rudenko, V. M. 97
Sasaki, T. 348, 163
Sato, M. 163
Saunders, B. D. 317
Schefflert, P. 379
Schemmel, K.-P. 300

Seese, D.	379
Serras, H.	208
Shabat, A. B.	81
Shablygin, E.	186
Shikalov, V. B.	216
Shirikov, V. P.	1
Smedley, T. J.	313
Smit, J.	149
Smith, G. C.	26
Spiridonova, M.	136
Stifter, S.	258
Suzuki, M.	163
Svinolupov, S. I.	81
Taranov, A. Yu.	192
Tasche, M.	161
Tomov, V.	52
Tonev, T.	52
Valibouze, A.	323
Vallée, B.	376
Vasiliev, N. N.	118
Vojtek, V.	412
Wallis, A. F.	244
Watt, S. M.	138
Weispfenning, V.	336
Werner, G.	454
Wolf, T.	479
Wright, F. J.	71
Wu, W.	44
Yamamoto, T.	134
Zharkov, A. Yu.	81, 93
Zhuravlev, V. F.	97
Zima, E. V.	223
Zimányi, M.	450
Zimmer, H. G.	iii

Vol. 324: M.P. Chytil, L. Janiga, V. Koubek (Eds.), Mathematical Foundations of Computer Science 1988. Proceedings. IX, 562 pages. 1988.

Vol. 325: G. Brassard, Modern Cryptology. VI, 107 pages. 1988.

Vol. 326: M. Gyssens, J. Paredaens, D. Van Gucht (Eds.), ICDT '88. 2nd International Conference on Database Theory. Proceedings, 1988. VI, 409 pages. 1988.

Vol. 327: G.A. Ford (Ed.), Software Engineering Education. Proceedings, 1988. V, 207 pages. 1988.

Vol. 328: R. Bloomfield, L. Marshall, R. Jones (Eds.), VDM '88. VDM – The Way Ahead. Proceedings, 1988. IX, 499 pages. 1988.

Vol. 329: E. Börger, H. Kleine Büning, M.M. Richter (Eds.), CSL '87. 1st Workshop on Computer Science Logic. Proceedings, 1987. VI, 346 pages. 1988.

Vol. 330: C.G. Günther (Ed.), Advances in Cryptology – EUROCRYPT '88. Proceedings, 1988. XI, 473 pages. 1988.

Vol. 331: M. Joseph (Ed.), Formal Techniques in Real-Time and Fault-Tolerant Systems. Proceedings, 1988. VI, 229 pages. 1988.

Vol. 332: D. Sannella, A. Tarlecki (Eds.), Recent Trends in Data Type Specification. V, 259 pages. 1988.

Vol. 333: H. Noltemeier (Ed.), Computational Geometry and its Applications. Proceedings, 1988. VI, 252 pages. 1988.

Vol. 334: K.R. Dittrich (Ed.), Advances in Object-Oriented Database Systems. Proceedings, 1988. VII, 373 pages. 1988.

Vol. 335: F.A. Vogt (Ed.), CONCURRENCY 88. Proceedings, 1988. VI, 401 pages. 1988.

Vol. 336: B.R. Donald, Error Detection and Recovery in Robotics. XXIV, 314 pages. 1989.

Vol. 337: O. Günther, Efficient Structures for Geometric Data Management. XI, 135 pages. 1988.

Vol. 338: K.V. Nori, S. Kumar (Eds.), Foundations of Software Technology and Theoretical Computer Science. Proceedings, 1988. IX, 520 pages. 1988.

Vol. 339: M. Rafanelli, J.C. Klensin, P. Svensson (Eds.), Statistical and Scientific Database Management. Proceedings, 1988. IX, 454 pages. 1989.

Vol. 340: G. Rozenberg (Ed.), Advances in Petri Nets 1988. VI, 439 pages. 1988.

Vol. 341: S. Bittanti (Ed.), Software Reliability Modelling and Identification. VII, 209 pages. 1988.

Vol. 342: G. Wolf, T. Legendi, U. Schendel (Eds.), Parcella '88. Proceedings, 1988. 380 pages. 1989.

Vol. 343: J. Grabowski, P. Lescanne, W. Wechler (Eds.), Algebraic and Logic Programming. Proceedings, 1988. 278 pages. 1988.

Vol. 344: J. van Leeuwen, Graph-Theoretic Concepts in Computer Science. Proceedings, 1988. VII, 459 pages. 1989.

Vol. 345: R.T. Nossum (Ed.), Advanced Topics in Artificial Intelligence. VII, 233 pages. 1988 (Subseries LNAI).

Vol. 346: M. Reinfrank, J. de Kleer, M.L. Ginsberg, E. Sandewall (Eds.), Non-Monotonic Reasoning. Proceedings, 1988. XIV, 237 pages. 1989 (Subseries LNAI).

Vol. 347: K. Morik (Ed.), Knowledge Representation and Organization in Machine Learning. XV, 319 pages. 1989 (Subseries LNAI).

Vol. 348: P. Deransart, B. Lorho, J. Małuszyński (Eds.), Programming Languages Implementation and Logic Programming. Proceedings, 1988. VI, 299 pages. 1989.

Vol. 349: B. Monien, R. Cori (Eds.), STACS 89. Proceedings, 1989. VIII, 544 pages. 1989.

Vol. 350: A. Törn, A. Žilinskas, Global Optimization. X, 255 pages. 1989.

Vol. 351: J. Díaz, F. Orejas (Eds.), TAPSOFT '89. Volume 1. Proceedings, 1989. X, 383 pages. 1989.

Vol. 352: J. Díaz, F. Orejas (Eds.), TAPSOFT '89. Volume 2. Proceedings, 1989. X, 389 pages. 1989.

Vol. 354: J.W. de Bakker, W.-P. de Roever, G. Rozenberg (Eds. Linear Time, Branching Time and Partial Order in Logics and Mode for Concurrency. VIII, 713 pages. 1989.

Vol. 355: N. Dershowitz (Ed.), Rewriting Techniques and Applic tions. Proceedings, 1989. VII, 579 pages. 1989.

Vol. 356: L. Huguet, A. Poli (Eds.), Applied Algebra, Algebra Algorithms and Error-Correcting Codes. Proceedings, 1987. VI, 41 pages. 1989.

Vol. 357: T. Mora (Ed.), Applied Algebra, Algebraic Algorithms ar Error-Correcting Codes. Proceedings, 1988. IX, 481 pages. 1989.

Vol. 358: P. Gianni (Ed.), Symbolic and Algebraic Computatio Proceedings, 1988. XI, 545 pages. 1989.

Vol. 359: D. Gawlick, M. Haynie, A. Reuter (Eds.), High Performanc Transaction Systems. Proceedings, 1987. XII, 329 pages. 1989.

Vol. 360: H. Maurer (Ed.), Computer Assisted Learning – ICCAL '8 Proceedings, 1989. VII, 642 pages. 1989.

Vol. 361: S. Abiteboul, P.C. Fischer, H.-J. Schek (Eds.), Neste Relations and Complex Objects in Databases. VI, 323 pages. 1989.

Vol. 362: B. Lisper, Synthesizing Synchronous Systems by Stat Scheduling in Space-Time. VI, 263 pages. 1989.

Vol. 363: A.R. Meyer, M.A. Taitslin (Eds.), Logic at Botik '8 Proceedings, 1989. X, 289 pages. 1989.

Vol. 364: J. Demetrovics, B. Thalheim (Eds.), MFDBS 89. Procee ings, 1989. VI, 428 pages. 1989.

Vol. 365: E. Odijk, M. Rem, J.-C. Syre (Eds.), PARLE '89. Parall Architectures and Languages Europe. Volume I. Proceedings, 198 XIII, 478 pages. 1989.

Vol. 366: E. Odijk, M. Rem, J.-C. Syre (Eds.), PARLE '89. Paral Architectures and Languages Europe. Volume II. Proceedings, 198 XIII, 442 pages. 1989.

Vol. 367: W. Litwin, H.-J. Schek (Eds.), Foundations of Data Orga zation and Algorithms. Proceedings, 1989. VIII, 531 pages. 1989.

Vol. 368: H. Boral, P. Faudemay (Eds.), IWDM '89, Database M chines. Proceedings, 1989. VI, 387 pages. 1989.

Vol. 369: D. Taubner, Finite Representations of CCS and TCS Programs by Automata and Petri Nets. X. 168 pages. 1989.

Vol. 370: Ch. Meinel, Modified Branching Programs and Their Co putational Power. VI, 132 pages. 1989.

Vol. 371: D. Hammer (Ed.), Compiler Compilers and High Spe Compilation. Proceedings, 1988. VI, 242 pages. 1989.

Vol. 372: G. Ausiello, M. Dezani-Ciancaglini, S. Ronchi Della Roc (Eds.), Automata, Languages and Programming. Proceedings, 198 XI, 788 pages. 1989.

Vol. 373: T. Theoharis, Algorithms for Parallel Polygon Renderin VIII, 147 pages. 1989.

Vol. 374: K.A. Robbins, S. Robbins, The Cray X-MP/Model 24. 165 pages. 1989.

Vol. 375: J.L.A. van de Snepscheut (Ed.), Mathematics of Progra Construction. Proceedings, 1989. VI, 421 pages. 1989.

Vol. 376: N.E. Gibbs (Ed.), Software Engineering Education. P ceedings, 1989. VII, 312 pages. 1989.

Vol. 377: M. Gross, D. Perrin (Eds.), Electronic Dictionaries and Automa in Computational Linguistics. Proceedings, 1987. V, 110 pages. 1989.

Vol. 378: J.H. Davenport (Ed.), EUROCAL '87. Proceedings, 1987. V 499 pages. 1989.